Student's Solutions Manual

Matthew Hudock
Denyse Kerr
Jamie Smolko

Brief Calculus
The Study of Rates of Change

Bill Armstrong
Don Davis

Upper Saddle River, NJ 07458

Executive Editor: Kathy Boothby Sestak
Supplement Editor: Joanne Wendelken
Special Projects Manager: Barbara A. Murray
Production Editor: Benjamin St. Jacques
Supplement Cover Manager: Paul Gourhan
Supplement Cover Designer: PM Workshop Inc.
Manufacturing Buyer: Lisa McDowell

© 2000 by Prentice Hall
Upper Saddle River, NJ 07458

All rights reserved. No part of this book may be
reproduced, in any form or by any means,
without permission in writing from the publisher.

Printed in the United States of America

10 9 8 7 6 5 4 3 2 1

ISBN 0-13-085882-X

Prentice-Hall International (UK) Limited, London
Prentice-Hall of Australia Pty. Limited, Sydney
Prentice-Hall Canada, Inc., Toronto
Prentice-Hall Hispanoamericana, S.A., Mexico
Prentice-Hall of India Private Limited, New Delhi
Pearson Education Asia Pte. Ltd., Singapore
Prentice-Hall of Japan, Inc., Tokyo
Editora Prentice-Hall do Brazil, Ltda., Rio de Janeiro

Table of Contents

Chapter 1 1
   Section 1.1 1
   Section 1.2 5
   Section 1.3 10
   Section 1.4 15
   Section 1.5 22
   Section 1.6 29
   Section 1.7 36
   Section 1.8 39
   Review Exercises 42
Chapter 2 53
   Section 2.1 53
   Section 2.2 64
   Section 2.3 70
   Section 2.4 77
   Section 2.5 83
   Section 2.6 90
   Review Exercises 94
Chapter 3 104
   Section 3.1 104
   Section 3.2 108
   Section 3.3 111
   Review Exercises 114
Chapter 4 122
   Section 4.1 122
   Section 4.2 128
   Section 4.3 131
   Section 4.4 134
   Section 4.5 141
   Review Exercises 145
Chapter 5 153
   Section 5.1 153
   Section 5.2 162
   Section 5.3 175
   Section 5.4 194
   Section 5.5 204
   Review Exercises 214
Chapter 6 233
   Chapter 6.1 Exercises 233
   Chapter 6.2 Exercises 237
   Chapter 6.3 Exercises 240
   Chapter 6.4 Exercises 245
   Chapter 6.5 Exercises 251
   Chapter 6.6 Exercises 256

Chapter 6.7 Exercises 260
Review Exercises 264
Chapter 7 270
Chapter 7.1 Exercises 270
Chapter 7.2 Exercises 273
Chapter 7.3 Exercises 277
Chapter 7.4 Exercises 281
Chapter 7.5 Exercises 287
Chapter 7.6 Exercises 290
Review Exercises 295
Chapter 8 302
Chapter 8.1 Exercises 302
Chapter 8.2 Exercises 306
Chapter 8.3 Exercises 309
Chapter 8.4 Exercises 314
Chapter 8.5 Exercises 319
Chapter 8.6 Exercises 324
Review Exercises 329
Appendix A 337

Chapter 1 Functions, Modeling, and Average Rate of Change

Section 1.1 Coordinate Systems and Functions

1) The ordered pairs are (-2, 3), (1, 5), and (4, 4). The scatterplot of the points is:

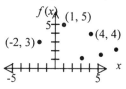

3) The ordered pairs are (-4, -5), (-3.5, -6), (-2, -8) and (2, -11). The scatterplot is:

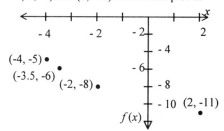

5)

x	y
2	6
4	5
6	4
8	2
10	0

The ordered pairs are (2, 6), (4, 5), (6, 4), (8, 2), and (10, 0).

7)

x	y
10	60
20	50
30	40
40	20
50	10

The ordered pairs are (10, 60), (20, 50), (30, 40), (40, 20), and (50, 10).

9)

x	y
-3	-30
-2	-10
-1	0
0	10
2	10

The ordered pairs are (−3, −30), (−2, −10), (−1, 0), (0, 10), and (2, 10).

11)

Years since 1980	Public Expenditures (in billion of dollars)
0	41
3	44
4	46
5	49
8	51

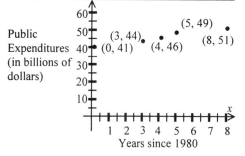

13)

Time	Speed (in miles per hour)
0	0
1	40
2	120
3	160
4	180

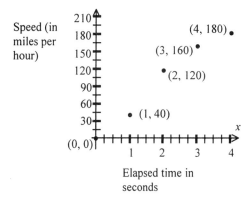

15) This relation is a function since each element of the domain is assigned to exactly one element of the range. We can have two different elements (in this case, 0 & 2) in the domain being assigned to the same value in the range (in this case, 1), but not the other way around.

17) This relation is not a function since we have an element of the domain (− 2) being assigned to two different values in the range (0 and 3).

19) Since every vertical line intersects at most one point of the graph of the relation, the relation is a function.

21) Since there exists a vertical line that intersects the graph of the relation at more than one point (the y-axis for example), the relation is not a function.

23) Since every vertical line intersects the graph of the relation only at one point, the relation is a function.

25) $f(2) = 2(2) - 1 = 4 - 1 = 3$. The ordered pair is (2, 3).

27) $f(1) = 2(1) - 1 = 2 - 1 = 1$. The ordered pair is (1, 1).

29) $g(3) = (3)^2 = 9$. The ordered pair is (3, 9).

31) $g(0) = (0)^2 = 0$. The ordered pair is (0, 0).

33) $g(\frac{1}{2}) = (\frac{1}{2})^2 = \frac{1}{4}$. The ordered pair is $(\frac{1}{2}, \frac{1}{4})$.

35) $g(-0.25) = (-0.25)^2 = 0.0625$. The ordered pair is (− 0.25, 0.0625).

37a) t is the independent variable.

37b) $y = f(t)$ is the dependent variable.

37c) $f(10) \approx 10$.

37d) $f(30) \approx 20$.

37e) Since t is greater than or equal to 0, then the domain of f is $[0, \infty)$.

37f) Since y is greater than or equal to 0, then the range of f is $[0, \infty)$.

39)

Interval Notation	Inequality Notation	Number Line
$[-\frac{1}{2}, 5)$	$-\frac{1}{2} \leq x < 5$	
$[-2, \infty)$	$x \geq -2$	
$(-\infty, -3)$	$x < -3$	
$(-1, 10)$	$-1 < x < 10$	
$(-\infty, 3) \cup (3, \infty)$	$x < 3$ or $x > 3$	

41) Since there are no values that make the denominator of a fraction zero or produce a negative result under an even-indexed radical, the domain of g is all real numbers or $(-\infty, \infty)$.

43) Since there are no values that make the denominator of a fraction zero or produce a negative result under an even-indexed radical, the domain of g is all real numbers or $(-\infty, \infty)$.

45) There are no values that produce a negative result under an even-indexed radical, but if $x = 5$, the denominator of y is 0. Hence, 5 must be excluded and the domain of y is all real numbers except 5 or $(-\infty, 5) \cup (5, \infty)$.

47) There are no values that make the denominator of a fraction zero, but if $x - 6$ is negative, then we have a negative result under the radical. Therefore, $6 - x$ has to be greater than or equal to zero. Solving yields:
$$6 - x \geq 0$$
$$-x \geq -6$$
$$x \leq 6$$
Thus, the domain of f is all real numbers less than or equal to 6 or $(-\infty, 6]$.

49) There are no values that make the denominator of a fraction zero, but if $x^2 + 3$ is negative, then we have a negative result under the radical. Since $x^2 \geq 0$, then $x^2 + 3$ will always be greater than zero for any real number x. Thus, the domain of y is all real numbers or $(-\infty, \infty)$.

51) There are no values that produce a negative result under an even-indexed radical, but if $2x^2 - 4x = 0$, the denominator of $f(x)$ is 0. Solving yields: $2x^2 - 4x = 0$
$$2x(x - 2) = 0$$
$$2x = 0 \text{ or } x - 2 = 0$$
$$x = 0 \text{ or } x = 2$$
Therefore, 0 and 2 must be excluded and the domain of f is all real numbers except 2 and 0 or $(-\infty, 0) \cup (0, 2) \cup (2, \infty)$.

53) The graph $f(x)$ in viewing window $[-3, 3]$ by $[0, 2]$ is displayed below:

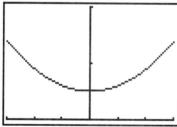

Note: The textbook writes the viewing window on the sides of the graph.
The domain is $(-\infty, \infty)$.
The range is $[\frac{1}{2}, \infty)$.

55) The graph y in viewing window $[-10, 10]$ by $[-0.7, 0.3]$ is displayed below:

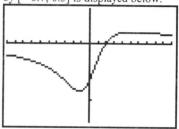

The domain is $(-\infty, \infty)$.
To determine the range is not quite as easy. In looking at the picture it seems that the lower endpoint about -0.4 and the higher endpoint about 0.1. We can use the maximum and minimum commands of the calculator to get a more precise answer:

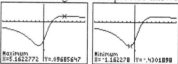

Rounding to the nearest hundredth, we get the range is $[-0.43, 0.1]$.

57a) $p(0) = \frac{165-(0)}{4} = \frac{165}{4} = 41.25$

$p(5) = \frac{165-(5)}{4} = \frac{160}{4} = 40$

$p(10) = \frac{165-(10)}{4} = \frac{155}{4} = 38.75$

$p(15) = \frac{165-(15)}{4} = \frac{150}{4} = 37.5$

$p(20) = \frac{165-(20)}{4} = \frac{145}{4} = 36.25$

$p(25) = \frac{165-(25)}{4} = \frac{140}{4} = 35$

$p(30) = \frac{165-(30)}{4} = \frac{135}{4} = 33.75$

$p(35) = \frac{165-(35)}{4} = \frac{130}{4} = 32.5$

x	0	5	10	15	20	25	30	35
$p(x)$	41.25	40	38.75	37.5	36.25	35	33.75	32.5

57b)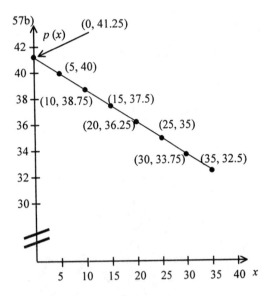

57c) Since $p(0) = 41.25$ and $p(80) = 32.5$, the range is $[32.5, 41.25]$.

57d) $p(22) = 35.75$. If 22 units of Teddy Bear designer lingerie are sold per day, the price will be $35.75 per unit.

59) Since $p(x) = \frac{165-x}{4}$, then
$p(18) = \frac{165-(18)}{4} = \frac{147}{4} = 36.75$ or $36.75.

61) The domain is $(-\infty, \infty)$.
The range is $(-\infty, \infty)$.

63) The domain is $(-5, \infty)$.
The range is about $(-\infty, 7]$.

65) The domain is $(-50, 100]$.
The range is $[0, 10)$.

Section 1.2 Linear Functions and Average Rate of Change

1) $m = \frac{y_2 - y_1}{x_2 - x_1} = \frac{3-8}{5-4} = \frac{-5}{1} = -5.$

3) $m = \frac{y_2 - y_1}{x_2 - x_1} = \frac{8-3}{-4-2} = \frac{5}{-6} = -\frac{5}{6}.$

5) $m = \frac{y_2 - y_1}{x_2 - x_1} = \frac{8.3-6.1}{7-5} = \frac{2.2}{2} = 1.1.$

7) $m = \frac{y_2 - y_1}{x_2 - x_1} = \frac{b-(b-1)}{a-(a-1)} = \frac{b-b+1}{a-a+1} = \frac{1}{1} = 1.$

9) $m = \frac{\text{Change in } y}{\text{Change in } x} = \frac{-5}{3} = -\frac{5}{3}.$

11) $m = \frac{y_2 - y_1}{x_2 - x_1} = \frac{0-40}{10-0} = \frac{-40}{10} = -4.$

13) $m = \frac{y_2 - y_1}{x_2 - x_1} = \frac{22-5}{5-1} = \frac{17}{4}.$

15) The slope is $m = \frac{y_2 - y_1}{x_2 - x_1} = \frac{-3-6}{5-2} = \frac{-9}{3} = -3.$
The equation is $y - y_1 = m(x - x_1)$, so
$y - 6 = -3(x - 2)$ or $y + 3 = -3(x - 5)$

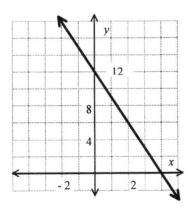

17) The slope is $m = \frac{y_2 - y_1}{x_2 - x_1} = \frac{13-5}{5-3} = \frac{8}{2} = 4.$
The equation is $y - y_1 = m(x - x_1)$, so
$y - 5 = 4(x - 3)$ or $y - 13 = 4(x - 5)$

17) Continued

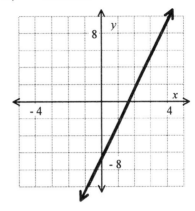

19) The origin is the point (0, 0) and the y-intercept at -3 is the point $(0, -3)$. Hence, $m = \frac{y_2 - y_1}{x_2 - x_1} = \frac{-3-0}{0-0} = \frac{-3}{0}$ is undefined. We have a vertical line of $x = 0$ or the y-axis.

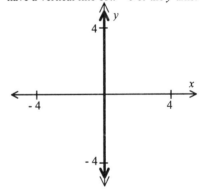

21) A vertical line passing through the point $(-2, 9)$ will have an equation of $x = -2$.

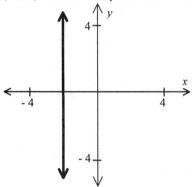

23) The slope is $m = 3$; $(x_1, y_1) = (8, 4)$.
The equation is $y = m(x - x_1) + y_1$, so
$y = 3(x - 8) + 4$.

25) The slope is $m = 5$; $(x_1, y_1) = (0, -13)$.
The equation is $y = m(x - x_1) + y_1$, so
$y = 5(x - 0) - 13$.

27) The slope is $m = 0.3$; $(x_1, y_1) = (-4, 0)$.
The equation is $y = m(x - x_1) + y_1$, so
$y = 0.3(x - (-4)) + 0$. Simplifying yields:
$y = 0.3(x + 4)$.

29) The slope is $m = \frac{y_2 - y_1}{x_2 - x_1} = \frac{1 - 2}{3 - (-1)} = \frac{-1}{4} = -\frac{1}{4}$.
The equation is $y = m(x - x_1) + y_1$, so
$y = -\frac{1}{4}(x - (-1)) + 2$ or $y = -\frac{1}{4}(x - 3) + 1$.
Simplifying yields:
$y = -\frac{1}{4}(x + 1) + 2$ or $y = -\frac{1}{4}(x - 3) + 1$.

31) The slope is $m = \frac{y_2 - y_1}{x_2 - x_1} = \frac{\frac{3}{2} - 0}{0 - (-12)} = \frac{\frac{3}{2}}{12}$
$= \frac{3}{2} \cdot \frac{1}{12} = \frac{3}{24}$.
The equation is $y = m(x - x_1) + y_1$, so
$y = \frac{3}{24}(x - (-12)) + 0$ or $y = \frac{3}{24}(x - 0) + \frac{3}{2}$.
Simplifying yields:
$y = \frac{3}{24}(x + 12)$ or $y = \frac{3}{24}x + \frac{3}{2}$.

33) Let $(x_1, y_1) = (0, 6)$ & $(x_2, y_2) = (4, 1.3)$.
The slope $m = \frac{y_2 - y_1}{x_2 - x_1} = \frac{1.3 - 6}{4 - 0} = \frac{-4.7}{4} = -\frac{47}{40}$.
The equation is $y = m(x - x_1) + y_1$, so
$y = -\frac{47}{40}(x - 0) + 6$ or $y = -\frac{47}{40}(x - 4) + 1.3$.
Simplifying yields:
$y = -\frac{47}{40}x + 6$ or $y = -\frac{47}{40}(x - 4) + 1.3$.

35) y-int: $f(0) = -4(0 - 1) + 1 = 4 + 1 = 5$.
x-int: $f(x) = -4(x - 1) + 1 = 0$; solve for x
$-4x + 4 + 1 = -4x + 5 = 0$
$-4x = -5$
$x = 1.25$
y-intercept is $(0, 5)$ & x-intercept is $(1.25, 0)$.

37) y-int.: $g(0) = 120 - 0.3(0) = 120$.
x-int.: $g(x) = 120 - 0.3x = 0$; solve for x
$120 = 0.3x$
$400 = x$
y-intercept is $(0, 120)$ & x-intercept is $(400, 0)$.

39) y-int.: $g(0) = 0.2((0) - 5.1) + 10$
$= -1.02 + 10 = 8.98$.
x-int.: $g(x) = 0.2(x - 5.1) + 10 = 0$; solve
$0.2x - 1.02 + 10 = 0$
$0.2x + 8.98 = 0$
$0.2x = -8.98$
$x = -44.9$
y-int. is $(0, 8.98)$ & x-int. is $(-44.9, 0)$.

41a) Using the linear function in the form of $f(x) = mx + b$ and the fact that the machine costs $25,000 new, we can replace b by $25,000$. Since the machine depreciates at a rate of $1,250 per year, we can replace m by -1250. Thus, the equation of the depreciation function is:
$f(x) = -1250x + 25,000$.

41b) Setting $f(x) = 10,000$ and solving for x, we get:
$-1250x + 25,000 = 10,000$
$-1250x = -15,000$
$x = 12$
So, when the machine is twelve years old, it will be worth $10,000.

43a) Using the linear function in the form of $f(x) = mx + b$ and the fact that the machine costs \$90,000 new, we can replace b by 90,000. Since the machine depreciates at a rate of \$6,000 per year, we can replace m by -6000. Thus, the equation of the depreciation function is:
$f(x) = -6000x + 90,000$.

43b) Half of \$90,000 is \$45,000.
Setting $f(x) = 45,000$ and solving for x, we get:
$-6000x + 90,000 = 45,000$
$-6000x = -45,000$
$x = 7.5$
So, after 7.5 years, the machine will be worth half of its original value.

45) We can rewrite $f(x)$ in slope-intercept form:
$f(x) = -4.5(x - 2) = -4.5x + 9$. Since the slope of the line is -4.5, f is decreasing.

47) We can rewrite $f(x)$ in slope-intercept form:
$f(x) = \frac{500 + 60x}{15} = 4x + \frac{100}{3}$. Since the slope of the line is $+4$, f is increasing.

49) $f(x)$ is already written in slope-intercept form. The slope of the line is $+4$, so f is increasing.

51) The slope is $m = \frac{y_2 - y_1}{x_2 - x_1} = \frac{2 - 5}{1 - 10} = \frac{-3}{-9} = \frac{1}{3}$.
Since $m = \frac{1}{3}$, f is increasing.

53) The slope is $m = \frac{y_2 - y_1}{x_2 - x_1} = \frac{700 - 0}{50 - 0} = \frac{700}{50} = 14$.
Since $m = 14$, f is increasing.

55) m will be the variable costs and b will be the fixed costs in $y = mx + b$. So, $m = 147.75$ and $b = 2700$. Thus, the linear cost function will be $C(x) = 147.75x + 2700$.

57) First find the slope and then plug into the point-slope formula and solve for y.
The slope is m $= \frac{y_2 - y_1}{x_2 - x_1} = \frac{30000 - 7000}{50 - 10} = \frac{23000}{40}$
$= 575$. Using (10, 7000) for (x_1, y_1), the equation $y - y_1 = m(x - x_1)$ becomes:
$y - 7000 = 575(x - 10)$. Solving for y yields:
$y - 7000 = 575x - 5750$
$y = 575x + 1250$
Thus, the linear cost function will be $C(x) = 575x + 1250$.

59) Plug $m = 80$ and $(x_1, y_1) = (40, 3850)$ into the point-slope formula and solve for y.
$y - y_1 = m(x - x_1)$
$y - 3850 = 80(x - 40)$
$y - 3850 = 80x - 3200$
$y = 80x + 650$
Thus, the linear cost function will be $C(x) = 80x + 650$.

61a) m will be the variable costs and b will be the fixed costs in $y = mx + b$. So, since $C(x) = 1.25x + 550$ the fixed costs is \$550 and the variable costs is \$1.25 per bottle.

61b) $C(70) = 1.25(70) + 550 = 87.5 + 550$
$= 637.5$. Hence, it costs \$637.50 to produce 70 bottles.

61c) If $C(x) = \$693.75$, then
$693.75 = 1.25x + 550$.
Solving for x yields:
$693.75 = 1.25x + 550$
$143.75 = 1.25x$
$115 = x$
Hence, 115 bottles can be produced for \$693.75.

61d) In a linear function, the marginal cost is the slope of the line. Hence, the marginal cost is \$1.25 per bottle.

61e) Since the marginal cost is constant, it will cost \$1.25 to produce the 71st bottle.

63)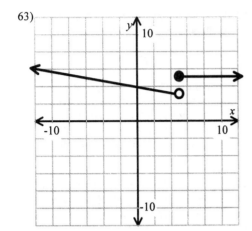

65)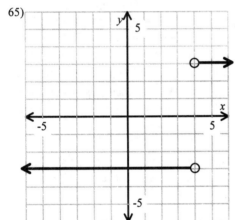

67)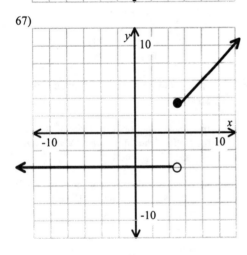

69)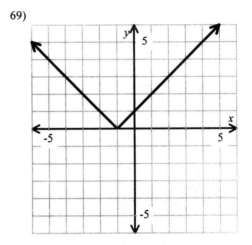

$$f(x) = |x+1| = \begin{cases} -x-1 & x < -1 \\ x+1 & x \geq -1 \end{cases}$$

71)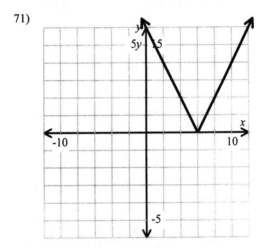

$$g(x) = |6-2x| = \begin{cases} 6-2x & x < 3 \\ -6+2x & x \geq 3 \end{cases}$$

73)

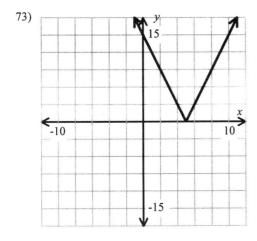

$$g(x) = |3x - 15| = \begin{cases} -3x + 15, & x < 5 \\ 3x - 15, & x \geq 5 \end{cases}$$

75a) The fixed costs are $20. The variable costs are $0.15 per mile.

75b) $C(x)$ = variable costs $\cdot x$ + fixed costs
Thus, $C(x) = 0.15x + 20$.

75c) A realistic domain would be $(0, \infty)$.

75d) $C(120) = 0.15(120) + 20 = 38$. It costs $38 to rent a truck and drive 120 miles.

75e) Since $C(x)$, the cost of driving the 121st mile is equal to the slope of the line. Thus, the cost of driving the 121st mile is $0.15. This is called the marginal cost.

77a) Since the fixed costs are equal to $C(0)$, then the fixed costs is $50. This corresponds to when 0 hours of labor on Table 1.2.10.

77b) $m = \frac{y_2 - y_1}{x_2 - x_1} = \frac{65 - 50}{1 - 0} = \frac{15}{1} = 15$. It costs $15 per hour for each additional hour of labor. This is called the variable or marginal cost.

77c) $C(x)$ = variable costs $\cdot x$ + fixed costs
Thus, $C(x) = 15x + 50$.

77d) $C(4.5) = 15(4.5) + 50 = 67.5 + 50 = 117.5$. It costs $117.50 to repair the microwave.

79a) The rate of change is equal to the slope. Since $m = 0.19$, then the amount of spending is increasing at a rate of $0.19 billion per year.

79b) Since $x = 3$ corresponds to 1985, then $f(3) = 0.19(3) + 1.67 = 0.57 + 1.67 = 2.24$. So, $2.24 billion was spent in college bookstores in 1985.

79c) Since $x = 8$ corresponds to 1990, then $f(8) = 0.19(8) + 1.67 = 1.52 + 1.67 = 3.19$. So, $3.19 billion was spent in college bookstores in 1990.

79d) Using the results from parts b and c, we compute: $3.19 - 2.24 = 0.95$. So $0.95 billion more was spent in 1990 than in 1985.

Section 1.3 Quadratic Functions and Average Rate of Change

1a) Since $f(x) = x^2 + 6x + 5$, a = 1 & b = 6. The x-coordinate of the vertex is $x = \frac{-b}{2a}$ so $x = \frac{-6}{2(1)} = \frac{-6}{2} = -3$. The y-coordinate is $f\left(\frac{-b}{2a}\right) = f(-3) = (-3)^2 + 6(-3) + 5 = 9 - 18 + 5 = -4$. Hence, the vertex is $(-3, -4)$.

1b) Since a > 0, the vertex is the minimum value of f. This means that f is increasing on $(-3, \infty)$ and decreasing on $(-\infty, -3)$.

3a) Since $g(x) = -x^2 + 6x + 6$, a = -1 and b = 6. The x-coordinate of the vertex is $x = \frac{-b}{2a}$ so $x = \frac{-6}{2(-1)} = \frac{-6}{-2} = 3$. The y-coordinate is $g\left(\frac{-b}{2a}\right) = g(3) = -(3)^2 + 6(3) + 6 = -9 + 18 + 6 = 15$. Hence, the vertex is $(3, 15)$.

3b) Since a < 0, the vertex is the maximum value of g. This means that g is increasing on $(-\infty, 3)$ and decreasing on $(3, \infty)$.

5a) Since $g(x) = 5x^2 + 6x - 3$, a = 5 & b = 6. The x-coordinate of the vertex is $x = \frac{-b}{2a}$ so $x = \frac{-(6)}{2(5)} = \frac{-6}{10} = -0.6$. The y-coordinate is $g\left(\frac{-b}{2a}\right) = g(-0.6) = 5(-0.6)^2 + 6(-0.6) - 3 = 1.8 - 3.6 - 3 = -4.8$. Hence, the vertex is $(-0.6, -4.8)$.

5b) Since a > 0, the vertex is the minimum value of g. This means that g is increasing on $(-0.6, \infty)$ and decreasing on $(-\infty, -0.6)$.

7a) Since $f(x) = 0.14x^2 + 0.5x - 0.3$, a = 0.14 & b = 0.5. The x-coordinate of the vertex is $x = \frac{-b}{2a}$ so $x = \frac{-0.5}{2(0.14)} = \frac{-0.5}{0.28} = \frac{-50}{28} = \frac{-25}{14} \approx -1.786$. The y-coordinate is $f\left(\frac{-b}{2a}\right) = f\left(\frac{-25}{14}\right) = 0.14\left(\frac{-25}{14}\right)^2 + 0.5\left(\frac{-25}{14}\right) - 0.3 = \frac{25}{56} - \frac{25}{28} - 0.3 = \frac{-209}{280} \approx -0.746$. Hence, the vertex is $\left(\frac{-25}{14}, \frac{-209}{280}\right) \approx (-1.79, -0.75)$.

7b) Since a > 0, the vertex is the minimum value of f. This means that f is increasing on $(\frac{-25}{14}, \infty)$ and decreasing on $(-\infty, \frac{-25}{14})$.

9a) Since $f(x) = -0.9x^2 - 1.8x + 0.5$, a = -0.9 and b = -1.8. The x-coordinate of the vertex is $x = \frac{-b}{2a}$ so $x = \frac{-(-1.8)}{2(-0.9)} = \frac{-1.8}{1.8} = -1$. The y-coordinate is $f\left(\frac{-b}{2a}\right) = f(-1) = -0.9(-1)^2 - 1.8(-1) + 0.5 = -0.9 + 1.8 + 0.5 = 1.4$. Hence, the vertex is $(-1, 1.4)$.

9b) Since a < 0, the vertex is the maximum value of f. This means that f is increasing on $(-\infty, -1)$ and decreasing on $(-1, \infty)$.

11) Setting $f(x) = 0$ and solving yields:
$x^2 - 16 = 0$
$(x - 4)(x + 4) = 0$
$(x - 4) = 0$ or $(x + 4) = 0$
$x = 4$ or $x = -4$.
Hence, the zeros of f are -4 and 4.

13) Setting $f(x) = 0$ and solving yields:
$$-2x^2 - 4x = 0$$
$$-2x(x+2) = 0$$
$$-2x = 0 \text{ or } (x+2) = 0$$
$$x = 0 \text{ or } x = -2.$$
Hence, the zeros of f are -2 and 0.

15) Setting $f(x) = 0$ and solving yields:
$$x^2 - 5x + 6 = 0$$
$$(x-2)(x-3) = 0$$
$$(x-2) = 0 \text{ or } (x-3) = 0$$
$$x = 2 \text{ or } x = 3.$$
Hence, the zeros of f are 2 and 3.

17) Setting $g(x) = 0$ and solving yields:
$$6x^2 - 5x - 50 = 0$$
$$(3x - 10)(2x + 5) = 0$$
$$(3x - 10) = 0 \text{ or } (2x + 5) = 0$$
$$x = \tfrac{10}{3} \text{ or } x = -2.5.$$
Hence, the zeros of g are -2.5 and $\tfrac{10}{3}$.

19) Since $f(x) = x^2 - x - 1$, $a = 1$, $b = -1$, and $c = -1$. Plugging into the quadratic formula yields:
$$x = \tfrac{-b \pm \sqrt{b^2 - 4ac}}{2a} = \tfrac{-(-1) \pm \sqrt{(-1)^2 - 4(1)(-1)}}{2(1)}$$
$$= \tfrac{1 \pm \sqrt{1+4}}{2} = \tfrac{1 \pm \sqrt{5}}{2}.$$ Therefore, there are two real roots: $x = \tfrac{1+\sqrt{5}}{2}$ and $x = \tfrac{1-\sqrt{5}}{2}$.

21) Since $y = 11x^2 - 7x + 1$, $a = 11$, $b = -7$, and $c = 1$. Plugging into the quadratic formula yields:
$$x = \tfrac{-b \pm \sqrt{b^2 - 4ac}}{2a} = \tfrac{-(-7) \pm \sqrt{(-7)^2 - 4(11)(1)}}{2(11)}$$
$$= \tfrac{7 \pm \sqrt{49-44}}{22} = \tfrac{7 \pm \sqrt{5}}{22}.$$ Therefore, there are two real roots: $x = \tfrac{7+\sqrt{5}}{22}$ and $x = \tfrac{7-\sqrt{5}}{22}$.

23) Since $f(x) = 2x^2 + 2x + 1$, $a = 2$, $b = 2$, and $c = 1$. Plugging into the quadratic formula yields:
$$x = \tfrac{-b \pm \sqrt{b^2 - 4ac}}{2a} = \tfrac{-(2) \pm \sqrt{(2)^2 - 4(2)(1)}}{2(2)}$$
$$= \tfrac{-2 \pm \sqrt{4-8}}{4} = \tfrac{-2 \pm \sqrt{-4}}{4}.$$ Since the discriminant is negative, there are *no real roots*.

25) The following graphs are graphed in viewing window $[-10, 10]$ by $[-30, 30]$

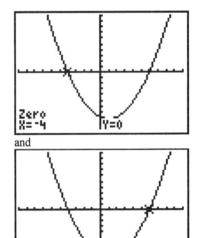

and

Note: The textbook writes the viewing window on the sides of the graph.
So the zeros are $x = -4$ and $x = 6$.

27) The following graph is graphed in viewing window $[-10, 10]$ by $[-30, 30]$

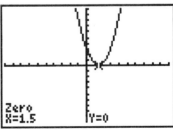

So the zero is $x = 1.5$.

12 Chapter 1 Functions, Modeling, and Average Rate of Change

29) The following graphs are graphed in viewing window [– 10, 10] by [– 10, 10]

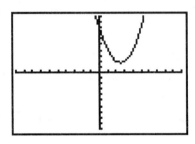

Since the graph does not intersect the *x*-axis, there are *no real zeros*. If you try to find a zero on your calculator, you will get an error message like this one:

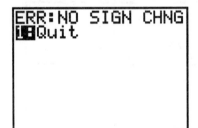

31a) $m_{sec} = \frac{f(x+\Delta x) - f(x)}{\Delta x} = \frac{f(6)-f(0)}{6-0}$

$= \frac{-5(6)+1 - [-5(0)+1]}{6} = \frac{-29-1}{6} = \frac{-30}{6} = -5.$

31b) $m_{sec} = \frac{f(x+\Delta x) - f(x)}{\Delta x} = \frac{f(3)-f(0)}{3-0}$

$= \frac{-5(3)+1 - [-5(0)+1]}{3} = \frac{-14-1}{3} = \frac{-15}{3} = -5.$

31c) $m_{sec} = \frac{f(x+\Delta x) - f(x)}{\Delta x} = \frac{f(4)-f(3)}{4-3}$

$= \frac{-5(4)+1 - [-5(3)+1]}{1} = \frac{-19+14}{1} = \frac{-5}{1} = -5.$

33a) $m_{sec} = \frac{f(x+\Delta x) - f(x)}{\Delta x} = \frac{f(4)-f(0)}{4-0}$

$= \frac{(4)^2 + (4) - [(0)^2 + (0)]}{4} = \frac{20-0}{4} = \frac{20}{4} = 5.$

33b) $m_{sec} = \frac{f(x+\Delta x) - f(x)}{\Delta x} = \frac{f(2)-f(0)}{2-0}$

$= \frac{(2)^2 + (2) - [(0)^2 + (0)]}{2} = \frac{6-0}{2} = \frac{6}{2} = 3.$

33c) $m_{sec} = \frac{f(x+\Delta x) - f(x)}{\Delta x} = \frac{f(4)-f(2)}{4-2}$

$= \frac{(4)^2 + (4) - [(2)^2 + (2)]}{2} = \frac{20-6}{2} = \frac{14}{2} = 7$

35a) $m_{sec} = \frac{f(x+\Delta x) - f(x)}{\Delta x} = \frac{f(2)-f(0)}{2-0}$

$= \frac{(2)^2 - (2)+3 - [(0)^2 - (0)+3]}{2} = \frac{5-3}{2} = \frac{2}{2} = 1.$

35b) $m_{sec} = \frac{f(x+\Delta x) - f(x)}{\Delta x} = \frac{f(1)-f(0)}{1-0}$

$= \frac{(1)^2 - (1)+3 - [(0)^2 - (0)+3]}{1} = \frac{3-3}{1} = \frac{0}{1} = 0.$

35c) $m_{sec} = \frac{f(x+\Delta x) - f(x)}{\Delta x} = \frac{f(2)-f(1)}{2-1}$

$= \frac{(2)^2 - (2)+3 - [(1)^2 - (1)+3]}{1} = \frac{5-3}{1} = \frac{2}{1} = 2.$

37) $\frac{f(5)-f(0)}{5-0} = \frac{\frac{1}{5}(5)^2 + 100(5)+30 - \left[\frac{1}{5}(0)^2 + 100(0)+30\right]}{5}$

$= \frac{535-30}{5} = \frac{505}{5} = 101.$ Over a five-year period, the number of citations was increasing at an average rate of 101 citations per year.

39a) $s(3) = -16(3)^2 + 80(3) = -144 + 240 = 96$. After 3 seconds, the rock is 96 feet above the ground.

39b) $m_{sec} = \frac{s(t+\Delta t)-s(t)}{\Delta t} = \frac{s(3)-s(1)}{3-1} =$

$\frac{-16(3)^2 + 80(3) - [-16(1)^2 + 80(1)]}{2} = \frac{96-64}{2} = \frac{32}{2}$

$= 16.$ Between 1 and 3 seconds, the rock was rising at an average speed of 16 feet per second.

41a) $g(3) = (3)^2 + 8(3) + 2000$
       $= 9 + 24 + 2000 = 2033$.
       This means that after 3 hours, there were 2,033 bacteria in the colony.

41b) $m_{sec} = \frac{g(t+\Delta t) - g(t)}{\Delta t} = \frac{g(6)-g(3)}{6-3}$
       $= \frac{(6)^2 + 8(6) + 2000 - [(3)^2 + 8(3) + 2000]}{3}$
       $= \frac{2084 - 2033}{3} = \frac{51}{3} = 17$.
       This means that from 3 to 6 hours after the colony was started, the colony was growing at an average rate of 17 bacteria per hour.

43a) $f(2) = -3.08(2)^2 + 40.35(2) + 305.89$
       $= -12.32 + 80.7 + 305.89 = 374.27$.
       This means that in 1988, there were 374.27 aggravated assaults per 100,000 people. The graph will open down since this is a quadratic equation and the coefficient of the squared term is negative.

43b) The graph is displayed in viewing window [0, 10] by [0, 500]

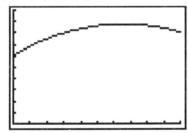

The greatest assault rate will occur at the vertex since this is a quadratic function and a < 0. Using $x = \frac{-b}{2a}$, we get:
$x = \frac{-40.35}{2(-3.08)} = \frac{-40.35}{-6.16} \approx 6.55$. The greatest assault rate occurred sometime during the middle of 1992.

43c) $m_{sec} = \frac{f(x+\Delta x) - f(x)}{\Delta x} = \frac{f(9)-f(4)}{9-4} =$
       $\frac{-3.08(9)^2 + 40.35(9) + 305.89 - [-3.08(4)^2 + 40.35(4) + 305.89]}{5}$
       $= \frac{419.56 - 418.01}{5} = \frac{1.55}{5} = 0.31$. The number of assaults per 100,000 was increasing at an average rate of 0.31 people per year between 1990 and 1995.

45a) $f(6) = 0.89(6)^2 - 1.93(6) + 3306.27$
       $= 32.04 - 11.58 + 3306.27 = 3326.73$.
       This means that in 1980, 3,326.73 calories were consumed each day per person in the U.S.

45b) The graph is displayed in viewing window [0, 15] by [3200, 3500].

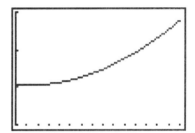

45c) Setting $f(x) = 3400$ and solving yields:
       $0.89x^2 - 1.93x + 3306.27 = 3400$
       $0.89x^2 - 1.93x - 93.73 = 0$.
       Therefore, a = 0.89, b = − 1.93, and c = − 93.73. Plugging into the quadratic formula yields:
       $x = \frac{-b \pm \sqrt{b^2 - 4ac}}{2a} =$
       $\frac{-(-1.93) \pm \sqrt{(-1.93)^2 - 4(0.89)(-93.73)}}{2(0.89)}$
       $= \frac{1.93 \pm \sqrt{3.7249 + 333.6788}}{1.78} = \frac{1.93 \pm \sqrt{337.4037}}{1.78}$
       ≈ 11.4037 or -16.4386. The negative solution is thrown out. So, when $x = 11$ or in 1985, the number of calories consumed exceeded 3400 calories per day.

45d) $m_{sec} = \dfrac{f(x+\Delta x) - f(x)}{\Delta x} = \dfrac{f(11)-f(1)}{11-1} =$

$= \dfrac{0.89(11)^2 - 1.93(11) + 3306.27 - [0.89(1)^2 - 1.93(1) + 3306.27]}{10}$

$= \dfrac{3392.73 - 3305.23}{10} = \dfrac{87.5}{10} = 8.75$

From 1975 to 1985, the average number of calories consumed per person per day was increasing at an average rate of 8.75 calories per year.

Section 1.4 Operations on Functions

1a) $(f+g)(x) = f(x) + g(x) = 6x + 3 + 4x - 1$
$= 10x + 2$. The domain is $(-\infty, \infty)$.

1b) $(f-g)(x) = f(x) - g(x)$
$= 6x + 3 - (4x - 1) = 2x + 4$. The domain is $(-\infty, \infty)$.

1c) $(f \cdot g)(x) = f(x) \cdot g(x) = (6x+3)(4x-1)$
$= 24x^2 + 6x - 3$. The domain is $(-\infty, \infty)$.

1d) $\left(\dfrac{f}{g}\right)(x) = \dfrac{f(x)}{g(x)} = \dfrac{6x+3}{4x-1}$. Setting the denominator equal to zero and solving yields:
$\quad 4x - 1 = 0$
$\quad 4x = 1$
$\quad x = 0.25$
This value must be excluded so the domain is $(-\infty, 0.25) \cup (0.25, \infty)$.

3a) $(f+g)(x) = f(x) + g(x)$
$= \sqrt{x+5} + x^2 + 5 = x^2 + \sqrt{x+5} + 5$.
Since $x + 5 \geq 0$, then $x \geq -5$. So, the domain is $[-5, \infty)$.

3b) $(f-g)(x) = f(x) - g(x)$
$= \sqrt{x+5} - (x^2 + 5) = -x^2 + \sqrt{x+5} - 5$.
Since $x + 5 \geq 0$, then $x \geq -5$. So, the domain is $[-5, \infty)$.

3c) $(f \cdot g)(x) = f(x) \cdot g(x)$
$= (\sqrt{x+5})(x^2+5) = x^2\sqrt{x+5} + 5\sqrt{x+5}$.
Since $x + 5 \geq 0$, then $x \geq -5$. So, the domain is $[-5, \infty)$.

3d) $\left(\dfrac{f}{g}\right)(x) = \dfrac{f(x)}{g(x)} = \dfrac{\sqrt{x+5}}{x^2+5}$. Since $x + 5 \geq 0$, then $x \geq -5$. Also, setting the denominator equal to zero and solving yields:
$\quad x^2 + 5 = 0$
$\quad x^2 = -5$
$\quad$ No Solution
There are no additional values that must be excluded so the domain is $[-5, \infty)$.

5) $(f+g)(0) = f(0) + g(0)$
$= 2(0)^2 - 4(0) + 5(0) + 1$
$= 0 - 0 + 0 + 1 = 1$.

7) $(f \cdot g)(4) = f(4) \cdot g(4)$
$= [2(4)^2 - 4(4)] \cdot [5(4) + 1]$
$= 16 \cdot 21 = 336$.

9) $(f-g)(2) = f(2) - g(2)$
$= 2(2)^2 - 4(2) - [5(2) + 1]$
$= 8 - 8 - [11] = -11$.

11) $\left(\dfrac{f}{g}\right)(5) = \dfrac{f(5)}{g(5)} = \dfrac{2(5)^2 - 4(5)}{5(5)+1} = \dfrac{50-20}{25+1} = \dfrac{30}{26} = \dfrac{15}{13}$.

13a) $(f-g)(20) = f(20) - g(20)$
$= 9.75(20) - [\frac{4}{5}(20)^2 + 6(20)]$
$= 195 - [320 + 120] = 195 - 440 = -245$.
The Handi-Neighbor Hardware Store makes $245 less than the U-Do-It Store when they both sell 20 hammers.

13b) $(f+g)(20) = f(20) + g(20)$
$= 9.75(20) + [\frac{4}{5}(20)^2 + 6(20)]$
$= 195 + [320 + 120] = 195 + 440 = 635$.
Both stores will make $635 combined if each sells 20 hammers.

15a) The conviction rate can be found by taking 100% times the number of convictions divided by the number of arrests. Since $x = 7$ corresponds to 1967, the conviction rate is: $\dfrac{g(7)}{f(7)} \cdot 100\%$
$= \dfrac{(7)+35}{4(7)+45} \cdot 100\% = \dfrac{42}{73} \cdot 100\% \approx 57.53\%$.

15b) The conviction rate can be found by taking 100% times the number of convictions divided by the number of arrests:
$\dfrac{g(x)}{f(x)} \cdot 100\% = \left(\dfrac{x+35}{4x+45}\right) \cdot 100\%$.

16   Chapter 1 Functions, Modeling, and Average Rate of Change

17a) The graph is displayed in viewing window [1, 11] by [0, 1].

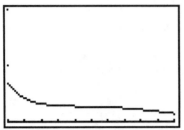

Note: The textbook writes the viewing window on the sides of the graph.

17b) We can display a table of values for $x = 1$, 3, 5, 7, 9, and 11 where Y_3 is the conviction rate:

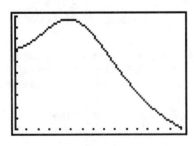

The decrease in the conviction rate over [1, 3] was $0.15921 - 0.33939 \approx -0.18018$. The decrease in the conviction rate over [5, 7] was $0.12817 - 0.13699 \approx -0.00882$. The decrease in the conviction rate over [9, 11] was $0.07179 - 0.10723 \approx -0.0354$. The decrease was the largest over [1, 3].

19a) The graph is displayed in viewing window [0, 15] by [400, 1400].

19b)

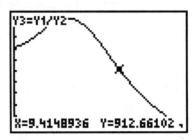

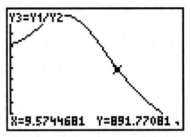

We see that at $x \approx 9.5$, the payment dropped below $900 per recipient. So, by 1989, the amount paid to each recipient had dropped below $900.

21) $R(x) = x \bullet p(x) = x(87.1) = 87.1x$.

23) $R(x) = x \bullet p(x) = x(-0.19x) = -0.19x^2$.

25) $R(x) = x \bullet p(x) = x\left(\frac{-x}{2000} + 3\right)$
   $= \frac{-x^2}{2000} + 3x$.

27) $R(x) = x \bullet p(x) = x(-0.12x + 30)$
   $= -0.12x^2 + 30x$.

29a) The graph is displayed in viewing window [0, 800] by [0, 8000]

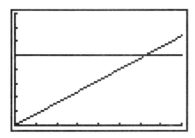

29b) The break-even point is when $R(x) = C(x)$. Solving $8x = 5000$ yields $x = 625$. $R(625) = 8(625) = 5000$. So, the break-even point is (625, 5000).

29c) The amount of revenue generated is $5,000 in order to break even.

31a) The graph is displayed in viewing window [0, 100] by [0, 3000].

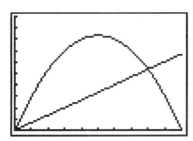

31b) The break-even point is when $R(x) = C(x)$. Using the INTERSECT command, and adjusting the viewing window to [0, 100] by [−800, 3000], we find that the functions intersect at:

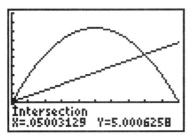

31b) Continued and

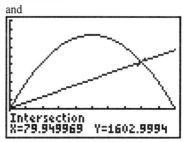

Rounding to the nearest hundredth, the break-even points are (0.05, 5.00) and (79.95, 1603.00).

31c) The amount of revenue generated is about $5.00 or $1603.00 in order to break even.

33) $P(x) = R(x) - C(x)$
$= 1200x - 37x^2 - (4300 + 148x)$
$= -37x^2 + 1052x - 4300$.
The x-coordinate of the vertex is $x = \frac{-b}{2a}$ so $x = \frac{-(1052)}{2(-37)} = \frac{-1052}{-74} = \frac{526}{37} \approx 14.22$. Plugging $x = \frac{526}{37}$ into the function, we find the y-coordinate to be:
$-37(\frac{526}{37})^2 + 1052(\frac{526}{37}) - 4300$
$= -\frac{276676}{37} + \frac{553352}{37} - 4300 = 3177\frac{27}{37} \approx$
3177.73. Hence, the vertex is about (14.22, 3177.73). Since $a < 0$, we have a maximum profit of about $3177.73 at $x \approx 14.22$.

35) $P(x) = R(x) - C(x)$
$= -2.1x^2 + 500x - (80x + 9500)$
$= -2.1x^2 + 420x - 9500$.
The x-coordinate of the vertex is $x = \frac{-b}{2a}$ so $x = \frac{-(420)}{2(-2.1)} = \frac{-420}{-4.2} = 100$. Plugging $x = 100$ into the function, we find the y-coordinate to be:
$-2.1(100)^2 + 420(100) - 9500$
$= -21,000 + 42,000 - 9500 = 11,500$.
Hence, the vertex is (100, 11,500). Since $a < 0$, we have a maximum profit of $11,500 at $x = 100$.

37a) $R(x) = x \cdot p(x) = x(37.8 - x)$
$= 37.8x - x^2$.
$C(x) = \begin{pmatrix} \text{Variable} \\ \text{Costs} \end{pmatrix} x + \begin{pmatrix} \text{Fixed} \\ \text{Costs} \end{pmatrix}$
$= 29x + 10.15$
$P(x) = R(x) - C(x)$
$= 37.8x - x^2 - (29x + 10.15)$
$= -x^2 + 8.8x - 10.15$.

37b) The graphs are displayed in viewing window [0, 10] by [−5, 10]

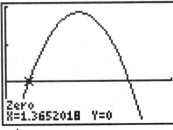

and

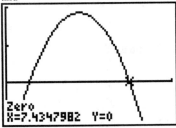

The zeros are approximately 1.37 and 7.43.

37c) $P(x) = -x^2 + 8.8x - 10.15$.
The x-coordinate of the vertex is $x = \frac{-b}{2a}$ so
$x = \frac{-(8.8)}{2(-1)} = \frac{-8.8}{-2} = 4.4$. Plugging $x = 4.4$ into the function, we find the y-coordinate to be:
$-(4.4)^2 + 8.8(4.4) - 10.15$
$= -19.36 + 38.72 - 10.15 = 9.21$. Hence, the vertex is (4.4, 9.21). Since $a < 0$, we have a maximum profit of $9.21 at $x = 4.4$.

39a) $R(x) = x \cdot p(x) = x(-10x + 1040)$
$= -10x^2 + 1040x$.
$C(x) = \begin{pmatrix} \text{Variable} \\ \text{Costs} \end{pmatrix} x + \begin{pmatrix} \text{Fixed} \\ \text{Costs} \end{pmatrix}$
$= 500x + 6650$
$P(x) = R(x) - C(x)$
$= -10x^2 + 1040x - (500x + 6650)$
$= -10x^2 + 540x - 6650$.

39b) The graphs are displayed in viewing window [0, 50] by [−250, 1000]

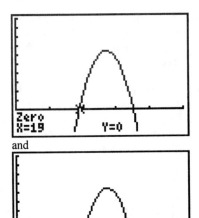

and

The zeros are 19 and 35.

39c) $P(x) = -10x^2 + 540x - 6650$.
The x-coordinate of the vertex is $x = \frac{-b}{2a}$ so
$x = \frac{-(540)}{2(-10)} = \frac{-540}{-20} = 27$. Plugging $x = 27$ into the function, we find the y-coordinate to be:
$-10(27)^2 + 540(27) - 6650$
$= -7290 + 14580 - 6650 = 640$. Hence, the vertex is (27, 640). Since $a < 0$, we have a maximum profit of $640 at $x = 27$.

41a) $R(x) = x \cdot p(x) = x(-x+30)$
$= -x^2 + 30x$.
$C(x) = \begin{pmatrix} \text{Variable} \\ \text{Costs} \end{pmatrix} x + \begin{pmatrix} \text{Fixed} \\ \text{Costs} \end{pmatrix}$
$= 5x + 60$
$P(x) = R(x) - C(x)$
$= -x^2 + 30x - (5x + 60)$
$= -x^2 + 25x - 60$.

41b) The graphs are displayed in viewing window [0, 25] by [−25, 100]

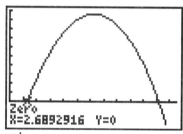

and

The zeros are approximately 2.69 and 22.31.

41c) $P(x) = -x^2 + 25x - 60$.
The x-coordinate of the vertex is $x = \frac{-b}{2a}$ so
$x = \frac{-(25)}{2(-1)} = \frac{-25}{-2} = 12.5$. Plugging $x = 12.5$ into the function, we find the y-coordinate to be:
$-(12.5)^2 + 25(12.5) - 60$
$= -156.25 + 312.5 - 60 = 96.25$. Hence, the vertex is (12.5, 96.25). Since $a < 0$, we have a maximum profit of \$96.25 at $x = 12.5$

43) $f(x) = -30$ is a polynomial (the exponent of x is the whole number 0). It is a linear function.

45) $g(x) = \frac{4}{x^2} + 3x + \frac{1}{2}$ is not a polynomial since the exponent of x in the first term is -2.

47) $f(x) = x^3 + x + 5$ is a polynomial since all the exponents of the variables in the terms are whole numbers. Since it has degree of 3, it is a cubic polynomial.

49) Since $f(x) = -17x$ is an odd degree function and the leading coefficient, -17, is negative, then as $x \to -\infty$, $f(x) \to \infty$ and as $x \to \infty$, $f(x) \to -\infty$.

51) Since $g(x) = 3.8x$ is an odd degree function and the leading coefficient, 3.8, is positive, then as $x \to -\infty$, $g(x) \to -\infty$ and as $x \to \infty$, $g(x) \to \infty$.

53) Since $g(x) = 5.5x^2 + 4$ is an even degree function and the leading coefficient, 5.5, is positive, then as $x \to -\infty$, $g(x) \to \infty$ and as $x \to \infty$, $g(x) \to \infty$.

55) Since $f(x) = 11x + 0.1x^3$ is an odd degree function and the leading coefficient, 0.1, is positive, then as $x \to -\infty$, $f(x) \to -\infty$ and as $x \to \infty$, $f(x) \to \infty$.

57) Since $g(x) = x^6 - 2x^3 - x^2$ is an even degree function and the leading coefficient, 1, is positive, then as $x \to -\infty$, $g(x) \to \infty$ and as $x \to \infty$, $g(x) \to \infty$.

59a) $s(2) = -16(2)^2 + 30(2) + 48$
$= -64 + 60 + 48 = 44$. Two seconds after the ball is thrown straight up, it is 44 feet above the ground.

59b) $m_{sec} = \dfrac{s(t+\Delta t) - s(t)}{\Delta t} = \dfrac{s(2.5) - s(0)}{2.5}$

$= \dfrac{-16(2.5)^2 + 30(2.5) + 48 - [-16(0)^2 + 30(0) + 48]}{2.5}$

$= \dfrac{23 - 48}{2.5}$

$= -10$

This means between zero and 2.5 seconds after the ball was thrown, the ball was falling to the ground at an average speed of 10 feet per second.

61a) $R(250) = 11(250) - \dfrac{(250)^2}{730}$

$= 2750 - \dfrac{62500}{730} = \dfrac{194500}{73} = 2664\dfrac{28}{73}$

≈ 2664.38.

This means that the BackStreet Company will generate about $2664.38 in revenue from the sale of 250 headphones.

61b) $m_{sec} = \dfrac{R(x+\Delta x) - R(x)}{\Delta x} = \dfrac{R(300) - R(100)}{200}$

$= \dfrac{11(300) - \dfrac{(300)^2}{730} - \left[11(100) - \dfrac{(100)^2}{730}\right]}{200}$

$= \dfrac{\dfrac{2319000}{730} - \dfrac{793000}{730}}{200} = \dfrac{\dfrac{1526000}{730}}{200} = \dfrac{763}{73} = 10\dfrac{33}{73}$

≈ 10.45

This means that when 100 to 300 headphones sold, the revenue was increasing at an average rate of about $10.45 per headphone.

63a) $f(7) = -0.09(7)^2 + 1.83(7) + 11.55$
$= -4.41 + 12.81 + 11.55 = 19.95$
This means that in 1982, 19.95% of the people taking the SAT intended to study business.

63b) $m_{sec} = \dfrac{f(x+\Delta x) - f(x)}{\Delta x} = \dfrac{f(10) - f(0)}{10}$

Since $f(0) = -0.09(0)^2 + 1.83(0) + 11.55$
$= 11.55$

& $f(10) = -0.09(10)^2 + 1.83(10) + 11.55$
$= 20.85$,

then $m_{sec} = \dfrac{20.85 - 11.55}{10} = \dfrac{9.3}{10} = 0.93$.

This means that between 1975 and 1985, the number of people taking the SAT who intended to study business was increasing at an average rate of 0.93% per year.

63c) $m_{sec} = \dfrac{f(x+\Delta x) - f(x)}{\Delta x} = \dfrac{f(20) - f(10)}{20 - 10}$

Since $f(10) = -0.09(10)^2 + 1.83(10)$
$+ 11.55 = 20.85$

and $f(20) = -0.09(20)^2 + 1.83(20)$
$+ 11.55 = 12.15$,

then $m_{sec} = \dfrac{12.15 - 20.85}{10} = \dfrac{-8.7}{10} = -0.87$.

This means that between 1985 and 1995, the number of people taking the SAT who intended to study engineering was decreasing at an average rate of 0.87% per year.

63d) We can see that the percentage of people taking the SAT who intended to study business was increasing between 1975 to 1985, but then that figure fell between 1985 and 1995.

65a) Since $f(x) = -3x + 6$ is an odd function and the leading coefficient, -3, is negative, then as $x \to -\infty$, $f(x) \to \infty$ and as $x \to \infty$, $f(x) \to -\infty$.

65b) Since $f(x)$ is a linear function, the average rate of change is constant and so it has no peaks and valleys.

65c) Since $m = -3$ is constant, the function is decreasing on $(-\infty, \infty)$ and increasing no where.

67a) Since $y = -x^2 - 10x + 21$ is an even function and the leading coefficient, -1, is negative, then as $x \to -\infty$, $y \to -\infty$ and as $x \to \infty$, $y \to -\infty$.

67b) The graph is displayed in viewing window $[-20, 20]$ by $[-50, 50]$

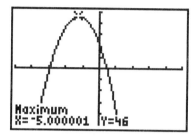

We have a "peak" at $(-5, 46)$.

67c) The function is decreasing on $(-5, \infty)$ and increasing on $(-\infty, -5)$.

69a) Since $g(x) = -2x^3 - 7x^2 + 4x - 2$ is an odd degree function and the leading coefficient, -2, is negative, then as $x \to -\infty$, $g(x) \to \infty$ and as $x \to \infty$, $g(x) \to -\infty$.

69b) The graphs are displayed in viewing window $[-20, 20]$ by $[-100, 100]$

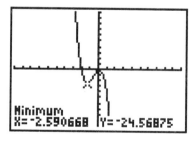

69b) Continued

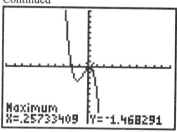

We have a "peak" at $(0.26, -1.47)$ and a "valley" at $(-2.59, -24.57)$.

69c) The function is decreasing on $(-\infty, -2.59) \cup (0.26, \infty)$ and increasing on $(-2.59, 0.26)$.

71a) Since $f(x) = -x^4 + 8x + 12$ is an even degree function and the leading coefficient, -1, is negative, then as $x \to -\infty$, $f(x) \to -\infty$ and as $x \to \infty$, $f(x) \to -\infty$.

71b) The graph is displayed in viewing window $[-10, 10]$ by $[-20, 20]$

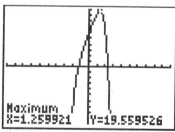

There is a peak at about $(1.26, 19.56)$

71c) The function is increasing on $(-\infty, 1.26)$ and decreasing on $(1.26, \infty)$.

Section 1.5 Rational, Radical and Power Functions

1a) Setting the denominator equal to zero and solving we get:
$$3x = 0$$
$$x = 0$$
Thus, the value of 0 must be excluded so the domain is $(-\infty, 0) \cup (0, \infty)$.

1b) Since $f(x)$ can be rewritten as: $4x^2, x \neq 0$, we have a hole at $x = 0$.

3a) Setting the denominator equal to zero and solving we get:
$$x - 2 = 0$$
$$x = 2$$
Thus, the value of 2 must be excluded so the domain is $(-\infty, 2) \cup (2, \infty)$.

3b) Since the numerator of $f(x)$ is not zero at $x = 2$, we have a vertical asymptote at $x = 2$.

5a) Setting the denominator equal to zero and solving we get:
$$x + 2 = 0$$
$$x = -2$$
Thus, the value of -2 must be excluded so the domain is $(-\infty, -2) \cup (-2, \infty)$.

5b) Since y can be rewritten as $\frac{3(x+2)}{x+2}$
$= 3, x \neq -2$, we have a hole at $x = -2$.

7a) Setting the denominator equal to zero and solving we get:
$$x^2 + 4x + 3 = 0$$
$$(x + 1)(x + 3) = 0$$
$$x + 1 = 0 \text{ or } x + 3 = 0$$
$$x = -1 \text{ or } x = -3$$
Thus, the values of -1 and -3 must be excluded so the domain is
$(-\infty, -3) \cup (-3, -1) \cup (-1, \infty)$.

7b) Since $g(x)$ can be rewritten as $\frac{(x+1)(x+1)}{(x+1)(x+3)}$
$= \frac{x+1}{x+3}, x \neq -1$, we have a hole at $x = -1$ and a vertical asymptote at $x = -3$.

9a) Setting the denominator equal to zero and solving we get:
$$2x^2 - x - 6 = 0$$
$$(x - 2)(2x + 3) = 0$$
$$x - 2 = 0 \text{ or } 2x + 3 = 0$$
$$x = 2 \text{ or } 2x = -3$$
$$x = 2 \text{ or } x = -1.5$$
Thus, the values of -1.5 and 2 must be excluded so the domain is
$(-\infty, -1.5) \cup (-1.5, 2) \cup (2, \infty)$.

9b) Since $f(x)$ can be rewritten as
$\frac{(2x+3)(2x+1)}{(x-2)(2x+3)} = \frac{2x+1}{x-2}, x \neq -1.5$,
we have a hole at $x = -1.5$ and a vertical asymptote at $x = 2$.

11) Since the degree of the polynomial in the numerator is greater than the degree of the polynomial in the denominator, then the function has no horizontal asymptote.

13) Since the degree of the polynomial in the numerator is the same as the degree of the polynomial in the denominator, then the function has a horizontal asymptote of $y = \frac{3}{1}$ or $y = 3$.

15) Since the degree of the polynomial in the numerator is the same as the degree of the polynomial in the denominator, then the function has a horizontal asymptote of $y = \frac{3}{1}$ or $y = 3$.

17) Since the degree of the polynomial in the numerator is the same as the degree of the polynomial in the denominator, then the function has a horizontal asymptote of $y = \frac{1}{1}$ or $y = 1$.

19) Since the degree of the polynomial in the numerator is the same as the degree of the polynomial in the denominator, then the function has a horizontal asymptote of $y = \frac{4}{2}$ or $y = 2$.

21) To find the y-intercept, we evaluate the function at $x = 0$:
$$g(0) = \frac{(0)}{2(0)+6} = 0.$$
So, the y-intercept is $(0, 0)$.
To find the x-intercept, we first need to find the domain of g. We then set the function equal to zero, solve for x and check to see if the answer is in the domain of g. To find the domain, we set the denominator equal to zero and solve:
$$2x + 6 = 0$$
$$2x = -6$$
$$x = -3$$
Thus, the domain is $(-\infty, -3) \cup (-3, \infty)$. Now, we set the function equal to 0 and solve:
$$g(x) = \frac{x}{2x+6} = 0$$
$$x = 0$$
Since 0 is in the domain of g, then the x-intercept is $(0, 0)$. So, $(0, 0)$ is both the y- and x-intercept.

23) To find the y-intercept, we evaluate the function at $x = 0$:
$$f(0) = \frac{6(0)-3}{2(0)+4} = \frac{-3}{4}.$$
So, the y-intercept is $(0, \frac{-3}{4})$.
To find the x-intercept, we first need to find the domain of f. We then set the function equal to zero, solve for x and check to see if the answer is in the domain of f. To find the domain, we set the denominator equal to zero and solve:
$$2x + 4 = 0$$
$$2x = -4$$
$$x = -2$$
Thus, the domain is $(-\infty, -2) \cup (-2, \infty)$.

23) Continued
Now, we set the function equal to 0 and solve:
$$f(x) = \frac{6x-3}{2x+4} = 0$$
$$6x - 3 = 0$$
$$6x = 3, \text{ so } x = 0.5$$
Since 0.5 is in the domain of f, then the x-intercept is $(0.5, 0)$.

25) To find the y-intercept, we evaluate the function at $x = 0$:
$$f(0) = \frac{(0)^2 - (0) - 6}{(0)^2 - 7(0) + 12} = \frac{-6}{12} = -0.5.$$
So, the y-intercept is $(0, -0.5)$.
To find the x-intercept, we first need to find the domain of f. We then set the function equal to zero, solve for x and check to see if the answer is in the domain of f. To find the domain, we set the denominator equal to zero and solve:
$$x^2 - 7x + 12 = 0$$
$$(x - 3)(x - 4) = 0$$
$$x - 3 = 0 \text{ or } x - 4 = 0$$
$$x = 3 \text{ or } x = 4$$
Thus, the domain is $(-\infty, 3) \cup (3, 4) \cup (4, \infty)$. Now, we set the function equal to 0 and solve:
$$f(x) = \frac{x^2 - x - 6}{x^2 - 7x + 12} = 0$$
$$x^2 - x - 6 = 0$$
$$(x - 3)(x + 2) = 0$$
$$x - 3 = 0 \text{ or } x + 2 = 0$$
$$x = 3 \text{ or } x = -2$$
Since -2 is in the domain of f, but 3 is not in the domain of f, then the x-intercept is $(-2, 0)$.

27) To find the y-intercept, we evaluate the function at $x = 0$:
$$g(0) = \frac{(0)-2}{(0)^2 + 6(0) + 9} = \frac{-2}{9} = -\frac{2}{9}.$$
So, the y-intercept is $(0, -\frac{2}{9})$.

27) Continued
To find the x-intercept, we first need to find the domain of g. We then set the function equal to zero, solve for x and check to see if the answer is in the domain of g. To find the domain, we set the denominator equal to zero and solve:
$$x^2 + 6x + 9 = 0$$
$$(x + 3)(x + 3) = 0$$
$$x + 3 = 0 \text{ or } x + 3 = 0$$
$$x = -3$$
Thus, the domain is $(-\infty, -3) \cup (-3, \infty)$. Now, we set the function equal to 0 and solve:
$$g(x) = \frac{x-2}{x^2+6x+9} = 0$$
$$x - 2 = 0$$
$$x = 2$$
Since 2 is in the domain of g, then the x-intercept is (2, 0).

29) To find the y-intercept, we evaluate the function at $x = 0$:
$$y = \frac{(0)^2 - 4}{(0)^2 + 2(0) - 3} = \frac{-4}{-3} = \frac{4}{3}.$$
So, the y-intercept is $(0, \frac{4}{3})$.
To find the x-intercept, we first need to find the domain of y. We then set the function equal to zero, solve for x and check to see if the answer is in the domain of y. To find the domain, we set the denominator equal to zero and solve:
$$x^2 + 2x - 3 = 0$$
$$(x + 3)(x - 1) = 0$$
$$x + 3 = 0 \text{ or } x - 1 = 0$$
$$x = -3 \text{ or } x = 1$$
Thus, the domain is $(-\infty, -3) \cup (-3, 1) \cup (1, \infty)$. Now, we set the function equal to 0 and solve:
$$y = \frac{x^2 - 4}{x^2 + 2x - 3} = 0$$
$$x^2 - 4 = 0$$
$$(x - 2)(x + 2) = 0$$
$$x = 2 \text{ or } x = -2$$
Since -2 and 2 is in the domain of y, then the x-intercepts are $(-2, 0)$ and $(2, 0)$.

31) The graph is displayed in viewing window $[-10, 10]$ by $[-20, 20]$.

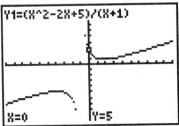

Note: The textbook writes the viewing window on the sides of the graph.
The y-intercept is (0, 5).
There are no x-intercepts since if $f(x) = 0$, then $x^2 - 2x + 5 = 0$. Checking the discriminant,
$b^2 - 4ac = (2)^2 - 4(1)(5) = 4 - 20$
$= -16$, we get a negative number.
Hence, there are no real solutions.

33) The graphs are displayed in viewing window $[-10, 10]$ by $[-10, 10]$.

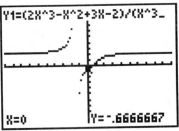

The y-intercept is $(0, -\frac{2}{3})$.

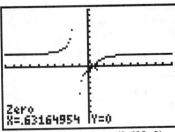

The x-intercept is about (0.632, 0).

Section 1.5 Rational, Radical and Power Functions   25

35) The graphs are displayed in viewing window [−10, 10] by [−10, 10].

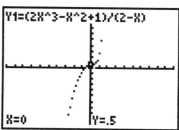

The y-intercept is (0, 0.5).

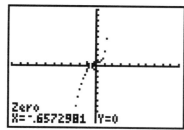

The x-intercept is (≈ −0.657, 0).

37)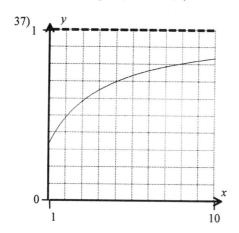

There is a horizontal asymptote at $y = 1$.

39)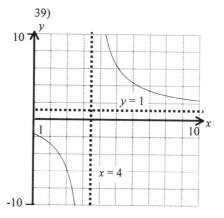

There is a horizontal asymptote at $y = 1$ and a vertical asymptote at $x = 4$.

41)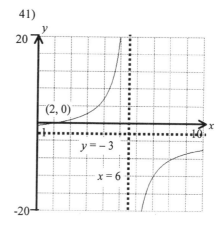

There is a horizontal asymptote at $y = -3$, a vertical asymptote at $x = 6$ and x-intercept at (2, 0).

43a) $f(85) = \frac{30(85)}{100-(85)} = \frac{2550}{15} = 170$.
It will cost $170 million to remove 85% of the pollutants.

43b) $f(5) = \frac{30(5)}{100-(5)} = \frac{150}{95} = \frac{30}{19} \approx 1.58$.

$f(50) = \frac{30(50)}{100-(50)} = \frac{1500}{50} = 30$.

$f(70) = \frac{30(70)}{100-(70)} = \frac{2100}{30} = 70$.

$f(90) = \frac{30(90)}{100-(90)} = \frac{2700}{10} = 270$.

$f(95) = \frac{30(95)}{100-(95)} = \frac{2850}{5} = 570$.

$f(100) = \frac{30(100)}{100-(100)} = \frac{3000}{0} =$ undefined.

x	5	50	70	90	95
f(x)	≈ 1.6	30	70	270	570

43c) $f(99.9) - f(95) = \frac{30(99.9)}{100-(99.9)} - \frac{30(95)}{100-(95)}$

$= \frac{2997}{0.1} - \frac{2850}{5} = 29{,}970 - 570 = 29{,}400$. It will approximately cost $29,400 million to remove the final 5% of the pollutant.

43d) $f(100)$ is undefined. We can interpret this to mean that it is impossible to remove all the pollutants.

45a) $f(10) = \frac{20(10)-18}{(10)} = \frac{182}{10} = 18.2$
After 10 days, the person had about 18 out of 20 items remembered.

45b) $f(100) = \frac{20(100)-18}{(100)} = \frac{1982}{100} = 19.82$
After 100 days, the person had about 20 out of 20 items remembered.

47a) $S(10) = \frac{120(10)^2 - 600(10) + 3}{2(10)^2 - 10(10) + 1} = \frac{12000 - 6000 + 3}{200 - 100 + 1}$

$= \frac{6003}{101} \approx 59.44$.
If the company spends $10,000 on advertising, the income from the sales will be about $5,944,000.

47b) The graph is displayed in viewing window [5, 10] by [50, 70].

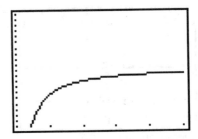

47c) Since the degree of the polynomial in the numerator is the same as the degree of the polynomial in the denominator, then the function has a horizontal asymptote of $y = \frac{120}{2}$ or $y = 60$. This means that the most income the company can generate by advertising is $6,000,000.

49) $f(x) = \sqrt{2x+3} = (2x+3)^{\frac{1}{2}}$
The index is even. Setting the radicand ≥ 0 and solving, we find:
$2x + 3 \geq 0$
$2x \geq -3$
$x \geq -1.5$
Thus, the domain of the function is $[-1.5, \infty)$.

51) $f(x) = \sqrt[3]{5x-8} = (5x-8)^{\frac{1}{3}}$
The index is odd, so the domain of the function is $(-\infty, \infty)$.

53) $g(x) = \sqrt[4]{(6x+1)^5} = (6x+1)^{\frac{5}{4}}$
The index is even. Setting the radicand ≥ 0 and solving, we find:
$(6x+1)^5 \geq 0$
$(6x+1) \geq 0$
$6x \geq -1$
$x \geq -\frac{1}{6}$
Thus, the domain of the function is $[-\frac{1}{6}, \infty)$.

55) $f(x) = \frac{3}{\sqrt{7x-2}} = 3(7x-2)^{-\frac{1}{2}}$

The index is even, but the denominator also cannot be zero. Setting the radicand > 0 and solving, we find:
$$7x - 2 > 0$$
$$7x > 2$$
$$x > \tfrac{2}{7}$$
Thus, the domain of the function is $(\tfrac{2}{7}, \infty)$.

57) $f(x) = (8x-9)^{\frac{1}{2}} = \sqrt{8x-9}$

The index is even. Setting the radicand ≥ 0 and solving, we find:
$$8x - 9 \geq 0$$
$$8x \geq 9$$
$$x \geq 1.125$$
Thus, the domain of the function is $[1.125, \infty)$.

59) $f(x) = (7x+9)^{\frac{1}{3}} = \sqrt[3]{7x+9}$

The index is odd, so the domain of the function is $(-\infty, \infty)$.

61) $y = (4x-5)^{\frac{3}{2}} = \sqrt{(4x-5)^3}$

The index is even. Setting the radicand ≥ 0 and solving, we find:
$$4x - 5 \geq 0$$
$$4x \geq 5$$
$$x \geq 1.25$$
Thus, the domain of the function is $[1.25, \infty)$.

63) $f(x) = (7x-2)^{-\frac{1}{3}} = \frac{1}{\sqrt[3]{7x-2}}$

The index is odd, but the denominator also cannot be zero. Setting the radicand = 0 and solving, we find:
$$7x - 2 = 0$$
$$7x = 2$$
$$x = \tfrac{2}{7}$$
This value of x must be excluded. Thus, the domain of the function is
$(-\infty, \tfrac{2}{7}) \cup (\tfrac{2}{7}, \infty)$.

65a) $g(70) = \tfrac{7}{5}\sqrt{70} \approx 11.71$

If the tower is 70 feet tall, an observer can see about 11.71 miles into the forest.

65b) Setting $g(x)$ equal to 29 and solving yields:
$$g(x) = \tfrac{7}{5}\sqrt{x} = 29$$
$$\sqrt{x} = \tfrac{145}{7}$$
$$x = \tfrac{21025}{49} \approx 429.08$$
The tower should be about 429.08 feet high.

67a) $g(300) = 28.1\sqrt[3]{300} \approx 188$

There are about 188 plant species in 300 square miles of rainforest.

67b) $m_{sec} = \frac{g(x+\Delta x) - g(x)}{\Delta x} = \frac{g(2197) - g(1728)}{469}$
$= \frac{28.1\sqrt[3]{2197} - 28.1\sqrt[3]{1728}}{469} \approx \frac{365.3 - 337.2}{469} = \frac{28.1}{469}$
$= 0.0599$

From 1728 square feet to 2197 square feet, the average increase in plant species is about 0.0599 species per square foot.

69a) $f(3) = 17311(3)^{-0.028} \approx 16,787$

In 1992, there were about 16,787 grocery stores in the United States.

69b) $\dfrac{f(6)-f(3)}{6-3} = \dfrac{17311(6)^{-0.028} - 17311(3)^{-0.028}}{3}$

$\approx \dfrac{16463.95 - 16786.60}{3} \approx \dfrac{-322.65}{3} = -107.55$

Between 1992 and 1995, the number of grocery stores were decreasing by an average rate of 107.55 stores per year.

71a) $f(11) = 19.14\,(11)^{0.112} \approx 25.04$

In 1990, there were about 25,040,000 retired workers receiving social security benefits.

71b) The graph is displayed in viewing window [1, 20] by [0, 30]

The number of retired workers receiving social security benefits is increasing since the graph is rising from left to right.

71c) $m_{sec} = \dfrac{f(x+\Delta x) - f(x)}{\Delta x} = \dfrac{f(7) - f(2)}{5}$

$= \dfrac{19.14(7)^{0.112} - 19.14(2)^{0.112}}{5}$

$\approx \dfrac{23.80 - 20.69}{5} = \dfrac{3.11}{5} = 0.623$

Between 1981 and 1986, the number of retired workers receiving benefits was growing at an average rate of 623,000 people per year.

Section 1.6 Exponential Functions

1a)

x	-5	-4	-3	-2	-1	0
$\left(\frac{1}{3}\right)^x$	243	81	27	9	3	1

x	1	2	3	4	5
$\left(\frac{1}{3}\right)^x$	$\frac{1}{3}$	$\frac{1}{9}$	$\frac{1}{27}$	$\frac{1}{81}$	$\frac{1}{243}$

1b)

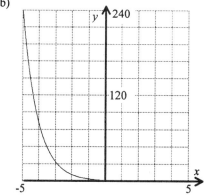

The function $\left(\frac{1}{3}\right)^x$ is decreasing on its domain, i.e., as $x \to -\infty, f(x) \to \infty$. Hence, the function is an exponential decay function.

3a)

x	-5	-4	-3	-2	-1	0
4^x	$\frac{1}{1024}$	$\frac{1}{256}$	$\frac{1}{64}$	$\frac{1}{16}$	$\frac{1}{4}$	1

x	1	2	3	4	5
4^x	4	16	64	256	1024

3b)

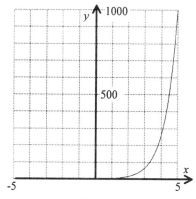

The function 4^x is increasing on its domain, i.e., as $x \to \infty, f(x) \to \infty$. Hence, the function is an exponential growth function.

5a) All values are rounded to the nearest hundredth or to two significant digits.

x	-5	-4	-3	-2	-1	0
$(2.3)^x$	0.016	0.036	0.082	0.19	0.43	1

x	1	2	3	4	5
$(2.3)^x$	2.3	5.29	12.17	27.98	64.36

5b)

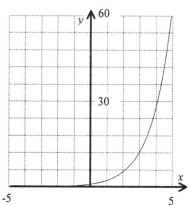

The function $(2.3)^x$ is increasing on its domain, i.e., as $x \to \infty, f(x) \to \infty$. Hence, the function is an exponential growth function.

7a) All values are rounded to the nearest hundredth.

x	-5	-4	-3	-2	-1	0
$(0.7)^x$	5.95	4.16	2.92	2.04	1.43	1

x	1	2	3	4	5
$(0.7)^x$	0.7	0.49	0.34	0.24	0.17

7b)

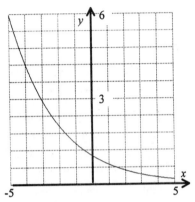

The function $(0.7)^x$ is decreasing on its domain, i.e., as $x \to -\infty, f(x) \to \infty$. Hence, the function is an exponential decay function.

9a) All values are rounded to the nearest hundredth or to two significant digits.

x	-5	-4	-3	-2
e^{2x}	0.000045	0.00034	0.0025	0.018

x	-1	0	1	2
e^{2x}	0.14	1	7.39	54.60

x	3	4	5
e^{2x}	403.43	2980.96	22026.47

9b)

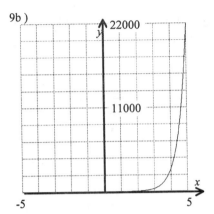

The function e^{2x} is increasing on its domain, i.e., as $x \to \infty, f(x) \to \infty$. Hence, the function is an exponential growth function.

11a) All values are rounded to the nearest hundredth.

x	-5	-4	-3	-2	-1	0
$e^{0.3x}$	0.22	0.30	0.41	0.55	0.74	1

x	1	2	3	4	5
$e^{0.3x}$	1.35	1.82	2.46	3.32	4.48

11b)

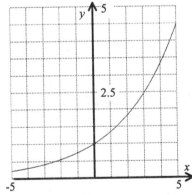

The function $e^{0.3x}$ is increasing on its domain, i.e., as $x \to \infty, f(x) \to \infty$. Hence, the function is an exponential growth function.

Section 1.6 Exponential Functions   31

13a) All values are rounded to the nearest hundredth or two significant digits.

x	-5	-4	-3	-2
$e^{-1.6x}$	2980.96	601.85	121.51	24.53

x	-1	0	1	2
$e^{-1.6x}$	4.95	1	0.20	0.041

x	3	4	5
$e^{-1.6x}$	0.0082	0.0017	0.00034

13b)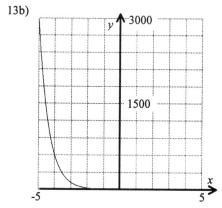

The function $e^{-1.6x}$ is decreasing on its domain, i.e., as $x \to -\infty, f(x) \to \infty$. Hence, the function is an exponential decay function.

15a) $f(7) = 200\left(\frac{4}{5}\right)^{(7)} = 200\,(0.2097152)$
$= 41.94304 \approx 41.94$.
This means that on the 7th day, the wound is about 41.94 cm² in size.

15b) Evaluating the function for $x = 0, 1, 2, 3,$ and 4, we find:
$f(0) = 200\left(\frac{4}{5}\right)^{(0)} = 200\,(1) = 200$
$f(1) = 200\left(\frac{4}{5}\right)^{(1)} = 200\,(0.8) = 160$
$f(2) = 200\left(\frac{4}{5}\right)^{(2)} = 200\,(0.64) = 128$
$f(3) = 200\left(\frac{4}{5}\right)^{(3)} = 200\,(0.512) = 102.4$

15b) Continued
$f(4) = 200\left(\frac{4}{5}\right)^{(4)} = 200\,(0.4096) = 81.92$
We can see that on the third day the wound is approximately half its original size. Later in this chapter, we will use a more precise technique involving logarithms for solving this problem.

15c) $m_{sec} = \frac{f(x+\Delta x) - f(x)}{\Delta x} = \frac{f(8) - f(1)}{7}$.
$f(1) = 200\left(\frac{4}{5}\right)^{(1)} = 200\,(0.8) = 160$
$f(8) = 200\left(\frac{4}{5}\right)^{(8)} = 200\,(0.16777216)$
$= 33.554432$
Therefore, $\frac{f(8)-f(1)}{7} = \frac{33.554432 - 160}{7}$
$= \frac{-126.445568}{7} \approx -18.06$
Between the first and eighth day, the wound was shrinking at an average rate of 18.06 square centimeters per day.

17a) $g(5) = 8000 - 8000\left(\frac{43}{50}\right)^{(5)}$
$\approx 8000 - 3763.42 = 4236.58$
By the fifth day, about 4,237 students had heard the rumor.

17b) $m_{sec} = \frac{g(x+\Delta x) - g(x)}{\Delta x} = \frac{g(5) - g(1)}{4}$.
$g(1) = 8000 - 8000\left(\frac{43}{50}\right)^{(1)}$
$= 8000 - 6880 = 1120$
$g(5) = 8000 - 8000\left(\frac{43}{50}\right)^{(5)}$
$\approx 8000 - 3763.42 = 4236.58$
Therefore, $\frac{g(5)-g(1)}{4} \approx \frac{4236.58 - 1120}{4}$
$= \frac{3116.58}{4} \approx 779.15$
Between the first and fifth day, the rumor was spreading at an average rate of about 779.15 students per day.

19a) $f(2) = 19.12\,(0.944)^2 \approx 17.04$
In 1992, about 17.04% of the waste generated in the U.S. was yard waste.

19b) The graph is displayed in viewing window [1, 6] by [10, 20]

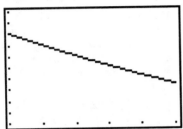

Note: The textbook writes the viewing window on the sides of the graph.
This is an exponential decay function since it is decreasing from left to right.

19c) $m_{sec} = \frac{f(x+\Delta x) - f(x)}{\Delta x} = \frac{f(6) - f(1)}{5}$
$= \frac{19.12(0.944)^6 - 19.12(0.944)^1}{5} \approx \frac{13.53069 - 18.04928}{5} \approx$
$\frac{-4.5186}{5} \approx -0.904$

Between the 1991 and 1996, the percentage of waste that was due to yard waste was shrinking at an average rate of 0.904% per year.

21) $I = Prt = (500)(0.04)(6) = 120$. So the total interest earned is $120.

23) $A = P(1 + \frac{r}{n})^{nt}$
Since $P = 3000, r = 0.06, n = 1$, and $t = 3$,
then $A = (3000)(1 + \frac{0.06}{1})^{(1)(3)}$
$A = (3000)(1.06)^3 \approx 3573.05$
After three years, about $3,573.05 will be in the account.

25) $A = P(1 + \frac{r}{n})^{nt}$
Since $P = 1800, r = 0.071, n = 12$, & $t = 5$,
then $A = (1800)(1 + \frac{0.071}{12})^{(12)(5)}$
$A = (1800)(1.00596\overline{6})^{60} \approx 2564.44$
After five years, about $2,564.44 will be in the account.

27a) $A = P(1 + \frac{r}{n})^{nt}$
Since $P = 2000, r = 0.08, n = 1$, & $t = 6$,
then $A = (2000)(1 + \frac{0.08}{1})^{(1)(6)}$
$A = (2000)(1.08)^6 \approx 3173.75$
After six years, about $3,173.75 will be in the account.

27b) $A = P(1 + \frac{r}{n})^{nt}$
Since $P = 2000, r = 0.08, n = 2$, & $t = 6$,
then $A = (2000)(1 + \frac{0.08}{2})^{(2)(6)}$
$A = (2000)(1.04)^{12} \approx 3202.06$
After six years, about $3,202.06 will be in the account.

27c) $A = P(1 + \frac{r}{n})^{nt}$
Since $P = 2000, r = 0.08, n = 4$, & $t = 6$,
then $A = (2000)(1 + \frac{0.08}{4})^{(4)(6)}$
$A = (2000)(1.02)^{24} \approx 3216.87$
After six years, about $3,216.87 will be in the account.

29a) $A = P(1 + \frac{r}{n})^{nt}$
Since $P = 15,000, r = 0.059, n = 1$, & $t = 16$, then $A = (15,000)(1 + \frac{0.059}{1})^{(1)(16)}$
$A = (15,000)(1.059)^{16} \approx 37,534.15$
After sixteen years, about $37,534.15 will be in the account.

29b) $A = P(1 + \frac{r}{n})^{nt}$
Since $P = 15,000, r = 0.059, n = 12$, & $t = 16$, then $A = (15,000)(1 + \frac{0.059}{12})^{(12)(16)}$
$A = (15,000)(1.0491\overline{6})^{192} \approx 38,464.55$
After sixteen years, about $38,464.55 will be in the account.

29c) $A = P(1 + \frac{r}{n})^{nt}$
Since $P = 15,000, r = 0.059, n = 52$, & $t = 16$, then $A = (15,000)(1 + \frac{0.059}{52})^{(52)(16)}$
$A \approx (15,000)(1.001134615)^{832} \approx 38,533.00$. After sixteen years, about $38,533.00 will be in the account.

Section 1.6 Exponential Functions   33

31) Since $P = 5500$, $r = 0.055$, & $n = 4$, then
$A(5) = (5500)(1 + \frac{0.055}{4})^{(4)(5)}$
$= (5500)(1.01375)^{20} \approx 7227.37$
If $5500 is invested at a 5.5% interest rate compounded quarterly, there will be about $7,227.37 in the account after five years. The interest made will be:
$7,227.37 − $5,500 = $1,727.37.

33) Since $P = 10,000$, $r = 0.0725$, & $n = 12$, then
$A(5) = (10,000)(1 + \frac{0.0725}{12})^{(12)(5)}$
$= (10,000)(1.006041\overline{6})^{60} \approx 14,353.51$
If $10,000 is invested at a 7.25% interest rate compounded monthly, there will be about $14,353.51 in the account after five years. The interest made will be:
$14,353.51 − $10,000 = $4,353.

35a) Since $P = 4000$, $r = 0.0575$, & $n = 12$,
then $A(t) = P(1 + \frac{r}{n})^{nt}$
$= (4000)(1 + \frac{0.0575}{12})^{12t}$
$= 4000(1.004791\overline{6})^{12t}$.

35b) The graph is displayed in viewing window [0, 16] by [4000, 10,000].

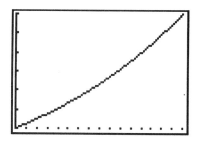

35c) The graph is displayed in viewing window [0, 16] by [4000, 10,000].

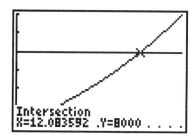

It will take approximately twelve years for the money to double.

37a) $f(5) = 1,200,000e^{0.23(5)} = 1,200,000e^{1.15}$
$\approx 1,200,000(3.15819291) \approx 3,789,831$.
After five minutes, there are approximately 3,789,831 bacteria in the colony.

37b) $m_{sec} = \frac{f(t+\Delta t) - f(t)}{\Delta t} = \frac{f(10) - f(0)}{10}$
$= \frac{1200000e^{0.23(10)} - 1200000e^{0.23(0)}}{10}$
$\approx \frac{11969018.95 - 1200000}{10} = \frac{10769018.95}{10}$
$\approx 1,076,901.90$
In the first ten minutes, the colony was growing at an average rate of about 1,076,901.90 bacteria per minute.

39a) $S(5) = 10,000(2 - e^{-0.2(5)})$
$= 10,000(2 - e^{-1}) \approx 10,000(1.63212056)$
$\approx 16,321$.
After five years on the market, approximately 16,321 Flashfast Cigarette Lighters had been sold..

39b) $m_{sec} = \dfrac{S(t+\Delta t) - S(t)}{\Delta t} = \dfrac{S(5)-S(0)}{5}$

$= \dfrac{10000(2-e^{-0.2(5)}) - 10000(2-e^{-0.2(0)})}{5}$

$\approx \dfrac{10000(1.632120559) - 10000(1)}{5}$

$\approx \dfrac{16321.20559 - 10000}{5} = \dfrac{6321.20559}{5}$

≈ 1264.24

During the first five years, the average rate of change in sales was about 1,264.24 lighters per year.

41a) $f(12) = 1285.34e^{-0.037(12)} = 1285.34e^{-0.444}$
$= 1285.34(0.6414654208) \approx 824.50$
In 1991, there were approximately 824.5 thousand W.W.II veterans receiving compensation for service-connected disabilities.

41b) The graph is displayed in viewing window [1, 16] by [600, 1300].

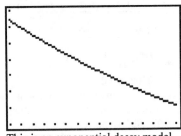

This is an exponential decay model.

41c) $m_{sec} = \dfrac{f(x+\Delta x) - f(x)}{\Delta x} = \dfrac{f(11) - f(1)}{10}$

$= \dfrac{1285.34e^{-0.037(11)} - 1285.34e^{-0.037(1)}}{10}$

$\approx \dfrac{855.5791036 - 1238.651484}{10} = \dfrac{-383.0723802}{10}$

≈ -38.31

Between 1980 and 1990, the number of W.W. II veterans receiving compensation was decreasing at an average rate of about 38,310 veterans per year.

43a) Since the exponent of e is positive, the function is an exponential growth model.

43b) $f(21) = 37.52e^{0.084(21)} = 37.52e^{1.764}$
$\approx 37.52(5.479424077) \approx 218.96$
In 1970, the Federal Government outlays were about $218.96 billion.

43c) The graph is displayed in viewing window [1, 46] by [0, 1500].

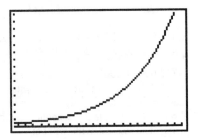

43d) $m_{sec} = \dfrac{f(x+\Delta x) - f(x)}{\Delta x} = \dfrac{f(41) - f(31)}{10}$

$= \dfrac{37.52e^{0.084(41)} - 37.52e^{0.084(31)}}{10}$

$\approx \dfrac{1174.824583 - 507.1841355}{10} = \dfrac{667.6404472}{10}$

≈ 66.76

Between 1980 and 1990, the Federal Government outlays were increasing at an average rate of about $66.76 billion per year.

45) $P = \$1000, r = 0.055, \& t = 7$.
$A(7) = Pe^{rt} = 1000e^{0.055(7)} = 1000e^{0.385}$
$\approx 1000(1.469614321) \approx 1469.61$.
After seven years, there will be about $1,469.61 in the account.

47) $P = \$20,000, r = 0.059, \& t = 7$.
$A(7) = Pe^{rt} = 20,000e^{0.059(7)} = 20,000e^{0.413}$
$\approx 20,000(1.511345026) \approx 30,226.90$.
After seven years, there will be about $30,226.90 in the account.

49) $P = \$8000, r = 0.081, \& t = 7$.
$A(7) = Pe^{rt} = 8000e^{0.081(7)} = 1000e^{0.567}$
$\approx 8000(1.7629702) \approx 14,103.76$.
After seven years, there will be about $14,103.76 in the account.

51a) $f(3) = \frac{0.8}{1+6e^{-0.3(3)}} = \frac{0.8}{1+6e^{-0.9}} \approx \frac{0.8}{3.439417958}$

$\approx 0.2325974946 \approx 0.2326$

In 2000, about 23.26% of the households in the U.S. owned a DVD player.

51b) $m_{sec} = \frac{f(x+\Delta x) - f(x)}{\Delta x} = \frac{f(5) - f(3)}{2}$

$f(5) = \frac{0.8}{1+6e^{-0.3(5)}} = \frac{0.8}{1+6e^{-1.5}} \approx \frac{0.8}{2.338780961}$

$\approx 0.3420585396 \approx 0.34206$

$f(3) = \frac{0.8}{1+6e^{-0.3(3)}} = \frac{0.8}{1+6e^{-0.9}} \approx \frac{0.8}{3.439417958}$

$\approx 0.2325974946 \approx 0.23260$

So, $\frac{f(5)-f(3)}{2} \approx \frac{0.34206 - 0.23260}{2} = \frac{0.10946}{2}$

≈ 0.0547.

Between the years 2000 and 2002, the number of households that owns a DVD player will be increasing by an average rate of 5.47% per year.

53a) $f(0) = \frac{105}{1+20.08e^{-(0)}} = \frac{105}{1+20.08} \approx \frac{105}{21.08}$

$\approx 4.981024668 \approx 5$

When the influenza initially broke out, 5 students were infected.

53b) $m_{sec} = \frac{f(t+\Delta t) - f(t)}{\Delta t} = \frac{f(7) - f(3)}{4}$

$f(7) = \frac{105}{1+20.08e^{-(7)}} = \frac{105}{1+.0183105899}$

$\approx \frac{105}{1.0183105899} \approx 103.1119592 \approx 103.11$

$f(3) = \frac{105}{1+20.08e^{-(3)}} = \frac{105}{1+.9997243328}$

$\approx \frac{105}{1.9997243328} \approx 52.50723726 \approx 52.51$

So, $\frac{f(7)-f(3)}{4} \approx \frac{103.11 - 52.51}{4} = \frac{50.6}{4} = 12.65$

Between the 3rd and 7th day after the outbreak, the number of infected students was increasing by an average rate of about 12.65 students per day.

55a) The initial size corresponds when t = 0. Thus,

$f(0) = \frac{1000}{1+121.51e^{-0.72(0)}} = \frac{1000}{1+121.51} = \frac{1000}{122.51}$

$\approx 8.162598972 \approx 8$

The initial size of the seed colony was 8 lizards.

55b) $f(8) = \frac{1000}{1+121.51e^{-0.72(8)}} \approx \frac{1000}{1+0.3828915703}$

$= \frac{1000}{1.3828915703} \approx 723.1224931 \approx 723$

After 8 years, there are 723 lizards in the colony.

55c) $m_{sec} = \frac{f(t+\Delta t) - f(t)}{\Delta t} = \frac{f(8) - f(0)}{8}$

$\approx \frac{723.12 - 8.16}{8} = \frac{714.96}{8} \approx 89.37$

In the first eight years, the population grew at an average rate of about 89.37 lizards per year.

Section 1.7 Composite, Inverse and Logarithmic Function

1) Since $g(x) = 6x$, then $g(3) = 6(3) = 18$.
   Replacing $g(3)$ by 18 we get:
   $f(g(3)) = f(18) = 3(18) = 54$.

   Since $g(x) = 6x$, then $g(-2) = 6(-2) = -12$. Replacing $g(-2)$ by -12 we get:
   $f(g(-2)) = f(-12) = 3(-12) = -36$.

3) Since $g(x) = 6x + 1$, then $g(0) = 6(0) + 1 = 1$. Replacing $g(0)$ by 1 we get:
   $f(g(0)) = f(1) = 3(1)^2 - (1) = 2$.

   Since $f(x) = 3x^2 - x$, then $f(0) = 3(0)^2 - (0) = 0$. Replacing $f(0)$ by 0 we get:
   $g(f(0)) = g(0) = 6(0) + 1 = 1$.

5) $g(f(x)) = g(3x - 1) = 8(3x - 1) + 2$
   $= 24x - 8 + 2 = 24x - 6$.

   $f(g(x)) = f(8x + 2) = 3(8x + 2) - 1$
   $= 24x + 6 - 1 = 24x + 5$.

7) $g(f(x)) = g(5x - 3)$
   $= (5x - 3)^2 + 3(5x - 3) + 4$
   $= 25x^2 - 30x + 9 + 15x - 9 + 4$
   $= 25x^2 - 15x + 4$.

   $f(g(x)) = f(x^2 + 3x + 4) = 5(x^2 + 3x + 4) - 3$
   $= 5x^2 + 15x + 20 - 3 = 5x^2 + 15x + 17$.

9) $g(f(x)) = g(x^3) = \frac{1}{(x^3)} = \frac{1}{x^3}$.
   $f(g(x)) = f(\frac{1}{x}) = (\frac{1}{x})^3 = \frac{1}{x^3}$.

11) $g(f(x)) = g(\sqrt{x+1}) = (\sqrt{x+1})^2 + 2$
    $= x + 1 + 2 = x + 3$.

    $f(g(x)) = f(x^2 + 2) = \sqrt{(x^2 + 2) + 1}$
    $= \sqrt{x^2 + 3}$.

13) Let $g(x) = x + 3$ and $f(x) = x^3$. Then $h(x) = f(g(x)) = f(x + 3) = (x + 3)^3$.

15) Let $g(x) = \frac{1}{x+3}$ and $f(x) = x^2$. Then $h(x) = f(g(x)) = f(\frac{1}{x+3}) = (\frac{1}{x+3})^2 = \frac{1}{(x+3)^2}$.

17) Let $g(x) = x - 2$ and $f(x) = \sqrt[4]{x}$. Then $h(x) = f(g(x)) = f(x - 2) = \sqrt[4]{x - 2}$.

19) Let $g(x) = \sqrt{x}$ and $f(x) = 2 - 3x$. Then
    $h(x) = f(g(x)) = f(\sqrt{x}) = 2 - 3\sqrt{x}$.

21a) $r(t) = 1.3t$. Note that if $t = 0$, then $r(t) = 0$.

21b) $V(t) = V(r(t)) = V(1.3t) = \frac{4}{3}\pi(1.3t)^3$
   $= \frac{4}{3}\pi \cdot 2.197t^3 = \frac{8.788}{3}\pi t^3 = 2.929\overline{3}\pi t^3$
   $V(t)$ is the volume of the sphere as a function of time.

21c) $V(6) = 2.929\overline{3}\pi (6)^3 = 2.929\overline{3}\pi (216)$
   $= 632.736\pi \approx 1987.80$
   After 6 seconds, the volume of the sphere is about 1987.80 in^3.

23) $f(g(x)) = f(\frac{1}{8}x) = 8(\frac{1}{8}x) = x$.

25) $f(g(x)) = f(\frac{x+10}{7}) = 7(\frac{x+10}{7}) - 10$
    $= x + 10 - 10 = x$.

27) $f(g(x)) = f(\sqrt[3]{x-6}) = (\sqrt[3]{x-6})^3 + 6$
    $= (x - 6) + 6 = x$.

29) $f(g(x)) = f(x^2 + 1) = \sqrt{(x^2 + 1) - 1}$
    $= \sqrt{x^2} = |x| = x$ since $x \geq 0$.

31) $5 = \log_2(32)$

33) $-3 = \log_2(\frac{1}{8})$

35) $2 = \log_{\frac{1}{2}}(\frac{1}{4})$

37) $\frac{3}{4} = \log_{81}(27)$

39) $5 = \log(100{,}000)$

41) $1 = \ln(e)$

43) $\log_2(\frac{3}{5}) = \log_2(3) - \log_2(5)$

45) $\log(8 \bullet 20) = \log(8) + \log(20)$

47) $\ln(\sqrt{26}) = \ln([26]^{\frac{1}{2}}) = \frac{1}{2}\ln(26)$

49) $\log_3(\frac{4\sqrt{3}}{9}) = \log_3(4\sqrt{3}) - \log_3(9)$
$= \log_3(4) + \log_3(\sqrt{3}) - \log_3(9)$
$= \log_3(4) + \log_3(3^{\frac{1}{2}}) - \log_3(3^2)$
$= \log_3(4) + \frac{1}{2}\log_3(3) - 2\log_3(3)$
$= \log_3(4) + \frac{1}{2} - 2$ (since $\log_3(3) = 1$)
$= \log_3(4) - 1.5$

51) In solving $6 = 3^x$, we begin by taking the natural log of both sides:
$\ln(6) = \ln(3^x)$
$\ln(6) = x\ln(3)$
Dividing both sides by $\ln(3)$, we get:
$x = \frac{\ln(6)}{\ln(3)} \approx 1.630929754 \approx 1.63$.

53) In solving $8 \bullet 2^x = 51$, we begin by dividing both sides by 8 and then taking the natural log of both sides:
$2^x = \frac{51}{8} = 6.375$
$\ln(2^x) = \ln(6.375)$
$x\ln(2) = \ln(6.375)$
Dividing both sides by $\ln(2)$, we get:
$x = \frac{\ln(6.375)}{\ln(2)} \approx 2.672425342 \approx 2.67$.

55) In solving $2e^x = 62$, we begin by dividing both sides by 2 and then taking the natural log of both sides:
$e^x = \frac{62}{2} = 31$
$\ln(e^x) = \ln(31)$
$x = \ln(31) \approx 3.433987204 \approx 3.43$.

57) In solving $1.21(0.3)^x = 42$, we begin by dividing both sides by 1.21 and then taking the natural log of both sides:
$(0.3)^x = \frac{42}{1.21}$
$\ln[(0.3)^x] = \ln(\frac{42}{1.21})$
$x\ln(0.3) = \ln(42) - \ln(1.21)$
Dividing both sides by $\ln(0.3)$, we get:
$x = \frac{\ln(42) - \ln(1.21)}{\ln(0.3)} \approx -2.946120748$
≈ -2.95.

59a) $f(4) = 11.39(0.94)^{(4)} = 11.39(0.78074896)$
$= 8.892730654 \approx 8.89$.
In 1996, 8.89% of the unemployed workers in the U.S. labor force were Hispanics.

59b) In solving $11.39(0.94)^x = 10$, we begin by dividing both sides by 11.39 and then taking the natural log of both sides:
$(0.94)^x = \frac{10}{11.39}$
$\ln[(0.94)^x] = \ln(\frac{10}{11.39})$
$x\ln(0.94) = \ln(10) - \ln(11.39)$
Dividing both sides by $\ln(0.94)$, we get:
$x = \frac{\ln(10) - \ln(11.39)}{\ln(0.94)} \approx 2.103431681 \approx 2$.

$x \approx 2$ corresponds to 1994, hence after 1994, the number of unemployed workers in the U.S. labor force that were Hispanics dropped below 10%.

61a) $f(1) = 2.05 + 1.3\ln(1)$
$= 2.05 + 1.3(0) = 2.05$.
$f(3) = 2.05 + 1.3\ln(3)$
$\approx 2.05 + 1.428 \approx 3.48$.
$f(5) = 2.05 + 1.3\ln(5)$
$\approx 2.05 + 2.092 \approx 4.14$.
$f(7) = 2.05 + 1.3\ln(7)$
$\approx 2.05 + 2.530 \approx 4.58$.

61a) Continued
$f(9) = 2.05 + 1.3 \ln(9)$
$\approx 2.05 + 2.856 \approx 4.91$.

x	$f(x)$
1	2.05
3	3.48
5	4.14
7	4.58
9	4.91

61b) $f(6) = 2.05 + 1.3 \ln(6)$
$\approx 2.05 + 2.329 \approx 4.38$.
In 1990, about $4.38 billion was spent on admission to spectator sports.

61c) $m_{sec} = \frac{f(x+\Delta x) - f(x)}{\Delta x} = \frac{f(9)-f(1)}{9-1}$
$\approx \frac{4.906392 - 2.05}{8} = \frac{2.856392}{8} \approx 0.357$
Between 1985 and 1993, the personal expenditure for admissions to spectator sports was increasing at an average rate of about $0.357 billion per year.

61d) $m_{sec} = \frac{f(x+\Delta x) - f(x)}{\Delta x} = \frac{f(3)-f(1)}{3-1}$
$= \frac{3.48 - 2.05}{2} = \frac{1.43}{2} \approx 0.72$
Between 1985 and 1987, the personal expenditure for admissions to spectator sports was increasing at an average rate of about $0.72 billion per year.

63a) $f(10) = 16.49 + 2.85 \ln(10)$
$\approx 16.49 + 6.562 \approx 23.05$.
In 1989, the Federal Government spent about $23.05 billion of constant 1993-1994 dollars on higher education.

63b) $f(15) = 16.49 + 2.85 \ln(15)$
$\approx 16.49 + 7.718 \approx 24.21$
$f(10) = 16.49 + 2.85 \ln(10)$
$\approx 16.49 + 6.562 \approx 23.05$.
$m_{sec} = \frac{f(x+\Delta x) - f(x)}{\Delta x} = \frac{f(15)-f(10)}{5}$
$\approx \frac{24.21 - 23.05}{5} = \frac{1.16}{5} \approx 0.23$.

63b) Continued
Between 1989 and 1994, the amount of money the Federal Government spent on higher education was increasing at an average rate of about $0.23 billion per year.

Section 1.8 Regression and Mathematical Modeling

1) The graph seems to resemble a parabola, so we should choose a quadratic model.

3) The graph seems to resemble a third degree polynomial, so we should choose a cubic model.

5) This seems to resemble either a logarithmic function or a power function, so we should choose a logarithmic model first and then a power model as a second choice.

7) This seems to be an example of exponential growth function. We should try an exponential model.

9) The almost backward "s – shaped" graph suggests a logistic model would be good to try.

11) The Curve Fitting Process
Step #1
Standardize the independent variable by letting x be the number of years since 1980.

Year	x	Multiple Births per 1000 live births
1980	0	19.3
1985	5	21
1986	6	21.6
1987	7	22
1988	8	22.4
1989	9	23
1990	10	23.3
1991	11	23.9
1992	12	24.4
1993	13	25.2
1994	14	25.7

11) Continued
Step #2
The scatterplot of the data is displayed in viewing window [- 1, 15] by [19, 26].

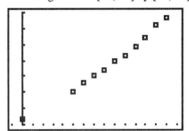

Note: The textbook writes the viewing window on the sides of the graph.

Step #3
The graph does suggest that a linear model would be good to try.

Step #4
Using the graphing calculator on the data, we get:

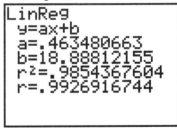

Rounding off to the nearest thousandths, we get: $y = 0.463x + 18.888$ as our model.

11) Continued
    Step #5
    Since r and r^2 are very close to one, we have a great fit. Here's the model graphed with the data points in viewing window [-1, 15] by [19, 26].

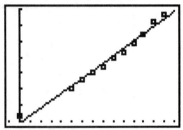

13) The Curve Fitting Process
    Step #1
    Standardize the independent variable by letting x be the number of years since 1990.

Year	x	Average Expenditure for a New Car
1990	1	$15,926
1991	2	$16,650
1992	3	$17,825
1993	4	$18,585
1994	5	$19,463
1995	6	$19,757

    Step #2
    The scatterplot of the data is displayed in viewing window [0, 7] by [15,800, 19,800].

    Step #3
    The graph does suggest that a logarithmic model would be good to try.

13) Continued
    Step #4
    Using the graphing calculator on the data, we get:

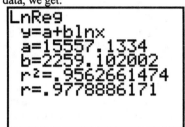

    Rounding off to the nearest thousandths, we get:
    $y = 15{,}557.133 + 2259.102 \ln(x)$
    as our model.

    Step #5
    Since r and r^2 are very close to one, we have a great fit. Here's the model graphed with the data points in viewing window [0, 7] by [15,800, 19,800].

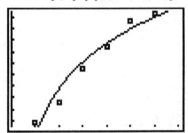

15a) $f(10) = 0.02(10)^3 - 0.23(10)^2 + 2.03(10) + 20.22$
    $= 20 - 23 + 20.3 + 20.22 = 37.52$
    In the tenth year, 37.52% of births were to unwed mothers.

15b) $f(20) = 0.02(20)^3 - 0.23(20)^2 + 2.03(20) + 20.22$
    $= 160 - 92 + 40.6 + 20.22 = 128.82$
    In the twentieth year, 128.82% of births were to unwed mothers.

15c) The answer in part b) does not make sense since the percentage cannot be larger than 100%. This would say that there were more births by unwed mothers than the total number of births which is impossible.

42 Chapter 1 Functions, Modeling, and Average Rate of Change

Chapter 1 Review Exercises

1) The ordered pairs are (−1, 3.5), (2, 1), (2.8, −1.7), and (4, 2).

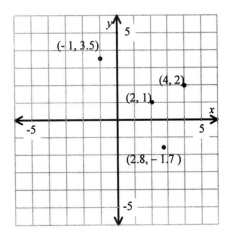

3) For the function $A = \pi r^2$, r is the independent variable, A is the dependent

5) This relation is a function since each element of the domain is assign to exactly one element of the range.

7) $f(3) = (3)^2 − 4 = 9 − 4 = 5$. The ordered pair is (3, 5).

9a) x is the independent variable.

9b) y is the dependent variable.

9c) $f(3) = 2$ on the graph.

9d) $f(1) = 2$ on the graph.

9e) The domain is [1, 4].

9f) The range is [1, 5].

11) There are no values that make the denominator of a fraction zero, but if −4 + 2x is negative, then we have a negative result under the radical. Therefore, −4 + 2x has to be greater than or equal to zero. Solving yields:
$$-4 + 2x \geq 0$$
$$-4 \geq -2x$$
$$2 \leq x \text{ or } x \geq 2$$
Thus, the domain of $f(x)$ is all real numbers greater than or equal to 2 or [2, ∞).

13) The graph is displayed in viewing window [−5, 5] by [−5, 5]:

Note: The textbook writes the viewing window on the sides of the graph.

We have a vertical asymptote of $x = -2$ and $x = 2$. There are no other values of x that make the function undefined so the domain is (−∞, −2) U (−2, 2) U (2, ∞). Finding the range is a bit more challenging. We will need to use the maximum and minimum functions on the calculator. Since the function has a horizontal asymptote of $y = 0$, we begin by examining the "left part" of the function ($x < -2$) to find its minimum value.

Chapter 1 Review Exercises 43

13) Continued
The graph is displayed in viewing window $[-20, -2]$ by $[-1, 1]$

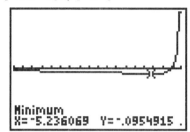

For the "left part" of the graph rounded the nearest thousandths, the range is $y \geq -0.095$. The "right part" of the function $(x > 2)$ has a range of $y > 0$ since there is no x-intercept in that part. Thus, $y \geq -0.095$ or $(-0.095, \infty)$.

Next, we need to examine the "middle part" of the function $(-2 < x < 2)$ and find the maximum value. The graph is displayed in viewing window $[-2, 2]$ by $[-10, 1]$:

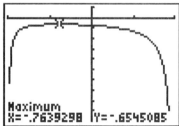

The range of the "middle part" rounded to the nearest thousandth is $y \leq -0.655$ or $(-\infty, -0.655)$. Thus the range of the entire function is the union of $(-\infty, -0.655)$ and $(-0.095, \infty)$.
Domain: $(-\infty, -2) \cup (-2, 2) \cup (2, \infty)$.
Range: $(-\infty, -0.655) \cup (-0.095, \infty)$.

15) The domain is all real number greater than -2 or $(-2, \infty)$. Since $f(1) = -3$, then the range is all real numbers greater than or equal to -3 or $[-3, \infty)$.

17) Two points that are on the line are $(6, 3)$ and $(3, 1)$ so the average rate of change is:
$m_{sec} = \frac{y_2 - y_1}{x_2 - x_1} = \frac{3-1}{6-3} = \frac{2}{3}$.

19) First, we have the average rate of change or slope of the line using $(-4, 0)$ & $(0, 5)$:
$m_{sec} = \frac{y_2 - y_1}{x_2 - x_1} = \frac{5-0}{0-(-4)} = \frac{5}{4}$.

Using $m = \frac{5}{4}$ and $(0, 5)$, the equation to the line in point-slope form becomes:
$y - y_1 = m(x - x_1)$
$y - 5 = \frac{5}{4}(x - 0)$

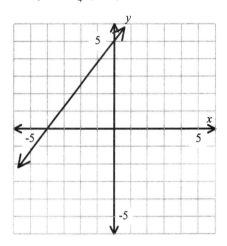

21) We first find the average rate of change or slope of the line using $(-4, -10)$ & $(12, 2)$.
$m_{sec} = \frac{y_2 - y_1}{x_2 - x_1} = \frac{2-(-10)}{12-(-4)} = \frac{12}{16} = \frac{3}{4}$.

Then, we replace m by $\frac{3}{4}$ and (x_1, y_1) by $(12, 2)$ in the C-F form and solve for y:
$y = m(x - x_1) + y_1$
$y = \frac{3}{4}(x - 12) + 2$
$y = \frac{3}{4}x - 9 + 2$.
$y = \frac{3}{4}x - 7$.

23a) Here, $m = -40$ and $b = 800$, so
$C(x) = -40x + 800$

23b) $C(x) = -40x + 800 = 600$; solve for x
$$-40x = -200$$
$$x = 5$$
In five years, the bike will be worth \$600.

25) The fixed costs are \$2600 so $b = 2600$. Thus, $C(x) = mx + 2600$. Now, we need to use the fact that $C(18) = 3950$ to find m.
$$C(18) = m(18) + 2600 = 3950;$$
$$18m + 2600 = 3950$$
$$18m = 1350$$
$$m = 75.$$
So, $C(x) = 75x + 2600$.

27)

29a) Assuming that the cost function is linear, the y-coordinate of the y-intercept will correspond to the fixed costs. Since $(0, 2500)$ is the y-intercept, then the fixed costs are \$2,500.

29b) Assuming the cost function is linear, cost to produce each additional ceiling fan will correspond to the average rate of change or the slope of the line. So,
$$m_{sec} = \frac{y_2 - y_1}{x_2 - x_1} = \frac{5500 - 2500}{50 - 0} = \frac{3000}{50} = 60.$$
The marginal cost, the cost to produce one additional ceiling fan, is \$60 per fan. This is also the variable cost.

29c) $C(x) = mx + b = 60x + 2500$.

29d) $C(84) = 60(84) + 2500 = 5040 + 2500 = 7540$. It will cost \$7,540 to produce 84 ceiling fans.

31a) Since $f(x) = -2x^2 - 6x + 12$, $a = -2$ & $b = -6$. The x-coordinate of the vertex is $x = \frac{-b}{2a}$ so $x = \frac{-(-6)}{2(-2)} = \frac{6}{-4} = -\frac{3}{2}$. The y-coordinate is $f\left(\frac{-b}{2a}\right) = f(-\frac{3}{2})$
$$= -2(-\frac{3}{2})^2 - 6(-\frac{3}{2}) + 12$$
$$= -4.5 + 9 + 12 = 16.5.$$ Hence, the vertex is $(-\frac{3}{2}, 16.5)$.

31b) Since $a < 0$, the vertex is the maximum value of f. So, the range is all real numbers less than or equal to 16.5 or $(-\infty, 16.5]$.

33a) Since $y = -x^2 + 3.2x - 7.8$, $a = -1$ and $b = 3.2$. The x-coordinate of the vertex is $x = \frac{-b}{2a}$ so $x = \frac{-(3.2)}{2(-1)} = \frac{-3.2}{-2} = 1.6$.
Plugging $x = 1.6$ into the function, we find the y-coordinate to be:
$-(1.6)^2 + 3.2(1.6) - 7.8$
$= -2.56 + 5.12 - 7.8 = -5.24$. Hence, the vertex is $(1.6, -5.24)$.

33b) Since $a < 0$, the vertex is the maximum value of y. This means that the function is increasing on $(-\infty, 1.6)$ and decreasing on $(1.6, \infty)$.

35) Setting $y = 0$ and solving yields:
$$2x^2 + 3x - 20 = 0$$
$$(2x - 5)(x + 4) = 0$$
$$2x - 5 = 0 \text{ or } x + 4 = 0$$
$$2x = 5 \text{ or } x = -4$$
$$x = 2.5 \text{ or } x = -4$$
The zeros are -4 and 2.5.

37) The graphs are displayed in viewing window [−5, 5] by [−5, 5].

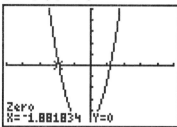

and

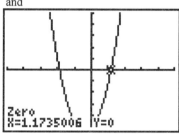

So, rounding off to the nearest hundredth, the two zeros are $x = -1.88$ and $x = 1.17$.

39a) $m_{sec} = \frac{g(x+\Delta x)-g(x)}{\Delta x} = \frac{g(1)-g(0)}{1}$

$= \frac{2(1)^2 - 3(1) - [2(0)^2 - 3(0)]}{2} = \frac{-1-0}{1} = \frac{-1}{1}$

$= -1.$

39b) $m_{sec} = \frac{g(x+\Delta x)-g(x)}{\Delta x} = \frac{g(2)-g(0)}{2-0}$

$= \frac{2(2)^2 - 3(2) - [2(0)^2 - 3(0)]}{2} = \frac{2-0}{2} = \frac{2}{2}$

$= 1.$

39c) $m_{sec} = \frac{g(x+\Delta x)-g(x)}{\Delta x} = \frac{g(4)-g(0)}{4}$

$= \frac{2(4)^2 - 3(4) - [2(0)^2 - 3(0)]}{4} = \frac{20-0}{4} = \frac{20}{4}$

$= 5.$

41a) $f(4) = 0.49(4)^2 - 3.41(4) + 88.50$
$= 7.84 - 13.64 + 88.50 = 82.7$
In 1994, the average American spent 82.7 hours a year watching premium cable television programming.

41b) The graph is displayed in viewing window [0, 10] by [75, 110]

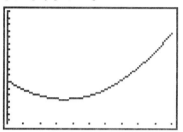

41c) First, we need to locate the x-coordinate of the vertex. Here, $a = 0.49$ and $b = -3.41$ so the x-coordinate of the vertex is

$x = \frac{-b}{2a}$ so $x = \frac{-(-3.41)}{2(-0.49)} = \frac{3.41}{0.98} \approx 3.480.$

Since a is positive and the x-coordinate of the vertex is in the domain of f, this function has a minimum value. The maximum value will occur at one of the endpoints. We therefore need to evaluate the function at $x = 0$, $x \approx 3.48$, and $x = 10$.
$f(0) = 0.49(0)^2 - 3.41(0) + 88.50$
$= 0 - 0 + 88.50 = 88.5$
$f(3.48) = 0.49(3.48)^2 - 3.41(3.48) + 88.5$
$\approx 5.934 - 11.971 + 88.50 \approx 82.57$
$f(10) = 0.49(10)^2 - 3.41(10) + 88.50$
$= 49 - 34.1 + 88.50 = 103.4$
So the range is [82.57, 103.4].

41d) $f(0) = 0.49(0)^2 - 3.41(0) + 88.50$
$= 0 - 0 + 88.50 = 88.5$
$f(1) = 0.49(1)^2 - 3.41(1) + 88.50$
$= 0.49 - 3.41 + 88.50 = 85.58$
$f(2) = 0.49(2)^2 - 3.41(2) + 88.50$
$= 1.96 - 6.82 + 88.50 = 83.64$
$f(3) = 0.49(3)^2 - 3.41(3) + 88.50$
$= 4.41 - 10.23 + 88.50 = 82.7$
$f(4) = 0.49(4)^2 - 3.41(4) + 88.50$
$= 7.84 - 13.64 + 88.50 = 82.7$
$f(5) = 0.49(5)^2 - 3.41(5) + 88.50$
$= 12.25 - 17.05 + 88.50 = 82.7$
$f(6) = 0.49(6)^2 - 3.41(6) + 88.50$
$= 17.64 - 20.46 + 88.50 = 82.7$
$f(7) = 0.49(7)^2 - 3.41(7) + 88.50$
$= 24.01 - 23.87 + 88.50 = 82.7$

41d) Continued
$f(8) = 0.49(8)^2 - 3.41(8) + 88.50$
$= 31.36 - 27.28 + 88.50 = 82.7$
$f(9) = 0.49(9)^2 - 3.41(8) + 88.50$
$= 39.69 - 30.69 + 88.50 = 82.7$
$f(10) = 0.49(10)^2 - 3.41(10) + 88.50$
$= 49 - 34.1 + 88.50 = 103.4$

x	f(x)
0	88.5
1	85.58
2	83.64
3	82.68
4	82.7
5	83.7
6	85.68
7	88.64
8	92.58
9	97.5
10	103.4

41e) In 1992 through 1995 ($x = 2, 3, 4, \& 5$), the average American spent less than 85 hours watching premium cable television programming.

43a) $(f + g)(x) = f(x) + g(x) = \sqrt{x+2} + x - 2$.
Since $x + 2 \geq 0$, then $x \geq -2$. So, the domain is $[-2, \infty)$.

43b) $(f - g)(x) = f(x) - g(x) = \sqrt{x+2} - x + 2$.
Since $x + 2 \geq 0$, then $x \geq -2$. So, the domain is $[-2, \infty)$.

43c) $(f \bullet g)(x) = f(x) \bullet g(x) = (\sqrt{x+2})(x-2)$
$= x\sqrt{x+2} - 2\sqrt{x+2}$.
Since $x + 2 \geq 0$, then $x \geq -2$. So, the domain is $[-2, \infty)$.

43d) $\left(\frac{f}{g}\right)(x) = \frac{f(x)}{g(x)} = \frac{\sqrt{x+2}}{x-2}$. Since $x + 2 \geq 0$,
then $x \geq -2$. Also, setting the denominator equal to zero and solving:
$x - 2 = 0$
$x = 2$
This value must be excluded so the domain is $[-2, 2) \cup (2, \infty)$.

45) $(f - g)(4) = f(4) - g(4)$
$= 3(4) - 5 - ((4) + 2)^2 = 7 - 36 = -29$.

47) $\left(\frac{f}{g}\right)(3) = \frac{f(3)}{g(3)} = \frac{3(3)-5}{((3)+2)^2} = \frac{4}{25} = 0.16$.

49) $R(x) = x \bullet p(x) = x(145 - 0.15x)$
$= 145x - 0.15x^2$.

51) $P(x) = R(x) - C(x)$
$= 32x - 0.08x^2 - 16x - 400$
$= -0.08x^2 + 16x - 400$
We need to find the x-coordinate of the vertex. Here, $a = -0.08$ and $b = 16$ so the x-coordinate of the vertex is
$x = \frac{-b}{2a}$ so $x = \frac{-(16)}{2(-0.08)} = \frac{-16}{-0.16} = 100$.
The y-coordinate then is:
$f\left(\frac{-b}{2a}\right) = f(100)$
$= -0.08(100)^2 + 16(100) - 400 = 400$.
So the vertex is at (100, 400). Since a is negative, P has a maximum profit of $400 when $x = 100$ units are produced.

53a) $f(x) = x^2 + 5x - 3$ is a polynomial since all the exponents of the variables in the terms are whole numbers. Since the degree of the polynomial is 2, this is a quadratic function.

53b) $y = 17$ is a polynomial since all the exponents of the variables in the terms are whole numbers. Since the degree of the polynomial is 0, this is a linear function.

53c) $g(x) = \frac{1}{3}x - 4$ is a polynomial since all the exponents of the variables in the terms are whole numbers. Since the degree of the polynomial is 1, this is a linear function.

53d) $f(x) = \frac{3}{4}x^4 - \frac{2}{3}x^3 + \frac{1}{2}x^2 + x$ is a polynomial since all the exponents of the variables in the terms are whole numbers. Since the degree of the polynomial is 4, this is a quartic function.

55) Since $g(x) = 3x^2 - 5x$ is an even degree function and the leading coefficient, 3, is positive, then as $x \to -\infty$, $g(x) \to \infty$ and as $x \to \infty$, $g(x) \to \infty$.

57a) Since $f(x) = x^3 - 6x^2 + 6x + 4$ is an odd degree function and the leading coefficient, 1, is positive, then as $x \to -\infty$, $f(x) \to -\infty$ and as $x \to \infty$, $f(x) \to \infty$.

57b) The graphs are displayed in viewing window $[-5, 5]$ by $[-10, 10]$.

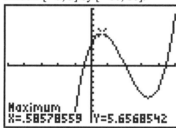

and

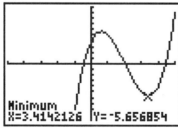

Thus, we have a "peak" at $(0.59, 5.66)$ and a "valley" at $(3.41, -5.66)$.

57c) $f(x)$ is decreasing on $(0.59, 3.41)$ and increasing on $(-\infty, 0.59) \cup (3.41, \infty)$.

59a) Setting the denominator equal to zero and solving we get:
$$2x + 3 = 0$$
$$2x = -3$$
$$x = -1.5$$
The value of -1.5 must be excluded so the domain is $(-\infty, -1.5) \cup (-1.5, \infty)$.

59b) Since the numerator of $f(x)$ is not zero at $x = -1.5$, we have a vertical asymptote at $x = -1.5$.

59c) Since the degree of the polynomial in the numerator is greater than the degree of the polynomial in the denominator, then the function has no horizontal asymptote.

61a) Setting the denominator equal to zero and solving we get:
$$x^2 - 9 = 0$$
$$(x + 3)(x - 3) = 0$$
$$x + 3 = 0 \text{ or } x - 3 = 0$$
$$x = -3 \text{ or } x = 3$$
Thus, the values of -3 and 3 must be excluded so the domain is
$(-\infty, -3) \cup (-3, 3) \cup (3, \infty)$.

61b) Since the numerator of $f(x)$ is not zero at $x = -3$, we have a vertical asymptote at $x = -3$. Since the numerator of $f(x)$ is zero at $x = 3$, we have a hole at $x = 3$.

61c) Since the degree of the polynomial in the numerator is the same as the degree of the polynomial in the denominator, then the function has a horizontal asymptote of
$y = \frac{2}{1}$ or $y = 2$.

63) To find the y-intercept, we evaluate the function at $x = 0$:
$$y = \frac{(0)^2 - 8(0) + 16}{(0)^2 - 9} = \frac{16}{-9} = -\frac{16}{9}.$$
So, the y-intercept is $(0, -\frac{16}{9})$.
To find the x-intercept, we first need to find the domain of y. We then set the function equal to zero, solve for x and check to see if the answer is in the domain of y. To find the domain, we set the denominator equal to zero and solve:
$x^2 - 9 = 0$
$(x + 3)(x - 3) = 0$
$x + 3 = 0$ or $x - 3 = 0$
$x = -3$ or $x = 3$
Hence, the domain is
$(-\infty, -3) \cup (-3, 3) \cup (3, \infty)$.
Now, we set the function equal to 0 and solve:
$$y = \frac{x^2 - 8x + 16}{x^2 - 9} = 0$$
$x^2 - 8x + 16 = 0$
$(x - 4)^2 = 0$
$x = 4$
Since 4 is in the domain of f, then the x-intercept is $(4, 0)$.

65) To find the y-intercept, we evaluate the function at $x = 0$:
$$g(x) = \frac{12}{(0)^2 + 6} = \frac{12}{6} = 2.$$
So, the y-intercept is $(0, 2)$.
To find the x-intercept, we first need to find the domain of g. We then set the function equal to zero, solve for x and check to see if the answer is in the domain of g. To find the domain, we set the denominator equal to zero and solve:
$x^2 + 6 = 0$
But since $x^2 \geq 0$, then $x^2 + 6 \geq 6$ so $x^2 + 6 = 0$ has no real solution. Hence, the domain is $(-\infty, \infty)$.

65) Continued
Now, we set the function equal to 0 and solve:
$$f(x) = \frac{12}{x^2 + 6} = 0$$
$12 = 0$
No solution
This means that $g(x)$ has no x-intercepts

67)

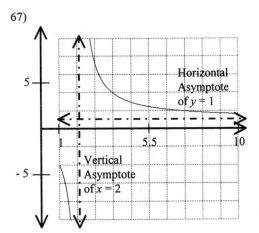

69) $\sqrt[4]{(x-2)^3} = [(x-2)^3]^{\frac{1}{4}} = (x-2)^{\frac{3}{4}}$.
Since the radical function has an even index, $x - 2 \geq 0$ or $x \geq 2$. Hence the domain is $[2, \infty)$.

71a) $f(64) = \frac{1}{4}\sqrt{64} = \frac{1}{4} \cdot 8 = 2$.
It will take 2 seconds for an object to fall 64 feet.

71b) $f(144) = \frac{1}{4}\sqrt{144} = \frac{1}{4} \cdot 12 = 3$.
It will take 3 seconds for an object to fall 144 feet.

73a) $f(12) = 19.66(12)^{0.74} \approx 19.66(6.2861841)$
≈ 123.65.
In 1987, about $123.65 billion was spent on research and development in the U.S.

73b) The graph is displayed in viewing window [5, 21] by [0, 200].

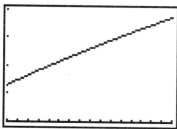

73c) $m_{sec} = \frac{f(x+\Delta x)-f(x)}{\Delta x} = \frac{f(12)-f(8)}{4}$

$= \frac{19.66(12)^{0.74}-19.66(8)^{0.74}}{4}$

$\approx \frac{123.6453-91.5945}{4} \approx \frac{32.0507}{4} \approx 8.01$.

Between 1983 and 1987, the amount spent on research and development was increasing at an average rate of about $8.01 billion per year.

75a) All values are rounded to the nearest hundredth or two significant digits.

x	−5	−4	−3	−2	−1	0
f(x)	0.016	0.036	0.082	0.19	0.43	1

x	1	2	3	4	5
f(x)	2.3	5.29	12.17	27.98	64.36

75b)

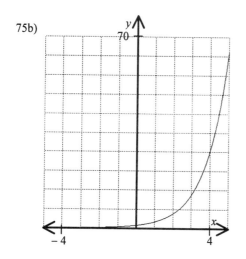

75c) This is an exponential growth function.

77a)

x	f(x)	x	f(x)
−5	≈ 729416	1	≈ 0.0672
−4	≈ 49021	2	≈ 0.0045
−3	≈ 3294.5	3	≈ 3 × 10⁻⁴
−2	≈ 221.41	4	≈ 2 × 10⁻⁵
−1	≈ 14.88	5	≈ 1 × 10⁻⁶
0	1		

77b)

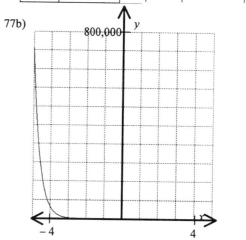

77c) This is an exponential decay function.

79) $A = P(1+\frac{r}{n})^{nt}$

Since $P = 1700$, $r = 0.053$, $n = 12$, & $t = 7$, then $A = (1700)(1+\frac{0.053}{12})^{(12)(7)}$

$A = (1700)(1.004416)^{84} \approx 2461.60$

After seven years, about $2,461.60 will be in the account.

81a) $A(t) = 8000(1+\frac{.0675}{4})^{4t}$

$= 8000(1.016875)^{4t}$

This is the amount in the account as a function of time.

81b) $A(6) = 8000(1.016875)^{4(6)}$

$= 8000(1.016875)^{24} \approx 11{,}953.96$.

After six years, about $11,953.96 will be in the account.

81c) The amount of interest earned is $A(6) - P$
= 11,953.96 − 8000 = $3,953.96.

83a) $f(4) = 64{,}000e^{0.04(4)} = 64{,}000e^{0.16}$
$\approx 75{,}105$. In 1984, there were about 75,105 people in Granaco City.

83b) $m_{\text{sec}} = \frac{f(x+\Delta x) - f(x)}{\Delta x} = \frac{f(8) - f(5)}{3}$
$= \frac{64000e^{0.04(8)} - 64000e^{0.04(5)}}{3} \approx \frac{88136.18 - 78169.78}{3}$
$\approx \frac{9966.4}{3} \approx 3322.13$

Between 1985 and 1988, the population was growing at an average rate of about 3322.13 people per year.

83c) Setting $f(x) = 110{,}000$ and solving yields:
$110{,}000 = 64{,}000e^{0.04x}$
$1.71875 = e^{0.04x}$
$\ln(1.71875) = 0.04x$
$x = \frac{\ln(1.71875)}{0.04} \approx 13.54 \approx 14$

In 1994, the population exceeded 110,000 people.

85a) Effective rate $= \left(1 + \frac{r}{n}\right)^n - 1$
$= \left(1 + \frac{0.082}{12}\right)^{12} - 1 = (1.00683)^{12} - 1$
$\approx 1.085153122 - 1 \approx 0.0852$ or 8.52%

So, investing in an account that pays 8.2% compounded monthly is equivalent to investing in an account that pays 8.52% compounded annually.

85b) Effective rate $= e^r - 1$
$= e^{0.067} - 1 = 1.069295478 - 1$
≈ 0.0693 or 6.93%

So, investing in an account that pays 6.7% compounded continuously is equivalent to investing in an account that pays 6.93% compounded annually.

87) $g(4) = 2(4) - 5 = 8 - 5 = 3$.
So, $f(g(4)) = f(3) = (3)^2 - 3(3)$
$= 9 - 9 = 0$.

89a) Since both f and g are polynomials, their domains are all real numbers or $(-\infty, \infty)$.

89b) $f(g(x)) = f(x^2) = 5(x^2) + 3 = 5x^2 + 3$.
Since this is a polynomial, the domain is all real numbers or $(-\infty, \infty)$.

89c) $g(f(x)) = g(5x + 3) = (5x + 3)^2$
$= 25x^2 + 30x + 9$.
Since this is a polynomial, the domain is all real numbers or $(-\infty, \infty)$.

91) Let $f(x) = x^5$ and $g(x) = x^2 + 2$. Then $h(x) = f(g(x)) = f(x^2 + 2) = (x^2 + 2)^5$.

93) Let $f(x) = \sqrt{x}$ and $g(x) = 6 - x$. Then
$h(x) = f(g(x)) = f(6 - x) = \sqrt{(6-x)}$
$= \sqrt{6-x}$.

95) The graph is displayed in viewing window [−5, 5] by [−5, 5].

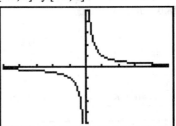

Since every horizontal line passes through at most one point on the graph, the function is one-to-one.

97a) $f(g(x)) = f(x^3 - 5) = \sqrt[3]{(x^3 - 5) + 5}$
$= \sqrt[3]{x^3} = x$.

97b)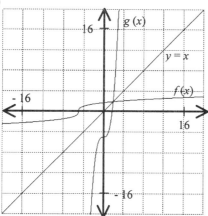

99) The base is 10, then exponent is −4 and the answer is 0.0001 so,
    −4 = log (0.0001)

101) $\log_2\left(\frac{5^2}{7}\right) = \log_2(5^2) - \log_2(7)$
     $= 2\log_2(5) - \log_2(7)$

103) In solving $6.3 \cdot (1.7)^x = 24$, we begin by dividing both sides by 6.3 and then taking the natural log of both sides:
     $6.3 \cdot (1.7)^x = 24$
     $(1.7)^x = \frac{24}{6.3} = \frac{240}{63} = \frac{80}{21}$
     $\ln[(1.7)^x] = \ln\left(\frac{80}{21}\right)$
     $x \ln(1.7) = \ln(80) - \ln(21)$
     $x = \frac{\ln(80) - \ln(21)}{\ln(1.7)} \approx 2.52.$

105) This seems to resemble either a logarithmic function or a power function, so we should choose a logarithmic model first and then a power model as a second choice.

107) The Curve Fitting Process
   Step #1
   Standardize the independent variable by letting x be the number of years since 1985.

Year	x	# of Americans that are covered by Health Insurance (millions)
1987	2	241
1989	4	246
1991	6	251
1993	8	260
1995	10	264
1997	12	269

   Step #2
   The scatterplot of the data is displayed in viewing window [−1, 13] by [240, 270].

   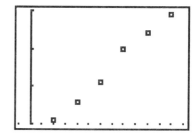

   Step #3
   The graph does suggest that a linear model would be good to try.

   Step #4
   Using the graphing calculator on the data, we get:

   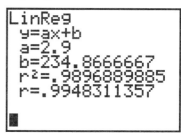

   Rounding off to the nearest thousandths, we get: $y = 2.9x + 234.867$ as our model.

107) Continued
Step #5
Since r and r^2 are very close to one, we have a great fit. Here's the model graphed with the data points in viewing window [-1, 13] by [240, 270].

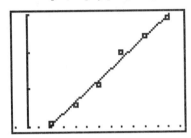

109) The Curve Fitting Process
Step #1
Standardize the independent variable by letting x be the number of years since 1900.

Year	x	Number of U.S. Workers who Worked at Home (millions)
1963	63	19.3
1968	68	26.6
1973	73	35.5
1978	78	62.5
1983	83	89.8
1988	88	138.3
1993	93	147.7
1998	98	181.9

Step #2
The scatterplot of the data is displayed in viewing window [-1, 100] by [19, 190].

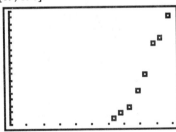

109) Continued
Step #3
The graph does suggest that a power model would be good to try.

Step #4
Using the graphing calculator on the data, we get:

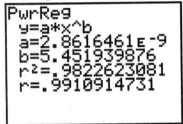

Rounding off, we get:
$y = (2.862 \times 10^{-9}) x^{5.452}$ as our model.

Step #5
Since r and r^2 are very close to one, we have a great fit. Here's the model graphed with the data points in viewing window [-1, 100] by [19, 190].

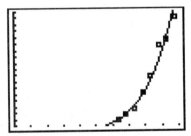

CHAPTER 2 LIMITS, INSTANTANEOUS RATE OF CHANGE AND THE DERIVATIVE

Section 2.1 Limits

1) We begin by evaluating the function for the various values in the table:
   $f(0) = 2 - 3(0) = 2 - 0 = 2$ $f(2) = 2 - 3(2) = 2 - 6 = -4$
   $f(0.9) = 2 - 3(0.9) = 2 - 2.7 = -0.7$ $f(1.1) = 2 - 3(1.1) = 2 - 3.3 = -1.3$
   $f(0.99) = 2 - 3(0.99) = 2 - 2.97 = -0.97$ $f(1.01) = 2 - 3(1.01) = 2 - 3.03 = -1.03$
   $f(0.999) = 2 - 3(0.999) = 2 - 2.997 = -0.997$ $f(1.001) = 2 - 3(1.001) = 2 - 3.003 = -1.003$
   From this side, the values approach -1.   From this side, the values approach -1.

x	0	0.9	0.99	0.999	1	1.001	1.01	1.1	2
$f(x)$	2	-0.7	-0.97	-0.997	?	-1.003	-1.03	-1.3	-4

Coming in from either direction, we see that the function values are approaching -1. Thus,
   a) $\lim_{x \to 1^-} f(x) = -1$ b) $\lim_{x \to 1^+} f(x) = -1$ c) $\lim_{x \to 1} f(x) = -1$

3) We begin by evaluating the function for the various values in the table:
   $f(-3) = 2(-3)^4 - 3(-3)^3 + 2(-3) - 4 = 162 + 81 - 6 - 4 = 233$
   $f(-2.1) = 2(-2.1)^4 - 3(-2.1)^3 + 2(-2.1) - 4 = 38.8962 + 27.783 - 4.2 - 4 \approx 58.479$
   $f(-2.01) = 2(-2.01)^4 - 3(-2.01)^3 + 2(-2.01) - 4 \approx 32.6448 + 24.3618 - 4.02 - 4 \approx 48.987$
   $f(-2.001) = 2(-2.001)^4 - 3(-2.001)^3 + 2(-2.001) - 4 \approx 16.0320 + 24.0360 - 4.002 - 4 = 48.098$
   From this side, the values approach 48.
   $f(-1) = 2(-1)^4 - 3(-1)^3 + 2(-1) - 4 = 2 + 3 - 2 - 4 = -1$
   $f(-1.9) = 2(-1.9)^4 - 3(-1.9)^3 + 2(-1.9) - 4 = 26.0642 + 20.577 - 3.8 - 4 \approx 38.841$
   $f(-1.99) = 2(-1.99)^4 - 3(-1.99)^3 + 2(-1.99) - 4 \approx 31.3648 + 23.6418 - 3.98 - 4 \approx 47.027$
   $f(-1.999) = 2(-1.999)^4 - 3(-1.999)^3 + 2(-1.999) - 4 \approx 31.9360 + 23.9640 - 3.998 - 4 \approx 47.902$
   From this side, the values approach 48.

x	-3	-2.1	-2.01	-2.001	-2	-1.999	-1.99	-1.9	-1
$f(x)$	233	≈ 58.48	≈ 48.99	≈ 48.10	?	≈ 47.90	≈ 47.03	≈ 38.84	-1

Coming in from either direction, we see that the function values are approaching 48. Thus,
   a) $\lim_{x \to -2^-} f(x) = 48$ b) $\lim_{x \to -2^+} f(x) = 48$ c) $\lim_{x \to -2} f(x) = 48$

5) We begin by evaluating the function for the various values in the table:
   $f(0) = \frac{(0)^2 - 1}{(0) - 1} = \frac{0 - 1}{0 - 1} = \frac{-1}{-1} = 1$ $f(2) = \frac{(2)^2 - 1}{(2) - 1} = \frac{4 - 1}{2 - 1} = \frac{3}{1} = 3$
   $f(0.9) = \frac{(0.9)^2 - 1}{(0.9) - 1} = \frac{0.81 - 1}{0.9 - 1} = \frac{-0.19}{-0.1} = 1.9$ $f(1.1) = \frac{(1.1)^2 - 1}{(1.1) - 1} = \frac{1.21 - 1}{1.1 - 1} = \frac{0.21}{0.1} = 2.1$
   $f(0.99) = \frac{(0.99)^2 - 1}{(0.99) - 1} = \frac{0.9801 - 1}{0.99 - 1} = \frac{-0.0199}{-0.01} = 1.99$ $f(1.01) = \frac{(1.01)^2 - 1}{(1.01) - 1} = \frac{1.0201 - 1}{1.01 - 1} = \frac{0.0201}{0.01} = 2.01$
   $f(0.999) = \frac{(0.999)^2 - 1}{(0.999) - 1} \approx \frac{0.998001 - 1}{0.999 - 1} = \frac{-0.001999}{-0.001} = 1.999$ $f(1.001) = \frac{(1.001)^2 - 1}{(1.001) - 1} = \frac{1.002001 - 1}{1.001 - 1} = \frac{0.002001}{0.001} = 2.001$
   From this side, the values approach 2.   From this side, the values approach 2.

54 Chapter 2 Limits, Instantaneous Rate of Change and the Derivative

11) We begin by graphing the function and zooming in. The display window is listed below the graph.

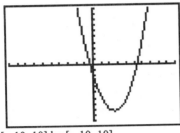

[−10, 10] by [−10, 10]

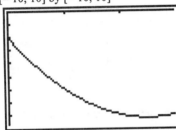

[0, ≈ 2.98] by [≈ −8.71, 0]
Note: The textbook writes the viewing window on the sides of the graph.
By using the trace features, we find:

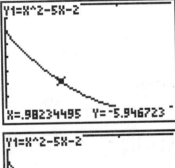

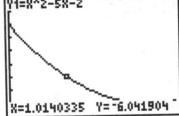

that $\lim_{x \to 1} f(x) = -6$.

11) Continued
We can verify our results numerically:

x	0	0.9	0.99	0.999
$f(x)$	−2	−5.69	−5.97	−5.997
x	2	1.1	1.01	1.001
$f(x)$	−8	−6.29	−6.03	−6.003

As x approaches 1, $f(x)$ approaches −6.

13) We begin by graphing the function and zooming in. The display window is listed below the graph.

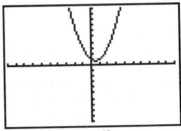

[−10, 10] by [−10, 10]

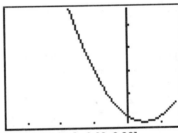

[1.75, 4.25] by [−2.22, 2.28]

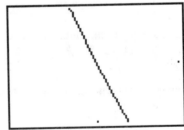

[−1.64, −0.39] by [2.36, 3.61]

5) Continued

x	0	0.9	0.99	0.999	1	1.001	1.01	1.1	2
$f(x)$	1	1.9	1.99	1.999	?	2.001	2.01	2.1	3

Coming in from either direction, we see that the function values are approaching 2. Thus,
a) $\lim_{x \to 1^-} f(x) = 2$ b) $\lim_{x \to 1^+} f(x) = 2$ c) $\lim_{x \to 1} f(x) = 2$

7) We begin by evaluating the function for the various values in the table:

$f(-3) = \frac{(-3)^3 + 8}{(-3)+2} = \frac{-27+8}{-3+2} = \frac{-19}{-1} = 19$ $f(-1) = \frac{(-1)^3 + 8}{(-1)+2} = \frac{-1+8}{-1+2} = \frac{7}{1} = 7$

$f(-2.1) = \frac{(-2.1)^3 + 8}{(-2.1)+2} = \frac{-9.261+8}{-2.1+2} = \frac{-1.261}{-0.1} = 12.61$ $f(-1.9) = \frac{(-1.9)^3 + 8}{(-1.9)+2} = \frac{-6.859+8}{-1.9+2} = \frac{1.141}{0.1} = 11.41$

$f(-2.01) = \frac{(-2.01)^3 + 8}{(-2.01)+2} = \frac{-8.120601+8}{-2.01+2}$ $f(-1.99) = \frac{(-1.99)^3 + 8}{(-1.99)+2} = \frac{-7.880599+8}{-1.99+2}$

$= \frac{-0.120601}{-0.01} = 12.0601$ $= \frac{0.119401}{0.01} = 11.9401$

$f(-2.001) = \frac{(-2.001)^3 + 8}{(-2.001)+2} = \frac{-8.012006001+8}{-2.001+2}$ $f(-1.999) = \frac{(-1.999)^3 + 8}{(-1.999)+2} = \frac{-7.988005999+8}{-1.999+2}$

$= \frac{-0.012006001}{-0.001} = 12.006001$ $= \frac{0.011994001}{0.001} = 11.994001$

From this side, the values approach 12.   From this side, the values approach 12.

x	-3	-2.1	-2.01	-2.001	-2	-1.999	-1.99	-1.9	-1
$f(x)$	19	12.61	≈ 12.06	≈ 12.006	?	≈ 11.99	≈ 11.94	11.41	7

Coming in from either direction, we see that the function values are approaching 12. Thus,
a) $\lim_{x \to -2^-} f(x) = 12$ b) $\lim_{x \to -2^+} f(x) = 12$ c) $\lim_{x \to -2} f(x) = 12$

9) $f(-1) = (-1)^2 + 1 = 1 + 1 = 2$ $f(1) = 3(1) + 1 = 3 + 1 = 4$
$f(-0.1) = (-0.1)^2 + 1 = 0.01 + 1 = 1.01$ $f(0.1) = 3(0.1) + 1 = 0.3 + 1 = 1.3$
$f(-0.01) = (-0.01)^2 + 1 = 0.0001 + 1 = 1.0001$ $f(0.01) = 3(0.01) + 1 = 0.03 + 1 = 1.03$
$f(-0.001) = (-0.001)^2 + 1 = 0.000001 + 1 = 1.000001$ $f(0.001) = 3(0.001) + 1 = 0.003 + 1 = 1.003$
From this side, the values approach 1.   From this side, the values approach 1.

x	-1	-0.1	-0.01	-0.001	0	0.001	0.01	0.1	1
$f(x)$	2	1.01	1.0001	1.000001	?	1.003	1.03	1.3	4

Coming in from either direction, we see that the function values are approaching 1. Thus,
a) $\lim_{x \to 0^-} f(x) = 1$ b) $\lim_{x \to 0^+} f(x) = 1$ c) $\lim_{x \to 0} f(x) = 1$

Section 2.1 Limits 55

13) Continued
By using the trace features, we find:

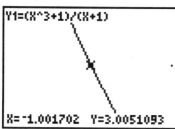

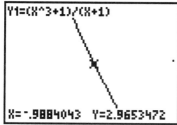

that $\lim_{x \to -1} g(x) = 3$.

We can verify our results numerically:

x	-2	-1.1	-1.01	-1.001
$g(x)$	7	3.31	3.03	3.003
x	0	-0.9	-0.99	-0.999
$g(x)$	1	2.71	2.97	2.997

As x approaches -1, $g(x)$ approaches 3.

15) We begin by graphing the function and zooming in. The display window is listed below the graph.

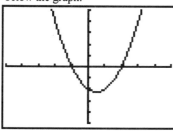

$[-5, 5]$ by $[-5, 5]$

15) Continued

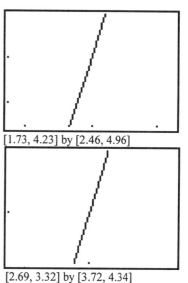

$[1.73, 4.23]$ by $[2.46, 4.96]$

$[2.69, 3.32]$ by $[3.72, 4.34]$

By using the trace features, we find:

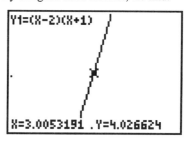

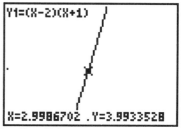

that $\lim_{x \to 3} f(x) = 4$.

56 Chapter 2 Limits, Instantaneous Rate of Change and the Derivative

15) Continued

We can verify our results numerically:

x	2	2.9	2.99	2.999
$f(x)$	0	3.51	3.95	3.995
x	4	3.1	3.01	3.001
$f(x)$	10	4.51	4.05	4.005

As x approaches 3, $f(x)$ approaches 4.

17) We begin by graphing the function and zooming in. The display window is listed below the graph.

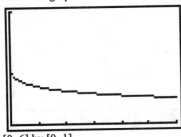

[0, 6] by [0, 1]

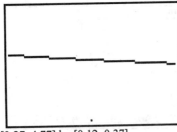

[3.27, 4.77] by [0.12, 0.37]

By using the trace features, we find:

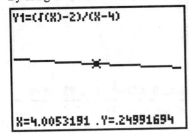

17) Continued

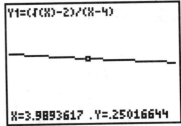

that $\lim_{x \to 4} f(x) = 0.25$.

We can verify our results numerically:

x	3	3.9	3.99	3.999
$f(x)$	0.268	0.252	0.2502	0.25002
x	5	4.1	4.01	4.001
$f(x)$	0.236	0.248	0.2498	0.24998

As x approaches 4, $f(x)$ approaches 0.25.

19) Using the Substitution Principle, we find that
$\lim_{x \to -2} (3x + 1) = 3(-2) + 1 = -6 + 1 = -5$.

We can verify this numerically and graphically:

x	-3	-2.1	-2.01	-2.001
$f(x)$	-8	-5.3	-5.03	-5.003
x	-1	-1.9	-1.99	-1.999
$f(x)$	-2	-4.7	-4.97	-4.997

The graphs are displayed in viewing window $[-2.1, -1.9]$ by $[-6, -4]$.

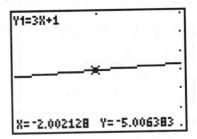

Section 2.1 Limits 57

19) Continued
and

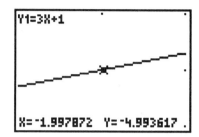

21) Using the Substitution Principle, we find that
$\lim_{x \to 2} (2x^2 + 3x - 1) = 2(2)^2 + 3(2) - 1$
$= 8 + 6 - 1 = 13$

We can verify this numerically and graphically:

x	1	1.9	1.99	1.999
$f(x)$	4	11.92	12.89	12.989
x	3	2.1	2.01	2.001
$f(x)$	26	14.12	13.11	13.011

The graphs are displayed in viewing window [1.9, 2.1] by [11, 15].

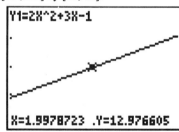
and
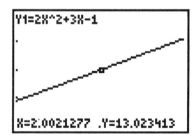

23) We need to simplify the expression before using the Substitution Principle.

$\lim_{x \to 3} \frac{x^2 - 9}{x - 3} = \lim_{x \to 3} \frac{(x+3)(x-3)}{x-3} = \lim_{x \to 3} (x + 3)$

(since $x \neq 3$) We can now use the Substitution Principle:
$\lim_{x \to 3} (x + 3) = (3) + 3 = 6.$

Thus, $\lim_{x \to 3} \frac{x^2 - 9}{x - 3} = 6.$

We can verify this numerically and graphically:

x	2	2.9	2.99	2.999
$f(x)$	5	5.9	5.99	5.999
x	4	3.1	3.01	3.001
$f(x)$	7	6.1	6.01	6.001

The graphs are displayed in viewing window [2.9, 3.1] by [5.8, 6.2].

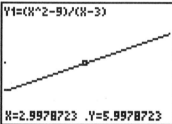
and
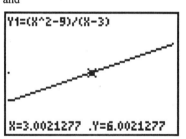

25) Since $|x| = x$ if $x > 0$ and $|x| = -x$ if $x < 0$ then using the Substitution Principle, we find that $\lim_{x \to 0^-} |x| = \lim_{x \to 0^-} (-x) = -(0) = 0$ and
$\lim_{x \to 0^+} |x| = \lim_{x \to 0^+} (x) = (0) = 0.$

25) Continued
Therefore, coming in from either direction, we see that the function values are approaching 0. Thus,
$\lim_{x \to 0} |x| = 0$.

We can verify this numerically and graphically:

x	−1	−0.1	−0.01	−0.001
f(x)	1	0.1	0.01	0.001
x	1	0.1	0.01	0.001
f(x)	1	0.1	0.01	0.001

The graphs are displayed in viewing window [−0.1, 0.1] by [−0.1, 0.1]

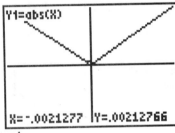

and

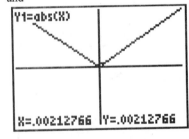

27) Since $|x+1| = x+1$ if $x > -1$ and $|x+1| = -(x+1)$ if $x < -1$, then using the Substitution Principle, we find that
$\lim_{x \to -1^-} |x+1| = \lim_{x \to -1^-} -(x+1) =$
$= -((-1)+1) = -(0) = 0$
and
$\lim_{x \to -1^+} |x+1| = \lim_{x \to -1^+} (x+1)$
$= ((-1)+1) = (0) = 0$.

27) Continued
Therefore, coming in from either direction, we see that the function values are approaching 0. Thus,
$\lim_{x \to -1} |x+1| = 0$.

We can verify this numerically and graphically:

x	−2	−1.1	−1.01	−1.001
f(x)	1	0.1	0.01	0.001
x	0	−0.9	−0.99	−0.999
f(x)	1	0.1	0.01	0.001

The graphs are displayed in viewing window [−1.1, −0.9] by [−0.1, 0.1]

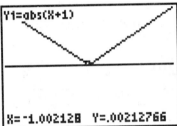

and

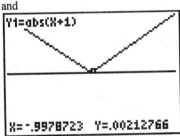

29) We need to simplify the expression before using the Substitution Principle.
$\lim_{x \to -1} \frac{x^2-1}{x+1} = \lim_{x \to -1} \frac{(x+1)(x-1)}{x+1} = \lim_{x \to -1} (x-1)$
$= (-1) - 1 = -2$.

We can verify this numerically and graphically:

x	−2	−1.1	−1.01	−1.001
f(x)	−3	−2.1	−2.01	−2.001
x	0	−0.9	−0.99	−0.999
f(x)	−1	−1.9	−1.99	−1.999

Section 2.1 Limits 59

29) Continued
The graphs are displayed in viewing window
$[-1.1, -0.9]$ by $[-2.1, -1.9]$

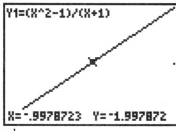

and

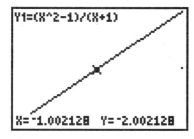

31) Using the Substitution Principle, we find that
$$\lim_{x \to 0} \sqrt{2x+3} = \sqrt{2(0)+3} = \sqrt{0+3}$$
$$= \sqrt{3} \approx 1.732.$$

We can verify this numerically and graphically:

x	−1	−0.1	−0.01	−0.001
f(x)	1	1.67	1.726	1.731
x	1	0.1	0.01	0.001
f(x)	2.24	1.79	1.738	1.733

The graphs are displayed in viewing window
$[-0.3, 0.3]$ by $[1.1, 2.4]$

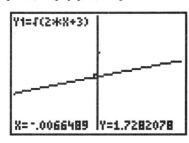

31) Continued
and

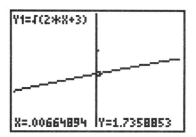

33) Using the Substitution Principle, we find that
$$\lim_{x \to -5} \frac{x^2 - 25}{x - 5} = \frac{(-5)^2 - 25}{(-5) - 5} = \frac{0}{-10} = 0.$$

We can verify this numerically and graphically:

x	−6	−5.1	−5.01	−5.001
f(x)	−1	−0.1	−0.01	−0.001
x	−4	−4.9	−4.99	−4.999
f(x)	1	0.1	0.01	0.001

The graphs are displayed in viewing window
$[-5.1, -4.9]$ by $[-0.1, 0.1]$

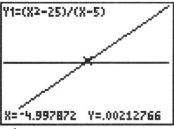

and

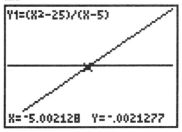

35) Using the Substitution Principle, we find that

$$\lim_{x \to 1} \frac{x^2-1}{x+1} = \frac{(1)^2-1}{(1)+1} = \frac{0}{2} = 0.$$

We can verify this numerically and graphically:

x	2	1.1	1.01	1.001
$f(x)$	1	0.1	0.01	0.001
x	0	0.9	0.09	0.009
$f(x)$	-1	-0.1	-0.01	-0.001

The graphs are displayed in viewing window $[0.9, 1.1]$ by $[-0.1, 0.1]$

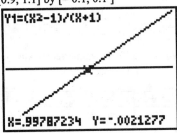

and

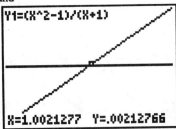

37) Using the Substitution Principle, we find that:

$$\lim_{x \to 2} (x+1)^2 \cdot (3x-1)^3$$
$$= (2+1)^2 \cdot (3(2)-1)^3 = (3)^2 \cdot (5)^3$$
$$= 9 \cdot 125 = 1125.$$

We can verify this numerically and graphically:

x	1	1.9	1.99	1.999
$f(x)$	32	873.15	1097.52	1122.23
x	3	2.1	2.01	2.001
$f(x)$	8192	1430.71	1153.02	1127.78

37) Continued

The graphs are displayed in viewing window $[1.9, 2.1]$ by $[1100, 1150]$

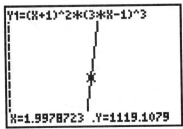

and

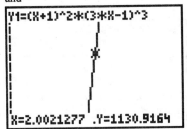

39) We need to simplify the expression before using the Substitution Principle.

$$\lim_{x \to 3} \frac{x^2-x-6}{x-3} = \lim_{x \to 3} \frac{(x+2)(x-3)}{x-3} = \lim_{x \to 3} (x+2)$$
$$= 3 + 2 = 5.$$

We can verify this numerically and graphically:

x	2	2.9	2.99	2.999
$f(x)$	4	4.9	4.99	4.999
x	4	3.1	3.01	3.001
$f(x)$	6	5.1	5.01	5.001

The graphs are displayed in viewing window $[2.9, 3.1]$ by $[4.9, 5.1]$

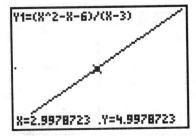

39) Continued
and

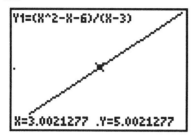

41) We need to simplify the expression before using the Substitution Principle.

$$\lim_{x \to -2} \frac{x+2}{x^2+5x+6} = \lim_{x \to -2} \frac{x+2}{(x+3)(x+2)}$$
$$= \lim_{x \to -2} \frac{1}{x+3} = \frac{1}{-2+3} = \frac{1}{1} = 1.$$

We can verify this numerically and graphically:

x	-3	-2.1	-2.01	-2.001
$f(x)$	$\emptyset$	1.1111	1.0101	1.001
x	-1	-1.9	-1.99	-1.999
$f(x)$	0.5	0.909	0.99	0.999

The graphs are displayed in viewing window $[-2.1, -1.9]$ by $[0.9, 1.1]$

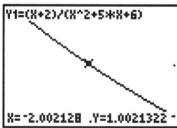

and

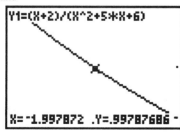

43) We need to simplify the expression before using the Substitution Principle.

$$\lim_{x \to 4} \frac{\sqrt{x}-2}{x-4} = \lim_{x \to 4} \frac{\sqrt{x}-2}{x-4} \cdot \frac{\sqrt{x}+2}{\sqrt{x}+2}$$
$$= \lim_{x \to 4} \frac{(\sqrt{x})^2 - 2^2}{(x-4)(\sqrt{x}+2)} = \lim_{x \to 4} \frac{(x-4)}{(x-4)(\sqrt{x}+2)}$$
$$= \lim_{x \to 4} \frac{1}{\sqrt{x}+2} = \frac{1}{\sqrt{4}+2} = \frac{1}{2+2} = \frac{1}{4}.$$

We can verify this numerically and graphically:

x	3	3.9	3.99	3.999
$f(x)$	0.268	0.2516	0.25016	0.25002
x	5	4.1	4.01	4.001
$f(x)$	0.236	0.2485	0.24984	0.24998

The graphs are displayed in viewing window $[3.9, 4.1]$ by $[0.2, 0.3]$

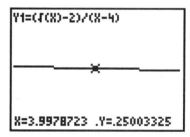

and

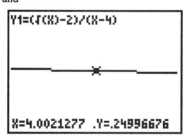

45a) Since $\lim_{x \to -1^-} f(x) = 1$ & $\lim_{x \to -1^+} f(x) = 1$ then $\lim_{x \to -1} f(x) = 1$.

45b) $f(-1) = 1$.

45c) $\lim_{x \to 3^-} f(x) = 1$

45d) $\lim_{x \to 3^+} f(x) = 2$

45e) Since the results from part c and d are not equal, then $\lim_{x \to 3} f(x)$ does not exist.

45f) $f(1) = 2$

45g) Since $\lim_{x \to 1^-} f(x) = 3$ & $\lim_{x \to 1^+} f(x) = 3$
then $\lim_{x \to 1} f(x) = 3$.

47) Using the Substitution Principle, we find that:
$\lim_{h \to 0} (2x + h) = 2x + (0) = 2x$.

49) We need to simplify the expression before using the Substitution Principle.
$\lim_{h \to 0} \frac{3xh + h^2}{h} = \lim_{h \to 0} \frac{h(3x+h)}{h} = \lim_{h \to 0} (3x + h)$
$= 3x + (0) = 3x$.

51) We need to simplify the expression before using the Substitution Principle.
$\lim_{h \to 0} \frac{4x^2 h + 2h^2}{h} = \lim_{h \to 0} \frac{h(4x^2 + 2h)}{h}$
$= \lim_{h \to 0} (4x^2 + 2h) = 4x^2 + 2(0) = 4x^2$.

53) We need to simplify the expression before using the Substitution Principle.
$\lim_{h \to 0} \frac{-3x^2 h + 6h^2}{h} = \lim_{h \to 0} \frac{h(-3x^2 + 6h)}{h}$
$= \lim_{h \to 0} (-3x^2 + 6h) = -3x^2 + 6(0) = -3x^2$.

55a) $f(1 + h) = 3(1 + h) - 1 = 3 + 3h - 1$
$= 3h + 2$.

55b) $f(1) = 3(1) - 1 = 3 - 1 = 2$.
So, $\lim_{h \to 0} \frac{f(1+h) - f(1)}{h} = \lim_{h \to 0} \frac{3h + 2 - 2}{h}$
$= \lim_{h \to 0} \frac{3h}{h} = \lim_{h \to 0} 3 = 3$.

57a) $f(1 + h) = (1 + h)^2 + 2 = 1 + 2h + h^2 + 2$
$= h^2 + 2h + 3$.

57b) $f(1) = (1)^2 + 2 = 1 + 2 = 3$.
So, $\lim_{h \to 0} \frac{f(1+h) - f(1)}{h} = \lim_{h \to 0} \frac{h^2 + 2h + 3 - 3}{h}$
$= \lim_{h \to 0} \frac{h^2 + 2h}{h} = \lim_{h \to 0} \frac{h(h+2)}{h} = \lim_{h \to 0} (h + 2)$
$= 0 + 2 = 2$.

59a) $f(1 + h) = (1 + h)^2 + 3(1 + h) - 1$
$= 1 + 2h + h^2 + 3 + 3h - 1$
$= h^2 + 5h + 3$.

59b) $f(1) = (1)^2 + 3(1) - 1 = 1 + 3 - 1 = 3$.
So, $\lim_{h \to 0} \frac{f(1+h) - f(1)}{h} = \lim_{h \to 0} \frac{h^2 + 5h + 3 - 3}{h}$
$= \lim_{h \to 0} \frac{h^2 + 5h}{h} = \lim_{h \to 0} \frac{h(h+5)}{h} = \lim_{h \to 0} (h + 5)$
$= 0 + 5 = 5$.

61a) $f(1 + h) = \frac{1}{1+h}$.

61b) $f(1) = \frac{1}{1} = 1$.
So, $\lim_{h \to 0} \frac{f(1+h) - f(1)}{h} = \lim_{h \to 0} \frac{\frac{1}{1+h} - 1}{h}$
$= \lim_{h \to 0} \frac{\frac{1}{1+h} - \frac{1+h}{1+h}}{h} = \lim_{h \to 0} \frac{\frac{1-1-h}{1+h}}{h}$
$= \lim_{h \to 0} \frac{-h}{1+h} \div h = \lim_{h \to 0} \frac{-h}{1+h} \cdot \frac{1}{h} = \lim_{h \to 0} \frac{-1}{1+h}$
$= \frac{-1}{1+0} = \frac{-1}{1} = -1$.

63a) The graph is displayed in viewing window [0, 20] by [0, 2500]

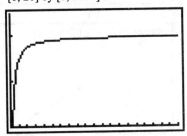

63b) $N(10) = 2000 - \frac{520}{10} = 2000 - 52 = 1948$.
This means that 194,800 items are sold when $10,000 is spent on advertising.

63c) Using the Substitution Principle, we find that:
$\lim_{x \to 10} 2000 - \frac{520}{x} = 2000 - \frac{520}{10}$
$= 2000 - 52 = 1948$.
This means that the limit of the number of items sold as advertising approaches $10,000 is 194,800 items.

65a) The graph is displayed in viewing window [0, 25] by [0, 20,000]

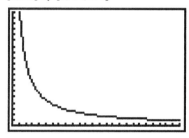

65b) $AC(10) = \frac{22500 + 7.35(10)}{10} = \frac{22500 + 73.5}{10}$
$= \frac{22573.5}{10} = 2257.35$.
When ten units are produced, the average cost per unit is $2257.35.

65c) Using the Substitution Principle, we find that:
$\lim_{x \to 10} \frac{22500 + 7.35x}{10} = \frac{22500 + 7.35(10)}{10}$
$= \frac{22500 + 73.5}{10} = \frac{22573.5}{10} = 2257.35$
This means that the limit of the average cost per item produced as the production approaches 10 units is $2257.35.

67a) $c(100) = 25 + 0.05(100) = 25 + 5 = 30$.
It will cost $30 to rent a medium-size car and drive it 100 miles.

67b) Since $\lim_{m \to 100^-} c(m) = 30$ and
$\lim_{m \to 100^+} c(m) = 30$,
then $\lim_{m \to 100} c(m) = 30$.
The limit of the cost of the rental as the miles driven approaches 100 miles is $30.

67c) $c(200) = 25(2) + 0.05(200) = 50 + 10 = 60$.
It will cost $60 to rent a medium-size car and drive it 200 miles.

67d) Since $\lim_{m \to 200^-} c(m) = 35$ and
$\lim_{m \to 200^+} c(m) = 60$,
then $\lim_{m \to 100} c(m)$ does not exist.

67e) The jump in the graph represents the end of the first day and the beginning of the second day.

64 Chapter 2 Limits, Instantaneous Rate of Change and the Derivative

Section 2.2 Limits and Asymptotes

1) We begin by evaluating the function for the various values in the table:

$f(-0.1) = \frac{1}{(-0.1)^3} = \frac{1}{-0.001} = -1000$ $\qquad$ $f(0.1) = \frac{1}{(0.1)^3} = \frac{1}{0.001} = 1000$

$f(-0.01) = \frac{1}{(-0.01)^3} = \frac{1}{-0.000001} = -1000000$ $\qquad$ $f(0.01) = \frac{1}{(0.01)^3} = \frac{1}{0.000001} = 1000000$

$f(-0.001) = \frac{1}{(-0.001)^3} = \frac{1}{-1\times10^{-9}} = -1\times10^9$ $\qquad$ $f(0.001) = \frac{1}{(0.001)^3} = \frac{1}{1\times10^{-9}} = 1\times10^9$

$f(-0.00001) = \frac{1}{(-0.00001)^3} = \frac{1}{-1\times10^{-15}} = -1\times10^{15}$ $\qquad$ $f(0.00001) = \frac{1}{(0.00001)^3} = \frac{1}{1\times10^{-15}} = 1\times10^{15}$

From this side, the values approach $-\infty$. $\qquad$ From this side, the values approach ∞.

x	-0.1	-0.01	-0.001	-0.00001	0	0.00001	0.001	0.01	0.1
$f(x)$	-1000	-1×10^6	-1×10^9	-1×10^{15}	?	1×10^{15}	1×10^9	1×10^6	1000

Coming in from different directions, we see that the function does not approach the same value. Thus,

a) $\lim\limits_{x\to 0^-} f(x) = -\infty$ $\qquad$ b) $\lim\limits_{x\to 0^+} f(x) = \infty$ $\qquad$ c) $\lim\limits_{x\to 0} f(x)$ does not exist

3) We begin by evaluating the function for the various values in the table:

$f(0.9) = \frac{1}{(0.9-1)^2} = \frac{1}{(-0.1)^2} = \frac{1}{0.01} = 100$ $\qquad$ $f(1.1) = \frac{1}{(1.1-1)^2} = \frac{1}{(0.1)^2} = \frac{1}{0.01} = 100$

$f(0.99) = \frac{1}{(0.99-1)^2} = \frac{1}{(-0.01)^2} = \frac{1}{0.0001} = 10000$ $\qquad$ $f(1.01) = \frac{1}{(1.01-1)^2} = \frac{1}{(0.01)^2} = \frac{1}{0.0001} = 10000$

$f(0.999) = \frac{1}{(0.999-1)^2} = \frac{1}{(-0.001)^2} = \frac{1}{1\times10^{-6}} = 1\times10^6$ $\qquad$ $f(1.001) = \frac{1}{(1.001-1)^2} = \frac{1}{(0.001)^2} = \frac{1}{1\times10^{-6}} = 1\times10^6$

$f(0.99999) = \frac{1}{(0.99999-1)^2} = \frac{1}{(-0.00001)^2} = \frac{1}{1\times10^{-10}} = 1\times10^{10}$ $\qquad$ $f(1.00001) = \frac{1}{(1.00001-1)^2} = \frac{1}{(0.00001)^2}$

$\qquad\qquad\qquad\qquad\qquad\qquad\qquad\qquad\qquad\qquad\qquad\qquad\qquad\qquad\qquad = \frac{1}{1\times10^{-10}} = 1\times10^{10}$

From this side, the values approach ∞. $\qquad$ From this side, the values approach ∞.

x	0.9	0.99	0.999	0.99999	1	1.00001	1.001	1.01	1.1
$f(x)$	100	1×10^4	1×10^6	1×10^{10}	?	1×10^{10}	1×10^6	1×10^4	100

Coming in from either direction, we see that the function values are approaching ∞. Thus,

a) $\lim\limits_{x\to 1^-} f(x) = \infty$ $\qquad$ b) $\lim\limits_{x\to 1^+} f(x) = \infty$ $\qquad$ c) $\lim\limits_{x\to 1} f(x) = \infty$

5) We begin by evaluating the function for the various values in the table:

$f(-2.1) = \frac{-2.1+2}{(-2.1)^2-(-2.1)-6} = \frac{-0.1}{4.41+2.1-6} = \frac{-0.1}{0.51} \approx -0.1961$

$f(-2.001) = \frac{-2.001+2}{(-2.001)^2-(-2.001)-6} = \frac{-0.001}{4.004001+2.001-6} = \frac{-0.001}{0.005001} \approx -0.19996$

$f(-2.0001) = \frac{-2.0001+2}{(-2.0001)^2-(-2.0001)-6} = \frac{-0.0001}{4.00040001+2.0001-6} = \frac{-0.0001}{0.00050001} \approx -0.199996$

From this side, the values approach -0.2.

5) Continued

$$f(-1.9) = \frac{-1.9+2}{(-1.9)^2-(-1.9)-6} = \frac{0.1}{3.61+1.9-6} = \frac{0.1}{-0.49} \approx -0.2041$$

$$f(-1.999) = \frac{-1.999+2}{(-1.999)^2-(-1.999)-6} = \frac{0.001}{3.996001+1.999-6} = \frac{0.001}{-0.004999} \approx -0.20004$$

$$f(-1.9999) = \frac{-1.9999+2}{(-1.9999)^2-(-1.9999)-6} = \frac{0.0001}{3.99960001+1.9999-6} = \frac{0.0001}{-0.00049999} \approx -0.200004$$

From this side, the values approach -0.2.

x	-2.1	-2.001	-2.0001	-2	-1.9999	-1.999	-1.9
$f(x)$	≈ -0.1961	≈ -0.19996	≈ -0.199996	?	≈ -0.200004	≈ -0.20004	≈ -0.2041

Coming in from either direction, we see that the function values are approaching -0.2. Thus,

a) $\lim_{x \to -2^-} f(x) = -0.2$ b) $\lim_{x \to -2^+} f(x) = -0.2$ c) $\lim_{x \to -2} f(x) = -0.2$

7) We begin by evaluating the function for the various values in the table:

$$f(-0.1) = \frac{13250+2.35(-0.1)}{(-0.1)} = \frac{13250-0.235}{-0.1} = \frac{13249.765}{-0.1} = -132497.65 \approx -1.3 \times 10^5$$

$$f(-0.01) = \frac{13250+2.35(-0.01)}{(-0.01)} = \frac{13250-0.0235}{-0.01} = \frac{13249.9765}{-0.01} = -1324997.65 \approx -1.3 \times 10^6$$

$$f(-0.001) = \frac{13250+2.35(-0.001)}{(-0.001)} = \frac{13250-0.00235}{-0.001} = \frac{13249.99765}{-0.001} = -13249997.65 \approx -1.3 \times 10^7$$

$$f(-0.00001) = \frac{13250+2.35(-0.00001)}{(-0.00001)} = \frac{13250-0.0000235}{-0.00001} = \frac{13249.9999765}{-0.00001} = -1324999997.65 \approx -1.3 \times 10^9$$

From this side, the values approach $-\infty$.

$$f(0.1) = \frac{13250+2.35(0.1)}{(0.1)} = \frac{13250+0.235}{0.1} = \frac{13250.235}{0.1} = 132502.35 \approx 1.3 \times 10^5$$

$$f(0.01) = \frac{13250+2.35(0.01)}{(0.01)} = \frac{13250+0.0235}{0.01} = \frac{13250.0235}{0.01} = 1325002.35 \approx 1.3 \times 10^6$$

$$f(0.001) = \frac{13250+2.35(0.001)}{(0.001)} = \frac{13250+0.00235}{0.001} = \frac{13250.00235}{0.001} = 13250002.35 \approx 1.3 \times 10^7$$

$$f(0.00001) = \frac{13250+2.35(0.00001)}{(0.00001)} = \frac{13250+0.0000235}{0.00001} = \frac{13250.0000235}{0.00001} = 1325000002.35 \approx 1.3 \times 10^9$$

From this side, the values approach ∞.

x	-0.1	-0.01	-0.001	-0.00001	0	0.00001	0.001	0.01	0.1
$f(x)$	-1.3×10^5	-1.3×10^6	-1.3×10^7	-1.3×10^9	?	1.3×10^9	1.3×10^7	1.3×10^5	1.3×10^5

Coming in from different directions, we see that the function does not approach the same value. Thus,

a) $\lim_{x \to 0^-} f(x) = -\infty$ b) $\lim_{x \to 0^+} f(x) = \infty$ c) $\lim_{x \to 0} f(x)$ does not exist

9) Since $\lim\limits_{x \to 0^-} \frac{2}{x^3} = -\infty$ and $\lim\limits_{x \to 0^+} \frac{2}{x^3} = \infty$, then $\lim\limits_{x \to 0} \frac{2}{x^3}$ does not exist.

11) We need to simplify the expression before evaluating the limit.
$$\lim_{x \to -2} \frac{x+2}{x^2-x-6} = \lim_{x \to -2} \frac{x+2}{(x+2)(x-3)}$$
$$= \lim_{x \to -2} \frac{1}{x-3} = \frac{1}{-2-3} = -\frac{1}{5}.$$

13) We need to simplify the expression before evaluating the limit.
$$\lim_{x \to 3} \frac{x-3}{x^2-9} = \lim_{x \to 3} \frac{x-3}{(x-3)(x+3)} = \lim_{x \to 3} \frac{1}{x+3}$$
$$= \frac{1}{3+3} = \frac{1}{6}.$$

15) We need to simplify the expression before evaluating the limit.
$$\lim_{x \to -1} \frac{x^2-x+1}{x^3+1} = \lim_{x \to -1} \frac{x^2-x+1}{(x+1)(x^2-x+1)}$$
$$= \lim_{x \to -1} \frac{1}{x+1}. \text{ But, } \lim_{x \to -1^-} \frac{1}{x+1} = -\infty \text{ and}$$
$$\lim_{x \to -1^+} \frac{1}{x+1} = \infty, \text{ so } \lim_{x \to -1} \frac{1}{x+1} \text{ does not exist.}$$

17a) The function 3^x is increasing on its domain, i.e., as $x \to \infty, f(x) \to \infty$. Hence, the function is an exponential growth function.

17b) Since f is an exponential growth function, $\lim\limits_{x \to \infty} f(x) = \infty$ & $\lim\limits_{x \to -\infty} f(x) = 0$.

19a) The function e^x is increasing on its domain, i.e., as $x \to \infty, f(x) \to \infty$. Hence, the function is an exponential growth function.

19b) Since f is an exponential growth function, $\lim\limits_{x \to \infty} f(x) = \infty$ & $\lim\limits_{x \to -\infty} f(x) = 0$.

21a) The function $e^{-0.215x}$ is decreasing on its domain ($e^{-0.215} < 1$), i.e., as $x \to -\infty$, $f(x) \to \infty$. Hence, the function is an exponential decay function.

21b) Since f is an exponential decay function, $\lim\limits_{x \to \infty} f(x) = 0$ & $\lim\limits_{x \to -\infty} f(x) = \infty$.

23a) The function $(0.987)^x$ is decreasing on its domain, i.e., as $x \to -\infty, f(x) \to \infty$. Hence, the function is an exponential decay function.

23b) Since f is an exponential decay function, $\lim\limits_{x \to \infty} f(x) = 0$ & $\lim\limits_{x \to -\infty} f(x) = \infty$.

25) Since the degree of the polynomial in denominator is 1, then we need to multiply top and bottom by $\frac{1}{x}$.

Thus, $\lim\limits_{x \to -\infty} \frac{2x+5}{x-1} = \lim\limits_{x \to -\infty} \left(\frac{2x+5}{x-1}\right) \cdot \frac{\frac{1}{x}}{\frac{1}{x}}$
$$= \lim_{x \to -\infty} \frac{2+\frac{5}{x}}{1-\frac{1}{x}} = \frac{2+0}{1-0} = 2.$$

We can verify our results numerically:

x	-1×10^3	-1×10^6	-1×10^9
$f(x)$	≈ 1.993	≈ 1.999993	≈ 2.000000

As x approaches $-\infty$, $f(x)$ approaches 2.

27) Since the degree of the polynomial in the denominator is 2, then we need to multiply top and bottom by $\frac{1}{x^2}$.

Thus, $\lim\limits_{x \to \infty} \frac{3x^2-x+2}{2x^2+x-5} = \lim\limits_{x \to \infty} \left(\frac{3x^2-x+2}{2x^2+x-5}\right) \cdot \frac{\frac{1}{x^2}}{\frac{1}{x^2}}$
$$= \lim_{x \to \infty} \frac{3-\frac{1}{x}+\frac{2}{x^2}}{2+\frac{1}{x}-\frac{5}{x^2}} = \frac{3-0+0}{2+0-0} = \frac{3}{2} = 1.5.$$

We can verify our results numerically:

x	1×10^3	1×10^6	1×10^9
$f(x)$	≈ 1.4988	≈ 1.499999	≈ 1.500000

As x approaches ∞, $f(x)$ approaches 1.5.

29) Since the degree of the polynomial in the denominator is 3, then we need to multiply top and bottom by $\frac{1}{x^3}$.

Thus, $\lim\limits_{x\to\infty} \frac{2x^2+2x+1}{5x^3+3x-5} = \lim\limits_{x\to\infty} \left(\frac{2x^2+2x+1}{5x^3+3x-5}\right)\frac{\frac{1}{x^3}}{\frac{1}{x^3}}$

$= \lim\limits_{x\to\infty} \frac{\frac{2}{x}+\frac{2}{x^2}+\frac{1}{x^3}}{5+\frac{3}{x^2}-\frac{5}{x^3}} = \frac{0+0+0}{5+0-0} = \frac{0}{5} = 0$.

We can verify our results numerically:

x	1×10^3	1×10^6	1×10^9
$f(x)$	$\approx 4\times 10^{-4}$	$\approx 4\times 10^{-7}$	$\approx 4\times 10^{-10}$

As x approaches ∞, $f(x)$ approaches 0.

31) Since the degree of the polynomial in denominator is 3, then we need to multiply top and bottom by $\frac{1}{x^3}$.

Thus, $\lim\limits_{x\to-\infty} \frac{2x^2+2x+1}{5x^3+3x-5}$

$= \lim\limits_{x\to-\infty} \left(\frac{2x^2+2x+1}{5x^3+3x-5}\right)\frac{\frac{1}{x^3}}{\frac{1}{x^3}}$

$= \lim\limits_{x\to-\infty} \frac{\frac{2}{x}+\frac{2}{x^2}+\frac{1}{x^3}}{5+\frac{3}{x^2}-\frac{5}{x^3}} = \frac{0+0+0}{5+0-0} = \frac{0}{5} = 0$.

We can verify our results numerically:

x	-1×10^3	-1×10^6	-1×10^9
$f(x)$	$\approx -4\times 10^{-4}$	$\approx -4\times 10^{-7}$	$\approx -4\times 10^{-10}$

As x approaches $-\infty$, $f(x)$ approaches 0.

33) Using the results from problems #25 and #26, we see that:

$\lim\limits_{x\to\pm\infty} \frac{2x+5}{x-1} = 2$. Thus, the function has a horizontal asymptote of $y = 2$.

35) Using the results from problems #29 and #31, we see that:

$\lim\limits_{x\to\pm\infty} \frac{2x^2+2x+1}{5x^3+3x-5} = 0$. Thus, the function has a horizontal asymptote of $y = 0$.

37) As x approaches 2 from the right, the function values decrease without bound. Thus, $\lim\limits_{x\to 2^+} f(x) = -\infty$.

39) Since $\lim\limits_{x\to 2^+} f(x) \neq \lim\limits_{x\to 2^-} f(x)$,

then $\lim\limits_{x\to 2} f(x)$ does not exist.

41) As x approaches -3 from the right, the function values approach 1.
Thus, $\lim\limits_{x\to -3^+} f(x) = 1$.

As x approaches -3 from the left, the function values approach 1.
Thus, $\lim\limits_{x\to -3^-} f(x) = 1$.

Since $\lim\limits_{x\to -3^+} f(x) = 1 = \lim\limits_{x\to -3^-} f(x)$,

then $\lim\limits_{x\to -3} f(x) = 1$.

43) As x decreases without bound, the function values approach 0. Thus, $\lim\limits_{x\to -\infty} f(x) = 0$.

45a) Since x represents the percentage of pollutants remove, then it needs to be a number between 0% and 100%. x can be equal to 0%, but not equal to 100% since the function is undefined at that value. Thus the domain is [0, 100).

The graph is displayed in viewing window [0, 100] by [0, 1500]

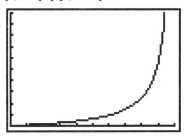

Note: The textbook writes the viewing window on the sides of the graph.

45b) $C(40) = \frac{93(40)}{100-40} = \frac{3720}{60} = 62$.

It will cost $62 thousand or $62,000 to clean up 40% of the pollutants.

45c) $\lim_{x \to 100^-} C(x) = \lim_{x \to 100^-} \frac{93x}{100-x} = \infty$. As the removal of pollutants nears 100%, the cost increases without bound.

47a) $\lim_{x \to \infty} N(x) = \lim_{x \to \infty} 3200 - \frac{750}{x} = 3200 - 0$

= 3200. As the amount of money spent increases without bound, the number of items purchased approaches 3200 hundred items or 320,000 items.

47b) The graph is displayed in viewing window [1, 50] by [2600, 3200]

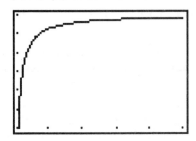

Using the trace feature, we find that

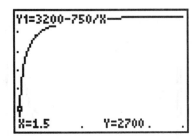

47b) Continued
and

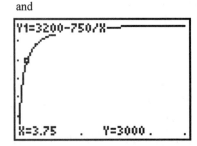

So, an interval for x that would result in function values between 2700 and 3000 is [1.5, 3.75].

49) $\lim_{q \to \infty} AC(q) = \lim_{q \to \infty} \frac{15200 + 4.85q}{q}$

$= \lim_{q \to \infty} \left(\frac{15200 + 4.85q}{q} \right) \frac{\frac{1}{q}}{\frac{1}{q}} = \lim_{q \to \infty} \frac{\frac{15200}{q} + 4.85}{1}$

$= \frac{0 + 4.85}{1} = \frac{4.85}{1} = 4.85$.

As production increases without bound, the average production cost approaches $4.85 per unit.

51a) $A(0) = 2.5e^{-0.2(0)} = 2.5(1) = 2.5$.
This means that initially, 2.5 milliliters of the medication is administered.

51b) $\lim_{x \to \infty} 2.5e^{-0.2t} = 0$. As time (after the medication is taken) increases without bound, the amount of medication in the body approaches zero.

51c) The graph is displayed in viewing window [0, 5] by [0, 5]

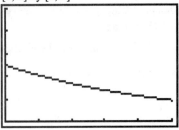

51d) Letting $y_2 = 1$ and using the intersect command on the calculator, we find that

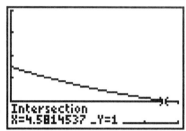

Approximately 4.58 hours after the initial dose, the amount present drops below 1 milliliter. Hence, the medication (a dose 1 milliliter less than the initial dose) should be taken every 4 ½ hours.

53a) The graph is displayed in viewing window [0, 60] by [0, 3000]

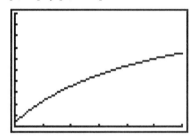

53b) One year corresponds to $t = 12$ months.
$P(12) = \frac{10(10 + 7(12))}{1 + 0.02(12)} = \frac{10(94)}{1 + 0.24} = \frac{940}{1.24} \approx 758$.

After one year, there are about 758 bass in the lake.

53c) $\lim\limits_{t \to \infty} P(t) = \lim\limits_{t \to \infty} \frac{10(10 + 7t)}{1 + 0.02t}$

$= \lim\limits_{t \to \infty} \left(\frac{100 + 70t}{1 + 0.02t} \right) \cdot \frac{\frac{1}{t}}{\frac{1}{t}} = \lim\limits_{t \to \infty} \frac{\frac{100}{t} + 70}{\frac{1}{t} + 0.02}$

$= \frac{0 + 70}{0 + 0.02} = \frac{70}{0.02} = 3500$.

As time increases without bound, the bass population in the lake approaches 3,500 bass.

55a) The graph is displayed in viewing window [0, 50] by [0, 1500]

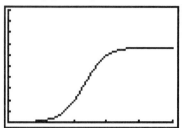

55b) $D(3) = \frac{1000}{1 + 999e^{-0.3(3)}} = \frac{1000}{1 + 999e^{-0.9}}$

$= \frac{1000}{1 + 406.1630901} = \frac{1000}{407.1630901} \approx 2.456 \approx 2$

After 3 days, approximately two people have been infected.

55c) $\lim\limits_{t \to \infty} \frac{1000}{1 + 999e^{-0.3t}} = \frac{1000}{1 + 999(0)} = \frac{1000}{1 + 0} = \frac{1000}{1}$

$= 1000$.

As time increases without bound, the number of people infected approaches 1000 people or the entire population of the town.

Section 2.3 Instantaneous Rate of Change and the Derivative

1) $m_{sec} = \dfrac{f(x+\Delta x) - f(x)}{\Delta x} = \dfrac{f(1) - f(-2)}{1-(-2)}$

$= \dfrac{2(1)^2 + (1) - 3 - [2(-2)^2 + (-2) - 3]}{1+2} = \dfrac{0-3}{3} = \dfrac{-3}{3} = -1$.

3) $m_{sec} = \dfrac{g(x+\Delta x) - g(x)}{\Delta x} = \dfrac{g(9) - g(1)}{9-1} = \dfrac{3\sqrt{9} - 3\sqrt{1}}{8}$

$= \dfrac{9-3}{8} = \dfrac{6}{8} = \dfrac{3}{4}$.

5) $m_{sec} = \dfrac{f(x+\Delta x) - f(x)}{\Delta x} = \dfrac{f(20) - f(0)}{20-0}$

$= \dfrac{2.1e^{0.2(20)} - 2.1e^{0.2(0)}}{20-0} \approx \dfrac{114.65612 - 2.1}{20} = \dfrac{112.55612}{20}$

≈ 5.63.

7) $m_{sec} = \dfrac{f(x+\Delta x) - f(x)}{\Delta x} = \dfrac{f(15) - f(0)}{15-0}$

$= \dfrac{2.1e^{-0.2(15)} - 2.1e^{-0.2(0)}}{15-0} \approx \dfrac{0.10455284 - 2.1}{15}$

$= \dfrac{-1.99544716}{15} \approx -0.13$.

9) $m_{sec} = \dfrac{f(x+\Delta x) - f(x)}{\Delta x} = \dfrac{f(10) - f(1)}{10-1}$

$= \dfrac{1 + \ln(10) - [1 + \ln(1)]}{9} = \dfrac{3.302585093 - 1}{9} = \dfrac{2.302585093}{9}$

≈ 0.26.

11a) Looking from left to right, $f(A) < f(B)$. Therefore, the average rate of change between the points A and B is positive.

Looking from left to right, $f(B) > f(C)$. Therefore, the average rate of change between the points B and C is negative.

Since $f(C) = f(D)$, then the average rate of change between the points C and D is zero.

11b) Looking from left to right, $f(B) < f(C)$. Therefore, the average rate of change between the points B and C is positive.

Looking from left to right, $f(C) > f(D)$. Therefore, the average rate of change between the points C and D is negative.

Since $f(A) = f(B)$, then the average rate of change between the points A and B is zero.

13)

h	-1	-0.1	-0.01	-0.001	-0.0001
$\dfrac{f(x+h) - f(x)}{h}$	3	3	3	3	3

h	1	0.1	0.01	0.001	0.0001
$\dfrac{f(x+h) - f(x)}{h}$	3	3	3	3	3

At $x = 2$, $\lim\limits_{h \to 0} \dfrac{f(x+h) - f(x)}{h} = 3$.

15)

h	-1	-0.1	-0.01	-0.001	-0.0001
$\dfrac{f(x+h) - f(x)}{h}$	6	6.9	6.99	6.999	6.9999

h	1	0.1	0.01	0.001	0.0001
$\dfrac{f(x+h) - f(x)}{h}$	8	7.1	7.01	7.001	7.0001

At $x = 3$, $\lim\limits_{h \to 0} \dfrac{f(x+h) - f(x)}{h} = 7$.

17)

h	-1	-0.1	-0.01	-0.001	-0.0001
$\dfrac{f(x+h) - f(x)}{h}$	77	96.62	98.76	98.98	98.998

h	1	0.1	0.01	0.001	0.0001
$\dfrac{f(x+h) - f(x)}{h}$	125	101.42	99.24	99.02	99.002

At $x = 4$, $\lim\limits_{h \to 0} \dfrac{f(x+h) - f(x)}{h} = 99$.

19)

h	−1	−0.1	−0.01	−0.001	−0.0001
$\frac{f(x+h)-f(x)}{h}$	0.8	0.755	0.751	0.7501	0.75

h	1	0.1	0.01	0.001	0.0001
$\frac{f(x+h)-f(x)}{h}$	0.7	0.745	0.7495	0.74995	0.75

At $x = 4$, $\lim_{h \to 0} \frac{f(x+h)-f(x)}{h} = 0.75$.

21)

h	−1	−0.1	−0.01	−0.001	−0.0001
$\frac{f(x+h)-f(x)}{h}$	0.6	0.62	0.626	0.6265	0.62656

h	1	0.1	0.01	0.001	0.0001
$\frac{f(x+h)-f(x)}{h}$	0.7	0.63	0.627	0.6266	0.62657

At $x = 2$, $\lim_{h \to 0} \frac{f(x+h)-f(x)}{h} \approx 0.627$.

23) The slope of the line tangent to the graph of f at $x = 3$ is 7.

25) $f'(2) = \lim_{h \to 0} \frac{f(2+h)-f(2)}{h}$

$= \lim_{h \to 0} \frac{2(2+h)+1-[2(2)+1]}{h} = \lim_{h \to 0} \frac{4+2h+1-[5]}{h}$

$= \lim_{h \to 0} \frac{2h}{h} = \lim_{h \to 0} 2 = 2$.

27) $f'(-2) = \lim_{h \to 0} \frac{f(-2+h)-f(-2)}{h}$

$= \lim_{h \to 0} \frac{(-2+h)^2 - 2(-2+h) - [(-2)^2 - 2(-2)]}{h}$

$= \lim_{h \to 0} \frac{4 - 4h + h^2 + 4 - 2h - [8]}{h} = \lim_{h \to 0} \frac{h^2 - 6h}{h}$

$= \lim_{h \to 0} \frac{h(h-6)}{h} = \lim_{h \to 0} h - 6 = 0 - 6 = -6$.

29) $f'(4) = \lim_{h \to 0} \frac{f(4+h)-f(4)}{h}$

$= \lim_{h \to 0} \frac{-(4+h)^2 + 2(4+h) - [-(4)^2 + 2(4)]}{h}$

$= \lim_{h \to 0} \frac{-16 - 8h - h^2 + 8 + 2h - [-8]}{h} = \lim_{h \to 0} \frac{-h^2 - 6h}{h}$

$= \lim_{h \to 0} \frac{h(-h-6)}{h} = \lim_{h \to 0} -h - 6 = 0 - 6 = -6$.

31) $f'(4) = \lim_{h \to 0} \frac{f(4+h)-f(4)}{h}$

$= \lim_{h \to 0} \frac{2\sqrt{4+h} - [2\sqrt{4}]}{h} = \lim_{h \to 0} \frac{2\sqrt{4+h} - [4]}{h}$

$= \lim_{h \to 0} \frac{2(\sqrt{4+h} - 2)}{h} = \lim_{h \to 0} 2\left(\frac{\sqrt{4+h} - 2}{h}\right)$

We need to multiply top and bottom by the conjugate of the numerator, $\sqrt{4+h} + 2$

$= \lim_{h \to 0} 2\left(\frac{\sqrt{4+h} - 2}{h}\right) \cdot \frac{\sqrt{4+h} + 2}{\sqrt{4+h} + 2}$

$= \lim_{h \to 0} 2\left(\frac{4+h-4}{h(\sqrt{4+h}+2)}\right) = \lim_{h \to 0} 2\left(\frac{h}{h(\sqrt{4+h}+2)}\right)$

$= \lim_{h \to 0} 2\left(\frac{1}{\sqrt{4+h}+2}\right) = 2\left(\frac{1}{\sqrt{4+0}+2}\right)$

$= 2\left(\frac{1}{\sqrt{4}+2}\right) = 2\left(\frac{1}{2+2}\right) = 2\left(\frac{1}{4}\right) = \frac{1}{2}$.

33) $f'(7) = \lim_{h \to 0} \frac{f(7+h)-f(7)}{h}$

$= \lim_{h \to 0} \frac{-1.2(7+h)^2 + 3.2(7+h) - 2.1 - [-1.2(7)^2 + 3.2(7) - 2.1]}{h}$

$= \lim_{h \to 0} \frac{-58.8 - 16.8h - 1.2h^2 + 22.4 + 3.2h - 2.1 + 58.8 - 22.4 + 2.1}{h}$

$= \lim_{h \to 0} \frac{-13.6h - 1.2h^2}{h} = \lim_{h \to 0} \frac{h(-13.6 - 1.2h)}{h}$

$= \lim_{h \to 0} -13.6 - 1.2h = -13.6 - 1.2(0)$

$= -13.6$.

35a) From #27, the slope of the tangent line at $x = -2$ is $m = f'(-2) = -6$. Since $f(-2) = (-2)^2 - 2(-2) = 4 + 4 = 8$, then the equation of the tangent line can by found by plugging $m = -6$ and $(-2, 8)$ into:
$$y - y_1 = m(x - x_1)$$
$$y - 8 = -6(x - (-2))$$
$$y - 8 = -6x - 12$$
$$y = -6x - 4$$

35b) The graph is displayed in viewing window $[-5, 5]$ by $[-5, 25]$

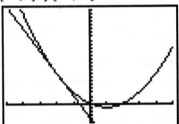

Note: The textbook writes the viewing window on the sides of the graph.

37a) From #29, the slope of the tangent line at $x = 4$ is $m = f'(4) = -6$. Since $f(4) = -(4)^2 + 2(4) = -16 + 8 = -8$, then the equation of the tangent line can by found by plugging $m = -6$ and $(4, -8)$ into:
$$y - y_1 = m(x - x_1)$$
$$y - (-8) = -6(x - 4)$$
$$y + 8 = -6x + 24$$
$$y = -6x + 16$$

37b) The graph is displayed in viewing window $[-10, 10]$ by $[-25, 5]$

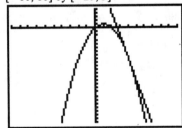

39a) From #33, the slope of the tangent line at $x = 7$ is $m = f'(7) = -13.6$. Since $f(7) = -1.2(7)^2 + 3.2(7) - 2.1$
$= -58.8 + 22.4 - 2.1 = -38.5$, then the equation of the tangent line can by found by plugging $m = -13.6$ and $(7, -38.5)$ into:
$$y - y_1 = m(x - x_1)$$
$$y - (-38.5) = -13.6(x - 7)$$
$$y + 38.5 = -13.6x + 95.2$$
$$y = -13.6x + 56.7$$

39b) The graph is displayed in viewing window $[0, 15]$ by $[-100, 50]$

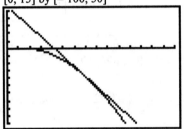

41) $f'(x) = \lim\limits_{h \to 0} \dfrac{f(x+h) - f(x)}{h}$
$= \lim\limits_{h \to 0} \dfrac{2(x+h) - 5 - [2x - 5]}{h} = \lim\limits_{h \to 0} \dfrac{2x + 2h - 5 - 2x + 5}{h}$
$= \lim\limits_{h \to 0} \dfrac{2h}{h} = \lim\limits_{h \to 0} 2 = 2.$

43) $g'(x) = \lim\limits_{h \to 0} \dfrac{g(x+h) - g(x)}{h}$
$= \lim\limits_{h \to 0} \dfrac{-2(x+h) + 3 - [-2x + 3]}{h}$
$= \lim\limits_{h \to 0} \dfrac{-2x - 2h + 3 + 2x - 3}{h}$
$= \lim\limits_{h \to 0} \dfrac{-2h}{h} = \lim\limits_{h \to 0} -2 = -2.$

45) $f'(x) = \lim\limits_{h \to 0} \dfrac{f(x+h) - f(x)}{h} = \lim\limits_{h \to 0} \dfrac{(x+h)^2 - [x^2]}{h}$
$= \lim\limits_{h \to 0} \dfrac{x^2 + 2xh + h^2 - x^2}{h} = \lim\limits_{h \to 0} \dfrac{2xh + h^2}{h}$
$= \lim\limits_{h \to 0} \dfrac{h(2x + h)}{h} = \lim\limits_{h \to 0} 2x + h = 2x + 0 = 2x.$

47) $h'(x) = \lim_{h \to 0} \frac{h(x+h) - h(x)}{h}$

$= \lim_{h \to 0} \frac{(x+h)^2 - 2(x+h) + 3 - [x^2 - 2x + 3]}{h}$

$= \lim_{h \to 0} \frac{x^2 + 2xh + h^2 - 2x - 2h + 3 - x^2 + 2x - 3}{h}$

$= \lim_{h \to 0} \frac{2xh + h^2 - 2h}{h} = \lim_{h \to 0} \frac{h(2x + h - 2)}{h}$

$= \lim_{h \to 0} 2x + h - 2 = 2x + (0) - 2 = 2x - 2.$

49) $g'(x) = \lim_{h \to 0} \frac{g(x+h) - g(x)}{h}$

$= \lim_{h \to 0} \frac{2.1(x+h)^2 + 3.2(x+h) - [2.1x^2 + 3.2x]}{h}$

$= \lim_{h \to 0} \frac{2.1x^2 + 4.2xh + 2.1h^2 + 3.2x + 3.2h - 2.1x^2 - 3.2x}{h}$

$= \lim_{h \to 0} \frac{4.2xh + 2.1h^2 + 3.2h}{h} = \lim_{h \to 0} \frac{h(4.2x + 2.1h + 3.2)}{h}$

$= \lim_{h \to 0} 4.2x + 2.1h + 3.2$

$= 4.2x + 2.1(0) + 3.2 = 4.2x + 3.2.$

51) $f'(x) = \lim_{h \to 0} \frac{f(x+h) - f(x)}{h}$

$= \lim_{h \to 0} \frac{-2(x+h)^2 + 3(x+h) - [-2x^2 + 3x]}{h}$

$= \lim_{h \to 0} \frac{-2x^2 - 4xh - 2h^2 + 3x + 3h + 2x^2 - 3x}{h}$

$= \lim_{h \to 0} \frac{-4xh - 2h^2 + 3h}{h} = \lim_{h \to 0} \frac{h(-4x - 2h + 3)}{h}$

$= \lim_{h \to 0} -4x - 2h + 3 = -4x - 2(0) + 3$

$= -4x + 3.$

53) $f'(x) = \lim_{h \to 0} \frac{f(x+h) - f(x)}{h}$

$= \lim_{h \to 0} \frac{-2.3(x+h)^2 + 3.1(x+h) - [-2.3x^2 + 3.1x]}{h}$

$= \lim_{h \to 0} \frac{-2.3x^2 - 4.6xh - 2.3h^2 + 3.1x + 3.1h + 2.3x^2 - 3.1x}{h}$

$= \lim_{h \to 0} \frac{-4.6xh - 2.3h^2 + 3.1h}{h} = \lim_{h \to 0} \frac{h(-4.6x - 2.3h + 3.1)}{h}$

$= \lim_{h \to 0} -4.6x - 2.3h + 3.1 = -4.6x - 0 + 3.1$

$= -4.6x + 3.1.$

55) $g'(x) = \lim_{h \to 0} \frac{g(x+h) - g(x)}{h}$

$= \lim_{h \to 0} \frac{(x+h)^3 + (x+h)^2 - [x^3 + x^2]}{h}$

$= \lim_{h \to 0} \frac{x^3 + 3x^2h + 3xh^2 + h^3 + x^2 + 2xh + h^2 - x^3 - x^2}{h}$

$= \lim_{h \to 0} \frac{3x^2h + 3xh^2 + h^3 + 2xh + h^2}{h}$

$= \lim_{h \to 0} \frac{h(3x^2 + 3xh + h^2 + 2x + h)}{h}$

$= \lim_{h \to 0} 3x^2 + 3xh + h^2 + 2x + h$

$= 3x^2 + 0 + 0 + 2x + 0 = 3x^2 + 2x.$

57) From Exercise 41, $f'(x) = 2$. Therefore, $f'(-1) = 2$ and $f'(2) = 2$.

59) From Exercise 47, $f'(x) = 2x - 2$. Therefore, $f'(-3) = 2(-3) - 2 = -6 - 2 = -8$ and $f'(2) = 2(2) - 2 = 4 - 2 = 2.$

61) From Exercise 49, $g'(x) = 4.2x + 3.2$. Hence, $g'(0) = 4.2(0) + 3.2 = 0 + 3.2 = 3.2$ and $g'(2) = 4.2(2) + 3.2 = 8.4 + 3.2 = 11.6.$

63) From Exercise 51, $f'(x) = -4x + 3$. Hence, $f'(3) = -4(3) + 3 = -12 + 3 = -9$ and $f'(6) = -4(6) + 3 = -24 + 3 = -21.$

65) From Exercise 55, $f'(x) = 3x^2 + 2x$. Hence, $f'(-1) = 3(-1)^2 + 2(-1) = 3 - 2 = 1$ and $f'(1) = 3(1)^2 + 2(1) = 3 + 2 = 5.$

67a) $P'(q) = \lim_{h \to 0} \frac{P(q+h) - P(q)}{h}$

$= \lim_{h \to 0} \frac{1000(q+h) - 2(q+h)^2 - [1000q - 2q^2]}{h}$

$= \lim_{h \to 0} \frac{1000q + 1000h - 2q^2 - 4qh - 2h^2 - 1000q + 2q^2}{h}$

$= \lim_{h \to 0} \frac{1000h - 4qh - 2h^2}{h} = \lim_{h \to 0} \frac{h(1000 - 4q - 2h)}{h}$

$= \lim_{h \to 0} 1000 - 4q - 2h = 1000 - 4q + 0$

$= 1000 - 4q.$

67b) $P'(200) = 1000 - 4(200) = 1000 - 800$
= 200. When 200 modems are produced and sold, the profit is increasing at a rate of $200 per modem.
$P'(300) = 1000 - 4(300) = 1000 - 1200$
= – 200. When 300 modems are produced and sold, the profit is decreasing at a rate of $200 per modem.

69a) $R(50) = 200(50) - (50)^2 = 10000 - 2500$
= 7500. The total revenue from the sale of 50 sets of Junior Golf Clubs is $7500.
$R(150) = 200(150) - (150)^2$
$= 30000 - 22500 = 7500$. The total revenue from the sale of 150 sets of Junior Golf Clubs is $7500.

69b) $R'(q) = \lim_{h \to 0} \frac{R(q+h) - R(q)}{h}$
$= \lim_{h \to 0} \frac{200(q+h) - (q+h)^2 - [200q - q^2]}{h}$
$= \lim_{h \to 0} \frac{200q + 200h - q^2 - 2qh - h^2 - 200q + q^2}{h}$
$= \lim_{h \to 0} \frac{200h - 2qh - h^2}{h} = \lim_{h \to 0} \frac{h(200 - 2q - h)}{h}$
$= \lim_{h \to 0} 200 - 2q - h = 200 - 2q + 0$
$= 200 - 2q.$

69c) $R'(50) = 200 - 2(50) = 200 - 100$
= 100. When 50 sets of Junior Golf Clubs are produced and sold, the revenue is increasing at a rate of $100 per set.
$R'(150) = 200 - 2(150) = 200 - 300$
= – 100. When 150 sets of Junior Golf Clubs are produced and sold, the revenue is decreasing at a rate of $100 per set.

71a) $C'(x) = \lim_{h \to 0} \frac{C(x+h) - C(x)}{h}$
$= \lim_{h \to 0} \frac{(x+h)^2 + 15(x+h) + 1500 - [x^2 + 15x + 1500]}{h}$
$= \lim_{h \to 0} \frac{x^2 + 2xh + h^2 + 15x + 15h + 1500 - x^2 - 15x - 1500}{h}$
$= \lim_{h \to 0} \frac{2xh + h^2 + 15h}{h} = \lim_{h \to 0} \frac{h(2x + h + 15)}{h}$
$= \lim_{h \to 0} 2x + h + 15 = 2x + 0 + 15$
$= 2x + 15.$

71b) $C'(40) = 2(40) + 15 = 80 + 15 = 95$
When 40 refrigerators are produced and sold, the total cost is increasing at a rate of $95 per refrigerator.
$C'(100) = 2(100) + 15 = 200 + 15 = 215$
When 100 refrigerators are produced and sold, the total cost is increasing at a rate of $215 per refrigerator.

73a) $f'(x) = \lim_{h \to 0} \frac{f(x+h) - f(x)}{h}$
$f(x + h) = 2.39(x + h)^2 + 47.05(x + h) + 432.2$
$= 2.39x^2 + 4.78xh + 2.39h^2 + 47.05x + 47.05h + 432.2$
Thus, $f(x + h) - f(x) =$
$= 2.39x^2 + 4.78xh + 2.39h^2 + 47.05x + 47.05h + 432.2 - 2.39x^2 - 47.05x - 432.2$
$= 4.78xh + 2.39h^2 + 47.05h$
Hence, $\lim_{h \to 0} \frac{f(x+h) - f(x)}{h}$
$= \lim_{h \to 0} \frac{4.78xh + 2.39h^2 + 47.05h}{h}$
$= \lim_{h \to 0} \frac{h(4.78x + 2.39h + 47.05)}{h}$
$= \lim_{h \to 0} 4.78x + 2.39h + 47.05$
$= 4.78x + 0 + 47.05 = 4.78x + 47.05.$

73b) $f'(3) = 4.78(3) + 47.05 = 14.34 + 47.05$
$= 61.39$. The U.S. imports were increasing at a rate of $61.39 billion per year in 1993.
$f'(4) = 4.78(4) + 47.05 = 19.12 + 47.05$
$= 66.17$. The U.S. imports were increasing at a rate of $66.17 billion per year in 1994.
$f'(5) = 4.78(5) + 47.05 = 23.9 + 47.05$
$= 70.95$. The U.S. imports were increasing at a rate of $70.95 billion per year in 1995.

73c) Since $f'(x)$ is greater than zero from part b, the tangent line is increasing and hence, the function is increasing.

75a) $f'(x) = \lim_{h \to 0} \frac{f(x+h) - f(x)}{h}$

$f(x+h) = -0.41(x+h)^2 + 13.53(x+h) + 330.69$
$= -0.41x^2 - 0.82xh - 0.41h^2 + 13.53x + 13.53h + 330.69$

Thus, $f(x+h) - f(x) =$
$= -0.41x^2 - 0.82xh - 0.41h^2 + 13.53x + 13.53h + 330.69 + 0.41x^2 - 13.53x - 330.69$
$= -0.82xh - 0.41h^2 + 13.53h$

Hence, $\lim_{h \to 0} \frac{f(x+h) - f(x)}{h}$
$= \lim_{h \to 0} \frac{-0.82xh - 0.41h^2 + 13.53h}{h}$
$= \lim_{h \to 0} \frac{h(-0.82x - 0.41h + 13.53)}{h}$
$= \lim_{h \to 0} -0.82x - 0.41h + 13.53$
$= -0.82x - 0 + 13.53 = -0.82x + 13.53$.

75b) $f'(6) = -0.82(6) + 13.53 = -4.92 + 13.53$
$= 8.61$. The U.S. water consumption was increasing at a rate of 8.61 gallons per day per capita in 1965.
$f'(16) = -0.82(16) + 13.53$
$= -13.12 + 13.53 = 0.41$. The U.S. water consumption was increasing at a rate of 0.41 gallons per day per capita in 1975.

75b) Continued
$f'(26) = -0.82(26) + 13.53$
$= -21.32 + 13.53 = -7.79$. The U.S. water consumption was decreasing at a rate of 7.79 gallons per day per capita in 1985.

75c) Since $f'(x) > 0$ between $x = 1$ (1960) and $x = 16$ (1975), the function is increasing at those values. So, water consumption was increasing between 1960 and 1975. Since $f'(x) < 0$ between $x = 17$ (1976) and $x = 31$ (1990), the function is decreasing at those values. In short, water consumption was decreasing between 1976 and 1990.

77a) $A(7) = 2000(1.065)^7 \approx 2000(1.553987)$
≈ 3107.97. After seven years, there will be $3,107.97 in the account.

77b) $A'(7) = \lim_{h \to 0} \frac{2000(1.065)^{7+h} - 2000(1.065)^7}{h}$
$= \lim_{h \to 0} \frac{2000(1.065)^7[(1.065)^h - 1]}{h}$
$= 2000(1.065)^7 \cdot \lim_{h \to 0} \frac{[(1.065)^h - 1]}{h}$
$\approx 3107.973092 \cdot \lim_{h \to 0} \frac{[(1.065)^h - 1]}{h}$.

We will create a table to numerically evaluate $\lim_{h \to 0} \frac{[(1.065)^h - 1]}{h}$:

77b) Continued

h	−1	−0.1	−0.01	−0.001	−0.0001
$\frac{[(1.065)^h - 1]}{h}$	0.06	0.063	0.0630	0.06297	0.062975

h	1	0.1	0.01	0.001	0.0001
$\frac{[(1.065)^h - 1]}{h}$	0.07	0.063	0.0630	0.06298	0.062975

Thus, $\lim_{h \to 0} \frac{[(1.065)^h - 1]}{h} \approx 0.062975$. Hence,

$A'(7) \approx 3107.973092 \cdot \lim_{h \to 0} \frac{[(1.065)^h - 1]}{h}$.

$\approx 3107.973092 \cdot 0.062975 \approx 195.72$. After seven years, the money in the account is increasing at a rate of $195.72 per year.

79a) $N(15) = 1500 - \frac{630}{(15)} = 1500 - 42 = 1458$.

When $15,000 is spent on advertising, 145,800 items are sold.

$N'(15) = \frac{630}{(15)^2} = \frac{630}{225} = 2.8$.

The number of items sold was increasing at a rate of 280 items per $1000 when $15,000 was spent on advertising.

79b) $N(25) = 1500 - \frac{630}{(25)} = 1500 - 25.2$

$= 1474.8$. When $25,000 is spent on advertising, 147,480 items are sold.

$N'(25) = \frac{630}{(25)^2} = \frac{630}{625} = 1.008$.

The number of items sold was increasing at a rate of about 101 items per $1000 when $25,000 was spent on advertising.

79c) As x gets larger, $N'(x)$ gets smaller so it is decreasing. In fact,

$\lim_{x \to \infty} N'(x) = \lim_{x \to \infty} \frac{630}{x^2} = 0$

79d) The graph of $N(x)$ is displayed in viewing window [0.42, 100] by [0, 1500]

The graph of $N'(x)$ is displayed in viewing window [0.26, 100] by [0, 100]

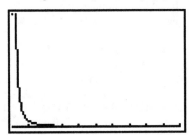

The graphs clearly support the answer in part c. Since $N'(x)$ is decreasing and going towards zero, we would expect the graph of $N(x)$ to level off as the picture indicates.

81) Since t_1 is steeper than t_2, it has a larger slope. Hence, the tree at $t = 25$ years old is growing at a faster rate that the tree at $t = 64$ years old,

Section 2.4 Derivatives of Constants, Powers, and Sums

1) $f'(x) = \frac{d}{dx}(3) = 0.$

3) $f'(t) = \frac{d}{dt}(-2) = 0.$

5) $f'(x) = \frac{d}{dx}(x^6) = 6x^{6-1} = 6x^5.$

7) $y' = \frac{d}{dx}(-3x^4) = -3 \cdot \frac{d}{dx}(x^4) = -3 \cdot 4x^{4-1}$
   $= -12x^3.$

9) $g'(x) = \frac{d}{dx}(2x^{2/3}) = 2 \cdot \frac{d}{dx}(x^{2/3}) = 2 \cdot \frac{2}{3} x^{2/3 - 1}$
   $= \frac{4}{3} x^{-1/3} = \frac{4}{3x^{\frac{1}{3}}} = \frac{4}{3\sqrt[3]{x}}.$

11) $f'(x) = \frac{d}{dx}(-3x^{-1/3}) = -3 \cdot \frac{d}{dx}(x^{-1/3})$
    $= -3 \cdot \frac{-1}{3} x^{-1/3 - 1} = 1x^{-4/3} = \frac{1}{x^{\frac{4}{3}}} = \frac{1}{\sqrt[3]{x^4}}$
    $= \frac{1}{x\sqrt[3]{x}}.$

13) $h'(x) = \frac{d}{dx}(\frac{2}{3} x^4) = \frac{2}{3} \cdot \frac{d}{dx}(x^4) = \frac{2}{3} \cdot 4x^{4-1}$
    $= \frac{8}{3} x^3.$

15) $y' = \frac{d}{dx}(-\frac{2}{5} x^{5/3}) = -\frac{2}{5} \cdot \frac{d}{dx}(x^{5/3})$
    $= -\frac{2}{5} \cdot \frac{5}{3} x^{5/3 - 1} = -\frac{2}{3} x^{2/3} = -\frac{2}{3} \sqrt[3]{x^2}.$

17) $f'(x) = \frac{d}{dx}[2x^3 + 4x^2 - 7x + 1]$
    $= \frac{d}{dx}[2x^3] + \frac{d}{dx}[4x^2] - \frac{d}{dx}[7x] + \frac{d}{dx}[1]$
    $= 6x^2 + 8x - 7 + 0 = 6x^2 + 8x - 7.$

19) $g'(x) = \frac{d}{dx}[3x^2 - 2x + 6]$
    $= \frac{d}{dx}[3x^2] - \frac{d}{dx}[2x] + \frac{d}{dx}[6] = 6x - 2 + 0$
    $= 6x - 2.$

21) $y' = \frac{d}{dx}[-5x^2 - 6x + 2]$
    $= \frac{d}{dx}[-5x^2] - \frac{d}{dx}[6x] + \frac{d}{dx}[2]$
    $= -10x - 6 + 0 = -10x - 6.$

23) $y' = \frac{d}{dx}[-5x^3 + 7x - 5]$
    $= \frac{d}{dx}[-5x^3] + \frac{d}{dx}[7x] - \frac{d}{dx}[5]$
    $= -15x^2 + 7 - 0 = -15x^2 + 7.$

25) $f'(x) = \frac{d}{dx}[\frac{1}{2} x^3 + \frac{3}{5} x^2 - \frac{2}{3} x + \frac{2}{5}]$
    $= \frac{d}{dx}[\frac{1}{2} x^3] + \frac{d}{dx}[\frac{3}{5} x^2] - \frac{d}{dx}[\frac{2}{3} x] + \frac{d}{dx}[\frac{2}{5}]$
    $= \frac{3}{2} x^2 + \frac{6}{5} x - \frac{2}{3} + 0 = \frac{3}{2} x^2 + \frac{6}{5} x - \frac{2}{3}.$

27) $g'(x) = \frac{d}{dx}[1.31x^2 + 2.05x - 3.9]$
    $= \frac{d}{dx}[1.31x^2] + \frac{d}{dx}[2.05x] - \frac{d}{dx}[3.9]$
    $= 2.62x + 2.05 - 0$
    $= 2.62x + 2.05.$

29) $d'(x) = \frac{d}{dx}[-0.2x^2 + 3.5x^3 - 0.4x^4]$
    $= \frac{d}{dx}[-0.2x^2] + \frac{d}{dx}[3.5x^3] - \frac{d}{dx}[0.4x^4]$
    $= -0.4x + 10.5x^2 - 1.6x^3.$

31) $y' = \frac{d}{dx}[1.15x^3 - 2.3x^2 + 2.53x - 7.1]$
    $= \frac{d}{dx}[1.15x^3] - \frac{d}{dx}[2.3x^2] + \frac{d}{dx}[2.53x]$
    $\quad - \frac{d}{dx}[7.1]$
    $= 3.45x^2 - 4.6x + 2.53 - 0$
    $= 3.45x^2 - 4.6x + 2.53.$

33) $f(x) = 3\sqrt{x} + \frac{1}{2} x - 5x^2 = 3x^{1/2} + \frac{1}{2} x - 5x^2.$
    Hence $f'(x) = \frac{d}{dx}[3x^{1/2} + \frac{1}{2} x - 5x^2]$
    $= \frac{d}{dx}[3x^{1/2}] + \frac{d}{dx}[\frac{1}{2} x] - \frac{d}{dx}[5x^2]$
    $= \frac{3}{2} x^{-1/2} + \frac{1}{2} - 10x = \frac{3}{2\sqrt{x}} + \frac{1}{2} - 10x.$

35) $y = \sqrt[3]{x} + x^2 - 3x^3 = x^{1/3} + x^2 - 3x^3.$
    Hence $y' = \frac{d}{dx}[x^{1/3} + x^2 - 3x^3]$
    $= \frac{d}{dx}[x^{1/3}] + \frac{d}{dx}[x^2] - \frac{d}{dx}[3x^3]$
    $= \frac{1}{3} x^{-2/3} + 2x - 9x^2 = \frac{1}{3\sqrt[3]{x^2}} + 2x - 9x^2.$

37) $g(x) = \sqrt[3]{x^2} - \frac{4}{\sqrt{x}} = x^{2/3} - 4x^{-1/2}$.

Hence, $g'(x) = \frac{d}{dx}[x^{2/3} - 4x^{-1/2}]$

$= \frac{d}{dx}[x^{2/3}] - \frac{d}{dx}[4x^{-1/2}] = \frac{2}{3}x^{-1/3} + 2x^{-3/2}$

$= \frac{2}{3\sqrt[3]{x}} + \frac{2}{\sqrt{x^3}} = \frac{2}{3\sqrt[3]{x}} + \frac{2}{x\sqrt{x}}$.

39) $f'(x) = \frac{d}{dx}[2.35x^{1.35}] = 3.1725x^{0.35}$.

41) $AC(x) = 2000 + \frac{5}{x^2} = 2000 + 5x^{-2}$.

Hence, $AC'(x) = \frac{d}{dx}[2000 + 5x^{-2}]$

$= \frac{d}{dx}[2000] + \frac{d}{dx}[5x^{-2}] = 0 - 10x^{-3} = -\frac{10}{x^3}$.

43) $y = 2x^2 + \frac{1}{x} = 2x^2 + x^{-1}$.

$y' = \frac{d}{dx}[2x^2 + x^{-1}] = \frac{d}{dx}[2x^2] + \frac{d}{dx}[x^{-1}]$

$= 4x - x^{-2} = 4x - \frac{1}{x^2}$.

45) $f'(x) = \frac{d}{dx}[3x^{-3/2} - 4x^{1/2} + 5]$

$= \frac{d}{dx}[3x^{-3/2}] - \frac{d}{dx}[4x^{1/2}] + \frac{d}{dx}[5]$

$= -\frac{9}{2}x^{-5/2} - 2x^{-1/2} + 0 = -\frac{9}{2\sqrt{x^5}} - \frac{2}{\sqrt{x}}$

$= -\frac{9}{2x^2\sqrt{x}} - \frac{2}{\sqrt{x}}$.

47) $y' = \frac{d}{dx}[2.35x^{-1/2} - 2.3x^{-2/3}]$

$= \frac{d}{dx}[2.35x^{-1/2}] - \frac{d}{dx}[2.3x^{-2/3}]$

$= -1.175x^{-3/2} + \frac{4.6}{3}x^{-5/3}$

$= -\frac{47}{40}x^{-3/2} + \frac{23}{15}x^{-5/3}$

$= -\frac{47}{40\sqrt{x^3}} + \frac{23}{15\sqrt[3]{x^5}} = -\frac{1.175}{\sqrt{x^3}} + \frac{23}{15\sqrt[3]{x^5}}$.

49a) Since there are no values that make the denominator zero and the numerator and denominator are polynomials, then f is defined for all real numbers or $(-\infty, \infty)$.

49b) $f(x) = \frac{3x^3 - 9x^2 + 4}{12} = \frac{3x^3}{12} - \frac{9x^2}{12} + \frac{4}{12}$

$= \frac{x^3}{4} - \frac{3x^2}{4} + \frac{1}{3} = 0.25x^3 - 0.75x^2 + \frac{1}{3}$.

49c) $f'(x) = \frac{d}{dx}[0.25x^3 - 0.75x^2 + \frac{1}{3}]$

$= \frac{d}{dx}[0.25x^3] - \frac{d}{dx}[0.75x^2] + \frac{d}{dx}[\frac{1}{3}]$

$= 0.75x^2 - 1.5x + 0 = 0.75x^2 - 1.5x$.

49d) Since $f'(x)$ is a polynomial, then the domain of f' is all real numbers or $(-\infty, \infty)$.

51a) Since there are no values except for zero that make the denominator zero and the numerator and denominator are polynomials, then f is defined for all real numbers except 0 or $(-\infty, 0) \cup (0, \infty)$.

51b) $f(x) = \frac{2x^3 + 3x^2 - x + 3}{x} = \frac{2x^3}{x} + \frac{3x^2}{x} - \frac{x}{x} + \frac{3}{x}$

$= 2x^2 + 3x - 1 + 3x^{-1}$.

51c) $f'(x) = \frac{d}{dx}[2x^2 + 3x - 1 + 3x^{-1}]$

$= \frac{d}{dx}[2x^2] + \frac{d}{dx}[3x] - \frac{d}{dx}[1] + \frac{d}{dx}[3x^{-1}]$

$= 4x + 3 - 0 - 3x^{-2} = 4x + 3 - \frac{3}{x^2}$.

51d) Since there are no values except for zero that make the denominator zero and the numerator and denominator are polynomials, then f' is defined for all real numbers except 0 or $(-\infty, 0) \cup (0, \infty)$.

53a) Since there are no values except for zero that make the denominator zero and the numerator and denominator are polynomials, then y is defined for all real numbers except 0 or $(-\infty, 0) \cup (0, \infty)$.

53b) $y = \frac{7x^4 - 50x^2 + x}{x^2} = \frac{7x^4}{x^2} - \frac{50x^2}{x^2} + \frac{x}{x^2}$

$= 7x^2 - 50 + x^{-1}$.

53c) $y' = \frac{d}{dx}[7x^2 - 50 + x^{-1}]$
$= \frac{d}{dx}[7x^2] - \frac{d}{dx}[50] + \frac{d}{dx}[x^{-1}]$
$= 14x - 0 - x^{-2} = 14x - \frac{1}{x^2}$.

53d) Since there are no values except for zero that make the denominator zero and the numerator and denominator are polynomials, then y' is defined for all real numbers except 0 or $(-\infty, 0) \cup (0, \infty)$.

55a) In looking at the denominator, h is only defined when $x > 0$. Thus, the domain is $(0, \infty)$.

55b) $h(x) = \frac{4x^3 - 14x^2 + 3}{2\sqrt{x}} = \frac{4x^3}{2\sqrt{x}} - \frac{14x^2}{2\sqrt{x}} + \frac{3}{2\sqrt{x}}$
$= \frac{4x^3}{2x^{1/2}} - \frac{14x^2}{2x^{1/2}} + \frac{3}{2x^{1/2}}$
$= 2x^{5/2} - 7x^{3/2} + 1.5x^{-1/2}$.

55c) $h'(x) = \frac{d}{dx}[2x^{5/2} - 7x^{3/2} + 1.5x^{-1/2}]$
$= \frac{d}{dx}[2x^{5/2}] - \frac{d}{dx}[7x^{3/2}] + \frac{d}{dx}[1.5x^{-1/2}]$
$= 5x^{3/2} - 10.5x^{1/2} - 0.75x^{-3/2}$
$= 5\sqrt{x^3} - 10.5\sqrt{x} - \frac{3}{4\sqrt{x^3}}$
$= 5x\sqrt{x} - 10.5\sqrt{x} - \frac{3}{4x\sqrt{x}}$.

55d) In looking at the denominator, h' is only defined when $x > 0$. Thus, the domain is $(0, \infty)$.

57a) $f'(x) = \frac{d}{dx}[x^3] = 3x^2$.

57b) The slope of the tangent line at $x = -1$ is equal to $f'(-1) = 3(-1)^2 = 3$.

57c) Since $f(-1) = (-1)^3 = -1$, we can find the equation of the tangent line by plugging $m = 3$ (from part b) & $(-1, -1)$ into:
$y - y_1 = m(x - x_1)$
$y - (-1) = 3(x - (-1))$
$y + 1 = 3x + 3$
$y = 3x + 2$.

57d) The function and the tangent line are graphed in viewing window $[-2, 2]$ by $[-4, 4]$

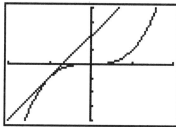

Note: The textbook writes the viewing window on the sides of the graph.

59a) $y' = \frac{d}{dx}[\frac{1}{x}] = \frac{d}{dx}[x^{-1}] = -x^{-2} = \frac{-1}{x^2}$.

59b) The slope of the tangent line at $x = 3$ is equal to $y' = \frac{-1}{(3)^2} = \frac{-1}{9}$.

59c) Since $f(3) = \frac{1}{3}$, we can find the equation of the tangent line by plugging $m = \frac{-1}{9}$ (from part b) & $(3, \frac{1}{3})$ into:
$y - y_1 = m(x - x_1)$
$y - (\frac{1}{3}) = \frac{-1}{9}(x - 3)$
$y - \frac{1}{3} = \frac{-1}{9}x + \frac{1}{3}$
$y = \frac{-1}{9}x + \frac{2}{3}$.

59d) The function and the tangent line are graphed in viewing window $[-3, 8]$ by $[-3, 3]$

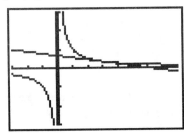

61a) $f'(x) = \frac{d}{dx}[x^{2/3}] = \frac{2}{3}x^{-1/3} = \frac{2}{3\sqrt[3]{x}}$.

61b) The slope of the tangent line at $x = 8$ is equal to $f'(8) = \frac{2}{3\sqrt[3]{8}} = \frac{2}{3 \cdot 2} = \frac{1}{3}$.

61c) Since $f(8) = (8)^{2/3} = \sqrt[3]{8^2} = \sqrt[3]{64} = 4$, we can find the equation of the tangent line by plugging $m = \frac{1}{3}$ (from part b) & $(8, 4)$ into:
$y - y_1 = m(x - x_1)$
$y - (4) = \frac{1}{3}(x - 8)$
$y - 4 = \frac{1}{3}x - \frac{8}{3}$
$y = \frac{1}{3}x + \frac{4}{3}$.

61d) The function and the tangent line are graphed in viewing window $[0, 20]$ by $[0, 8]$

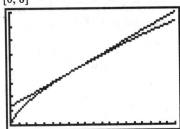

63a) Since $N(x) = 2500 - \frac{520}{x} = 2500 - 520x^{-1}$, then $N'(x) = \frac{d}{dx}[2500 - 520x^{-1}]$
$= 0 + 520x^{-2} = \frac{520}{x^2}$.

63b) $N(10) = 2500 - \frac{520}{10} = 2500 - 52 = 2448$.
If $10,000 is spent on advertising, then 244,800 CDs will be purchased.
$N'(10) = \frac{520}{10^2} = \frac{520}{100} = 5.2$.
When $10,000 is spent on advertising, the number of CDs sold is increasing at a rate of 520 CDs per thousand dollars.

65) $C(300) = 3000 + 11(300) - 7\sqrt{300} + 0.03(300)^{3/2}$
$\approx 3000 + 3300 - 121.24 + 155.88$
$= 6334.64$.
It will cost about $6,334.64 to make 300 coats.
$C'(x) = \frac{d}{dx}[3000 + 11x - 7x^{1/2} + 0.03x^{3/2}]$
$= 0 + 11 - 3.5x^{-1/2} + 0.045x^{1/2}$
$= 11 - \frac{3.5}{\sqrt{x}} + 0.045\sqrt{x}$.
Thus, $C'(300) = 11 - \frac{3.5}{\sqrt{300}} + 0.045\sqrt{300}$
$\approx 11 - 0.202 + 0.779 \approx 11.58$.
When 300 coats are produced, the total cost is increasing at a rate of about $11.58 per coat.

67) $SD(1) = \frac{93.21}{(1)^2} = 93.21$. At one mile downwind from the plant, the amount of sulfur dioxide in the air is 93.21 parts per million.
$SD'(x) = \frac{d}{dx}[93.21x^{-2}] = -186.42x^{-3}$
$= \frac{-186.42}{x^3}$.
$SD'(1) = \frac{-186.42}{(1)^3} = -186.42$.
At one mile downwind from the plant, the amount of sulfur dioxide in the air is decreasing at a rate of 186.42 parts per million per mile.

Section 2.4 Derivatives of Constants, Powers, and Sums

69a) $SD(3) = -0.02(3)^3 + 0.27(3)^2 - 1.22(3) + 24.12$
$= -0.54 + 2.43 - 3.66 + 24.12 = 22.35$.
In 1987, 22.35 million tons of sulfur dioxide was released into the air.
$SD(7) = -0.02(7)^3 + 0.27(7)^2 - 1.22(7) + 24.12$
$= -6.86 + 13.23 - 8.54 + 24.12 = 21.95$.
In 1991, 21.95 million tons of sulfur dioxide was released into the air.

69b) $SD'(t) = \frac{d}{dt}[-0.02t^3 + 0.27t^2 - 1.22t + 24.12]$
$= -0.06t^2 + 0.54t - 1.22$
$SD'(3) = -0.06(3)^2 + 0.54(3) - 1.22$
$= -0.54 + 1.62 - 1.22 = -0.14$.
In 1987, the amount of sulfur dioxide emitted into the atmosphere was decreasing at a rate of 0.14 million tons per year.
$SD'(7) = -0.06(7)^2 + 0.54(7) - 1.22$
$= -2.94 + 3.78 - 1.22 = -0.38$.
In 1991, the amount of sulfur dioxide emitted into the atmosphere was decreasing at a rate of 0.38 million tons per year.

71a) $m_{sec} = \frac{N(t+\Delta t) - f(t)}{\Delta t} = \frac{N(4) - N(1)}{4-1}$
$= \frac{6(4)^{\frac{5}{2}} - 6(1)^{\frac{5}{2}}}{3} = \frac{192 - 6}{3} = \frac{186}{3} = 62$.
Between 1 and 4 minutes, the culture was growing at an average rate of 62 milligrams of bacteria per minute.

71b) $N'(t) = \frac{d}{dt}[6t^{5/2}] = 15t^{3/2} = 15t\sqrt{t}$.
Hence, $N'(4) = 15(4)\sqrt{4} = 15(4)2 = 120$. After 4 minutes, the bacteria culture was growing at a rate of 120 milligrams of bacteria per minute.

73a) $v(t) = s'(t) = \frac{d}{dt}[150 - 16t^2] = -32t$.

73b) $s(1) = 150 - 16(1)^2 = 150 - 16 = 134$.
After 1 second, the egg is 134 feet above the ground.
$s(3) = 150 - 16(3)^2 = 150 - 144 = 6$.
After 3 seconds, the egg is 6 feet above the ground.

73c) $m_{sec} = \frac{s(t+\Delta t) - s(t)}{\Delta t} = \frac{s(3) - s(1)}{3-1} = \frac{6 - 134}{2}$
$= \frac{-128}{2} = -64$. The average velocity of the egg between one and three seconds was decreasing at a rate of 64 feet per second.

73d) $v(1) = -32(1) = -32$. After one second, the velocity of the egg was -32 feet per second.
$v(3) = -32(3) = -96$. After three seconds, the velocity of the egg was decreasing at a rate of 96 feet per second.

73e) When the egg hits the ground, its height above the ground will be zero. Setting $s(t) = 0$ and solving yields:
$s(t) = 150 - 16t^2 = 0$
$150 = 16t^2$
$9.375 = t^2$
$t \approx \pm 3.1$.
Rejecting the negative root, we get $t \approx 3.1$. Thus, after about 3.1 seconds, the egg will hit the ground.

75a) $f'(x) = \frac{d}{dx}[-0.02x^3 + 0.57x^2 - 4.25x + 31.47]$
$= -0.06x^2 + 1.14x - 4.25$.
$f'(12) = -0.06(12)^2 + 1.14(12) - 4.25$
$= -8.64 + 13.68 - 4.25 = 0.79$.
In 1991, the number of cases of tuberculosis was increasing at a rate of 790 cases per year.
$f'(15) = -0.06(15)^2 + 1.14(15) - 4.25$
$= -13.5 + 17.1 - 4.25 = -0.65$.
In 1994, the number of cases of tuberculosis was decreasing at a rate of 650 cases per year.

77a) $C(11) = 1.65(11)^2 - 9.14(11) + 3306.88$
$= 199.65 - 100.54 + 3306.88 = 3405.99$
In 1985, the number of calories consumed per capita was about 3406 calories per day.
$C(16) = 1.65(16)^2 - 9.14(16) + 3306.88$
$= 422.4 - 146.24 + 3306.88 = 3583.04$
In 1990, the number of calories consumed per capita was about 3583 calories per day.

77b) $C'(t) = \frac{d}{dt}[1.65t^2 - 9.14t + 3306.88]$
$= 3.3t - 9.14$.
$C'(11) = 3.3(11) - 9.14 = 36.3 - 9.14$
$= 27.16$. In 1985, the number of calories consumed per capita was increasing at a rate of 27.16 calories per day per year.
$C'(16) = 3.3(16) - 9.14 = 52.8 - 9.14$
$= 43.66$. In 1990, the number of calories consumed per capita was increasing at a rate of 43.66 calories per day per year.

79a) $S(23) = -0.51(23)^2 + 20.2(23) + 49.49 =$
$= -269.79 + 464.6 + 49.49 = 244.3$.
Twenty-three weeks after the ad campaign began, the total weekly sales were $244,300.

79b) $S'(t) = \frac{d}{dt}[-0.51t^2 + 20.2t + 49.49]$
$= -1.02t + 20.20$.
$S'(23) = -1.02(23) + 20.2$
$= -23.46 + 20.2 = -3.26$.
Twenty-three weeks after the ad campaign began, the total weekly sales were decreasing at a rate of $3,260 per week.

79c) Twenty-three weeks after the ad campaign began, weekly sales of the new cola were $244,300. Weekly sales were decreasing at a rate of $3,260 per week.

81a) $f(4) = 9.25(4)^3 - 123.81(4)^2 + 484.8(4)$
$+ 2882.57$
$= 592 - 1980.96 + 1939.2 + 2882.57$
$= 3432.81$.
In 1993, there were 3,432,810 members in the Girl Scouts of America.

81b) $f'(t) = \frac{d}{dt}[9.25t^3 - 123.81t^2 + 484.8t$
$+ 2882.57]$
$= 27.75t^2 - 247.62t + 484.8$.

81c) $f'(4) = 27.75(4)^2 - 247.62(4) + 484.8$
$= 444 - 990.48 + 484.8$
$= -61.68$.
In 1993, the membership in the Girl Scouts of America was decreasing at a rate of 61,680 members per year.

83) The Boys Scouts of America membership was increasing at a faster rate. In fact, the membership in the Girl Scouts of America was actually falling and not rising.

Section 2.5 Derivatives of Products and Quotients

1) $f'(x) = \frac{d}{dx}(x^2) \cdot (2x+1) + x^2 \cdot \frac{d}{dx}(2x+1)$
   $= 2x \cdot (2x+1) + x^2 \cdot (2)$
   $= 4x^2 + 2x + 2x^2 = 6x^2 + 2x$.

3) $f'(x) = \frac{d}{dx}(x^3) \cdot (3x^2 + 2x - 5)$
   $\quad\quad + x^3 \cdot \frac{d}{dx}(3x^2 + 2x - 5)$
   $= 3x^2 \cdot (3x^2 + 2x - 5) + x^3 \cdot (6x + 2)$
   $= 9x^4 + 6x^3 - 15x^2 + 6x^4 + 2x^3$
   $= 15x^4 + 8x^3 - 15x^2$.

5) $y' = \frac{d}{dx}(3x^4) \cdot (2x^2 - 9x + 1)$
   $\quad\quad + (3x^4) \cdot \frac{d}{dx}(2x^2 - 9x + 1)$
   $= 12x^3 \cdot (2x^2 - 9x + 1) + (3x^4) \cdot (4x - 9)$
   $= 24x^5 - 108x^4 + 12x^3 + 12x^5 - 27x^4$
   $= 36x^5 - 135x^4 + 12x^3$.

7) $y' = \frac{d}{dx}(-5x^2) \cdot (3x^3 + 5x - 7)$
   $\quad\quad + (-5x^2) \cdot \frac{d}{dx}(3x^3 + 5x - 7)$
   $= -10x \cdot (3x^3 + 5x - 7) + (-5x^2) \cdot (9x^2 + 5)$
   $= -30x^4 - 50x^2 + 70x - 45x^4 - 25x^2$
   $= -75x^4 - 75x^2 + 70x$.

9) $f'(x) = \frac{d}{dx}(3x+4) \cdot (2x-1)$
   $\quad\quad + (3x+4) \cdot \frac{d}{dx}(2x-1)$
   $= 3 \cdot (2x-1) + (3x+4) \cdot 2$
   $= 6x - 3 + 6x + 8$
   $= 12x + 5$.

11) $y' = \frac{d}{dx}(5x+3) \cdot (3x^3 + 2x^2 + 1)$
    $\quad\quad + (5x+3) \cdot \frac{d}{dx}(3x^3 + 2x^2 + 1)$
    $= 5 \cdot (3x^3 + 2x^2 + 1) + (5x+3) \cdot (9x^2 + 4x)$
    $= 15x^3 + 10x^2 + 5 + 45x^3 + 20x^2 + 27x^2 + 12x$
    $= 60x^3 + 57x^2 + 12x + 5$.
    $= 6x^2 - 10x + 8$.

13) $g'(x) = \frac{d}{dx}(3x^2 - 2x + 1) \cdot (2x^2 + 5x - 7)$
    $\quad\quad + (3x^2 - 2x + 1) \cdot \frac{d}{dx}(2x^2 + 5x - 7)$
    $= (6x - 2) \cdot (2x^2 + 5x - 7)$
    $\quad\quad + (3x^2 - 2x + 1) \cdot (4x + 5)$
    $= 12x^3 + 26x^2 - 52x + 14$
    $\quad\quad + 12x^3 + 7x^2 - 6x + 5$
    $= 24x^3 + 33x^2 - 58x + 19$.

15) $y' = \frac{d}{dx}(2x^{1/2} + 4x - 3) \cdot (3x - 4)$
    $\quad\quad + (2x^{1/2} + 4x - 3) \cdot \frac{d}{dx}(3x - 4)$
    $= (x^{-1/2} + 4) \cdot (3x - 4) + (2x^{1/2} + 4x - 3) \cdot 3$
    $= 3x^{1/2} - 4x^{-1/2} + 12x - 16 + 6x^{1/2} + 12x - 9$
    $= 24x + 9x^{1/2} - 25 - 4x^{-1/2}$
    $= 24x + 9\sqrt{x} - 25 - \frac{4}{\sqrt{x}}$.

17) $f'(x) = \frac{d}{dx}(3x^{6/5} - 5x) \cdot (4x^{5/3} + 2x - 5)$
    $\quad\quad + (3x^{6/5} - 5x) \cdot \frac{d}{dx}(4x^{5/3} + 2x - 5)$
    $= (3.6x^{1/5} - 5) \cdot (4x^{5/3} + 2x - 5)$
    $\quad\quad + (3x^{6/5} - 5x) \cdot (\frac{20}{3}x^{2/3} + 2)$
    $= 14.4x^{28/15} + 7.2x^{6/5} - 18x^{1/5} - 20x^{5/3} - 10x$
    $\quad\quad + 25 + 20x^{28/15} + 6x^{6/5} - \frac{100}{3}x^{5/3} - 10x$
    $= 34.4x^{28/15} - \frac{160}{3}x^{5/3} + 13.2x^{6/5} - 20x$
    $\quad\quad - 18x^{1/5} + 25$
    $= 34.4x\sqrt[15]{x^{13}} - \frac{160}{3}x\sqrt[3]{x^2} + 13.2x\sqrt[5]{x}$
    $\quad\quad - 20x - 18\sqrt[5]{x} + 25$.

19) $f'(x) = \frac{d}{dx}(3x^{1/2} - 5) \cdot (2x^{1/2} - x^{-3})$
    $\quad\quad + (3x^{1/2} - 5) \cdot \frac{d}{dx}(2x^{1/2} - x^{-3})$
    $= 1.5x^{-1/2} \cdot (2x^{1/2} - x^{-3})$
    $\quad\quad + (3x^{1/2} - 5) \cdot (x^{-1/2} + 3x^{-4})$
    $= 3 - 1.5x^{-7/2} + 3 + 9x^{-7/2} - 5x^{-1/2} - 15x^{-4}$
    $= 6 - 5x^{-1/2} + 7.5x^{-7/2} - 15x^{-4}$
    $= 6 - \frac{5}{\sqrt{x}} + \frac{7.5}{x^3\sqrt{x}} - \frac{15}{x^4}$.

21) $h'(x) = \frac{d}{dx}(x^{2/3} + x + 1) \bullet (x^{-1} + x^{-2})$
$\qquad + (x^{2/3} + x + 1) \bullet \frac{d}{dx}(x^{-1} + x^{-2})$
$= (\frac{2}{3}x^{-1/3} + 1) \bullet (x^{-1} + x^{-2})$
$\qquad + (x^{2/3} + x + 1) \bullet (-x^{-2} - 2x^{-3})$
$= \frac{2}{3}x^{-4/3} + \frac{2}{3}x^{-7/3} + x^{-1} + x^{-2} - x^{-4/3}$
$\qquad - 2x^{-7/3} - x^{-1} - 2x^{-2} - x^{-2} - 2x^{-3}$
$= -\frac{1}{3}x^{-4/3} - 2x^{-2} - \frac{4}{3}x^{-7/3} - 2x^{-3}$
$= -\frac{1}{3x\sqrt[3]{x}} - \frac{2}{x^2} - \frac{4}{3x^2\sqrt[3]{x}} - \frac{2}{x^3}$

23) $f'(x) = \frac{\frac{d}{dx}(x+2) \bullet (x+1) - (x+2) \bullet \frac{d}{dx}(x+1)}{(x+1)^2}$
$= \frac{1 \bullet (x+1) - (x+2) \bullet 1}{(x+1)^2} = \frac{x+1-x-2}{(x+1)^2} = \frac{-1}{(x+1)^2}$.

25) $y' = \frac{\frac{d}{dx}(4x-3) \bullet (2x+1) - (4x-3) \bullet \frac{d}{dx}(2x+1)}{(2x+1)^2}$
$= \frac{4 \bullet (2x+1) - (4x-3) \bullet 2}{(2x+1)^2} = \frac{8x+4-8x+6}{(2x+1)^2} = \frac{10}{(2x+1)^2}$.

27) $f'(x) =$
$\frac{\frac{d}{dx}(3x^2-5x+1) \bullet (5x^2+3x+2) - (3x^2-5x+1) \bullet \frac{d}{dx}(5x^2+3x+2)}{(5x^2+3x+2)^2}$
$= \frac{(6x-5) \bullet (5x^2+3x+2) - (3x^2-5x+1) \bullet (10x+3)}{(5x^2+3x+2)^2}$
$= \frac{30x^3 - 7x^2 - 3x - 10 - 30x^3 + 41x^2 + 5x - 3}{(5x^2+3x+2)^2}$
$= \frac{34x^2 + 2x - 13}{(5x^2+3x+2)^2}$.

29) $f'(x) = \frac{\frac{d}{dx}(3x^{\frac{1}{2}} - 5) \bullet (6x-1) - (3x^{\frac{1}{2}} - 5) \bullet \frac{d}{dx}(6x-1)}{(6x-1)^2}$
$= \frac{(1.5x^{-\frac{1}{2}}) \bullet (6x-1) - (3x^{\frac{1}{2}} - 5) \bullet 6}{(6x-1)^2}$
$= \frac{9x^{\frac{1}{2}} - 1.5x^{-\frac{1}{2}} - 18x^{\frac{1}{2}} + 30}{(6x-1)^2} = \frac{-9x^{\frac{1}{2}} + 30 - 1.5x^{-\frac{1}{2}}}{(6x-1)^2}$
$= \frac{x^{-\frac{1}{2}}(-9x + 30x^{\frac{1}{2}} - 1.5)}{(6x-1)^2} = \frac{-9x + 30\sqrt{x} - 1.5}{\sqrt{x}(6x-1)^2}$.

31) When differentiating the numerator, we will have to use the product rule. Let's calculate the derivative of the numerator first:
Let $h(x) = (x^2 + 2)(x - 3)$, then
$h'(x) = \frac{d}{dx}(x^2 + 2) \bullet (x - 3)$
$\qquad + (x^2 + 2) \bullet \frac{d}{dx}(x - 3)$
$= 2x(x-3) + (x^2+2) \bullet 1 = 2x^2 - 6x + x^2 + 2$
$= 3x^2 - 6x + 2$.
Hence, $y = \frac{(x^2+2)(x-3)}{x-1} = \frac{h(x)}{x-1}$. Thus,
$y' = \frac{\frac{d}{dx}(h(x)) \bullet (x-1) - (h(x)) \bullet \frac{d}{dx}(x-1)}{(x-1)^2}$
$= \frac{(3x^2 - 6x + 2) \bullet (x-1) - [(x^2+2)(x-3)] \bullet 1}{(x-1)^2}$
$= \frac{3x^3 - 9x^2 + 8x - 2 - x^3 + 3x^2 - 2x + 6}{(x-1)^2}$
$= \frac{2x^3 - 6x^2 + 6x + 4}{(x-1)^2}$.

33) When differentiating the numerator, we will have to use the product rule. Let's calculate the derivative of the numerator first:
Let $h(x) = (5x^4 + 2)(x^2 + 3)$, then
$h'(x) = \frac{d}{dx}(5x^4 + 2) \bullet (x^2 + 3)$
$\qquad + (5x^4 + 2) \bullet \frac{d}{dx}(x^2 + 3)$
$= 20x^3(x^2 + 3) + (5x^4 + 2)2x$
$= 20x^5 + 60x^3 + 10x^5 + 4x$
$= 30x^5 + 60x^3 + 4x$.
Hence, $f(x) = \frac{(5x^4+2)(x^2+3)}{x-4} = \frac{h(x)}{x-4}$. Thus,
$f'(x) = \frac{\frac{d}{dx}(h(x)) \bullet (x-4) - (h(x)) \bullet \frac{d}{dx}(x-4)}{(x-4)^2}$
$= \frac{(30x^5 + 60x^3 + 4x) \bullet (x-4) - [(5x^4+2)(x^2+3)] \bullet 1}{(x-4)^2}$
$= \frac{30x^6 - 120x^5 + 60x^4 - 240x^3 + 4x^2 - 16x - 5x^6 - 15x^4 - 2x^2 - 6}{(x-4)^2}$
$= \frac{25x^6 - 120x^5 + 45x^4 - 240x^3 + 2x^2 - 16x - 6}{(x-4)^2}$.

35) $g'(x) = \frac{d}{dx}(\frac{1}{2}) \bullet (4x^3 + 2x^2 - 3x - 5)$
$\qquad + (\frac{1}{2}) \bullet \frac{d}{dx}(4x^3 + 2x^2 - 3x - 5)$
$= (0) \bullet (4x^3 + 2x^2 - 3x - 5)$
$\qquad + \frac{1}{2} \bullet (12x^2 + 4x - 3)$
$= 0 + 6x^2 + 2x - \frac{3}{2} = 6x^2 + 2x - \frac{3}{2}$.

37a) $f'(x) = \frac{d}{dx}(x^2) \bullet (x^2 - 5)$
$\qquad + x^2 \bullet \frac{d}{dx}(x^2 - 5)$
$= 2x(x^2 - 5) + x^2(2x)$
$= 2x^3 - 10x + 2x^3 = 4x^3 - 10x$.

37b) $m = f'(1) = 4(1)^3 - 10(1)$
$= 4 - 10 = -6$.
Since $f(1) = (1)^2((1)^2 - 5) = 1(-4) = -4$,
we can find the equation of the tangent line by plugging $m = -6$ (from above) & $(1, -4)$ into:
$y - y_1 = m(x - x_1)$
$y - (-4) = -6(x - 1)$
$y + 4 = -6x + 6$
$y = -6x + 2$.

37c) The graph is displayed in viewing window $[-5, 5]$ by $[-10, 10]$

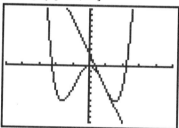

Note: The textbook writes the viewing window on the sides of the graph.

37d) The graph is displayed in viewing window $[-5, 5]$ by $[-10, 10]$

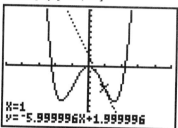

We can see that this matches our answer very well.

39a) $f'(x) = \frac{d}{dx}(x^2 + 1) \bullet (x^3 + 1)$
$\qquad + (x^2 + 1) \bullet \frac{d}{dx}(x^3 + 1)$
$= 2x(x^3 + 1) + (x^2 + 1) \bullet 3x^2$
$= 2x^4 + 2x + 3x^4 + 3x^2 = 5x^4 + 3x^2 + 2x$.

39b) $m = f'(1) = 5(1)^4 + 3(1)^2 + 2(1)$
$= 5 + 3 + 2 = 10$.
Since $f(1) = ((1)^2 + 1)((1)^3 + 1) = (2)(2) = 4$, we can find the equation of the tangent line by plugging $m = 10$ (from above) & $(1, 4)$ into:
$y - y_1 = m(x - x_1)$
$y - 4 = 10(x - 1)$
$y - 4 = 10x - 10$
$y = 10x - 6$.

39c) The graph is displayed in viewing window $[-3, 3]$ by $[-10, 10]$

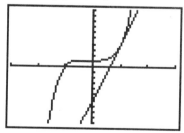

86 Chapter 2 Limits, Instantaneous Rate of Change and the Derivative

39d) The graph is displayed in viewing window [−3, 3] by [−10, 10]

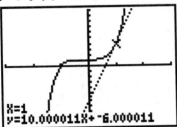

We can see that this matches our answer very well.

41a) $y' = \dfrac{\frac{d}{dx}(x+2)\cdot(x-1) - (x+2)\cdot\frac{d}{dx}(x-1)}{(x-1)^2}$

$= \dfrac{1\cdot(x-1) - (x+2)\cdot 1}{(x-1)^2} = \dfrac{x-1-x-2}{(x-1)^2} = \dfrac{-3}{(x-1)^2}$.

41b) $m =$ at $x = 2$, $y' = \dfrac{-3}{((2)-1)^2} = \dfrac{-3}{1} = -3$.

Since, at $x = 2$, $y = \dfrac{(2)+2}{(2)-1} = \dfrac{4}{1} = 4$, we can find the equation of the tangent line by plugging $m = -3$ (from above) & $(2, 4)$ into:

$y - y_1 = m(x - x_1)$
$y - 4 = -3(x - 2)$
$y - 4 = -3x + 6$
$y = -3x + 10$.

41c) The graph is displayed in viewing window [1, 5] by [0, 10]

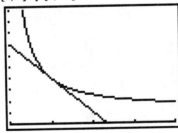

41d) The graph is displayed in viewing window [1, 5] by [0, 10]

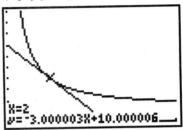

The graph matches our answer very well.

43a) $g'(x)$

$= \dfrac{\frac{d}{dx}(3x^2-2x)\cdot(-2x+3) - (3x^2-2x)\cdot\frac{d}{dx}(-2x+3)}{(-2x+3)^2}$

$= \dfrac{(6x-2)\cdot(-2x+3) - (3x^2-2x)\cdot(-2)}{(-2x+3)^2}$

$= \dfrac{-12x^2 + 22x - 6 + 6x^2 - 4x}{(-2x+3)^2} = \dfrac{-6x^2 + 18x - 6}{(-2x+3)^2}$.

43b) $m =$ at $g'(-1) = \dfrac{-6(-1)^2 + 18(-1) - 6}{(-2(-1)+3)^2}$

$= \dfrac{-6 - 18 - 6}{(5)^2} = \dfrac{-30}{25} = -1.2$.

Since $g(-1) = \dfrac{3(-1)^2 - 2(-1)}{-2(-1)+3} = \dfrac{3+2}{2+3} = \dfrac{5}{5}$

$= 1$, we can find the equation of the tangent line by plugging $m = -1.2$ (from above) & $(-1, 1)$ into:

$y - y_1 = m(x - x_1)$
$y - 1 = -1.2(x - (-1))$
$y - 1 = -1.2(x + 1)$
$y - 1 = -1.2x - 1.2$
$y = -\dfrac{6}{5}x - \dfrac{1}{5}$.

43c) The graph is displayed in viewing window [−5, 1] by [−5, 10]

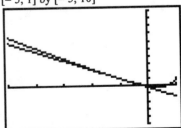

43d) The graph is displayed in viewing window [−5, 0] by [−10, 0]

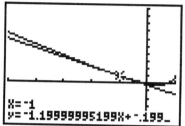

The graph matches our answer very well.

45) $f'(x) = \frac{d}{dx}(x+1) \cdot (x-2) \cdot (x+5)$
$\qquad + (x+1) \cdot \frac{d}{dx}(x-2) \cdot (x+5)$
$\qquad + (x+1) \cdot (x-2) \cdot \frac{d}{dx}(x+5)$
$= (1)(x-2)(x+5) + (x+1)(1)(x+5)$
$\qquad + (x+1)(x-2)(1)$
$= x^2 + 3x - 10 + x^2 + 6x + 5 + x^2 - x - 2$
$= 3x^2 + 8x - 7$.

47) $f'(x) = \frac{d}{dx}(x+1) \cdot (2x^2 - 3) \cdot (3x+4)$
$\qquad + (x+1) \cdot \frac{d}{dx}(2x^2 - 3) \cdot (3x+4)$
$\qquad + (x+1) \cdot (2x^2 - 3) \cdot \frac{d}{dx}(3x+4)$
$= (1)(2x^2 - 3)(3x+4) + (x+1)(4x)(3x+4)$
$\qquad + (x+1)(2x^2 - 3)(3)$
$= 6x^3 + 8x^2 - 9x - 12 + 12x^3 + 28x^2 + 16x$
$\qquad + 6x^3 + 6x^2 - 9x - 9$
$= 24x^3 + 42x^2 - 2x - 21$.

49) $f'(x) = \frac{d}{dx}(x^{1/2}) \cdot (2x-1) \cdot (3x^2 + 2)$
$\qquad + (x^{1/2}) \cdot \frac{d}{dx}(2x-1) \cdot (3x^2 + 2)$
$\qquad + (x^{1/2}) \cdot (2x-1) \cdot \frac{d}{dx}(3x^2 + 2)$
$= \frac{1}{2}x^{-1/2}(2x-1)(3x^2+2) + (x^{1/2})(2)(3x^2+2)$
$\qquad + (x^{1/2})(2x-1)(6x)$
$= 3x^{5/2} - 1.5x^{3/2} + 2x^{1/2} - x^{-1/2} + 6x^{5/2} + 4x^{1/2}$
$\qquad + 12x^{5/2} - 6x^{3/2}$
$= 21x^{5/2} - 7.5x^{3/2} + 6x^{1/2} - x^{-1/2}$
$= 21x^2\sqrt{x} - 7.5x\sqrt{x} + 6\sqrt{x} - \frac{1}{\sqrt{x}}$.

51) $q(3) = 30(3) - 0.5(3)^2 = 90 - 4.5 = 85.5$
Three months after a new computer hits the market, the monthly sales are 8,550 computers.
$q'(t) = \frac{d}{dt}(30t - 0.5t^2) = 30 - t$. Thus,
$q'(3) = 30 - 3 = 27$. Three months after a new computer hits the market, the monthly sales are increasing at a rate of 2700 computers per month.

53a) $R(t) = p(t) \cdot q(t)$
$= (2200 - 34t^2)(30t - 0.5t^2)$.
Since $q(t)$ is measured in hundreds of units, $R(t)$ then will be measured in hundreds of dollars.

53b) $R'(t) = \frac{d}{dt}(2200 - 34t^2) \cdot (30t - 0.5t^2)$
$\qquad + (2200 - 34t^2) \cdot \frac{d}{dt}(30t - 0.5t^2)$
$= (-68t)(30t - 0.5t^2)$
$\qquad + (2200 - 34t^2)(30 - t)$
$= -2040t^2 + 34t^3 + 66{,}000 - 2200t$
$\qquad - 1020t^2 + 34t^3$
$= 68t^3 - 3060t^2 - 2200t + 66{,}000$.

53c) $R(3) = (2200 - 34(3)^2)(30(3) - 0.5(3)^2)$
$= (1894)(85.5) = 161{,}937$.
After three months on the market, the total monthly revenue from the sales of a new computer was $16,193,700.
$R'(3) = 68(3)^3 - 3060(3)^2$
$\qquad - 2200(3) + 66{,}000$
$1836 - 27540 - 6600 + 66{,}000 = 33{,}696$.
After three months, the total monthly revenue was increasing at a rate of $3,369,600 per month.

55) $p(6) = 2200 - 34(6)^2 = 2200 - 1224 = 976$.
After six months on the market, the price of the computer was $ 976.
$p'(t) = \frac{d}{dt}(2200 - 34t^2) = -68t$. Thus,
$p'(6) = -68(6) = -408$. After six months on the market, the price of the computer was decreasing at a rate of $408 per month.

57a) Recall that
$R(t) = (2200 - 34t^2)(30t - 0.5t^2)$.
Multiplying and simplifying, we get:
$R(t) = (2200 - 34t^2)(30t - 0.5t^2)$
$= 66000t - 1100t^2 - 1020t^3 + 17t^4$
$= 17t^4 - 1020t^3 - 1100t^2 + 66000t$.
Then, $R'(t)$
$= \frac{d}{dt}[17t^4 - 1020t^3 - 1100t^2 + 66000t]$
$= 68t^3 - 3060t^2 - 2200t + 66000$.

57b) The graph is displayed in viewing window [0, 7] by [– 80,000, 80,000]

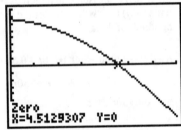

$R'(t) = 0$ when $t \approx 4.5129307$.

57c) The graph is displayed in viewing window [1, 7] by [0, 250,000]

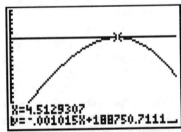

Thus, $R'(4.5129307) \approx -0.001015$.

57d) The revenue is maximized at $t \approx 4.5129307$ since the graph "peaks" at that value.

59) $p(3) = 220 - (3)^2 = 220 - 9 = 211$. After three months on the market, the price of the CD-ROM drive was $211.
$p'(t) = \frac{d}{dt}(220 - t^2) = -2t$. Thus,
$p'(3) = -2(3) = -6$. After three months on the market, the price of the CD-ROM drive was decreasing at a rate of $6 per month.

61a) $f(3) = 17.5(3) + 336.86$
$= \$389.36$ per week $= \frac{\$389.36}{\text{week}}$.
$g(3) = 0.48(3) + 9.44 = 10.88$
$= \$10.88$ per hour $= \frac{\$10.88}{\text{hour}}$

Thus, $\frac{f(3)}{g(3)} = \frac{\frac{\$389.36}{\text{week}}}{\frac{\$10.88}{\text{hour}}} = \frac{\$389.36}{\text{week}} \div \frac{\$10.88}{\text{hour}}$

$= \frac{\$389.36}{\text{week}} \cdot \frac{\text{hour}}{\$10.88} \approx 35.79$ hours per week.

In 1992, an employee in finance, insurance, and real estate was working an average of about 35.79 hours per week.

61b) $h(t) = \frac{f(t)}{g(t)} = \frac{17.5t + 336.86}{0.48t + 9.44}$. This represents the average number hours per week that an employee in finance, insurance, and real estate works.

61c) $h(3) = \frac{f(3)}{g(3)} \approx 35.79$ hours per week (from part a). In 1992, an employee in finance, insurance, and real estate was working an average of about 35.79 hours per week.

$f'(t) = \frac{d}{dt}(17.5t + 336.86) = 17.5$ and

$g'(t) = \frac{d}{dt}(0.48t + 9.44) = 0.48$.

Hence, $h'(t) = \frac{f'(t) \cdot g(t) - f(t) \cdot g'(t)}{[g(t)]^2}$

$= \frac{(17.5) \cdot (0.48t + 9.44) - (17.5t + 336.86) \cdot (0.48)}{(0.48t + 9.44)^2}$

$= \frac{8.4t + 165.2 - 8.4t - 161.6928}{(0.48t + 9.44)^2} = \frac{3.5072}{(0.48t + 9.44)^2}$.

Therefore, $h'(3) = \frac{3.5072}{(0.48(3) + 9.44)^2}$

$= \frac{3.5072}{(10.88)^2} = \frac{3.5072}{118.3744} \approx 0.03$.

Section 2.5 Derivatives of Products and Quotients 89

61c) Continued
In 1992, the amount of hours per week an employee in finance, insurance, and real estate was increasing at a rate of about 0.03 hours per week per year.

63) $C(50) = \frac{50(50)}{100-(50)} = \frac{2500}{50} = 50$. It will cost $50,000 to clean up 50% of the pollutants.

$$C'(x) = \frac{\frac{d}{dx}(50x) \cdot (100-x) - (50x) \cdot \frac{d}{dx}(100-x)}{(100-x)^2}$$

$$= \frac{50(100-x) - (50x)(-1)}{(100-x)^2} = \frac{5000 - 50x + 50x}{(100-x)^2}$$

$$= \frac{5000}{(100-x)^2}.$$

Thus, $C'(50) = \frac{5000}{(100-50)^2} = \frac{5000}{2500} = 2.$

When 50% of the pollutants is removed, the cost of removing the pollutants is increasing at a rate of $2,000 per 1%.

65) $P(5) = \frac{20(10+7(5))}{1+0.02(5)} = \frac{20(45)}{1+0.1} = \frac{900}{1.1} \approx 818.18.$

After five months, there are about 818 bass in the lake.

Since $P(t) = \frac{20(10+7t)}{1+0.02t} = \frac{200+140t}{1+0.02t}$, then $P'(t)$

$$= \frac{\frac{d}{dt}(200+140t) \cdot (1+0.02t) - (200+140t) \cdot \frac{d}{dt}(1+0.02t)}{(1+0.02t)^2}$$

$$= \frac{(140) \cdot (1+0.02t) - (200+140t) \cdot (0.02)}{(1+0.02t)^2}$$

$$= \frac{140 + 2.8t - 4 - 2.8t}{(1+0.02t)^2} = \frac{136}{(1+0.02t)^2}.$$

Hence, $P'(5) = \frac{136}{(1+0.02(5))^2} = \frac{136}{(1.1)^2} = \frac{136}{1.21}$

≈ 112.4. After five months, the bass population was increasing at a rate of about 112.4 bass per month.

67a) $SD(5) = -0.02(5)^3 + 0.27(5)^2 - 1.22(5) + 24.12$
$= -2.5 + 6.75 - 6.1 + 24.12 = 22.27.$
In 1989, 22.27 million tons of sulfur dioxide was released into the air.
$CP(5) = 2.57(5) + 232.91$
$= 12.85 + 232.91 = 245.76.$
In 1989, the U.S. civilian population was 245.76 million people.

67b) $\frac{SD(5)}{CP(5)} = \frac{22.27 \text{ million tons of sulfur dioxide}}{245.76 \text{ million people}}$
about 0.09 tons of sulfur dioxide per person.
In 1989, the amount of sulfur dioxide released into the air was about 0.09 tons per capita.

67c) $H(t) = \frac{SD(t)}{CP(t)} = \frac{-0.02t^3 + 0.27t^2 - 1.22t + 24.12}{2.57t + 232.91}$.

Let us first find $SD'(t)$ and $CP'(t)$:
$SD'(t) = \frac{d}{dt}[-0.02t^3 + 0.27t^2 - 1.22t + 24.12]$
$= -0.06t^2 + 0.54t - 1.22.$
$CP'(t) = \frac{d}{dt}[2.57t + 232.91] = 2.57.$
Now, $SD'(t) \cdot CP(t) - SD(t) \cdot CP'(t) =$
$(-0.06t^2 + 0.54t - 1.22) \cdot (2.57t + 232.91)$
$- (-0.02t^3 + 0.27t^2 - 1.22t + 24.12)(2.57)$
$= -0.1542t^3 - 12.5868t^2 + 122.636t$
$- 284.1502 + 0.0514t^3 - 0.6939t^2$
$+ 3.1354t - 61.9884$
$= -0.1028t^3 - 13.2807t^2 + 125.7714t$
$- 346.1386$

Hence, $H'(t) = \frac{SD'(t) \cdot CP(t) - SD(t) \cdot CP'(t)}{[CP(t)]^2}$

$= \frac{-0.1028t^3 - 13.2807t^2 + 125.7714t - 346.1386}{(2.57t + 232.91)^2}.$

67d) $H'(5)$
$= \frac{-0.1028(5)^3 - 13.2807(5)^2 + 125.7714(5) - 346.1386}{(2.57(5) + 232.91)^2}$
$= \frac{-12.85 - 332.0175 + 628.857 - 346.1386}{(245.76)^2} = \frac{-62.1491}{60397.9776}$
≈ -0.001

In 1989, the amount of sulfur dioxide released into the air was decreasing at a rate of about 0.001 tons per capita per year.

Section 2.6 Continuity and Nondifferentiability

1) Yes, f is continuous at $x = 1$ since all three of the conditions in the definition of continuity are satisfied.

3) No, f is not continuous at $x = 3$ since condition #2 of the definition is not satisfied ($\lim\limits_{x \to 3} f(x)$ does not exist).

5) Yes, f is continuous at $x = 1$ since all three of the conditions in the definition of continuity are satisfied.

7) Yes, f is continuous at $x = 3$ since all three of the conditions in the definition of continuity are satisfied.

9) We need to check each condition of the definition of continuity:
   1. Is $g(-2)$ defined? Yes, since $g(-2) = 3(-2) - 2 = -8$.
   2. Does $\lim\limits_{x \to -2} g(x)$ exist? Yes, since, by the Substitution Principle, $\lim\limits_{x \to -2} g(x) = \lim\limits_{x \to -2} 3x - 2 = 3(-2) - 2 = -8$.
   3. Does $\lim\limits_{x \to -2} g(x) = g(-2)$? Yes, since $-8 = -8$.

   Hence, g is continuous at $x = -2$.

11) We need to check each condition of the definition of continuity:
    1. Is $h(3)$ defined? Yes, since $h(3) = (3)^2 - (3) - 6 = 0$.
    2. Does $\lim\limits_{x \to 3} h(x)$ exist? Yes, since, by the Substitution Principle, $\lim\limits_{x \to 3} h(x) = \lim\limits_{x \to 3} x^2 - x - 6 = (3)^2 - (3) - 6 = 0$.
    3. Does $\lim\limits_{x \to 3} h(x) = h(3)$? Yes, since $0 = 0$.

    Hence, h is continuous at $x = 3$.

13) We need to check each condition of the definition of continuity:
    1. Is $g(3)$ defined? No, $g(3) = \frac{3+1}{3-3} = \frac{4}{0}$ which is undefined.

    Since condition #1 is not satisfied, g is not continuous at $x = 3$.

15) We need to check each condition of the definition of continuity:
    1. Is $f(5)$ defined? No, $f(5) = \frac{5^2 - 25}{5 - 5} = \frac{0}{0}$ which is undefined.

    Since condition #1 is not satisfied, f is not continuous at $x = 5$.

17) We need to check each condition of the definition of continuity:
    1. Is $f(1)$ defined? Yes, since $f(1) = (1) + 2 = 3$.
    2. Does $\lim\limits_{x \to 1} f(x)$ exist? No, since
    $$\lim\limits_{x \to 1^-} f(x) = \lim\limits_{x \to 1^-} x + 2 = 1 + 2 = 3$$
    & $\lim\limits_{x \to 1^+} f(x) = \lim\limits_{x \to 1^+} x^2 + 3 = 1 + 3 = 4$,
    then $\lim\limits_{x \to 1} f(x)$ does not exist.

    Since condition #2 is not satisfied, f is not continuous at $x = 1$.

19) By Theorem 2.1, since f is a polynomial, it is continuous for all real numbers or on the interval $(-\infty, \infty)$.

21) By Theorem 2.1, since g is a rational function, it is continuous for all real numbers in its domain. Since g is undefined only at $x = -5$, then g is continuous on the interval $(-\infty, -5) \cup (-5, \infty)$.

23) By Theorem 2.1, since f is a rational function, it is continuous for all real numbers in its domain. Since f is undefined only at $x = -1$ and $x = 3$, then f is continuous on the interval $(-\infty, -1) \cup (-1, 3) \cup (3, \infty)$.

25) By Theorem 2.1, since f is an exponential function, it is continuous for all real numbers or on the interval $(-\infty, \infty)$.

27) By Theorem 2.1, since f is a logarithmic function, it is continuous for all real numbers in its domain. f is defined for all values of $x > -3$, hence f is continuous on the interval $(-3, \infty)$.

29) By Theorem 2.1, since h is an exponential function, it is continuous for all real numbers or on the interval $(-\infty, \infty)$.

31) By Theorem 2.1, since f is a radical function, it is continuous for all real numbers in its domain. f is defined for all values of $x \geq -1.5$, hence f is continuous on the interval $[-1.5, \infty)$.

33) f is not differentiable at $x = -2$ (since it is not continuous at $x = -2$) and at $x = 2$ (since it is not continuous at $x = 2$).

35) f is not differentiable at $x = -2$ (since it has a corner at $x = -2$) and at $x = 1$ (since it has a corner at $x = 1$).

37) f is not differentiable at $x = 2$ (since it has a corner at $x = 2$).

39) f is differentiable for all values of x since there are no points were it is not continuous, there are no corners, and there are no vertical tangents.

41a) The graph is displayed in viewing window $[-10, 10]$ by $[-5, 5]$

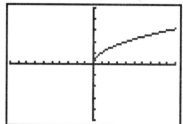

Note: The textbook writes the viewing window on the sides of the graph.
Note, the domain of f is $[0, \infty)$

41b) $f'(x) = \frac{d}{dx}\sqrt{x} = \frac{d}{dx} x^{1/2} = \frac{1}{2} x^{-1/2} = \frac{1}{2\sqrt{x}}$.

41c) f' is undefined at $x = 0$. (Also note that f' is undefined on $(-\infty, 0)$ since these values are not the domain of f.)

41d) f has a vertical tangent at $x = 0$.

43a) The graph is displayed in viewing window $[-10, 10]$ by $[-5, 5]$

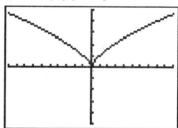

Note, the domain of f is $(-\infty, \infty)$.

43b) $f'(x) = \frac{d}{dx}(x^{2/3}) = \frac{2}{3} x^{-1/3} = \frac{2}{3\sqrt[3]{x}}$.

43c) f' is undefined at $x = 0$.

43d) f has a corner at $x = 0$.

45a) The graph is displayed in viewing window $[-6, 4]$ by $[-5, 5]$

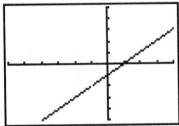

Note, the domain is $(-\infty, -1) \cup (-1, \infty)$ and hence, we have a hole at $x = -1$.

45b) $f'(x) = \dfrac{\dfrac{d}{dx}(x^2-1)\bullet(x+1)-(x^2-1)\bullet\dfrac{d}{dx}(x+1)}{(x+1)^2}$

$= \dfrac{2x\bullet(x+1)-(x^2-1)\bullet(1)}{(x+1)^2} = \dfrac{2x^2+2x-x^2+1}{(x+1)^2}$

$= \dfrac{x^2+2x+1}{(x+1)^2} = \dfrac{(x+1)^2}{(x+1)^2} = 1$ as long as $x \neq -1$.

45c) f' is undefined at $x = -1$.

45d) f is not continuous at $x = -1$.

47a) The graph is displayed in viewing window $[-4, 8]$ by $[-4, 8]$

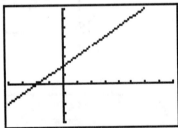

Note, the domain is $(-\infty, 2) \cup (2, \infty)$ and hence, we have a hole at $x = 2$.

47b) $f'(x) = \dfrac{\dfrac{d}{dx}(x^2-4)\bullet(x-2)-(x^2-4)\bullet\dfrac{d}{dx}(x-2)}{(x-2)^2}$

$= \dfrac{2x\bullet(x-2)-(x^2-4)\bullet(1)}{(x-2)^2} = \dfrac{2x^2-4x-x^2+4}{(x-2)^2}$

$= \dfrac{x^2-4x+4}{(x-2)^2} = \dfrac{(x-2)^2}{(x-2)^2} = 1$ as long as $x \neq 2$.

47c) f' is undefined at $x = 2$.

47d) f is not continuous at $x = 2$.

49) The jump in the graph represents the end of the first day and the beginning of the second day.

51a) For $x < 2$ lb, $C(x) = 1.5x$. For $x \geq 2$ lb, $C(x) = x$, thus:

$C(x) = \begin{cases} 1.5x, & 0 < x < 2 \\ 1x, & 2 \leq x \leq 6 \end{cases}$

51b)

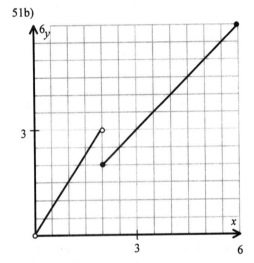

51c) $C(1.5) = 1.5(1.5) = 2.25$.
It will cost \$2.25 to buy 1.5-pound package of hamburger.

51d) $C(2) = 1(2) = 2$.
It will cost \$2 to buy a 2-pound package of hamburger.

51e) Since $\lim\limits_{x \to 2^-} C(x) = 1.5(2) = 3$ and

$\lim\limits_{x \to 2^+} C(x) = 1(2) = 2$, then

$\lim\limits_{x \to 2} C(x)$ does not exist.

51f) C is not continuous at $x = 2$ since $\lim_{x \to 2} C(x)$ does not exist (condition #2 of the definition of continuity).

Chapter 2 Review Exercises

1)

x	−0.1	−0.01	−0.001	0	0.001	0.01	0.1
$f(x)$	0.91	0.99	0.999	?	1.001	1.01	1.1

Coming in from either direction, we see that the function values are approaching 1. Thus,
a) $\lim_{x \to 0^-} f(x) = 1.$
b) $\lim_{x \to 0^+} f(x) = 1.$
c) $\lim_{x \to 0} f(x) = 1.$

3) We begin by graphing the function and zooming in. The display window is listed below the graph.

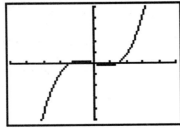

[− 5, 5] by [− 25, 25]
Note: The textbook writes the viewing window on the sides of the graph.

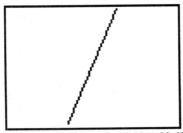

[≈ 3.095, ≈ 3.105] by [≈ 23.55, ≈ 23.63]

3) Continued
By using the trace features, we find:

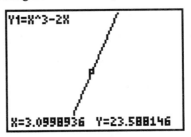

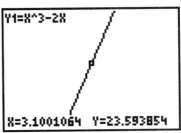

that $\lim_{x \to 3.1} f(x) \approx 23.59.$

We can verify our results numerically:

x	3	3.09	3.099	3.0999
$f(x)$	21	≈ 23.32	≈ 23.56	≈ 23.59
x	3.2	3.11	3.101	3.1001
$f(x)$	26.368	≈ 23.86	≈ 23.62	≈ 23.59

As x approaches 3.1, $f(x)$ approaches ≈ 23.59.

5) Using the Substitution Principle, we find that:
$\lim_{x \to 2} (7x^3 - 10x) = 7(2)^3 - 10(2) = 56 - 20$
$= 36.$

7) Using the Substitution Principle, we find that:
$\lim_{x \to -3} |x - 5| = |-3 - 5| = |-8| = 8.$

9) We need to simplify the expression before using the Substitution Principle.
$$\lim_{x\to 10} \frac{x+10}{x^2-100} = \lim_{x\to 10} \frac{x+10}{(x+10)(x-10)}$$
$$= \lim_{x\to 10} \frac{1}{x-10}.$$
But, since $\lim_{x\to 10^-} \frac{1}{x-10} = -\infty$ and
$\lim_{x\to 10^+} \frac{1}{x-10} = \infty$, then $\lim_{x\to 10} \frac{1}{x-10}$
does not exist.

11a) $\lim_{x\to -4} f(x) = 2$.

11b) $\lim_{x\to 0^-} f(x) = 0$.

11c) $f(0) = -\frac{3}{2}$.

11d) $\lim_{x\to 2} f(x)$ does not exist.

11e) $\lim_{x\to 3} f(x) = 1$.

11f) $f(3) = 1$.

13) We need to simplify the expression before using the Substitution Principle.
$$\lim_{h\to 0} \frac{2x^2h-9h}{h} = \lim_{h\to 0} \frac{h(2x^2-9)}{h} = \lim_{h\to 0} (2x^2-9)$$
$$= 2x^2 - 9.$$

15a) $f(2+h) = 3(2+h)^2 = 12 + 12h + 3h^2$.

15b) $f(2) = 3(2)^2 = 3(4) = 12$. Thus,
$$\lim_{h\to 0} \frac{f(2+h)-f(2)}{h} = \lim_{h\to 0} \frac{12+12h+h^2-12}{h}$$
$$= \lim_{h\to 0} \frac{12h+h^2}{h} = \lim_{h\to 0} \frac{h(12+h)}{h} = \lim_{h\to 0} 12 + h$$
$$= 12 + (0) = 12.$$

17a) $C(400) = 36,000 + \sqrt{10,000(400)}$
$= 36,000 + \sqrt{4,000,000} = 36,000 + 2000$
$= 38,000$. It will cost $38,000 to produce 400 teddy bears with glowing eyes.

17b) Using the Substitution Principle, we find that:
$$\lim_{x\to 100} C(x) = \lim_{x\to 100} 36,000 + \sqrt{10,000x}$$
$$= 36,000 + \sqrt{10,000(100)}$$
$$= 36,000 + \sqrt{1,000,000} = 36,000 + 1000$$
$$= 37,000.$$
As the production approaches 100 teddy bears, the production cost approaches $37,000.

17c) $AC(25) = \frac{C(25)}{25} = \frac{36000+\sqrt{10000(25)}}{25}$
$= \frac{36000+\sqrt{250000}}{25} = \frac{36000+500}{25} = \frac{36500}{25}$
$= 1460$.
When 25 teddy bears are produced, the average cost will be $1460 per bear.

19a) The graph is displayed in viewing window [0, 40] by [0, 10]

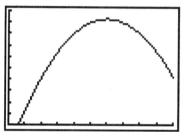

19b) $R(20) = 8 - \frac{(20)^2-48(20)+512}{50} = 8 - \frac{-48}{50}$
$= 8 + 0.96 = 8.96.$
$R(25) = 8 - \frac{(25)^2-48(25)+512}{50} = 8 - \frac{-63}{50}$
$= 8 + 1.26 = 9.26.$
$R(30) = 8 - \frac{(30)^2-48(30)+512}{50} = 8 - \frac{-28}{50}$
$= 8 + 0.56 = 8.56.$

19c) $R(24) = 8 - \frac{(24)^2 - 48(24) + 512}{50} = 8 - \frac{-64}{50}$
$= 8 + 1.28 = 9.28.$
$R(24 + h) = 8 - \frac{(24+h)^2 - 48(24+h) + 512}{50}$
$= 8 - \frac{576 + 48h + h^2 - 1152 - 48h + 512}{50}$
$= 8 - \frac{h^2 - 64}{50}$
$= 8 - 0.02h^2 + 1.28 = h^2 + 9.28.$
Thus,
$\lim_{h \to 0} \frac{R(24+h) - R(24)}{h} = \lim_{h \to 0} \frac{h^2 + 9.28 - 9.28}{h}$
$= \lim_{h \to 0} \frac{h^2}{h} = \lim_{h \to 0} h = 0.$

19d) From part c, since $R'(24) = 0$, twenty-four tablespoons of sugar seems to be optimal. We can verify this using the maximum command on a calculator. Here, the graph is displayed in viewing window [0, 40] by [0, 10]

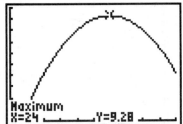

21)

x	−3.1	−3.01	−3.001	−3	−2.999	−2.99	−2.9
f(x)	200	20000	2 × 10⁶	?	2 × 10⁶	20000	200

Coming in from either direction, we see that the function values are approaching ∞.
Thus,
a) $\lim_{x \to 3^-} f(x) = \infty.$
b) $\lim_{x \to 3^+} f(x) = \infty.$
c) $\lim_{x \to 3} f(x) = \infty.$

23a) f is defined for all real numbers except 0. Thus, the domain of f is $(-\infty, 0) \cup (0, \infty)$.

23b) $\lim_{x \to 0} f(x) = \lim_{x \to 0} \frac{1250 + 3.2x}{x} = \lim_{x \to 0} \frac{1250}{x} + 3.2.$
But, $\lim_{x \to 0^-} \frac{1250}{x} + 3.2 = -\infty$ and
$\lim_{x \to 0^+} \frac{1250}{x} + 3.2 = \infty.$ Thus,
$\lim_{x \to 0} \frac{1250}{x} + 3.2$ does not exist.

25a) As x increases without bound, f increases without bound. Thus, f is an exponential growth function.

25b) Since f is an exponential growth function, $\lim_{x \to \infty} f(x) = \infty.$

25c) Since f is an exponential growth function, $\lim_{x \to -\infty} f(x) = 0.$

27a) As x increases without bound, f increases without bound. Thus, f is an exponential growth function.

27b) Since f is an exponential growth function, $\lim_{x \to \infty} f(x) = \infty.$

27c) Since f is an exponential growth function, $\lim_{x \to -\infty} f(x) = 0.$

29) Since the degree of the polynomial in the denominator is 2, then we need to multiply top and bottom by $\frac{1}{x^2}$.

Thus, $\lim_{x \to \infty} \frac{-2x^2 + 5x - 1}{x^2 - 13}$
$= \lim_{x \to \infty} \left(\frac{-2x^2 + 5x - 1}{x^2 - 13} \right) \cdot \frac{\frac{1}{x^2}}{\frac{1}{x^2}}$
$= \lim_{x \to \infty} \frac{-2 + \frac{5}{x} - \frac{1}{x^2}}{1 - \frac{13}{x^2}} = \frac{-2 + 0 - 0}{1 - 0} = \frac{-2}{1} = -2.$

We can verify our results numerically:

x	1 × 10³	1 × 10⁶	1 × 10⁹
f(x)	≈ −1.995	≈ −1.999995	≈ −2.000000

As x approaches ∞, $f(x)$ approaches −2.

31) Since the degree of the polynomial in the denominator is 1, then we need to multiply top and bottom by $\frac{1}{x}$.

Thus, $\lim\limits_{x\to-\infty} \frac{3x^4-27}{x+3} = \lim\limits_{x\to-\infty} \left(\frac{3x^4-27}{x+3}\right) \cdot \frac{\frac{1}{x}}{\frac{1}{x}}$

$= \lim\limits_{x\to-\infty} \frac{3x^3 - \frac{27}{x}}{1+\frac{3}{x}} = \frac{\left[\lim\limits_{x\to-\infty}(3x^3)\right]-27}{1+0} = -\infty.$

We can verify our results numerically:

x	1×10^3	1×10^6	1×10^9
$f(x)$	$\approx -3\times 10^9$	$\approx -3\times 10^{18}$	$\approx -3\times 10^{27}$

As x approaches $-\infty$, $f(x)$ approaches $-\infty$.

33) Since the highest degree of the polynomials in the numerator and denominator is 2, then we need to multiply top and bottom by $\frac{1}{x^2}$.

Thus, $\lim\limits_{x\to\infty} \frac{7x+9}{3x^2+2} = \lim\limits_{x\to\infty} \left(\frac{7x+9}{3x^2+2}\right) \cdot \frac{\frac{1}{x^2}}{\frac{1}{x^2}}$

$= \lim\limits_{x\to\infty} \frac{\frac{7}{x}+\frac{9}{x^2}}{3+\frac{2}{x^2}} = \frac{0+0}{3+0} = \frac{0}{3} = 0$

and $\lim\limits_{x\to-\infty} \frac{7x+9}{3x^2+2} = \lim\limits_{x\to-\infty} \left(\frac{7x+9}{3x^2+2}\right) \cdot \frac{\frac{1}{x^2}}{\frac{1}{x^2}}$

$= \lim\limits_{x\to-\infty} \frac{\frac{7}{x}+\frac{9}{x^2}}{3+\frac{2}{x^2}} = \frac{0+0}{3+0} = \frac{0}{3} = 0$

Therefore, f has a horizontal asymptote of $y = 0$.

35) No, f is not continuous at $x = 0$ since condition #2 of the definition is not satisfied ($\lim\limits_{x\to 0} f(x)$ does not exist).

37) For $x = 9$, we need to check each condition of the definition of continuity:
1. Is $g(9)$ defined? Yes, since $g(9) = \frac{(9)^2-9}{(9)+3} = \frac{72}{12} = 6$.
2. Does $\lim\limits_{x\to 9} g(x)$ exist? Yes, since, by the Substitution Principle, $\lim\limits_{x\to 9} g(x)$
$= \lim\limits_{x\to 9} \frac{x^2-9}{x+3} = \frac{(9)^2-9}{(9)+3} = \frac{72}{12} = 6.$
3. Does $\lim\limits_{x\to 9} g(x) = g(9)$? Yes, since $6 = 6$.

Hence, g is continuous at $x = 9$.

For $x = -3$, we need to check each condition of the definition of continuity:
1. Is $g(-3)$ defined? No, since $g(-3)$ gives us zero in the denominator.

Hence, since condition #1 is not satisfied, then g is not continuous at $x = -3$.

39) By Theorem 2.1, since f is a rational function, it is continuous for all real numbers in its domain. Since f is undefined only at $x = -3$ and $x = 5$, then f is continuous on the interval $(-\infty, -3) \cup (-3, 5) \cup (5, \infty)$.

41a) The graph is displayed in viewing window [0, 100] by [0, 100]

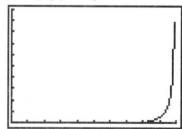

It does not cost anything to get the reliability rate anywhere between 0% and 80%. To get the rate above 80%, it does start to cost some money. In fact, the closer the rate gets to 100%, the more it will cost.

41b) Since $\lim_{x \to 80^-} C(x) = \lim_{x \to 80^-} 0 = 0$ and

$\lim_{x \to 80^+} C(x) = \lim_{x \to 80^+} \frac{400-5x}{x-100} = \frac{400-5(80)}{20}$

$= \frac{0}{20} = 0$, then $\lim_{x \to 80} C(x) = 0$.

41c) $\lim_{x \to 100^-} C(x) = \lim_{x \to 100^-} \frac{400-5x}{x-100} = \infty$.

43a) $N(0) = 2.4e^{0.008(0)} = 2.4e^0 = 2.4(1) = 2.4$.
On April 1st, the mosquito population was 2.4 million.

43b) Since $N(t)$ is an exponential growth function, then $\lim_{t \to \infty} N(t) = \infty$. This means that as time increases without bound, the mosquito population will increase without bound. This is very unrealistic since the population cannot increase without bound.

43c) The graph is displayed in viewing window [0, 100] by [0, 10]

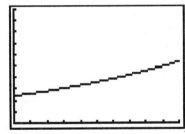

43d) We will set $N(t) = 5$, divide both sides by 2.4, and then take the natural log of both sides:
$N(t) = 2.4e^{0.008t} = 5$
$e^{0.008t} = \frac{5}{2.4} = \frac{25}{12}$
$\ln[e^{0.008t}] = \ln[\frac{25}{12}]$
$0.008t = \ln[\frac{25}{12}]$
Now, dividing both sides by 0.008, we get:
$t = \ln[\frac{25}{12}] / 0.008$
$t \approx 91.75 \approx 92$.
It will take about 92 days for the population to reach 5 million.

45a) The graph is displayed in viewing window [0, 10] by [0, 30]

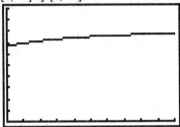

45b) $P(3) = \frac{25((3)+4)}{(3)+5} = \frac{25(7)}{8} = \frac{175}{8} = 21.875$.
After 3 years, the annual profit was $21,875.

45c) Since the degree of the polynomial in the denominator is 1, then we need to multiply top and bottom by $\frac{1}{t}$.

Thus, $\lim_{t \to \infty} \frac{25t+100}{t+5} = \lim_{t \to \infty} \left(\frac{25t+100}{t+5}\right) \cdot \frac{\frac{1}{t}}{\frac{1}{t}}$

$= \lim_{t \to \infty} \frac{25 + \frac{100}{t}}{1 + \frac{5}{t}} = \frac{25+0}{1+0} = 25$.

As time increases without bound, the annual profit approaches $25,000.

47) $m_{sec} = \dfrac{g(x+\Delta x)-g(x)}{\Delta x} = \dfrac{g(9)-g(4)}{9-4}$

$= \dfrac{2\sqrt{4(9)}-2\sqrt{4(4)}}{9-4} = \dfrac{12-8}{5} = \dfrac{4}{5}$.

49) $m_{sec} = \dfrac{f(x+\Delta x)-f(x)}{\Delta x} = \dfrac{f(7.9)-f(6.2)}{7.9-6.2}$

$= \dfrac{\frac{3}{7.9-2}-\frac{3}{6.2-2}}{1.7} = \dfrac{\frac{3}{5.9}-\frac{3}{4.2}}{1.7} = \dfrac{-50}{413}$

≈ -0.121.

51)

h	-1	-0.1	-0.01	-0.001	-0.0001
$\frac{f(x+h)-f(x)}{h}$	2.5	2.5	2.5	2.5	2.5

h	1	0.1	0.01	0.001	0.0001
$\frac{f(x+h)-f(x)}{h}$	2.5	2.5	2.5	2.5	2.5

At $x = 4$, $\lim\limits_{h\to 0} \dfrac{f(x+h)-f(x)}{h} = 2.5$.

53)

h	-1	-0.1	-0.01	-0.001	-0.0001
$\frac{f(x+h)-f(x)}{h}$	0.3	0.45	0.460	0.4609	0.46099

h	1	0.1	0.01	0.001	0.0001
$\frac{f(x+h)-f(x)}{h}$	0.6	0.48	0.463	0.4611	0.46102

At $x = -1$ $\lim\limits_{h\to 0} \dfrac{f(x+h)-f(x)}{h} \approx 0.461$.

55) $f'(7) = \lim\limits_{h\to 0} \dfrac{f(7+h)-f(7)}{h}$

$= \lim\limits_{h\to 0} \dfrac{(7+h)^2-2(7+h)-[(7)^2-2(7)]}{h}$

$= \lim\limits_{h\to 0} \dfrac{49+14h+h^2-14-2h-[35]}{h} = \lim\limits_{h\to 0} \dfrac{12h+h^2}{h}$

$= \lim\limits_{h\to 0} \dfrac{h(12+h)}{h} = \lim\limits_{h\to 0} 12 + h = 12 + 0 = 12$.

57) $f'(3) = \lim\limits_{h\to 0} \dfrac{f(3+h)-f(3)}{h}$

$= \lim\limits_{h\to 0} \dfrac{\frac{1}{(3+h)^2}-\frac{1}{(3)^2}}{h} = \lim\limits_{h\to 0} \left(\dfrac{1}{9+6h+h^2} - \dfrac{1}{9}\right)\cdot\dfrac{1}{h}$

$= \lim\limits_{h\to 0} \left(\dfrac{1}{9+6h+h^2}\cdot\dfrac{9}{9} - \dfrac{1}{9}\cdot\dfrac{9+6h+h^2}{9+6h+h^2}\right)\cdot\dfrac{1}{h}$

$= \lim\limits_{h\to 0} \left(\dfrac{9-9-6h-h^2}{9(9+6h+h^2)}\right)\cdot\dfrac{1}{h} = \lim\limits_{h\to 0} \dfrac{-6h-h^2}{81+54h+9h^2}\cdot\dfrac{1}{h}$

$= \lim\limits_{h\to 0} \dfrac{h(-6-h)}{81+54h+9h^2}\cdot\dfrac{1}{h} = \lim\limits_{h\to 0} \dfrac{-6-h}{81+54h+9h^2}$

$= \dfrac{-6-0}{81+0+0} = \dfrac{-6}{81} = -\dfrac{2}{27}$.

59) $f'(x) = \lim\limits_{h\to 0} \dfrac{f(x+h)-f(x)}{h}$

$= \lim\limits_{h\to 0} \dfrac{3(x+h)-7-[3x-7]}{h} = \lim\limits_{h\to 0} \dfrac{3x+3h-7-3x+7}{h}$

$= \lim\limits_{h\to 0} \dfrac{3h}{h} = \lim\limits_{h\to 0} 3 = 3$.

61) $h'(x) = \lim\limits_{h\to 0} \dfrac{h(x+h)-h(x)}{h}$

$= \lim\limits_{h\to 0} \dfrac{3.4(x+h)^2+1.9(x+h)-[3.4x^2+1.9x]}{h}$

$= \lim\limits_{h\to 0} \dfrac{3.4x^2+6.8xh+3.4h^2+1.9x+1.9h-3.4x^2-1.9x}{h}$

$= \lim\limits_{h\to 0} \dfrac{6.8xh+3.4h^2+1.9h}{h} = \lim\limits_{h\to 0} \dfrac{h(6.8x+3.4h+1.9)}{h}$

$= \lim\limits_{h\to 0} (6.8x+3.4h+1.9) = 6.8x+1.9$.

63) $f'(x) = \dfrac{d}{dx}[6x] = 6$.

The slope of the tangent line at $x = 4$ is equal to $f'(4) = 6$.

65) $h'(x) = \dfrac{d}{dx}\left[\dfrac{1}{x}\right] = \dfrac{d}{dx}[x^{-1}] = -x^{-2} = \dfrac{-1}{x^2}$.

The slope of the tangent line at $x = 7$ is equal to $h'(7) = \dfrac{-1}{(7)^2} = \dfrac{-1}{49}$.

67a) $A(6) = 3000(1.085)^6$
$\approx 3000(1.631467509) \approx 4894.40$.
After six years, there is about \$4,894.40 in the account.

67b) $A'(6) = \lim\limits_{h \to 0} \frac{3000(1.085)^{6+h} - 3000(1.085)^6}{h}$

$= \lim\limits_{h \to 0} \frac{3000(1.085)^6 [(1.085)^h - 1]}{h}$

$= 3000(1.085)^6 \cdot \lim\limits_{h \to 0} \frac{[(1.085)^h - 1]}{h}$

$\approx 4894.402527 \cdot \lim\limits_{h \to 0} \frac{[(1.085)^h - 1]}{h}$.

We will create a table to numerically evaluate $\lim\limits_{h \to 0} \frac{[(1.085)^h - 1]}{h}$:

h	-1	-0.1	-0.01	-0.001	-0.0001
$\frac{[(1.085)^h - 1]}{h}$	0.08	0.081	0.0815	0.08158	0.081580

h	1	0.1	0.01	0.001	0.0001
$\frac{[(1.085)^h - 1]}{h}$	0.09	0.082	0.0816	0.08158	0.081580

Thus, $\lim\limits_{h \to 0} \frac{[(1.085)^h - 1]}{h} \approx 0.08158$. Hence,

$A'(6) \approx 4894.402527 \cdot \lim\limits_{h \to 0} \frac{[(1.085)^h - 1]}{h}$.

$\approx 4894.402527 \cdot 0.08158 \approx 399.29$. After six years, the money in the account will be increasing at a rate of $399.29 per year.

69a)

h	-1	-0.1	-0.01	-0.001	-0.0001
$\frac{f(t+h) - f(t)}{h}$	5.4	5.03	5.003	5.0003	5.00003

h	1	0.1	0.01	0.001	0.0001
$\frac{f(t+h) - f(t)}{h}$	4.7	4.97	4.997	4.9997	4.99997

At $t = 4$, $\lim\limits_{h \to 0} \frac{f(t+h) - f(t)}{h} = 5$.

69b) The slope of the tangent line at $t = 4$ is equal to $f'(4) = 5$ (from part a).
Since $f(4) = 20 + \sqrt{400(4)} = 20 + 40 = 60$, we can find the equation of the tangent line by plugging $m = 5$ (from part b) & (4, 60) into:
$y - y_1 = m(t - t_1)$
$y - (60) = 5(t - (4))$
$y - 60 = 5t - 20$
$y = 5t + 40$.

69c) After 4 years, the fish population was growing at a rate of 5 fish per year.

71) $g'(x) = \frac{d}{dx}[3x^{1/5}] = 3 \cdot \frac{1}{5} x^{-4/5} = \frac{3}{5\sqrt[5]{x^4}}$.

73) $f'(x) = \frac{d}{dx}[-\frac{4}{5} x^{7/6}] = -\frac{4}{5} \cdot \frac{7}{6} x^{1/6}$
$= -\frac{14}{15\sqrt[6]{x}}$.

75) $h'(x) = \frac{d}{dx}[-9x^3 + x + 12] = -9 \cdot 3x^2 + 1 + 0$
$= -27x^2 + 1$.

77) $g'(x) = \frac{d}{dx}[6.23x^2 + 1.98x - 3.34]$
$= 6.23 \cdot 2x + 1.98 - 0 = 12.46x + 1.98$.

79) $f'(x) = \frac{d}{dx}[2\sqrt{x} + 7x^2 - \frac{1}{2}x^3]$
$= \frac{d}{dx}[2x^{1/2} + 7x^2 - \frac{1}{2}x^3]$
$= 2 \cdot \frac{1}{2} x^{-1/2} + 7 \cdot 2x - \frac{1}{2} \cdot 3x^2$
$= -1.5x^2 + 14x + \frac{1}{\sqrt{x}}$.

81) $h'(x) = \frac{d}{dx}[2.08x^{3.79}] = 2.08 \cdot 3.79 x^{2.79}$
$= 7.8832x^{2.79}$.

83a) Since there are no values except for zero that make the denominator zero and the numerator and denominator are polynomials, then f is defined for all real numbers except 0 or $(-\infty, 0) \cup (0, \infty)$.

83b) $f(x) = \frac{6x^4 + 25x^3 - 9x + 11}{x^2}$

$= \frac{6x^4}{x^2} + \frac{25x^3}{x^2} - \frac{9x}{x^2} + \frac{11}{x^2}$

$= 6x^2 + 25x - 9x^{-1} + 11x^{-2}$.

83c) $f'(x) = \frac{d}{dx}[6x^2 + 25x - 9x^{-1} + 11x^{-2}]$

$= \frac{d}{dx}[6x^2] + \frac{d}{dx}[25x] - \frac{d}{dx}[9x^{-1}]$
$\quad + \frac{d}{dx}[11x^{-2}]$

$= 12x + 25 + 9x^{-2} - 22x^{-3}$

$= 12x + 25 + \frac{9}{x^2} - \frac{22}{x^3}$.

83d) Since there are no values except for zero that make the denominator zero and the numerator and denominator are polynomials, then f' is defined for all real numbers except 0 or $(-\infty, 0) \cup (0, \infty)$.

85a) $f'(x) = \frac{d}{dx}[x^{3/4}] = \frac{3}{4}x^{-1/4} = \frac{3}{4\sqrt[4]{x}}$.

85b) The slope of the tangent line at $x = 16$ is equal to $f'(16) = \frac{3}{4\sqrt[4]{16}} = \frac{3}{4(2)} = \frac{3}{8}$.

85c) Since $f(16) = (16)^{3/4} = 8$, we can find the equation of the tangent line by plugging $m = \frac{3}{8}$ (from part b) & (16, 8) into:

$y - y_1 = m(x - x_1)$
$y - (8) = \frac{3}{8}(x - (16))$
$y - 8 = \frac{3}{8}x - 6$
$y = \frac{3}{8}x + 2$.

85d) The graph is displayed in viewing window [0, 32] by [0, 16]

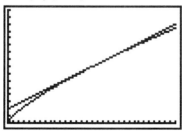

87) $h'(t) = \frac{d}{dt}[200t^2] = 400t$.

$h'(4) = 400(4) = 1600$. After four seconds, the rocket's height is increasing at a rate of 1600 feet per second.

89a) We need to set-up a function that doubles every 1.5 years. Let t be the time in years after 1985 and let k be a constant, then we need a function 2^{kt} such that:
For $t = 0$, $2^{k(0)} = 1$,
For $t = 1.5$, $2^{k(1.5)} = 2$, hence $1.5k = 1$,
For $t = 3$, $2^{k(3)} = 2^2 = 4$, hence $3k = 2$, Etc.
In each case, $k = 2/3$ will work. Now, multiply our function by 1.172 to get the desired function:

$S(t) = 1.172(2)^{\frac{2t}{3}}$

89b) $S'(13) = \lim_{h \to 0} \frac{1.172(2)^{\frac{2(13+h)}{3}} - 1.172(2)^{\frac{2(13)}{3}}}{h}$

$= \lim_{h \to 0} \frac{1.172(2)^{\frac{26}{3}}[(2)^{\frac{2h}{3}} - 1]}{h}$

$= 1.172(2)^{\frac{26}{3}} \cdot \lim_{h \to 0} \frac{[(2)^{\frac{2h}{3}} - 1]}{h}$

$\approx 476.2711124 \cdot \lim_{h \to 0} \frac{[(2)^{\frac{2h}{3}} - 1]}{h}$.

We will create a table to numerically evaluate $\lim_{h \to 0} \frac{[(2)^{\frac{2h}{3}} - 1]}{h}$:

h	-1	-0.1	-0.01	-0.001	-0.0001
$\frac{[(2)^{\frac{2h}{3}} - 1]}{h}$	0.37	0.452	0.4610	0.46199	0.462087

h	1	0.1	0.01	0.001	0.0001
$\frac{[(2)^{\frac{2h}{3}} - 1]}{h}$	0.59	0.473	0.4632	0.46220	0.462109

89b) Continued

Thus, $\lim_{h \to 0} \frac{[(2)^{\frac{2h}{3}} - 1]}{h} \approx 0.4621$. Hence,

$S'(13) \approx 476.2711124 \cdot \lim_{h \to 0} \frac{[(2)^{\frac{2h}{3}} - 1]}{h}$

$\approx 476.2711124 \cdot 0.4621 \approx 220.084$. In 1998, the clock speed was increasing at a rate of about 220.084 MHz per year.

91) $g'(x) = \frac{d}{dx}(-3x^2) \cdot (6x^3 - 2x^2 + 9x + 10)$
$\quad + (-3x^2) \cdot \frac{d}{dx}(6x^3 - 2x^2 + 9x + 10)$
$= -6x(6x^3 - 2x^2 + 9x + 10)$
$\quad + (-3x^2)(18x^2 - 4x + 9)$
$= -36x^4 + 12x^3 - 54x^2 - 60x - 54x^4$
$\quad + 12x^3 - 27x^2$
$= -90x^4 + 24x^3 - 81x^2 - 60x$.

93) $f'(x) = \frac{d}{dx}(3x^{\frac{1}{2}} + 5x) \cdot (-2x^{2/5} + x - 9)$
$\quad + (3x^{\frac{1}{2}} + 5x) \cdot \frac{d}{dx}(-2x^{2/5} + x - 9)$
$= (1.5x^{-\frac{1}{2}} + 5)(-2x^{2/5} + x - 9)$
$\quad + (3x^{1/2} + 5x)(-0.8x^{-3/5} + 1)$
$= -3x^{-1/10} + 1.5x^{\frac{1}{2}} - 13.5x^{-\frac{1}{2}} - 10x^{2/5} + 5x$
$\quad - 45 - 2.4x^{-1/10} + 3x^{\frac{1}{2}} - 4x^{2/5} + 5x$
$= 10x + 4.5x^{\frac{1}{2}} - 14x^{2/5} - 45 - 5.4x^{-1/10}$
$\quad - 13.5x^{-\frac{1}{2}}$
$= 10x + \frac{9\sqrt{x}}{2} - 14\sqrt[5]{x^2} - 45 - \frac{27}{5\sqrt[10]{x}} - \frac{27}{2\sqrt{x}}$.

95) $h'(x) = \frac{\frac{d}{dx}(3x+2) \cdot (x-1) - (3x+2) \cdot \frac{d}{dx}(x-1)}{(x-1)^2}$
$= \frac{3(x-1) - (3x+2)(1)}{(x-1)^2} = \frac{3x - 3 - 3x - 2}{(x-1)^2} = \frac{-5}{(x-1)^2}$.

97) $h'(x) = \frac{\frac{d}{dx}(3x^{\frac{1}{2}} - 2) \cdot (3x+1) - (3x^{\frac{1}{2}} - 2) \cdot \frac{d}{dx}(3x+1)}{(3x+1)^2}$
$= \frac{1.5x^{-\frac{1}{2}}(3x+1) - (3x^{\frac{1}{2}} - 2)3}{(3x+1)^2} = \frac{4.5x^{\frac{1}{2}} + 1.5x^{-\frac{1}{2}} - 9x^{\frac{1}{2}} + 6}{(3x+1)^2}$
$= \frac{-4.5x^{\frac{1}{2}} + 1.5x^{-\frac{1}{2}} + 6}{(3x+1)^2} = \frac{-4.5x^{\frac{1}{2}} + 6 + 1.5x^{-\frac{1}{2}}}{(3x+1)^2} \cdot \frac{2\sqrt{x}}{2\sqrt{x}}$
$= \frac{-9x + 12\sqrt{x} + 3}{2\sqrt{x}(3x+1)^2}$.

99a) $f'(x) = \frac{d}{dx}(-3x^2) \cdot (3x + 5)$
$\quad + (-3x^2) \cdot \frac{d}{dx}(3x + 5)$
$= -6x(3x + 5) + (-3x^2)(3)$
$= -18x^2 - 30x - 9x^2 = -27x^2 - 30x$.

99b) The slope of the tangent line at $x = 2$ is equal to $f'(2) = -27(2)^2 - 30(2)$
$= -108 - 60 = -168$. Since $f(2)$
$= -3(2)^2(3(2) + 5) = -12(11) = -132$,
we can find the equation of the tangent line by plugging $m = -168$ & $(2, -132)$ into:

$y - y_1 = m(x - x_1)$
$y - (-132) = -168(x - (2))$
$y + 132 = -168x + 336$
$y = -168x + 204$.

99c) The graph is displayed in viewing window $[0, 4]$ by $[-800, 0]$

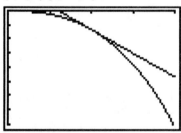

99d) The graph is displayed in viewing window $[0, 4]$ by $[-800, 0]$

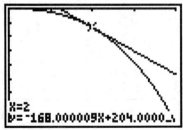

We can see that this matches our work very well.

101) The function is not differentiable at $x = 1$ since it is not continuous at $x = 1$.

103a) The graph is displayed in viewing window [− 10, 10] by [− 5, 5]

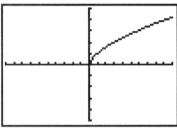

Note that the domain of f is $[0, \infty)$.

103b) $f'(x) = \frac{d}{dx}(x^{5/8}) = \frac{5}{8}x^{-3/8} = \frac{5}{8\sqrt[8]{x^3}}$.

103c) Since f' is only defined for $x > 0$, then it is undefined at $x = 0$. It is also undefined for $x < 0$ since the original function was not defined at those values.

103d) At $x = 0$, the function has a vertical tangent.

105) $f'(x) = \frac{d}{dx}(x^{3/4}) \bullet (3x - 1) \bullet (2x + 9)$
$\qquad + (x^{3/4}) \bullet \frac{d}{dx}(3x - 1) \bullet (2x + 9)$
$\qquad + (x^{3/4}) \bullet (3x - 1) \bullet \frac{d}{dx}(2x + 9)$
$= 0.75x^{-1/4}(3x - 1)(2x + 9) + (x^{3/4})(3)(2x + 9)$
$\qquad + (x^{3/4})(3x - 1)(2)$
$= 4.5x^{7/4} + 18.75x^{3/4} - 6.75x^{-1/4} + 6x^{7/4}$
$\qquad + 27x^{3/4} + 6x^{7/4} - 2x^{3/4}$
$= 16.5x^{7/4} + 43.75x^{3/4} - 6.75x^{-1/4}$
$= 16.5x\sqrt[4]{x^3} + 43.75\sqrt[4]{x^3} - \frac{27}{4\sqrt[4]{x}}$.

104 Chapter 3 Applications of the Derivative

CHAPTER 3 APPLICATIONS OF THE DERIVATIVE

Section 3.1 The Differential and Linear Approximations

1) $y' = \frac{d}{dx}(6x) = 6$. So, $dy = y' \, dx = 6 \, dx$.

3) $f'(x) = \frac{d}{dx}(-3x^2 + 2x) = -6x + 2$. Thus,
$dy = f'(x) \, dx = (-6x + 2) \, dx$.

5) $y' = \dfrac{\frac{d}{dx}(5) \cdot (x-1) - (5) \cdot \frac{d}{dx}(x-1)}{(x-1)^2} = \dfrac{(0)(x-1) - (5)(1)}{(x-1)^2}$
$= \dfrac{-5}{(x-1)^2}$. Hence, $dy = y' \, dx = \dfrac{-5}{(x-1)^2} \, dx$.

7) $f'(x) = \dfrac{\frac{d}{dx}(x) \cdot (x+1) - (x) \cdot \frac{d}{dx}(x+1)}{(x+1)^2}$
$= \dfrac{1(x+1) - (x)1}{(x+1)^2} = \dfrac{x+1-x}{(x+1)^2} = \dfrac{1}{(x+1)^2}$. Hence,
$dy = f'(x) \, dx = \dfrac{1}{(x+1)^2} \, dx$.

9) $y' = \frac{d}{dx}(x^{1/2} + 2x^{-1}) = \frac{1}{2}x^{-1/2} - 2x^{-2}$
$= \dfrac{1}{2\sqrt{x}} - \dfrac{2}{x^2}$.
Therefore, $dy = y' \, dx = \left(\dfrac{1}{2\sqrt{x}} - \dfrac{2}{x^2}\right) dx$.

11) $y' = \frac{d}{dx}(x^{-1/2} + x^{2/3}) = -\frac{1}{2}x^{-3/2} + \frac{2}{3}x^{-1/3}$
$= \dfrac{-1}{2x\sqrt{x}} + \dfrac{2}{3\sqrt[3]{x}}$.
Therefore, $dy = y' \, dx = \left(\dfrac{-1}{2x\sqrt{x}} + \dfrac{2}{3\sqrt[3]{x}}\right) dx$.

13) $f'(x) = \dfrac{\frac{d}{dx}(x^2+1) \cdot (x^2-1) - (x^2+1) \cdot \frac{d}{dx}(x^2-1)}{(x^2-1)^2}$
$= \dfrac{(2x)(x^2-1) - (x^2+1)(2x)}{(x^2-1)^2} = \dfrac{2x^3 - 2x - 2x^3 - 2x}{(x^2-1)^2}$
$= \dfrac{-4x}{(x^2-1)^2}$. Thus, $dy = f'(x) \, dx = \dfrac{-4x}{(x^2-1)^2} \, dx$.

15) $y' = \frac{d}{dx}(3x^{1.7} + 7x^{0.8} + 3) = 5.1x^{0.7} + 5.6x^{-0.2}$
$= 5.1\, x^{\frac{7}{10}} + 5.6\, x^{-\frac{1}{5}} = 5.1\sqrt[10]{x^7} + \dfrac{28}{5\sqrt[5]{x}}$. So,
$dy = y' \, dx = \left(5.1x^{0.7} + \dfrac{28}{5x^{0.2}}\right) dx$.

17) Let $f(x) = y = x^2 - 2x - 1$. Then,
$\Delta y = f(x + \Delta x) - f(x) = f(2.1) - f(2)$
$= (2.1)^2 - 2(2.1) - 1 - [(2)^2 - 2(2) - 1]$
$= -0.79 - [-1] = 0.21$.

Also, since $f'(x) = \frac{d}{dx}[x^2 - 2x - 1] = 2x - 2$,
then $dy = f'(x) \, dx = (2x - 2) \, dx$. For $x = 2$
and $dx = 0.1$, we get:
$dy = (2(2) - 2)(0.1) = (2)(0.1) = 0.2$.

19) Let $f(x) = y = 750 + 5x - 2x^2$. Then,
$\Delta y = f(x + \Delta x) - f(x) = f(52) - f(50)$
$= 750 + 5(52) - 2(52)^2$
$\qquad - [750 + 5(50) - 2(50)^2]$
$= -4398 - [-4000] = -398$.

Also, since $f'(x) = \frac{d}{dx}[750 + 5x - 2x^2]$
$= 5 - 4x$, then $dy = f'(x) \, dx = (5 - 4x) \, dx$.
For $x = 50$ and $dx = 2$, we get:
$dy = (5 - 4(50))(2) = (-195)(2) = -390$.

21) Let $f(x) = y = 100 - \dfrac{270}{x}$. Then,
$\Delta y = f(x + \Delta x) - f(x) = f(9.5) - f(9)$
$= 100 - \dfrac{270}{9.5} - [100 - \dfrac{270}{9}]$
$\approx 71.57894737 - [70] \approx 1.58$.

Also, since $f'(x) = \frac{d}{dx}[100 - 270x^{-1}]$
$= 270x^{-2} = \dfrac{270}{x^2}$, then $dy = f'(x) \, dx = \dfrac{270}{x^2} \, dx$.
For $x = 9$ and $dx = 0.5$, we get:
$dy = \dfrac{270}{(9)^2}(0.5) = (\dfrac{10}{3})(0.5) = \dfrac{5}{3} \approx 1.67$.

23) Let $f(x) = y = \sqrt{x}$. Then,
$\Delta y = f(x + \Delta x) - f(x) = f(2.1) - f(2)$
$= \sqrt{2.1} - \sqrt{2}$
$\approx 1.449137675 - 1.414213562 \approx 0.035$.

Also, since $f'(x) = \frac{d}{dx}[x^{1/2}] = 0.5x^{-1/2} = \frac{1}{2\sqrt{x}}$,
then $dy = f'(x)\, dx = \frac{1}{2\sqrt{x}}\, dx$. For $x = 2$ and $dx = 0.1$, we get:
$dy = \frac{1}{2\sqrt{2}}(0.1) \approx (0.3535533906)(0.1)$
≈ 0.035.

25) $\Delta y = f(x + \Delta x) - f(x) = f(2.1) - f(2)$
$= \frac{(2.1)^2 + 1}{(2.1)^2 - 1} - \frac{(2)^2 + 1}{(2)^2 - 1} = \frac{5.41}{3.41} - \frac{5}{3}$
$\approx 1.586510264 - 1.666666667 \approx -0.0802$.

Also, since $f'(x) = \frac{d}{dx}\left[\frac{x^2+1}{x^2-1}\right]$
$= \frac{\frac{d}{dx}(x^2+1)\bullet(x^2-1) - (x^2+1)\bullet\frac{d}{dx}(x^2-1)}{(x^2-1)^2}$
$= \frac{(2x)(x^2-1) - (x^2+1)(2x)}{(x^2-1)^2} = \frac{2x^3 - 2x - 2x^3 - 2x}{(x^2-1)^2}$
$= \frac{-4x}{(x^2-1)^2}$, then $dy = f'(x)\, dx = \frac{-4x}{(x^2-1)^2}\, dx$.
For $x = 2$ and $dx = 0.1$, we get:
$dy = \frac{-4(2)}{((2)^2-1)^2}(0.1) = \frac{-8}{(3)^2}(0.1) = \frac{-8}{9}(0.1)$
$\approx (-0.888888889)(0.1) \approx -0.0889$.

27) Let $f(x) = y = 2x^2(3x^2 - 2x)$. Then,
$\Delta y = f(x + \Delta x) - f(x) = f(1.1) - f(1)$
$= 2(1.1)^2(3(1.1)^2 - 2(1.1))$
$\qquad - [2(1)^2(3(1)^2 - 2(1))]$
$= 2.42(1.43) - [2(1)] = 3.4606 - 2 = 1.4606$.

27) Continued
Also, since $f'(x) = \frac{d}{dx}[2x^2(3x^2 - 2x)]$
$= \frac{d}{dx}[2x^2]\bullet(3x^2 - 2x) + 2x^2\bullet\frac{d}{dx}[(3x^2 - 2x)]$
$= 4x(3x^2 - 2x) + 2x^2(6x - 2)$
$= 12x^3 - 8x^2 + 12x^3 - 4x^2 = 24x^3 - 12x^2$,
then $dy = f'(x)\, dx = (24x^3 - 12x^2)\, dx$.
For $x = 1$ and $dx = 0.1$, we get:
$dy = (24(1)^3 - 12(1)^2)(0.1) = (12)(0.1)$
$= 1.2$.

29) Since $\sqrt{26} = \sqrt{25 + 1}$, we can approximate the answer using $f(x + dx) \approx f(x) + f'(x)\, dx$.
Let $f(x) = \sqrt{x}$, $x = 25$ and $dx = 1$. Then,
$f'(x) = \frac{d}{dx}[\sqrt{x}] = \frac{d}{dx}[x^{1/2}] = \frac{1}{2}x^{-1/2} = \frac{1}{2\sqrt{x}}$.
So now, $f(x + dx) \approx f(x) + f'(x)\, dx$ can be rewritten as $f(25 + 1) \approx f(25) + f'(25)(1)$
$= \sqrt{25} + \frac{1}{2\sqrt{25}}(1) = 5 + \frac{1}{10} = 5.1$.
Thus, by the linear approximation,
$\sqrt{26} \approx 5.1$. On a calculator,
$\sqrt{26} \approx 5.0990$, so our answer is off only by about one one-thousandth.

31) Since $\sqrt[3]{26} = \sqrt[3]{27 - 1}$, we can approximate the answer using $f(x + dx) \approx f(x) + f'(x)\, dx$.
Let $f(x) = \sqrt[3]{x}$, $x = 27$ and $dx = -1$. Then,
$f'(x) = \frac{d}{dx}[\sqrt[3]{x}] = \frac{d}{dx}[x^{1/3}] = \frac{1}{3}x^{-2/3}$
$= \frac{1}{3\sqrt[3]{x^2}}$. So, $f(x + dx) \approx f(x) + f'(x)\, dx$
becomes $f(27 + (-1)) \approx f(27) + f'(27)(-1)$
$= \sqrt[3]{27} + \frac{1}{3\sqrt[3]{(27)^2}}(-1) = 3 - \frac{1}{27} \approx 2.9630$.
Thus, by the linear approximation,
$\sqrt[3]{26} \approx 2.9630$. On a calculator,
$\sqrt[3]{26} \approx 2.9625$, so our answer is off only by about five ten-thousandths.

33) Since $\sqrt[4]{15.8} = \sqrt[4]{16-0.2}$, we can estimate the answer using $f(x+dx) \approx f(x) + f'(x)\,dx$. Let $f(x) = \sqrt[4]{x}$, $x = 16$ and $dx = -0.2$.
Then, $f'(x) = \frac{d}{dx}[\sqrt[4]{x}] = \frac{d}{dx}[x^{1/4}] = \frac{1}{4}x^{-3/4}$
$= \frac{1}{4\sqrt[4]{x^3}}$. So, $f(x+dx) \approx f(x) + f'(x)\,dx$
becomes $f(16 + (-0.2))$
$\approx f(16) + f'(16)(-0.2)$
$= \sqrt[4]{16} + \frac{1}{4\sqrt[4]{(16)^3}}(-0.2) = 2 - \frac{1}{160}$
≈ 1.9938.
Thus, by the linear approximation, $\sqrt[4]{15.8} \approx 1.9938$. On a calculator, $\sqrt[4]{15.8} \approx 1.9937$, so our answer is off only by about one ten-thousandth.

35) Since $\frac{2}{\sqrt{50}} = \frac{2}{\sqrt{49+1}}$, we can estimate the answer using $f(x+dx) \approx f(x) + f'(x)\,dx$. Let $f(x) = \frac{2}{\sqrt{x}}$, $x = 49$ and $dx = 1$. Then,
$f'(x) = \frac{d}{dx}[\frac{2}{\sqrt{x}}] = \frac{d}{dx}[2x^{-1/2}] = -x^{-3/2}$
$= \frac{-1}{x\sqrt{x}}$. So, $f(x+dx) \approx f(x) + f'(x)\,dx$
becomes $f(49 + 1) \approx f(49) + f'(49)(1)$
$= \frac{2}{\sqrt{49}} + \frac{-1}{49\sqrt{49}}(1) = \frac{2}{7} - \frac{1}{343} = \frac{97}{343}$
≈ 0.28280.
Thus, by the linear approximation, $\frac{2}{\sqrt{50}} \approx 0.28280$. On a calculator, $\frac{2}{\sqrt{50}} \approx 0.28284$, so our answer is off only by about four hundred-thousandths.

37) $A'(r) = \frac{d}{dr}(\pi r^2) = 2\pi r$. So, $dA = A'(r)\,dr$
$= 2\pi r\,dr$. For $r = 5$ and $dr = 0.2$, dA
$= A'(5)(0.2) = 2\pi(5)(0.2) = 2\pi \approx 6.28$.
We estimate that the area of the circle increased by about 6.28 square inches.

39) $y' = \frac{d}{dx}(1000 + 110x^{1/2}) = 55x^{-1/2} = \frac{55}{\sqrt{x}}$. So, $dy = y'\,dx = \frac{55}{\sqrt{x}}\,dx$. For $x = 250$ and $dx = 254 - 250 = 4$, $dy = \frac{55}{\sqrt{250}}(4) = \frac{220}{\sqrt{250}}$
≈ 13.91. As the number of employees increases from 250 to 254, the annual cost of covering the employees eye and dental insurance increases by about $13.91.

41) $y' = \frac{d}{dx}(0.22x^3 - 2.35x^2 + 14.32x + 10.22) = 0.66x^2 - 4.7x + 14.32$. So, $dy = y'\,dx$
$= (0.66x^2 - 4.7x + 14.32)\,dx$. For $x = 30$ and $dx = 33 - 30 = 3$,
$dy = (0.66(30)^2 - 4.7(30) + 14.32)(3)$
$\approx (594 - 141 + 14.32)(3) = 1401.96$. As the number of vehicles produces increases from 30 to 33 per week, the production cost increases by about $1,401,960.

43) $y' = \frac{d}{dx}(120x - 2.4x^2) = 120 - 4.8x$.
So, $dy = y'\,dx = (120 - 4.8x)\,dx$.
For $x = 10$ and $dx = 10 - 11 = 1$,
$dy = (120 - 4.8(10))(1) = (72)(1) = 72$. As the amount of advertising increases from $10,000 to $11,000, the total number of volleyballs sold increases by about 7,200 balls.

45) For $x = 11$, $y = 120(11) - 2.4(11)^2$
$= 1320 - 290.4 = 1029.6$.
For $x = 10$, $y = 120(10) - 2.4(10)^2$
$= 1200 - 240 = 960$.
Therefore $\Delta y = 1029.6 - 960 = 69.6$
As the amount of advertising increases from $10,000 to $11,000, the total number of volleyballs sold increases by 6,960 balls. The value of dy found in #43 is off by 240 balls.

47a) Since $f'(x) = \frac{d}{dx}[0.08x^2 - 0.11x + 65.1]$
$= 0.16x - 0.11$, then $f'(2)$
$= 0.16(2) - 0.11 = 0.21$. The slope of the tangent line at $x = 2$ is equal to $f'(2) = 0.21$.
Since $f(2) = 0.08(2)^2 - 0.11(2) + 65.1 = 65.2$, we can find the equation of the tangent line by plugging $m = 0.21$ & $(2, 65.2)$ into:
$y - y_1 = m(x - x_1)$
$y - (65.2) = 0.21(x - (2))$
$y - 65.2 = 0.21x - 0.42$
$y = 0.21x + 64.78$.

47b) Plugging in $x = 3$ into $y = 0.21x + 64.78$, we get:
$y = 0.21(3) + 64.78 = 65.41$. This means that if the number of people living alone in Florida continued at the 1992 rate, then 65.41% of the people would be living alone in 1993.

108 Chapter 3 Applications of the Derivative

Section 3.2 Marginal Analysis

1a) $MC(x) = \frac{d}{dx}[23x + 5200] = 23$.

1b) $MC(10) = 23$. The estimated cost of producing the 11th unit is $23.

1c) $C(11) = 23(11) + 5200 = 5453$.
$C(10) = 23(10) + 5200 = 5430$.
Hence, $C(11) - C(10) = 5453 - 5430 = 23$. The actual cost of producing the 11th unit is $23. This matches exactly our estimate in part b.

3a) $MC(x) = \frac{d}{dx}[\frac{1}{2}x^2 + 12.7x + 2100]$
$= x + 12.7$.

3b) $MC(11) = 11 + 12.7 = 23.7$. The estimated cost of producing the 12th unit is $23.70.

3c) $C(12) = \frac{1}{2}(12)^2 + 12.7(12) + 2100$
$= 2324.4$.
$C(11) = \frac{1}{2}(11)^2 + 12.7(11) + 2100$
$= 2300.2$.
Hence, $C(12) - C(11) = 2324.4 - 2300.2 = 24.2$. The actual cost of producing the 12th unit is $24.20. Our estimate in part b is off by $0.50.

5a) $MC(x) = \frac{d}{dx}[0.2x^3 - 3x^2 + 50x + 20]$
$= 0.6x^2 - 6x + 50$.

5b) $MC(30) = 0.6(30)^2 - 6(30) + 50 = 410$. The estimated cost of producing the 31st unit is $410.

5c) $C(31) = 0.2(31)^3 - 3(31)^2 + 50(31) + 20$
$= 4645.2$.
$C(30) = 0.2(30)^3 - 3(30)^2 + 50(30) + 20$
$= 4220$.
Hence, $C(31) - C(30) = 4645.2 - 4220 = 425.2$. The actual cost of producing the 31st unit is $425.20. Our estimate in part b is off by $15.20.

7a) $R(x) = x \bullet p(x) = x(6) = 6x$.
$P(x) = R(x) - C(x) = 6x - (5x + 500)$
$= x - 500$.

7b) $MC(x) = \frac{d}{dx}[5x + 500] = 5$.
$MP(x) = \frac{d}{dx}[x - 500] = 1$.

9a) $R(x) = x \bullet p(x) = x(\frac{-x}{20} + 15) = \frac{-x^2}{20} + 15x$.
$P(x) = R(x) - C(x) = \frac{-x^2}{20} + 15x$
$\quad - (\frac{x^2}{100} + 7x + 1000)$
$= \frac{-3x^2}{50} + 8x - 1000$.

9b) $MC(x) = \frac{d}{dx}[\frac{x^2}{100} + 7x + 1000] = \frac{x}{50} + 7$.
$MP(x) = \frac{d}{dx}[\frac{-3x^2}{50} + 8x - 1000] = \frac{-3x}{25} + 8$.

11a) $R(x) = x \bullet p(x) = x(-0.005x + 7)$
$= -0.005x^2 + 7x$.
$P(x) = R(x) - C(x)$
$= -0.005x^2 + 7x - (-0.001x^3 + 4x + 100)$
$= 0.001x^3 - 0.005x^2 + 3x - 100$.

11b) $MC(x) = \frac{d}{dx}[-0.001x^3 + 4x + 100]$
$= -0.003x^2 + 4$.
$MP(x) = \frac{d}{dx}[0.001x^3 - 0.005x^2 + 3x - 100]$
$= 0.003x^2 - 0.01x + 3$.

13) $MC(x) = \frac{d}{dx}[100 + 40x - 0.001x^2]$
$= 40 - 0.002x$.
Thus, $MC(200) = 40 - 0.002(200)$
$= 40 - 0.4 = 39.6$. The estimated daily cost of producing the 201st unit is $39.60.

15a) $AC(x) = \frac{C(x)}{x} = \frac{100 + 40x - 0.001x^2}{x}$

$= \frac{100}{x} + 40 - 0.001x.$

$AC(200) = \frac{100}{200} + 40 - 0.001(200)$

$= 0.5 + 40 - 0.2 = 40.3.$

When 200 tires are produced per day, the average cost per day to produce each tire is $40.30.

15b) $MAC(x) = \frac{d}{dx}[100x^{-1} + 40 - 0.001x]$

$= -100x^{-2} - 0.001 = -\frac{100}{x^2} - 0.001.$

$MAC(200) = -\frac{100}{(200)^2} - 0.001$

$= -0.0025 - 0.001 = -0.0035.$

When 200 tires are produced each day, the average cost per day to produce each tire is decreasing at a rate of 0.35¢ per tire.

17) $MP(x) = \frac{d}{dx}[5x + x^{1/2}] = 5 + \frac{1}{2}x^{-1/2}$

$= 5 + \frac{1}{2\sqrt{x}}.$

$MP(55) = 5 + \frac{1}{2\sqrt{55}} \approx 5.06742.$

The estimated monthly profit of selling the 56th subscription is about $5.07.

19a) $AP(x) = \frac{P(x)}{x} = \frac{5x + x^{\frac{1}{2}}}{x} = 5 + x^{-1/2} = 5 + \frac{1}{\sqrt{x}}.$

$AP(55) = 5 + \frac{1}{\sqrt{55}} \approx 5.13484.$

When 55 subscriptions are sold per month, the average profit per month for each subscription is about $5.13.

19b) $MAP(x) = \frac{d}{dx}[5 + x^{-1/2}] = -\frac{1}{2}x^{-3/2}$

$= -\frac{1}{2x\sqrt{x}}.$

$MAP(55) = -\frac{1}{2(55)\sqrt{55}} \approx -0.00123$

When 55 subscriptions are sold per month, the average profit per month for each subscription is decreasing at a rate of about 0.12¢ per subscription.

21a) $C(q)$ = Variable costs•q + fixed costs
$= 12q + 1200.$
$MC(q) = \frac{d}{dq}[12q + 1200] = 12.$

21b) $MC(100) = 12$ and $MC(150) = 12$. The estimated cost to produce 101st pager is $12. The estimated cost to produce 151st pager is also $12.

21c) Since the cost function is a linear equation, the marginal cost is constant and equal to the slope of the line.

23a) Since $R(x) = x \cdot p(x) = x \cdot 23 = 23x$, then
$P(x) = R(x) - C(x)$
$= 23x - (\frac{x^2}{95} + 3.5x + 5500)$
$= -\frac{x^2}{95} + 19.5x - 5500.$

23b) $P(500) = -\frac{(500)^2}{95} + 19.5(500) - 5500$
$= -\frac{50000}{19} + 9750 - 5500 \approx 1618.42.$

When 500 doll sets are produced and sold, the profit is about $1,618.42.

23c) $MP(x) = \frac{d}{dx}[-\frac{x^2}{95} + 19.5x - 5500]$
$= -\frac{2x}{95} + 19.5.$ Thus,
$MP(500) = -\frac{2(500)}{95} + 19.5 \approx 8.97.$

The estimated profit of producing and selling the 501st doll set is $8.97.

25a) $R(x) = x \cdot p(x) = x(\frac{-x}{30} + 300)$
$= \frac{-x^2}{30} + 300x.$
$C(x)$ = Variable costs•x + fixed costs
$= 30x + 150,000.$

25b) $P(x) = R(x) - C(x)$
$= \frac{-x^2}{30} + 300x - (30x + 150,000)$
$= \frac{-x^2}{30} + 270x - 150,000.$

Since P is a quadratic equation with $a < 0$, we know that it can intersect the x-axis at zero, one, or two points. If it does intersect the x-axis at two points, then the profit will be positive between the two points since the parabola points down. Setting $P(x) = 0$ and using the quadratic formula yields:

$\frac{-x^2}{30} + 270x - 150,000 = 0$

$x = \frac{-b \pm \sqrt{b^2 - 4ac}}{2a}$

$= \frac{-(270) \pm \sqrt{(270)^2 - 4\left(\frac{-1}{30}\right)(-150000)}}{2\left(\frac{-1}{30}\right)}$

$= \frac{-270 \pm \sqrt{72900 - 20000}}{\left(\frac{-1}{15}\right)}$

$= (-270 \pm \sqrt{52900}) \div \frac{-1}{15}$

$= (-270 \pm 230) \bullet (-15) = 600 \text{ or } 7500.$
For there to be a profit, more than 600 but less than 7500 hand-held computer devices must be produced and sold.

25c) $P'(x) = \frac{d}{dx}\left[\frac{-x^2}{30} + 270x - 150,000\right]$
$= \frac{-x}{15} + 270.$ So, $P'(1000) = \frac{-1000}{15} + 270$
≈ 203.33
The estimated profit of producing and selling the 1001st hand-held computer device is about $203.33.

27) $C(2001) = C(2000 + 1)$
$\approx C(2000) + MC(2000).$
But $C(2000) = 57,070$ from the table and $MC(2000) = 30$ from the graph.
Thus, $C(2000) + MC(2000) = 57,070 + 30$
$= 57100.$ Using the approximation, we estimate that the cost to produce 2001 units is $57,100.

Section 3.3 Measuring Rates and Errors

1a) The maximal error, dx, is equal to the difference of the highest value (x_b) and lowest value (x_a), divided by two. Hence, $dx = \frac{x_b - x_a}{2} = \frac{14 - 10}{2} = \frac{4}{2} = 2$.

1b) The forecast value is equal to the average of the highest value (x_b) and lowest value (x_a). Hence, $x = \frac{x_b + x_a}{2} = \frac{14 + 10}{2} = \frac{24}{2} = 12$.
Thus, the relative error $\varepsilon_x = \pm \frac{dx}{x} = \pm \frac{2}{12}$
$= \pm \frac{1}{6} \approx \pm 0.1667$.

1c) The percentage error $= \varepsilon_x \bullet 100\%$
$= \pm 16.67\%$.

3b) The forecast value is equal to the average of the highest value (x_b) and lowest value (x_a). Hence, $x = \frac{x_b + x_a}{2} = \frac{4300 + 4000}{2} = \frac{8300}{2}$
$= 4150$.
Thus, the relative error $\varepsilon_x = \pm \frac{dx}{x} = \pm \frac{150}{4150}$
$= \pm \frac{3}{83} \approx \pm 0.0361$.

3c) The percentage error $= \varepsilon_x \bullet 100\%$
$= \pm 3.61\%$.

5a) The maximal error, dx, is equal to the difference of the highest value (x_b) and lowest value (x_a), divided by two. Hence, $dx = \frac{x_b - x_a}{2} = \frac{45000 - 42000}{2} = \frac{3000}{2} = 1500$.

5b) The forecast value is equal to the average of the highest value (x_b) and lowest value (x_a). Hence, $x = \frac{x_b + x_a}{2} = \frac{45000 + 42000}{2}$
$= \frac{87000}{2} = 43500$.
Thus, the relative error $\varepsilon_x = \pm \frac{dx}{x} = \pm \frac{1500}{43500}$
$= \pm \frac{1}{29} \approx \pm 0.0345$.

5c) The percentage error $= \varepsilon_x \bullet 100\%$
$= \pm 3.45\%$.

7) Since $f'(x) = \frac{d}{dx}[6x^3] = 18x^2$, then $f'(1)$
$= 18(1)^2 = 18$. Also, $f(1) = 6(1)^3 = 6$ and
$\varepsilon_x = \pm 0.10$. Thus, $\varepsilon_y = \frac{f'(x) \bullet x}{f(x)} \bullet \varepsilon_x$
$= \frac{(18)(1)}{(6)}(\pm 0.10) = 3(\pm 0.10) = \pm 0.30$.

9) Since $f'(x) = \frac{d}{dx}[\frac{5}{x-1}]$
$= \frac{\frac{d}{dx}[5] \bullet (x-1) - 5 \bullet \frac{d}{dx}[x-1]}{(x-1)^2} = \frac{(0)(x-1) - 5(1)}{(x-1)^2} = \frac{-5}{(x-1)^2}$,
then $f'(-3) = \frac{-5}{(-3-1)^2} = \frac{-5}{16} = -0.3125$. Also,
$f(-3) = \frac{5}{(-3)-1} = \frac{5}{-4} = -1.25$ and
$\varepsilon_x = \pm 0.15$. Thus, $\varepsilon_y = \frac{f'(x) \bullet x}{f(x)} \bullet \varepsilon_x$
$= \frac{(-0.3125)(-3)}{(-1.25)}(\pm 0.15) = -0.75(\pm 0.15)$
$= \pm 0.1125$.

11) Since $f'(x) = \frac{d}{dx}[x^2 - 5x - 2] = 2x - 5$,
then $f'(2) = 2(2) - 5 = -1$. Also, $f(2)$
$= (2)^2 - 5(2) - 2 = 4 - 10 - 2 = -8$ and
$\varepsilon_x = \pm 0.05$. Thus, $\varepsilon_y = \frac{f'(x) \bullet x}{f(x)} \bullet \varepsilon_x$
$= \frac{(-1)(2)}{(-8)}(\pm 0.05) = 0.25(\pm 0.05) = \pm 0.0125$.

13) Since $f'(x) = \frac{d}{dx}[3\sqrt{x}] = \frac{d}{dx}[3x^{\frac{1}{2}}]$
$= 1.5x^{-\frac{1}{2}} = \frac{3}{2\sqrt{x}}$, then $f'(2) = \frac{3}{2\sqrt{2}}$
$= \frac{3}{2\sqrt{2}} \bullet \frac{\sqrt{2}}{\sqrt{2}} = \frac{3\sqrt{2}}{2(2)} = 0.75\sqrt{2} \approx 1.06066$.
Also, $f(2) = 3\sqrt{2} \approx 4.24264$ and
$\varepsilon_x = \pm 0.07$. Thus, $\varepsilon_y = \frac{f'(x) \bullet x}{f(x)} \bullet \varepsilon_x$
$= \frac{(0.75\sqrt{2})(2)}{(3\sqrt{2})}(\pm 0.07) = 0.5(\pm 0.07)$
$= \pm 0.035$.

112   Chapter 3   Applications of the Derivative

15) Since $f'(x) = \frac{d}{dx}[\frac{3-2x}{x}] = \frac{d}{dx}[\frac{3}{x} - 2]$
$= \frac{d}{dx}[3x^{-1} - 2] = -3x^{-2} = \frac{-3}{x^2}$, then $f'(1)$
$= \frac{-3}{(1)^2} = -3$. Also, $f(1) = \frac{3-2(1)}{1} = 1$ and
$\varepsilon_x = \pm 0.19$. Thus, $\varepsilon_y = \frac{f'(x) \cdot x}{f(x)} \cdot \varepsilon_x$
$= \frac{(-3)(1)}{(1)}(\pm 0.19) = -3(\pm 0.19) = \pm 0.57$.

17) Since the margin of error is $\pm 3\%$, the relative error ± 0.03. With the relative error of ± 0.03, the range of the fraction of people voting for the first and second candidate is [0.495, 0.555] and [0.445, 0.505] respectively. Here, you cannot be sure who is actually ahead because of the relative error involved in the poll. That is why it is called a statistical dead heat.

19) Since the actual range is [18000, 30000], the maximal error, dx, in the forecast is 30000 minus 24000 = 6000. Thus, the percentage error is $\varepsilon_x \cdot 100\% = \pm \frac{dx}{x} \cdot 100\%$
$= \pm \frac{6000}{24000} \cdot 100\% = \pm 25\%$.

21) Since the actual range is [55, 65], the maximal error, dx, in the forecast is 65 minus 60 = 5. Thus, the percentage error is $\varepsilon_x \cdot 100\% = \pm \frac{dx}{x} \cdot 100\%$
$= \pm \frac{5}{60} \cdot 100\% \approx \pm 8.3\%$.

23a)   Since $R'(x) = \frac{d}{dx}[200x^{3/2}] = 300x^{1/2}$
$= 300\sqrt{x}$, then $R'(60) = 300\sqrt{60}$
$= 600\sqrt{15}$. Also, $R(60) = 200\sqrt{60^3}$
$= 24000\sqrt{15}$ and $dx = \pm 5$. Thus,
$\varepsilon_y = \frac{f'(x) \cdot dx}{f(x)} \cdot 100\% = \frac{(600\sqrt{15})(\pm 5)}{(24000\sqrt{15})} \cdot 100\%$
$= \pm 0.125 \cdot 100\% \approx \pm 12.5\%$.

23b) $R(5) = 200\sqrt{5^3} = 1000\sqrt{5} \approx 2236.07$.
The company will receive \$2,236.07 in revenue from the sale of 5,000 bolts.
Since $MR(x) = \frac{d}{dx}[200x^{3/2}] = 300x^{1/2}$
$= 300\sqrt{x}$, then $MR(5) = 300\sqrt{5} \approx$
670.82. At a production level of 5,000 bolts, the approximate revenue from selling the next 1,000 bolts is about \$670.82.

23c) $R(6) = R(5+1) \approx R(5) + MR(5)$
$= 2236.07 + 670.82 = 2906.89$. We estimate that the company will receive \$2,906.89 in revenue from the sale of 6,000 bolts

23d) Since $R(6) = = 200\sqrt{6^3} = 1200\sqrt{6} \approx$
2939.39, the maximal error, dR, in the forecast is 2906.89 (from part c) minus 2939.39 $= -32.50$. Thus, the relative error is $\varepsilon_R = \pm \frac{dR}{R} = \pm \frac{32.50}{2939.39} \approx \pm 0.0111$ or $\pm 1.1\%$.

25) Since $f'(x) = \frac{d}{dx}[2x^2] = 4x$, then $Rel(x)$
$= \frac{f'(x)}{f(x)} = \frac{4x}{2x^2} = \frac{2}{x}$. Thus, $Rel(5) = \frac{2}{5} = 0.4$.

27) Since $f'(x) = \frac{d}{dx}[\frac{x-100}{10}] = \frac{d}{dx}[0.1x - 10]$
$= 0.1$, then $Rel(x) = \frac{f'(x)}{f(x)} = \frac{0.1}{0.1x - 10} = \frac{1}{x - 100}$.
Thus, $Rel(125) = \frac{1}{125-100} = \frac{1}{25}$.

29) Since $f'(x) = \frac{d}{dx}[20\sqrt{x}] = \frac{d}{dx}[20x^{1/2}] =$
$10x^{-1/2} = \frac{10}{\sqrt{x}}$, then $Rel(x) = \frac{f'(x)}{f(x)}$
$= f'(x) \div f(x) = \frac{10}{\sqrt{x}} \div 20\sqrt{x}$
$= \frac{10}{\sqrt{x}} \cdot \frac{1}{20\sqrt{x}} = \frac{1}{2x}$. Thus, $Rel(8) = \frac{1}{2(8)}$
$= \frac{1}{16} = 0.0625$.

31) Since $f'(x) = \frac{d}{dx}[3x^2 + 5x] = 6x + 5$, then

$Rel(x) = \frac{f'(x)}{f(x)} = \frac{6x+5}{3x^2+5x}$. Thus, $Rel(2)$

$= \frac{6(2)+5}{3(2)^2+5(2)} = \frac{17}{22} \approx 0.7727$.

33) Since $f'(x) = \frac{d}{dx}[0.004x^2 + 0.057x + 1.280]$

$= 0.008x + 0.057$, then $Rel(x) = \frac{f'(x)}{f(x)}$

$= \frac{0.008x + 0.057}{0.004x^2 + 0.057x + 1.280}$. Thus, $Rel(25)$

$= \frac{0.008(25) + 0.057}{0.004(25)^2 + 0.057(25) + 1.280} = \frac{0.257}{5.205} \approx 0.0494$.

The poverty threshold was growing at a rate of about 4.9% during the year in 1985.

Chapter 3 Review Exercises

1) $y' = \frac{d}{dx}(4x+2) = 4$. So, $dy = y'\, dx = 4\, dx$.

3) $y' = \frac{\frac{d}{dx}(x+3)\cdot(x-5)-(x+3)\cdot\frac{d}{dx}(x-5)}{(x-5)^2}$

$= \frac{(1)(x-5)-(x+3)(1)}{(x-5)^2} = \frac{x-5-x-3}{(x-5)^2} = \frac{-8}{(x-5)^2}$.

Hence, $dy = y'\, dx = \frac{-8}{(x-5)^2}\, dx$.

5) $y' = \frac{d}{dx}[\sqrt{x} - \frac{3}{x^4}] = \frac{d}{dx}[x^{1/2} - 3x^{-4}]$

$= \frac{1}{2}x^{-1/2} + 12x^{-5} = \frac{1}{2\sqrt{x}} + \frac{12}{x^5}$.

Hence, $dy = y'\, dx = (\frac{1}{2\sqrt{x}} + \frac{12}{x^5})\, dx$.

7) $y' = \frac{d}{dx}[x^4 - 2x + \sqrt[3]{x^2}]$

$= \frac{d}{dx}[x^4 - 2x + x^{2/3}] = 4x^3 - 2 + \frac{2}{3}x^{-1/3}$.

$= 4x^3 - 2 + \frac{2}{3\sqrt[3]{x}}$. Hence, $dy = y'\, dx$

$= (4x^3 - 2 + \frac{2}{3\sqrt[3]{x}})\, dx$.

9) $y' = \frac{d}{dx}[4x^5 - 21x + 4] = 20x^4 - 21$.

Hence, $dy = y'\, dx = (20x^4 - 21)\, dx$.

11) $y' = \frac{d}{dx}[2x^{1.7} - 5x^{0.8} + 4] = 3.4x^{0.7} - 4x^{-0.2}$

Hence, $dy = y'\, dx = (3.4x^{0.7} - \frac{4}{x^{0.2}})\, dx$.

13) Let $f(x) = y = 4x^2 - x + 6$. Then,
$\Delta y = f(x + \Delta x) - f(x) = f(3.1) - f(3)$
$= 4(3.1)^2 - (3.1) + 6 - [4(3)^2 - (3) + 6]$
$= 41.34 - [39] = 2.34$.

Also, since $f'(x) = \frac{d}{dx}[4x^2 - x + 6] = 8x - 1$,
then $dy = f'(x)\, dx = (8x - 1)\, dx$. For $x = 3$
and $dx = 0.1$, we get:
$dy = (8(3) - 1)(0.1) = (23)(0.1) = 2.3$.

15) Let $f(x) = y = \sqrt[3]{x}$. Then,
$\Delta y = f(x + \Delta x) - f(x) = f(9.261) - f(8)$
$= \sqrt[3]{9.261} - [\sqrt[3]{8}]$
$= 2.1 - [2] = 0.1$.

Also, since $f'(x) = \frac{d}{dx}[\sqrt[3]{x}] = \frac{d}{dx}[x^{1/3}]$
$= \frac{1}{3}x^{-2/3} = \frac{1}{3\sqrt[3]{x^2}}$, then $dy = f'(x)\, dx$
$= (\frac{1}{3\sqrt[3]{x^2}})\, dx$. For $x = 8$ and $dx = 1.261$, we

get:
$dy = (\frac{1}{3\sqrt[3]{8^2}})(1.261) = (\frac{1}{12})(1.261)$

≈ 0.1051.

17) Let $f(x) = y = \frac{x^2+2}{x^2-2}$. Then,

$\Delta y = f(x + \Delta x) - f(x) = f(2.1) - f(2)$
$= \frac{(2.1)^2+2}{(2.1)^2-2} - \frac{(2)^2+2}{(2)^2-2} \approx 2.65975 - 3 \approx -0.34$

Also, since
$y' = \frac{\frac{d}{dx}(x^2+2)\cdot(x^2-2)-(x^2+2)\cdot\frac{d}{dx}(x^2-2)}{(x^2-2)^2}$

$= \frac{2x(x^2-2)-(x^2+2)2x}{(x^2-2)^2} = \frac{2x^3-4x-2x^3-4x}{(x^2-2)^2}$

$= \frac{-8x}{(x^2-2)^2}$, then $dy = f'(x)\, dx$

$= (\frac{-8x}{(x^2-2)^2})\, dx$. For $x = 2$ and $dx = 0.1$, we

get:
$dy = (\frac{-8(2)}{((2)^2-2)^2})(0.1) = (\frac{-16}{4})(0.1) = -0.4$.

19) Since $\sqrt{65} = \sqrt{64+1}$, we can approximate the answer using $f(x + dx) \approx f(x) + f'(x)\, dx$. Let $f(x) = \sqrt{x}$, $x = 64$ and $dx = 1$. Then,
$f'(x) = \frac{d}{dx}[\sqrt{x}] = \frac{d}{dx}[x^{1/2}] = \frac{1}{2}x^{-1/2} = \frac{1}{2\sqrt{x}}$.
So now, $f(x + dx) \approx f(x) + f'(x)\, dx$ can be rewritten as $f(64 + 1) \approx f(64) + f'(64)(1)$
$= \sqrt{64} + \frac{1}{2\sqrt{64}}(1) = 8 + \frac{1}{16} = 8.0625$.
Thus, by the linear approximation, $\sqrt{65} \approx 8.0625$. On a calculator, $\sqrt{65} \approx 8.0623$, so our answer is off only by about two ten-thousandths.

21) Since $\sqrt[4]{16.3} = \sqrt[4]{16 + 0.3}$, we can estimate the answer using $f(x + dx) \approx f(x) + f'(x)\, dx$. Let $f(x) = \sqrt[4]{x}$, $x = 16$ and $dx = 0.3$. Then,
$f'(x) = \frac{d}{dx}[\sqrt[4]{x}] = \frac{d}{dx}[x^{1/4}] = \frac{1}{4}x^{-3/4}$
$= \frac{1}{4\sqrt[4]{x^3}}$. So, $f(x + dx) \approx f(x) + f'(x)\, dx$
becomes $f(16 + 0.3)$
$\approx f(16) + f'(16)(0.3)$
$= \sqrt[4]{16} + \frac{1}{4\sqrt[4]{(16)^3}}(0.3) = 2 + \frac{3}{320}$
≈ 2.0094.
Thus, by the linear approximation, $\sqrt[4]{16.3} \approx 2.0094$. On a calculator, $\sqrt[4]{16.3} \approx 2.0093$, so our answer is off only by about one ten-thousandth.

23a) $y' = \frac{d}{dx}(90x - 2.7x^2) = 90 - 5.4x$. So, $dy = y'\, dx = (90 - 5.4x)\, dx$.

23b) Here, $x = 4$ and $dx = 5 - 4 = 1$. So, $dy = (90 - 5.4(4))(1) = 68.4(1) = 68.4$. As the amount spent on advertising increases from $400 to $500, we estimate that the number of T-shirts sold increases by 6,840 shirts.

25) In #23, $y = 90x - 2.7x^2$.
For $x = 4$, $y = 90(4) - 2.7(4)^2 = 316.8$.
For $x = 5$, $y = 90(5) - 2.7(5)^2 = 382.5$.
Thus, $\Delta y = 382.5 - 316.8 = 65.7$. Thus, the actual change in sales was 6,570 T-shirts, so we were off by $6840 - 6570 = 270$ shirts.
In #24, $y = 82.76x - 1.87x^2$.
For $x = 4$, $y = 82.76(4) - 1.87(4)^2 = 301.12$.
For $x = 5$, $y = 82.76(5) - 1.87(5)^2 = 367.05$.
Thus, $\Delta y = 367.05 - 301.12 = 65.93$. Thus, the actual change in sales was 6,593 T-shirts, so we were off by $6780 - 6593 = 187$ shirts.

27a) $MC(x) = \frac{d}{dx}[18x + 642] = 18$.

27b) $MC(12) = 18$. Thus, it will cost about $18 to produce the 13th unit.

27c) Since $C(13) = 18(13) + 642 = 876$ and $C(12) = 18(12) + 642 = 858$, then $C(13) - C(12) = 876 - 858 = 18$. This matches our answer in part b exactly.

29a) $MC(x) = \frac{d}{dx}[26.7x + 87.4] = 26.7$.

29b) $MC(8) = 26.7$. Thus, it will cost about $26.70 to produce the 9th unit.

29c) Since $C(9) = 26.7(9) + 87.4 = 327.7$ and $C(8) = 26.7(8) + 87.4 = 301$, then $C(13) - C(12) = 327.7 - 301 = 26.7$. This matches our answer in part b exactly.

31a) $MC(x) = \frac{d}{dx}[\frac{1}{4}x^2 + 12x + 47] = \frac{1}{2}x + 12$.

31b) $MC(31) = \frac{1}{2}(31) + 12 = 27.5$. Thus, it will cost about $27.50 to produce the 32nd unit.

31c) Since $C(32) = \frac{1}{4}(32)^2 + 12(32) + 47 = 687$ and $C(31) = \frac{1}{4}(31)^2 + 12(31) + 47 = 659.25$, then $C(32) - C(31) = 687 - 659.25 = 27.75$. This is only $0.25 off from our answer in part b.

33a) $R(x) = p(x) \cdot x = 11x$.

33b) $P(x) = R(x) - C(x) = 11x - (7x + 250)$
$= 4x - 250$.

33c) $MP(x) = \frac{d}{dx}[4x - 250] = 4$.

33d) $MC(x) = \frac{d}{dx}[7x + 250] = 7$ and
$MR(x) = \frac{d}{dx}[11x] = 11$.

33e) $MR(x) - MC(x) = 11 - 7 = 4$. This is the same as part c.

35a) $R(x) = p(x) \cdot x = (-\frac{x}{15} + 50)x = \frac{-x^2}{15} + 50x$.

35b) $P(x) = R(x) - C(x)$
$= \frac{-x^2}{15} + 50x - (\frac{1}{10}x^2 + 3x + 850)$
$= \frac{-x^2}{6} + 47x - 850$.

35c) $MP(x) = \frac{d}{dx}[\frac{-x^2}{6} + 47x - 850] = \frac{-x}{3} + 47$.

35d) $MC(x) = \frac{d}{dx}[\frac{1}{10}x^2 + 3x + 850] = \frac{1}{5}x + 3$
and $MR(x) = \frac{d}{dx}[\frac{-x^2}{15} + 50x] = \frac{-2x}{15} + 50$.

35e) $MR(x) - MC(x) = [\frac{-2x}{15} + 50] - (\frac{1}{5}x + 3)$
$= \frac{-x}{3} + 47$. This is the same as part c.

37a) $R(x) = p(x) \cdot x = (-0.005x + 10)x$
$= -0.005x^2 + 10x$.

37b) $P(x) = R(x) - C(x) = -0.005x^2 + 10x$
$\quad -(-0.001x^3 + 8x + 100)$
$= 0.001x^3 - 0.005x^2 + 2x - 100$.

37c) $MP(x) = \frac{d}{dx}[0.001x^3 - 0.005x^2 + 2x - 100]$
$= 0.003x^2 - 0.01x + 2$.

37d) $MC(x) = \frac{d}{dx}[-0.001x^3 + 8x + 100]$
$= -0.003x^2 + 8$ and $MR(x)$
$= \frac{d}{dx}[-0.005x^2 + 10x.] = -0.01x + 10$.

37e) $MR(x) - MC(x)$
$= [-0.01x + 10] - [-0.003x^2 + 8]$
$= 0.003x^2 - 0.01x + 2$. This is the same as part c.

39a) $C(q)$ = Variable costs $\cdot q$ + fixed costs
$= 85q + 5600$.

39b) $MC(q) = \frac{d}{dq}[85q + 5600] = 85$.

41a) $MAC(q) = \frac{d}{dq}[85 + \frac{5600}{q}]$
$= \frac{d}{dq}[85 + 5600q^{-1}] = -5600q^{-2} = \frac{-5600}{q^2}$.

41b) $MAC(250) = \frac{-5600}{(250)^2} \approx -0.09$. When 250 bicycles are produced, then average cost per bike is decreasing at a rate of about $0.09 per bike.

43a) $P(q) = R(q) - C(q)$
$= -0.02q^2 + 150q - [85q + 5600]$
$= -0.02q^2 + 65q - 5600$.

43b) Since P is a quadratic equation with a < 0, we know that the maximum value will occur at the vertex. The q-coordinate of the vertex is equal to $\frac{-b}{2a} = \frac{-65}{2(-0.02)} = 1625$.
The two lot sizes closest to 1625 are 1500 bicycles and 1750 bicycles. Evaluate P at those values yields:
$P(1500) = -0.02(1500)^2 + 65(1500)$
$\quad - 5600$
$= 46,900$ and
$P(1750) = -0.02(1750)^2 + 65(1750)$
$\quad - 5600$
$= 46,900$.
Since both $P(1500)$ and $P(1750)$ are the same, this suggests that either 1500 or 1750 bicycles should be manufactured so that the profit can be as large as possible.

43b) Continued
We can verify this by looking at the table below:

PRODUCED, q	TOTAL PROFIT, $P(q)$
0	-5600
250	9400
500	21900
750	31900
1000	39400
1250	44400
1500	46900
1750	46900
2000	44400

45a) $MC(250) = 125$. It will cost $125 to produce the 251st.

$MC(500) = 125$. It will cost $125 to produce the 501st.

45b) Since the cost function is linear, the rate of change is constant and hence, $MC(250) = MC(500)$.

45c) $AC(q) = \frac{C(q)}{q} = \frac{125q + 7300}{q} = 125 + \frac{7300}{q}$.

45d) $AC(250) = 125 + \frac{7300}{250} = 125 + 29.2$
$= 154.2$. When 250 bicycles are produced, then average cost of each bike is $154.20.

47a) $R(q) = q \bullet p(q) = q(-0.022q + 180)$
$= -0.022q^2 + 180q$.

47b) $MR(q) = \frac{d}{dq}[-0.022q^2 + 180q]$
$= -0.044q + 180$

47c) $MR(500) = -0.044(500) + 180 = 158$. The approximate revenue from producing and selling the 501st bike is $158.

49a) Since $R(x) = x \bullet p(x) = x \bullet 6 = 6x$, then
$P(x) = R(x) - C(x)$
$= 6x - (0.0002x^2 + 2x + 1250)$
$= -0.0002x^2 + 4x - 1250$.

49b) $P(3000) = -0.0002(3000)^2 + 4(3000) - 1250$
$= 8950$. When 3000 watercolor sets are produced and sold, Colorama Company earns $8,950 profit.

49c) $MP(x) = \frac{d}{dx}[-0.0002x^2 + 4x - 1250]$
$= -0.0004x + 4$.

49d) $MP(3000) = -0.0004(3000) + 4 = 2.8$. The approximate profit from producing and selling the 3001st water color set is $2.80.

49e) $P(3001) = P(3000 + 1)$
$\approx P(3000) + MP(3000) = 8950 + 2.8$
$= 8952.80$. When 3001 sets are produced, we estimate that the profit will be $8,952.80.

49f) Since $P(3001) = -0.0002(3001)^2 + 4(3001) - 1250$
$= 8952.7998$, then the error in the approximation is two ten-thousandths.

51) $MAC(x) = \frac{d}{dx}\left[\frac{C(x)}{x}\right] = \frac{\frac{d}{dx}[C(x)] \bullet x - C(x) \bullet \frac{d}{dx}[x]}{x^2}$
$= \frac{C'(x) \bullet x - C(x)}{x^2}$. Factoring out an x in the numerator and simplifying, we get:
$= \frac{x(C'(x) - \frac{C(x)}{x})}{x^2} = \frac{C'(x) - \frac{C(x)}{x}}{x} = \frac{MC(x) - AC(x)}{x}$
since $MC(x) = C'(x)$ and $AC(x) = \frac{C(x)}{x}$.
Thus, $MAC(x) = \frac{MC(x) - AC(x)}{x}$.

53a) $C(x)$ = Variable costs $\bullet x$ + fixed costs
$= 150x + 10000$ and
$R(x) = p(x) \bullet x = \left(\frac{-x}{75} + 250\right)x$
$= \frac{-x^2}{75} + 250x$.

118   Chapter 3   Applications of the Derivative

53b) The graph is displayed in viewing window [0, 10,000] by [– 400,000, 1,500,000]

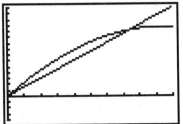

Note: The textbook writes the viewing window on the sides of the graph. The break-even point is when $R(x) = C(x)$. In other words, it is when the graphs of R and C intersect. Using the intersect command on the calculator, we find that the break even point is at

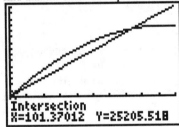

$x \approx 101.370$ and at

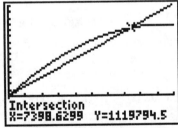

$x \approx 7398.630$. This means that 102 bookshelves is the smallest level of production that $R(x) > C(x)$ and 7398 bookshelves is the largest level of production that $R(x) > C(x)$.

53c) $P(x) = \frac{-x^2}{75} + 250x - (150x + 10,000)$
$= \frac{-x^2}{75} + 100x - 10,000.$

53d) From part b, $x = 102$ bookshelves produced is the smallest production level that $R(x) > C(x)$ and $x = 7,398$ bookshelves produced is the largest production level that $R(x) > C(x)$.

53e) $P'(x) = \frac{d}{dx}[\frac{-x^2}{75} + 100x - 10,000]$
$= \frac{-2x}{75} + 100.$

53f) $P'(2500) = \frac{-2(2500)}{75} + 100 \approx 33.3333.$
When 2500 bookshelves are produced, the profit is increasing at a rate of $\approx \$33.33$ per bookshelf.

55) $C(201) = C(200 + 1)$
$\approx C(200) + MC(200).$
But $C(200) = 30,830$ from the table and $MC(200) = 80$ from the graph.
Thus, $C(200) + MC(200) = 30,830 + 80$
$= 30,910.$ Using the approximation, we estimate that the cost to produce 201 units is \$30,910.

57) $R(101) = R(100 + 1)$
$\approx R(100) + MR(100).$
But $R(100) = 38000$ from the table and $MR(100) = 360$ from the graph.
Thus, $R(100) + MR(100) = 38000 + 360$
$= 38360.$ Using the approximation, we estimate that the revenue from selling 101 units is \$38,360.

59) $R(301) = R(300 + 1)$
$\approx R(300) + MR(300).$
But $R(300) = 102000$ from the table and $MR(300) = 280$ from the graph.
Thus, $R(300) + MR(300) = 102,000 + 280$
$= 102,280.$ Using the approximation, we estimate that the revenue from selling 301 units is \$102,280.

61a) $P(x) = -3000x^3 - 50x^2 + 300,000x$
$\quad\quad\quad - (119,890x + 123,450)$
$\quad\quad = -3000x^3 - 50x^2 + 180,110x - 123,450.$
Thus, $AP(x)$
$= \frac{P(x)}{x} = \frac{-3000x^3 - 50x^2 + 180110x - 123450}{x}$
$= -3000x^2 - 50x + 180,110 - \frac{123450}{x}$.
Hence,
$MAP(x)$
$= \frac{d}{dx}[-3000x^2 - 50x + 180,110 - \frac{123450}{x}]$
$= \frac{d}{dx}[-3000x^2 - 50x + 180,110$
$\quad\quad\quad - 123450x^{-1}]$
$= -6000x - 50 + 123,450x^{-2}$
$= -6000x - 50 + \frac{123450}{x^2}$.

61b) $MAP(1.5) = -6000(1.5) - 50 + \frac{123450}{(1.5)^2}$
$\approx 45,816.67.$ When 1500 printers are produced and sold, we estimate that the average profit per 1000 printers is increasing by about $45,816.67 for each additional 1000 printers.

63) Since $f'(x) = \frac{d}{dx}[x^3 - 2x + 3] = 3x^2 - 2$, then
$f'(1) = 3(1)^2 - 2 = 1.$ Also,
$f(1) = (1)^3 - 2(1) + 3 = 2$ and $\varepsilon_x = \pm 0.10.$
Thus, $\varepsilon_y = \frac{f'(x) \cdot x}{f(x)} \cdot \varepsilon_x = \frac{(1)(1)}{(2)}(\pm 0.10)$
$= 0.5(\pm 0.10) = \pm 0.05.$

65) Since $f'(x) = \frac{\frac{d}{dx}[2x+3] \cdot (3x-5) - (2x+3) \cdot \frac{d}{dx}[3x-5]}{(3x-5)^2}$
$= \frac{2(3x-5) - (2x+3)3}{(3x-5)^2} = \frac{6x - 10 - 6x - 9}{(3x-5)^2} = \frac{-19}{(3x-5)^2}$,
then $f'(-4) = \frac{-19}{(3(-4)-5)^2} = \frac{-19}{(-17)^2} = \frac{-19}{289}.$
Also, $f(-4) = \frac{2(-4)+3}{3(-4)-5} = \frac{-5}{-17} = \frac{5}{17}$ and
$\varepsilon_x = \pm 0.17.$ Thus, $\varepsilon_y = \frac{f'(x) \cdot x}{f(x)} \cdot \varepsilon_x$
$= \frac{(\frac{-19}{289})(-4)}{(\frac{5}{17})}(\pm 0.17) = \frac{76}{85}(\pm 0.17) = \pm 0.152.$

67) Since $f'(x) = \frac{d}{dx}[8\sqrt[4]{x^3}] = \frac{d}{dx}[8x^{3/4}]$
$= 6x^{-1/4} = \frac{6}{\sqrt[4]{x}}$, then $f'(2) = \frac{6}{\sqrt[4]{2}}.$ Also, $f(2)$
$= 8\sqrt[4]{2^3} = 8\sqrt[4]{8}$ and $\varepsilon_x = \pm 0.15.$ Thus,
$\varepsilon_y = \frac{f'(x) \cdot x}{f(x)} \cdot \varepsilon_x = \frac{(\frac{6}{\sqrt[4]{2}})(2)}{(8\sqrt[4]{8})}(\pm 0.15) = \frac{3}{2\sqrt[4]{16}}$
$= 0.75(\pm 0.15) = \pm 0.1125.$

69a) The forecast value is equal to the average of the highest value (x_b) and lowest value (x_a). Hence, $x = \frac{x_b + x_a}{2} = \frac{250 + 180}{2} = \frac{430}{2}$
$= 215.$

69b) The maximal error, dx, is equal to the difference of the highest value (x_b) and lowest value (x_a), divided by two. Hence, $dx = \frac{x_b - x_a}{2} = \frac{250 - 180}{2} = \frac{70}{2} = 35.$ Thus, the relative error $\varepsilon_x = \pm \frac{dx}{x} = \pm \frac{35}{215} \approx \pm 0.1628.$

69c) The percentage error $= \varepsilon_x \cdot 100\%$
$= \pm 16.28\%.$

71a) The forecast value is equal to the average of the highest value (x_b) and lowest value (x_a). Hence, $x = \frac{x_b + x_a}{2} = \frac{254 + 246}{2} = \frac{500}{2}$
$= 250.$

71b) The maximal error, dx, is equal to the difference of the highest value (x_b) and lowest value (x_a), divided by two. Hence, $dx = \frac{x_b - x_a}{2} = \frac{254 - 246}{2} = \frac{8}{2} = 4.$ Thus, the relative error $\varepsilon_x = \pm \frac{dx}{x} = \pm \frac{4}{250} \approx \pm 0.016.$

71c) The percentage error $= \varepsilon_x \cdot 100\%$
$= \pm 1.6\%.$

73a) The forecast value is equal to the average of the highest value (x_b) and lowest value (x_a). Hence, $x = \frac{x_b + x_a}{2} = \frac{94350 + 94150}{2}$
$= \frac{188500}{2} = 94,250.$

73b) The maximal error, dx, is equal to the difference of the highest value (x_b) and lowest value (x_a), divided by two. Hence, $dx = \frac{x_b - x_a}{2} = \frac{94350 - 94150}{2} = \frac{200}{2} = 100$.
Thus, the relative error $\varepsilon_x = \pm \frac{dx}{x}$
$= \pm \frac{100}{94250} \approx \pm 0.0011$.

73c) The percentage error $= \varepsilon_x \bullet 100\%$
$= \pm 0.11\%$.

75) Since 58% − 4% = 54% and 42% + 4% = 46%, the closest the race could be is 54% support for Evans and 46% support for Hawthorne.

77a) Since $MC(x) = C'(x)$
$= \frac{d}{dx} [350x^2 + 30,000x + 24,000]$
$= 700x + 30,000$, then $MC(88) = C'(88)$
$= 700(88) + 30,000 = 91,600$. Also,
$C(88) = 350(88)^2 + 30,000(88) + 24,000$
$= 5,374,400$ and $\varepsilon_x \approx \pm 0.0340909091$.
Thus, $\varepsilon_y \bullet 100\% = \frac{C'(x) \bullet x}{C(x)} \bullet \varepsilon_x \bullet 100\%$
$\approx \frac{91600(88)}{5374400} (\pm 0.0340909091)100\%$
$\approx 5.11\%$.

77b) $C(88) = 350(88)^2 + 30,000(88) + 24,000$
$= 5,374,400$. It will cost $5,374,400 to produce 88,000 modems.
$MC(x) = C'(x)$
$= \frac{d}{dx} [350x^2 + 30,000x + 24000]$
$= 700x + 30,000$. Thus, $MC(88) = C'(88)$
$= 700(88) + 30,000 = 91,600$.
When 88,000 modems are produced, it will cost the company about $91,600 to produce an additional 1000 modems.

77c) $C(89) \approx C(88) + MC(88)$
$= 5,374,400 + 91,600 = 5,466,000$. Thus, we estimate that it will cost $5,466,000 to produce 89,000 modems.

77d) $C(89) = 350(89)^2 + 30,000(89) + 24,000$
$= 5,466,350$. Thus, the marginal error is $5,466,000 - 5,466,350 = -350$. Hence, the relative error is $-\frac{350}{5466350} \approx -0.000064$
or -0.0064%.

79) The highest value is 78% + 2% = 80% and the lowest value is 78% − 2% = 76%.

81) Since $f'(x) = \frac{d}{dx}[7x^3 - 2x] = 21x^2 - 2$, then
$Rel(x) = \frac{f'(x)}{f(x)} = \frac{21x^2 - 2}{7x^3 - 2x}$.
Thus, $Rel(5) = \frac{21(5)^2 - 2}{7(5)^3 - 2(5)} = \frac{523}{865} \approx 0.6046$.

83) Since $f'(x) = \frac{d}{dx}[\frac{x-4}{x^2}] = \frac{d}{dx}[\frac{1}{x} - \frac{4}{x^2}]$
$= \frac{d}{dx}[x^{-1} - 4x^{-2}] = -x^{-2} + 8x^{-3} = \frac{-1}{x^2} + \frac{8}{x^3}$
$= \frac{-x+8}{x^3}$, then $Rel(x) = \frac{f'(x)}{f(x)} = f'(x) \div f(x)$
$\frac{-x+8}{x^3} \div \frac{x-4}{x^2} = \frac{-x+8}{x^3} \bullet \frac{x^2}{x-4} = \frac{-x+8}{x(x-4)}$.
Thus, $Rel(8) = \frac{-8+8}{8(8-4)} = \frac{0}{32} = 0$.

85) Since $f'(x) = \frac{d}{dx}[3\sqrt{x}] = \frac{d}{dx}[3x^{1/2}]$
$= 1.5x^{-1/2} = \frac{3}{2\sqrt{x}}$, then $Rel(x) = \frac{f'(x)}{f(x)}$
$= f'(x) \div f(x) = \frac{3}{2\sqrt{x}} \div (3\sqrt{x})$
$= \frac{3}{2\sqrt{x}} \bullet \frac{1}{3\sqrt{x}} = \frac{1}{2x}$.
Thus, $Rel(2) = \frac{1}{2(2)} = \frac{1}{4} = 0.25$.

87) Since $f'(x) = \frac{d}{dx}[7x^{-1} - 1] = -7x^{-2} = \frac{-7}{x^2}$,
then $Rel(x) = \frac{f'(x)}{f(x)} = f'(x) \div f(x)$
$= \frac{-7}{x^2} \div \frac{7-x}{x} = \frac{-7}{x^2} \bullet \frac{x}{7-x} = \frac{-7}{x(7-x)}$. Thus, $Rel(2)$
$= \frac{-7}{2(7-2)} = \frac{-7}{10} = -0.7$.

89a) $f(18) = 1.52(18)^2 + 19.9(18) + 437.52$
= 1288.2. In 1988, there were 1,288,200 households with a male householder and related children under 18 years of age.

89b) Since $f'(x) = \frac{d}{dx}[1.52x^2 + 19.9x + 437.52]$
= $3.04x + 19.9$, then $f'(18)$
= $3.04(18) + 19.9 = 74.62$.
In 1988, the number of households with a male householder and related children under 18 years of age was increasing at a rate of 74,620 households per year.

89c) $Rel(18) = \frac{f'(18)}{f(18)} = \frac{74.62}{1288.2} \approx 0.0579$ or 5.79%. Hence, the number of households with a male householder and related children under 18 years of age was increasing by about 5.79%.

Chapter 4 Additional Differentiation Techniques

Section 4.1 The Chain Rule

1) $f'(x) = \frac{d}{dx}[(x+1)^2] = 2(x+1)^1 \cdot \frac{d}{dx}(x+1)$
   $= 2(x+1)\cdot 1 = 2x + 2$.

3) $f'(x) = \frac{d}{dx}[(x-5)^3] = 3(x-5)^2 \cdot \frac{d}{dx}(x-5)$
   $= 3(x-5)^2 \cdot 1 = 3(x-5)^2$.

5) $f'(x) = \frac{d}{dx}[(2-x)^2] = 2(2-x)^1 \cdot \frac{d}{dx}(2-x)$
   $= 2(2-x)\cdot(-1) = 2x - 4$.

7) $g'(x) = \frac{d}{dx}[(2x+4)^3]$
   $= 3(2x+4)^2 \cdot \frac{d}{dx}(2x+4) = 3(2x+4)^2 \cdot 2$
   $= 6(2x+4)^2$ or $24(x+2)^2$.

9) $f'(x) = \frac{d}{dx}[(5-2x)^5]$
   $= 5(5-2x)^4 \cdot \frac{d}{dx}(5-2x) = 5(5-2x)^4 \cdot (-2)$
   $= -10(5-2x)^4$.

11) $f'(x) = \frac{d}{dx}[(3x^2+7)^5]$
    $= 5(3x^2+7)^4 \cdot \frac{d}{dx}(3x^2+7)$
    $= 5(3x^2+7)^4 \cdot (6x) = 30x(3x^2+7)^4$.

13) $f'(x) = \frac{d}{dx}[(x^3-2x^2+x)^2]$
    $= 2(x^3-2x^2+x)^1 \cdot \frac{d}{dx}(x^3-2x^2+x)$
    $= 2(x^3-2x^2+x)\cdot(3x^2-4x+1)$
    $= (6x^2-8x+2)(x^3-2x^2+x)$
    or $2x(3x^2-4x+1)(x^2-2x+1)$.

15) $g'(x) = \frac{d}{dx}[3(x^3-4)^3]$
    $= 9(x^3-4)^2 \cdot \frac{d}{dx}(x^3-4) = 9(x^3-4)^2 \cdot 3x^2$
    $= 27x^2(x^3-4)^2$.

17) $f'(x) = \frac{d}{dx}[5(5x^2-3x-1)^{10}]$
    $= 50(5x^2-3x-1)^9 \cdot \frac{d}{dx}(5x^2-3x-1)$
    $= 50(5x^2-3x-1)^9(10x-3)$.

19) $g'(x) = \frac{d}{dx}[(4x^2-x-4)^{55}]$
    $= 55(4x^2-x-4)^{54} \cdot \frac{d}{dx}(4x^2-x-4)$
    $= 55(4x^2-x-4)^{54}(8x-1)$.

21) $g'(x) = \frac{d}{dx}[(2x-4)^{1/2}]$
    $= \frac{1}{2}(2x-4)^{-1/2} \cdot \frac{d}{dx}(2x-4)$
    $= \frac{1}{2}(2x-4)^{-1/2} \cdot 2$
    $= (2x-4)^{-1/2} = \frac{1}{\sqrt{2x-4}}$.

23) $g'(x) = \frac{d}{dx}[(x^2+2x)^{1/3}]$
    $= \frac{1}{3}(x^2+2x)^{-2/3} \cdot \frac{d}{dx}(x^2+2x)$
    $= \frac{1}{3}(x^2+2x)^{-2/3} \cdot (2x+2)$
    $= \frac{2x+2}{3(x^2+2x)^{\frac{2}{3}}} = \frac{2x+2}{3\sqrt[3]{(x^2+2x)^2}}$.

25) $f'(x) = \frac{d}{dx}[(5x-2)^{-2}]$
    $= -2(5x-2)^{-3} \cdot \frac{d}{dx}(5x-2)$
    $= -2(5x-2)^{-3} \cdot (5)$
    $= -10(5x-2)^{-3}$
    $= \frac{-10}{(5x-2)^3}$.

27) $g'(x) = \frac{d}{dx}[(x^2 + 2x + 4)^{-\frac{1}{2}}]$

$= -\frac{1}{2}(x^2 + 2x + 4)^{-3/2} \bullet \frac{d}{dx}(x^2 + 2x + 4)$

$= -\frac{1}{2}(x^2 + 2x + 4)^{-3/2} \bullet (2x + 2)$

$= (-x - 1)(x^2 + 2x + 4)^{-3/2}$

$= \frac{-x-1}{(x^2+2x+4)\sqrt{x^2+2x+4}}$.

29) $g'(x) = \frac{d}{dx}[(3x^3 - x)^{-\frac{1}{4}}]$

$= -\frac{1}{4}(3x^3 - x)^{-5/4} \bullet \frac{d}{dx}(3x^3 - x)$

$= -\frac{1}{4}(3x^3 - x)^{-5/4}(9x^2 - 1)$

$= \frac{-9x^2+1}{(12x^3-4x)\sqrt[4]{3x^3-x}}$ or $\frac{-9x^2+1}{4x(3x^2-1)\sqrt[4]{3x^3-x}}$.

31a) $f'(x) = \frac{\frac{d}{dx}[1]\bullet(3x+4)-1\bullet\frac{d}{dx}[3x+4]}{(3x+4)^2}$

$= \frac{0\bullet(3x+4)-1\bullet(3)}{(3x+4)^2} = \frac{-3}{(3x+4)^2}$.

31b) $f'(x) = \frac{d}{dx}[(3x + 4)^{-1}]$

$= -1(3x + 4)^{-2} \bullet \frac{d}{dx}(3x + 4)$

$= -1(3x + 4)^{-2}(3) = -3(3x + 4)^{-2}$

$= \frac{-3}{(3x+4)^2}$.

33a) $f'(x) = \frac{\frac{d}{dx}[5]\bullet(x-2)^2-5\bullet\frac{d}{dx}[(x-2)^2]}{((x-2)^2)^2}$

$= \frac{0\bullet(x-2)^2-5\bullet2(x-2)\frac{d}{dx}[x-2]}{(x-2)^4} = \frac{-10(x-2)(1)}{(x-2)^4}$

$= \frac{-10}{(x-2)^3}$.

33b) $f'(x) = \frac{d}{dx}[5(x - 2)^{-2}]$

$= -10(x - 2)^{-3} \bullet \frac{d}{dx}(x - 2)$

$= -10(x - 2)^{-3}(1) = \frac{-10}{(x-2)^3}$.

35a) $f'(x) = \frac{\frac{d}{dx}[2]\bullet(x^2+2x+3)-2\bullet\frac{d}{dx}[x^2+2x+3]}{(x^2+2x+3)^2}$

$= \frac{0\bullet(x^2+2x+3)-2\bullet(2x+2)}{(x^2+2x+3)^2} = \frac{-4x-4}{(x^2+2x+3)^2}$.

35b) $f'(x) = \frac{d}{dx}[2(x^2 + 2x + 3)^{-1}]$

$= -2(x^2 + 2x + 3)^{-2} \bullet \frac{d}{dx}(x^2 + 2x + 3)$

$= -2(x^2 + 2x + 3)^{-2}(2x + 2)$

$= (-4x - 4)(x^2 + 2x + 3)^{-2}$

$= \frac{-4x-4}{(x^2+2x+3)^2}$.

37) $f'(x) = \frac{d}{dx}[(2x - 1)^3]$

$= 3(2x - 1)^2 \bullet \frac{d}{dx}(2x - 1) = 3(2x - 1)^2 \bullet 2$

$= 6(2x - 1)^2$. Since the slope of the tangent line at $x = 1$ is equal to $f'(1) = 6(2(1) - 1)^2 = 6(1)^2 = 6$, we can find the equation of the tangent line by plugging $m = 6$ & $(1, 1)$ into:
$y - y_1 = m(x - x_1)$
$y - 1 = 6(x - 1)$
$y - 1 = 6x - 6$
$y = 6x - 5$.

39) $f'(x) = \frac{d}{dx}[(2 - x)^4]$

$= 4(2 - x)^3 \bullet \frac{d}{dx}(2 - x) = 4(2 - x)^3 \bullet(-1)$

$= -4(2 - x)^3$. Since the slope of the tangent line at $x = 1$ is equal to $f'(1) = -4(2 - (1))^3 = -4(1)^3 = -4$, we can find the equation of the tangent line by plugging $m = -4$ & $(1, 1)$ into:
$y - y_1 = m(x - x_1)$
$y - 1 = -4(x - 1)$
$y - 1 = -4x + 4$
$y = -4x + 5$.

41) $f'(x) = \frac{d}{dx}[(x^3 - 4x + 2)^4]$
$= 4(x^3 - 4x + 2)^3 \bullet \frac{d}{dx}(x^3 - 4x + 2)$
$= 4(x^3 - 4x + 2)^3 \bullet (3x^2 - 4)$
$= (12x^2 - 16)(x^3 - 4x + 2)^3$. Since the slope of the tangent line at $x = 2$ is equal to $f'(2) = (12(2)^2 - 16)((2)^3 - 4(2) + 2)^3 = 32(2)^3 = 256$, we can find the equation of the tangent line by plugging $m = 256$ & $(2, 16)$ into:
$y - y_1 = m(x - x_1)$
$y - 16 = 256(x - 2)$
$y - 16 = 256x - 512$
$y = 256x - 496.$

43) $f'(x) = \frac{d}{dx}[(2x - 4)^{1/2}]$
$= \frac{1}{2}(2x - 4)^{-1/2} \bullet \frac{d}{dx}(2x - 4)$
$= \frac{1}{2}(2x - 4)^{-1/2} \bullet 2 = \frac{1}{\sqrt{2x-4}}$. Since the slope of the tangent line at $x = 2$ is equal to $f'(2) = \frac{1}{\sqrt{2(2)-4}} = \frac{1}{0}$ or undefined, we know that the equation of the tangent line a vertical line passing through $(2, 0)$. Hence, the equation of the tangent line is $x = 2$.

45) $f'(x) = \frac{d}{dx}[\sqrt{x^2 + 5}] = \frac{d}{dx}[(x^2 + 5)^{1/2}]$
$= \frac{1}{2}(x^2 + 5)^{-1/2} \bullet \frac{d}{dx}(x^2 + 5)$
$= \frac{1}{2}(x^2 + 5)^{-1/2}(2x) = x(x^2 + 5)^{-1/2}$
$= \frac{x}{\sqrt{x^2 + 5}}.$

47) $f'(x) = \frac{d}{dx}[\sqrt[3]{2x - 1}] = \frac{d}{dx}[(2x - 1)^{1/3}]$
$= \frac{1}{3}(2x - 1)^{-2/3} \bullet \frac{d}{dx}(2x - 1)$
$= \frac{1}{3}(2x - 1)^{-2/3}(2) = \frac{2}{3}(2x - 1)^{-2/3}$
$= \frac{2}{3\sqrt[3]{(2x-1)^2}}.$

49) $f'(x) = \frac{d}{dx}[\frac{5}{\sqrt{2x-8}}] = \frac{d}{dx}[5(2x - 8)^{-1/2}]$
$= -\frac{5}{2}(2x - 8)^{-3/2} \bullet \frac{d}{dx}(2x - 8)$
$= -\frac{5}{2}(2x - 8)^{-3/2}(2) = -5(2x - 8)^{-3/2}$
$= \frac{-5}{(2x-8)\sqrt{2x-8}}.$

51) $f'(x) = \frac{d}{dx}[\frac{64}{\sqrt[3]{5x^2 - 6x + 3}}]$
$= \frac{d}{dx}[64(5x^2 - 6x + 3)^{-1/3}]$
$= -\frac{64}{3}(5x^2 - 6x + 3)^{-4/3} \bullet \frac{d}{dx}(5x^2 - 6x + 3)$
$= -\frac{64}{3}(5x^2 - 6x + 3)^{-4/3}(10x - 6)$
$= -\frac{128}{3}(5x - 3)(5x^2 - 6x + 3)^{-4/3}$
$= \frac{-128(5x-3)}{3(5x^2 - 6x + 3)\sqrt[3]{5x^2 - 6x + 3}}.$

53) $g'(x) = \frac{d}{dx}[x] \bullet (x - 4)^3 + x \bullet \frac{d}{dx}[(x - 4)^3]$
$= 1(x - 4)^3 + x \bullet 3(x - 4)^2 \frac{d}{dx}(x - 4)$
$= (x - 4)^3 + 3x(x - 4)^2(1)$
$= (x - 4)^2 [(x - 4) + 3x]$
$= 4(x - 4)^2(x - 1).$

55) $g'(x) = \frac{d}{dx}[x] \bullet (x^2 + 3x)^{1/2} + x \bullet \frac{d}{dx}[(x^2 + 3x)^{1/2}]$
$= 1(x^2 + 3x)^{1/2} + x \bullet \frac{1}{2}(x^2 + 3x)^{-1/2} \frac{d}{dx}(x^2 + 3x)$
$= (x^2 + 3x)^{1/2} + \frac{1}{2}x(x^2 + 3x)^{-1/2}(2x + 3)$
$= \sqrt{x^2 + 3x} + \frac{x(2x+3)}{2\sqrt{x^2+3x}}$
$= \sqrt{x^2 + 3x} \bullet \frac{2\sqrt{x^2+3x}}{2\sqrt{x^2+3x}} + \frac{x(2x+3)}{2\sqrt{x^2+3x}}$
$= \frac{2(x^2+3x)}{2\sqrt{x^2+3x}} + \frac{x(2x+3)}{2\sqrt{x^2+3x}} = \frac{2x^2 + 6x + 2x^2 + 3x}{2\sqrt{x^2+3x}}$
$= \frac{4x^2 + 9x}{2\sqrt{x^2+3x}} = \frac{x(4x+9)}{2\sqrt{x(x+3)}}.$

57) $f'(x) = \dfrac{\frac{d}{dx}[x^3]\bullet(3x-8)^2 - x^3\bullet\frac{d}{dx}[(3x-8)^2]}{[(3x-8)^2]^2}$

$= \dfrac{3x^2(3x-8)^2 - x^3\bullet 2(3x-8)\frac{d}{dx}(3x-8)}{(3x-8)^4}$

$= \dfrac{3x^2(3x-8)^2 - 2x^3(3x-8)(3)}{(3x-8)^4}$

$= \dfrac{3x^2(3x-8)^2 - 6x^3(3x-8)}{(3x-8)^4} = \dfrac{3x^2(3x-8)[(3x-8)-2x]}{(3x-8)^4}$

$= \dfrac{3x^2(x-8)}{(3x-8)^3}$.

59) $f'(x) = \frac{d}{dx}[(x+3)^3]\bullet(2x-1)^2$
$\qquad + (x+3)^3 \bullet \frac{d}{dx}[(2x-1)^2]$

$= 3(x+3)^2 \frac{d}{dx}[x+3]\bullet(2x-1)^2$
$\qquad + (x+3)^3 \bullet 2(2x-1)\bullet \frac{d}{dx}[2x-1]$

$= 3(x+3)^2[1](2x-1)^2 + 2(x+3)^3(2x-1)[2]$
$= 3(x+3)^2(2x-1)^2 + 4(x+3)^3(2x-1)$
$= (x+3)^2(2x-1)[3(2x-1) + 4(x+3)]$
$= (x+3)^2(2x-1)(10x+9)$.

61) $g'(x) = \frac{d}{dx}\left[\sqrt{\frac{x+3}{x-3}}\right] = \frac{d}{dx}\left[\left(\frac{x+3}{x-3}\right)^{\frac{1}{2}}\right]$

$= \frac{1}{2}\left(\frac{x+3}{x-3}\right)^{-\frac{1}{2}} \bullet \frac{d}{dx}\left(\frac{x+3}{x-3}\right)$

$= \frac{1}{2}\sqrt{\frac{x-3}{x+3}} \bullet \frac{d}{dx}\left(\frac{x+3}{x-3}\right)$.

But, $\frac{d}{dx}\left(\frac{x+3}{x-3}\right) = \dfrac{\frac{d}{dx}[x+3]\bullet(x-3) - (x+3)\bullet\frac{d}{dx}[x-3]}{(x-3)^2}$

$= \dfrac{1\bullet(x-3) - (x+3)\bullet 1}{(x-3)^2} = \dfrac{x-3-x-3}{(x-3)^2} = \dfrac{-6}{(x-3)^2}$.

Therefore, $\frac{1}{2}\sqrt{\frac{x-3}{x+3}} \bullet \frac{d}{dx}\left(\frac{x+3}{x-3}\right) =$

$\frac{1}{2}\sqrt{\frac{x-3}{x+3}} \bullet \dfrac{-6}{(x-3)^2} = \dfrac{-3}{(x-3)\sqrt{(x-3)(x+3)}}$.

63) $y' = \frac{d}{dx}[(4x^2+5x+6)^{0.23}]$

$= 0.23(4x^2+5x+6)^{-0.77}\frac{d}{dx}(4x^2+5x+6)$

$= 0.23(4x^2+5x+6)^{-0.77}(8x+5)$

$= \dfrac{1.84x+1.15}{(4x^2+5x+6)^{0.77}}$.

65) Since $g(x) = \left(\frac{1}{x+3}\right)^{-1.03} = ([x+3]^{-1})^{-1.03}$

$= (x+3)^{1.03}$, then $g'(x) = \frac{d}{dx}[(x+3)^{1.03}]$

$= 1.03(x+3)^{0.03}\bullet \frac{d}{dx}(x+3)$

$= 1.03(x+3)^{0.03}(1) = 1.03(x+3)^{0.03}$.

67) $f'(x) = \frac{d}{dx}[1.44(x+1)^{1.22}]$

$= 1.22\bullet 1.44(x+1)^{0.22} \bullet \frac{d}{dx}(x+1)$

$= 1.7568(x+1)^{0.22}(1) = 1.7568(x+1)^{0.22}$.

69) Since $f'(x) = \frac{d}{dx}[400\sqrt{100-x}\,]$

$= \frac{d}{dx}[400(100-x)^{\frac{1}{2}}]$

$= \frac{1}{2}(400)(100-x)^{-\frac{1}{2}}\bullet\frac{d}{dx}(100-x)$

$= 200(100-x)^{-\frac{1}{2}}(-1) = \dfrac{-200}{\sqrt{100-x}}$, then $f'(x)$

$= \dfrac{-200}{\sqrt{100-70}} = \dfrac{-200}{\sqrt{30}} \approx -36.51$.

When the group is 70 years old, the number that are surviving is decreasing at a rate of about 36.51 people per year.

71) Since $f'(t) = \frac{d}{dt}\left[-\dfrac{10000}{\sqrt{1+0.18t}} + 11000\right]$

$= \frac{d}{dt}[-10000(1+0.18t)^{-\frac{1}{2}} + 11000]$

$= -\frac{1}{2}\bullet -10000(1+0.18t)^{-3/2}\bullet\frac{d}{dt}(1+0.18t)$

$= 5000(1+0.18t)^{-3/2}(0.18)$

$= \dfrac{900}{(1+0.18t)\sqrt{1+0.18t}}$, then $f'(10)$

$= \dfrac{900}{(1+0.18(10))\sqrt{1+0.18(10)}} = \dfrac{900}{(2.8)\sqrt{2.8}}$

≈ 192.09. In the tenth year of the study, the number of students enrolled in the Arts and Sciences was increasing by about 192.09 students per year.

73a) $MC(x) = \frac{d}{dx}[(3x+6)^{1.5} + 30]$

$= 1.5(3x+6)^{\frac{1}{2}} \bullet \frac{d}{dx}[3x+6]$

$= 1.5(3x+6)^{\frac{1}{2}}(3) = 4.5\sqrt{3x+6}$.

126   Chapter 4 Additional Differentiation Techniques

73b) $MC(5) = 4.5\sqrt{3(5)+6} = 4.5\sqrt{21}$
      ≈ 20.62. When 500 auto anti-theft devices are produced, the costs to produce the next 100 devices is about \$20,620.

75a) $MC(x) = \frac{d}{dx}[60 + (3x+5)^{\frac{1}{2}}]$
      $= 0.5(3x+5)^{-\frac{1}{2}} \cdot \frac{d}{dx}[3x+5]$
      $= 0.5(3x+5)^{-\frac{1}{2}}(3) = \frac{3}{2\sqrt{3x+5}}$.

75b) $MC(15) = \frac{3}{2\sqrt{3(15)+5}} = \frac{3}{2\sqrt{50}}$
      ≈ 0.2121. The cost of producing the 16th camera is about \$21.21.

77a) $p'(x) = \frac{d}{dx}[\sqrt{22500 - 50x}]$
      $= \frac{d}{dx}[(22500 - 50x)^{\frac{1}{2}}]$
      $= \frac{1}{2}(22500 - 50x)^{-\frac{1}{2}} \cdot \frac{d}{dx}[22500 - 50x]$
      $= \frac{1}{2}(22500 - 50x)^{-\frac{1}{2}}(-50)$
      $= \frac{-25}{\sqrt{22500-50x}} = \frac{-25}{\sqrt{25(900-2x)}} = \frac{-25}{5\sqrt{900-2x}}$
      $= \frac{-5}{\sqrt{900-2x}}$.

77b) $p'(25) = \frac{-5}{\sqrt{900-2(25)}} = \frac{-5}{\sqrt{850}} \approx -0.17$.
      When 2,500 dolls are produced, the price of the dolls is decreasing by \$0.17 per 100 dolls.

79a) $p'(x) = \frac{d}{dx}[\frac{125}{\sqrt{2x+5}}] = \frac{d}{dx}[125(2x+5)^{-\frac{1}{2}}]$
      $= -\frac{1}{2}125(2x+5)^{-3/2} \cdot \frac{d}{dx}[2x+5]$
      $= -\frac{125}{2}(2x+5)^{-3/2}(2)$
      $= \frac{-125}{(2x+5)\sqrt{2x+5}}$.

79b) $p'(20) = \frac{-125}{(2(20)+5)\sqrt{2(20)+5}} = \frac{-125}{(45)\sqrt{45}}$
      ≈ -0.41. When 2,000 spark plugs are produced, the price of the plugs is decreasing by \$0.41 per 100 spark plugs.

79c) $R(x) = x \cdot p(x) = x(\frac{125}{\sqrt{2x+5}}) = \frac{125x}{\sqrt{2x+5}}$.

79d) $MR(x) = \frac{d}{dx}[\frac{125x}{\sqrt{2x+5}}] = \frac{d}{dx}[\frac{125x}{(2x+5)^{\frac{1}{2}}}]$
      $= \frac{\frac{d}{dx}[125x] \cdot (2x+5)^{\frac{1}{2}} - (125x) \cdot \frac{d}{dx}[(2x+5)^{\frac{1}{2}}]}{((2x+5)^{\frac{1}{2}})^2}$
      $= \frac{125(2x+5)^{\frac{1}{2}} - (125x)\frac{1}{2}(2x+5)^{-\frac{1}{2}}\frac{d}{dx}(2x+5)}{(2x+5)}$
      $= \frac{125(2x+5)^{\frac{1}{2}} - (125x)\frac{1}{2}(2x+5)^{-\frac{1}{2}}(2)}{(2x+5)}$
      $= \frac{125(2x+5)^{\frac{1}{2}} - (125x)(2x+5)^{-\frac{1}{2}}}{(2x+5)} \cdot \frac{\sqrt{2x+5}}{\sqrt{2x+5}}$
      $= \frac{125(2x+5) - (125x)}{(2x+5)\sqrt{2x+5}} = \frac{125x+625}{(2x+5)\sqrt{2x+5}}$.

79e) $MR(20) = \frac{125(20)+625}{(2(20)+5)\sqrt{2(20)+5}} = \frac{3125}{45\sqrt{45}}$
      ≈ 10.35. When 2000 spark plugs are produced, the approximate revenue from selling the next 100 spark plugs is \$10.35.

81a) $f(12) = 2(12)\sqrt{12-3} = 24\sqrt{9} = 72$. It will take 72 minutes to learn 12 items on the list.

81b) $f'(x) = \frac{d}{dx}[2x\sqrt{x-3}] = \frac{d}{dx}[2.5x(x-6)^{\frac{1}{2}}]$
      $= \frac{d}{dx}[2x] \cdot (x-3)^{\frac{1}{2}} + 2x \cdot \frac{d}{dx}[(x-3)^{\frac{1}{2}}]$
      $= 2(x-3)^{\frac{1}{2}} + 2x \cdot \frac{1}{2}(x-3)^{-\frac{1}{2}}\frac{d}{dx}(x-3)$
      $= 2(x-3)^{\frac{1}{2}} + x \cdot (x-3)^{-\frac{1}{2}}(1)$
      $= 2\sqrt{x-3} + \frac{x}{\sqrt{x-3}}$
      $= 2\sqrt{x-3}\frac{\sqrt{x-3}}{\sqrt{x-3}} + \frac{x}{\sqrt{x-3}}$
      $= \frac{2x-6}{\sqrt{x-3}} + \frac{x}{\sqrt{x-3}} = \frac{3x-6}{\sqrt{x-3}}$.

81c) $f'(12) = \frac{3(12)-6}{\sqrt{(12)-3}} = \frac{30}{\sqrt{9}} = 10$.

After 12 items in the list have been learned, the time required to learn an item on the list is increasing at a rate of 10 minutes per item.

83a) $P(75) = 400\sqrt{101-75} = 400\sqrt{26}$

≈ 2040. There are still about 2,040 people alive at age 75 in the township.

83b) $P'(t) = \frac{d}{dt}[400\sqrt{101-t}\,]$

$= \frac{d}{dt}[400(101-t)^{1/2}]$

$= 400 \cdot \frac{1}{2}(101-t)^{-1/2} \cdot \frac{d}{dt}(101-t)$

$= 200(101-t)^{-1/2}(-1)$

$= \frac{-200}{\sqrt{101-t}}$.

83c) $P'(75) = \frac{-200}{\sqrt{101-75}} \approx -39.22$.

At age 75, the number of people still alive is decreasing at a rate about 39.22 people per year.

85a) $g(8) = 70(10 + 0.5(8))^{2.1} = 70(14)^{2.1}$
≈ 17864. Eight hours after the culture was first observed, there are about 17,864 bacteria in the culture.

85b) $g'(t) = \frac{d}{dt}[70(10 + 0.5t)^{2.1}]$

$= 70 \cdot 2.1(10 + 0.5t)^{1.1} \frac{d}{dt}[10 + 0.5t]$

$= 147(10 + 0.5t)^{1.1}(0.5)$

$= 73.5(10 + 0.5t)^{1.1}$

85c) $g'(8) = 73.5(10 + 0.5(8))^{1.1} = 73.5(14)^{1.1}$
≈ 1339.76. Eight hours after the culture was first observed, the number of bacteria was growing at a rate of about 1339.76 bacteria per hour.

128 Chapter 4 Additional Differentiation Techniques

Section 4.2 Derivatives of Logarithmic Functions

1) $f'(x) = [5 \ln(x)] = 5 \cdot \quad = \quad .$

3) Since $f(x) = \ln(x^6) = 6 \ln(x)$, then
$f'(x) = [6 \ln(x)] = 6 \cdot \quad = \quad .$

5) $f'(x) = [4x^3 \cdot \ln(x)]$
$= [4x^3] \cdot \ln(x) + 4x^3 \cdot [\ln(x)]$
$= 12x^2 \cdot \ln(x) + 4x^3 \cdot \quad = 12x^2 \ln(x) + 4x^2$
$= 4x^2 (3 \ln(x) + 1).$

7) $f'(x) = [\quad]$

$=$

$= \quad =$

$= \quad .$

9) $f'(x) = [10 - 12 \ln(x)] = 0 - 12 [\ln(x)]$
$= -12 \cdot \quad = \quad .$

11) $g'(x) = [\ln(x+7)] = \quad (x+7)$
$= \quad (1) = \quad .$

13) $g'(x) = [\ln(2x-5)] = \quad (2x-5)$
$= \quad (2) = \quad .$

15) $g'(x) = [\ln(x^2+3)] = \quad (x^2+3)$
$= \quad (2x) = \quad .$

17) $g'(x) = [\ln(\quad)]$
$= [\ln((2x+5)^{1/2})] = [\quad \ln(2x+5)]$
$= \quad \cdot \quad (2x+5) = \quad \cdot \quad (2)$
$= \quad .$

19) $g'(x) = \frac{dg}{dx}[(\ln(x))^6] = 6(\ln(x))^5 \cdot \frac{d}{dx}[\ln(x)]$
$= 6(\ln(x))^5 \cdot \frac{1}{x} = \quad .$

21) Since $g(x) = \sqrt{x} \cdot \ln(\sqrt{x})$
$= x^{1/2} \cdot \ln(x^{1/2}) = \frac{1}{2} x^{1/2} \cdot \ln(x)$, then
$g'(x) = [\quad x^{1/2} \cdot \ln(x)]$
$= [\quad x^{1/2}] \cdot \ln(x) + x^{1/2} \cdot [\ln(x)]$
$= \quad \cdot x^{-1/2} \cdot \ln(x) + x^{1/2} \cdot \quad$
$= \quad + \quad .$

23) $g'(x) = [\quad]$

$=$

$=$

$= \quad - \quad .$

25) $f'(x) = [\log_{10}(x)] = \quad \cdot \quad = \quad .$

27) $f'(x) = [6 \log_3(x)] = \quad \cdot \quad = \quad .$

29) $f'(x) = [x^2 \log_9(x)]$
$= \frac{dy}{dx}[x^2] \cdot \log_9(x) + x^2 \cdot \frac{dy}{dx}[\log_9(x)]$
$= 2x \log_9(x) + x^2 \cdot \frac{1}{\ln 9} \cdot \frac{1}{x}$
$= 2x \log_9 x + \frac{x}{x \ln 9} = 2x \log_9 x + \frac{x}{\ln 9}$

Section 4.2 Derivatives of Logarithmic Functions 129

31) $f'(x) =$ [$\log_2 (5x + 3)$]

= • • $(5x + 3)$

= • •(5) = .

33) $f'(x) =$ [$\log_{10} ($ $)$]

= [$\log_{10} (x + 3) - \log_{10} (x^2 + 1)$]

= • • $(x + 3)$

$-$ • • $(x^2 + 1)$

= • •(1) $-$ • •(2x)

= $-$

= • $-$ •

= =   .

35) $f'(x) = \frac{dy}{dx} [\ln (x)] = \frac{1}{x}$.

The slope of the tangent line at $x = 2$ is equal to $f'(2) = \frac{1}{2}$. We can find the equation of the tangent line by plugging $m = \frac{1}{2}$ & $(2, \ln (2))$ into:

$y - y_1 = m (x - x_1)$
$y - \ln (2) = \frac{1}{2} (x - 2)$
$y - \ln (2) = x - 1$
$y = x - 1 + \ln (2)$.

37) $f'(x) =$ [ln (   )]

= [ln $((2x - 1)^{1/2})$] = [ln $(2x - 1)$]

= • • $[2x - 1]$ = • •(2)

=   .

The slope of the tangent line at $x = 1$ is equal to $f'(1) =$ = 1.

37)   Continued
We can find the equation of the tangent line by plugging $m = 1$ & $(1, 0)$ into:
$y - y_1 = m (x - x_1)$
$y - 0 = 1(x - 1)$
$y = x - 1$.

39) $f'(x) =$ [$4x^3 \bullet \ln (x)$]

= [$4x^3$]•ln $(x) + 4x^3 \bullet$ [ln (x)]

= $12x^2 \bullet$ln $(x) + 4x^3 \bullet$ = $12x^2$ ln $(x) + 4x^2$

= $4x^2 (3 \ln (x) + 1)$.

The slope of the tangent line at $x = 1$ is equal to $f'(1) = 4(1)^2 (3 \ln (1) + 1) = 4$. We can find the equation of the tangent line by plugging $m = 4$ & $(1, 0)$ into:
$y - y_1 = m (x - x_1)$
$y - 0 = 4(x - 1)$
$y = 4x - 4$.

41) $f'(x) =$ [$(\ln (x))^6$] = $6(\ln (x))^5$ [ln (x)]

= $6(\ln (x))^5 \bullet$ =   .

The slope of the tangent line at $x = e$ is equal to $f'(e) =$ =   . We can find the equation of the tangent line by plugging $m =$ & $(e, 1)$ into:
$y - y_1 = m (x - x_1)$
$y - 1 =$ $(x - e)$
$y - 1 =$ $x - 6$.
$y =$ $x - 5$.

43a) $f'(t) =$ [$750 + 12 \ln (t)$] = $0 + 12 \bullet$

=   .

43b) $f(12) = 750 + 12 \ln (12) \approx 780$. Twelve hours after the start of the experiment, there were about 780 bacteria present. $f'(12) =$ $= 1$. Twelve hours after the experiment started, the culture was growing at a rate of 1 bacteria per hour.

130 Chapter 4 Additional Differentiation Techniques

45a) $f'(x) = [150 + 5 \log_2(x)]$
$= 0 + 5 \bullet \bullet = .$

45b) $f'(2) = \approx 3.61$. When Vectrum has been on the market for two years, the number of people using the drug was growing at a rate of about 3,610 people per year.
$f'(10) = \approx 0.721$. When Vectrum has been on the market for ten years, the number of people using the drug was growing at a rate of about 721 prescriptions per year.

47a) $f'(x) = [68.41 + 1.75 \ln(x)]$
$= 0 + 1.75 \bullet = .$

47b) $f(3) = 68.41 + 1.75 \ln(3) \approx 70.33$. In 1972, the life expectancy of African-American females in the US was 70.33 years.
$f'(3) = \approx 0.583$
In 1972, the life expectancy of African-American females in the US was increasing at a rate of 0.58 years per birth year.

47c) The slope of the tangent line at $x = 3$ is equal to $f'(3) \approx 0.583$. We can find the equation of the tangent line by plugging $m = 0.583$ & $(3, 70.33)$ into:
$y - y_1 = m(x - x_1)$
$y - 70.33 = 0.583(x - 3)$
$y - 70.33 = 0.583x - 1.75$
$y = 0.583x + 68.58$.
Plugging in $x = 15$, we get:
$y = 0.583(15) + 68.58 \approx 77.33$. Thus, using the tangent line, we estimate that in 1984, the life expectancy of African-American females in the US was about 77.33. But, $f(15) = 68.41 + 1.75 \ln(15) \approx 73.15$, so our estimate is more than four years off.

49a) The graph is displayed in viewing window [1, 16] by [10, 17]

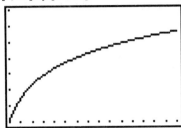

49b) $f'(x) = [10.12 + 2 \ln(x)]$
$= 0 + 2 \bullet = .$

49c) $f'(5) = = 0.4$. In 1984, the annual per capita consumption of light and skim milk was increasing at a rate of 0.4 gallons per year.
$f'(10) = = = 0.2$. In 1989, the annual per capita consumption of light and skim milk was increasing at a rate of 0.2 gallons per year. Notice that the annual per capita consumption was increasing twice as fast in 1984 than in 1989.

Section 4.3 Derivatives of Exponential Functions

1) $g'(x) = \frac{d}{dx}[7e^x] = 7e^x$.

3) $g'(x) = \frac{d}{dx}[2x(4 + e^x)]$
$= \frac{d}{dx}[2x] \bullet (4 + e^x) + 2x \frac{d}{dx}[4 + e^x]$
$= 2(4 + e^x) + 2x(e^x) = 2xe^x + 2e^x + 8$.

5) $g'(x) = \frac{d}{dx}\left[\frac{10}{5 - e^x}\right] = \frac{d}{dx}[10(5 - e^x)^{-1}]$
$= -10(5 - e^x)^{-2} \bullet \frac{d}{dx}[5 - e^x]$
$= -10(5 - e^x)^{-2}(-e^x)$
$= \frac{10e^x}{(5 - e^x)^2}$.

7) $g'(x) = \frac{d}{dx}[4x^2 e^x]$
$= \frac{d}{dx}[4x^2] \bullet e^x + 4x^2 \bullet \frac{d}{dx}[e^x]$
$= 8xe^x + 4x^2 e^x = 4xe^x(x + 2)$.

9) $g'(x) = \frac{d}{dx}[\sqrt{12 - e^x}] = \frac{d}{dx}[(12 - e^x)^{1/2}]$
$= \frac{1}{2}(12 - e^x)^{-1/2} \bullet \frac{d}{dx}(12 - e^x)$
$= \frac{1}{2}(12 - e^x)^{-1/2} \bullet (-e^x) = \frac{-e^x}{2\sqrt{12 - e^x}}$.

11) $g'(x) = \frac{d}{dx}\left[\frac{e^x - 10}{x^3 - 1}\right] =$
$\frac{\frac{d}{dx}[e^x - 10] \bullet (x^3 - 1) - (e^x - 10) \bullet \frac{d}{dx}[x^3 - 1]}{(x^3 - 1)^2}$
$= \frac{e^x(x^3 - 1) - (e^x - 10)3x^2}{(x^3 - 1)^2}$
$= \frac{x^3 e^x - e^x - 3x^2 e^x + 30x^2}{(x^3 - 1)^2}$.

13) $g'(x) = \frac{d}{dx}\left[\frac{e^x + 1}{e^x - 1}\right] =$
$\frac{\frac{d}{dx}[e^x + 1] \bullet (e^x - 1) - (e^x + 1) \bullet \frac{d}{dx}[e^x - 1]}{(e^x - 1)^2}$
$= \frac{e^x(e^x - 1) - (e^x + 1)e^x}{(e^x - 1)^2}$
$= \frac{e^{2x} - e^x - e^{2x} - e^x}{(e^x - 1)^2} = \frac{-2e^x}{(e^x - 1)^2}$.

15) $g'(x) = \frac{d}{dx}[2xe^x - x] = 2\frac{d}{dx}[xe^x] - \frac{d}{dx}[x]$
$= 2\frac{d}{dx}[x] \bullet e^x + 2x \bullet \frac{d}{dx}[e^x] - 1$
$= 2e^x + 2xe^x - 1$.

17) $f'(x) = \frac{d}{dx}[e^{2x-1}] = e^{2x-1} \frac{d}{dx}(2x - 1)$
$= e^{2x-1}(2) = 2e^{2x-1}$.

19) $f'(x) = \frac{d}{dx}[e^{\sqrt{x}}] = e^{\sqrt{x}} \bullet \frac{d}{dx}[x^{1/2}]$
$= e^{\sqrt{x}} \frac{1}{2} x^{-1/2} = \frac{e^{\sqrt{x}}}{2\sqrt{x}}$.

21) $f'(x) = \frac{d}{dx}[e^{\ln(x)}] = \frac{d}{dx}[x] = 1$.

23) $f'(x) = \frac{d}{dx}[5x \bullet e^{2x}] = \frac{d}{dx}[5x] \bullet e^{2x} + 5x \bullet \frac{d}{dx}[e^{2x}]$
$= 5e^{2x} + 5xe^{2x} \bullet \frac{d}{dx}[2x] = 5e^{2x} + 5xe^{2x}(2)$
$= 5e^{2x} + 10xe^{2x}$.

25) $f'(x) = \frac{d}{dx}\left[\frac{e^{x-1}}{e^{x+1}}\right] = \frac{d}{dx}[e^{x-1-(x+1)}]$
$= \frac{d}{dx}[e^{-2}] = 0$

27) $f'(x) = \frac{d}{dx} [\ln (x^2 + e^{-x})]$

$= \frac{1}{x^2 + e^{-x}} \cdot \frac{d}{dx} [x^2 + e^{-x}]$

$= \frac{1}{x^2 + e^{-x}} (2x - e^{-x}) = \frac{2x - e^{-x}}{x^2 + e^{-x}} \cdot \frac{e^x}{e^x}$

$= \frac{2xe^x - 1}{x^2 e^x + 1}.$

29) $g'(x) = \frac{d}{dx} [10^x] = 10^x \ln (10).$

31) $g'(x) = \frac{d}{dx} [\frac{5^x}{15^x}] = \frac{d}{dx} [(\frac{1}{3})^x] = (\frac{1}{3})^x \ln (\frac{1}{3})$

$= (\frac{1}{3})^x [\ln (1) - \ln (3)] = (\frac{1}{3})^x [0 - \ln (3)]$

$= -(\frac{1}{3})^x \ln (3).$

33) $g'(x) = \frac{d}{dx} [x^3 \cdot 0.3^x]$

$= \frac{d}{dx} [x^3] \cdot 0.3^x + x^3 \frac{d}{dx} [0.3^x]$

$= 3x^2 \cdot 0.3^x + x^3 \cdot 0.3^x \ln (0.3)$

$= x^2 \, 0.3^x \, (3 + x \ln (0.3)).$

35) $f'(x) = \frac{d}{dx} [10^{x+3}] = 10^{x+3} \ln (10) \cdot \frac{d}{dx} [x + 3]$

$= 10^{x+3} \ln (10) \cdot (1) = 10^{x+3} \ln (10)$

$= 10^x \cdot 10^3 \ln (10) = 10^{x+3} \ln (10).$

37) $f'(x) = \frac{d}{dx} [9^{1/x}] = 9^{1/x} \ln (9) \cdot \frac{d}{dx} [1/x]$

$= 9^{1/x} \ln (9) \cdot \frac{d}{dx} [x^{-1}] = 9^{1/x} \ln (9) \cdot (-1) x^{-2}$

$= \frac{-9^{1/x} \ln (9)}{x^2}.$

39) $f'(x) = \frac{d}{dx} [xe^x - 5^{2x}] = \frac{d}{dx} [xe^x] - \frac{d}{dx} [5^{2x}]$

$= \frac{d}{dx} [x] \cdot e^x + x \cdot \frac{d}{dx} [e^x] - 5^{2x} \ln (5) \frac{d}{dx} (2x)$

$= e^x + xe^x - (5^2)^x \ln (5) (2)$

$= e^x + xe^x - 2 \cdot 25^x \ln (5).$

41) $f'(x) = \frac{d}{dx} [\ln (5x) \cdot 5^{x^2}]$

$= \frac{d}{dx} [(\ln (5) + \ln (x)) \cdot 5^{x^2}]$

$= \frac{d}{dx} [(\ln (5) + \ln (x))] \cdot 5^{x^2}$

$\quad + (\ln (5) + \ln (x)) \cdot \frac{d}{dx} [5^{x^2}]$

$= \frac{1}{x} \cdot 5^{x^2} + (\ln (5x)) \cdot 5^{x^2} \cdot \ln (5) \cdot \frac{d}{dx} [x^2]$

$= \frac{1}{x} \cdot 5^{x^2} + (\ln (5x)) \cdot 5^{x^2} \ln (5) \cdot (2x)$

$= \frac{5^{x^2}}{x} + 2x \cdot 5^{x^2} \ln (5) \cdot \ln (5x).$

43a) $f'(t) = \frac{d}{dt} [2.1e^{0.2t}] = 2.1e^{0.2t} \cdot \frac{d}{dt} [0.2t]$

$= 2.1e^{0.2t} (0.2) = 0.42e^{0.2t}.$

43b) $f'(3) = 0.42e^{0.2(3)} = 0.42e^{0.6} \approx 0.77.$
Three days after an animal specimen is exposed to a new pesticide, the diameter of the tumor is increasing at a rate of about 0.77 millimeters per day.

45a) $p'(t) = \frac{d}{dt} [12(0.8)^t] = 12(0.8)^t \ln (0.8).$

45b) $p'(1) = 12(0.8)^{(1)} \ln (0.8) \approx -2.14.$
One year after the factory opened, the fish population was decreasing at a rate of about 214 fish per year.
$p'(8) = 12(0.8)^{(8)} \ln (0.8) \approx -0.45.$
Eight years after the factory opened, the fish population was decreasing at a rate of about 45 fish per year. Notice that decay rate one year after the factory opened was more than 4.75 times the decay rate nine years after the factory opened.

47a) The function $264.1e^{0.19x}$ is increasing on its domain, i.e., as $x \to \infty, f(x) \to \infty$. Hence, the function is an exponential growth function.

47b) $f'(x) = \frac{d}{dx} [264.1e^{0.19x}]$

$= 264.1e^{0.19x} \cdot \frac{d}{dx} [0.19x]$

$= 264.1e^{0.19x} (0.19) = 50.179e^{0.19x}.$

Section 4.3 Derivatives of Exponential Functions 133

47c) $f'(3) = 50.179e^{0.19(3)} \approx 88.73$.
In 1985, the average salary was increasing at a rate of about \$88,730 per year.
$f'(9) = 50.179e^{0.19(9)} \approx 277.438$.
In 1991, the average salary was increasing at a rate of about \$277,438 per year.
Notice that growth rate in 1991 was more than 3.1 times the growth rate in 1985.

49a) The graph is displayed in viewing window [0, 10] by [2000, 4000]

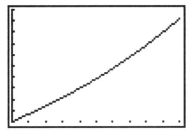

49b) $A'(t) = \frac{d}{dt}[2000e^{0.065t}]$
$= 2000e^{0.065t} \cdot \frac{d}{dt}[0.065t]$
$= 2000e^{0.065t} \cdot 0.065$
$= 130e^{0.065t}$.

49c) $A(5) = 2000e^{0.065(5)} \approx 2768.06$. After five years, there is about \$2,768.06 in the account.
$A'(5) = 130e^{0.065(5)} \approx 179.92$. After five years, amount in the account was growing at a rate of about \$179.92 per year.

51a) The function $28.69 \bullet (1.28)^x$ is increasing on its domain, i.e., as $x \to \infty, f(x) \to \infty$. Hence, the function is an exponential growth function.

51b) $f'(x) = \frac{d}{dx}[28.69 \bullet (1.28)^x]$
$= 28.69(1.28)^x \ln(1.28)$.
$f'(1) = 28.69(1.28)^{(1)} \ln(1.28) = 9.07$
In 1985, the total amount invested was growing at a rate of \$9.07 billion per year.
$f'(11) = 28.69(1.28)^{(11)} \ln(1.28) = 107.03$
In 1995, the total amount invested was growing at a rate of \$107.03 billion per year.
Notice that growth rate in 1995 was more than 11.8 times the growth rate in 1985.

51c) Let $k = \ln(1.28) \approx 0.25$ and $a = 28.69$. Then, $g(x) = a \bullet e^{kx} = 28.69e^{0.25x}$.

Section 4.4 Implicit Differentiation

1) $\frac{d}{dx}(2x) + \frac{d}{dx}(y) = \frac{d}{dx}(5)$

   $2 + \frac{dy}{dx} = 0$. Solving for $\frac{dy}{dx}$ yields:

   $2 + \frac{dy}{dx} = 0$

   $\frac{dy}{dx} = -2$.

3) $\frac{d}{dx}(x) + \frac{d}{dx}(3y^2) = \frac{d}{dx}(4)$

   $1 + 6y\frac{dy}{dx} = 0$. Solving for $\frac{dy}{dx}$ yields:

   $1 + 6y\frac{dy}{dx} = 0$

   $6y\frac{dy}{dx} = -1$

   $\frac{dy}{dx} = -\frac{1}{6y}$.

5) $\frac{d}{dx}(2x^2) + \frac{d}{dx}(2y^2) = \frac{d}{dx}(32)$

   $4x + 4y\frac{dy}{dx} = 0$. Solving for $\frac{dy}{dx}$ yields:

   $4x + 4y\frac{dy}{dx} = 0$

   $4y\frac{dy}{dx} = -4x$

   $\frac{dy}{dx} = -\frac{x}{y}$.

7) $\frac{d}{dx}(5x^3) + \frac{d}{dx}(y^3) - \frac{d}{dx}(x^4) = \frac{d}{dx}(0)$

   $15x^2 + 3y^2\frac{dy}{dx} - 4x^3 = 0$. Solving for $\frac{dy}{dx}$ yields:

   $15x^2 + 3y^2\frac{dy}{dx} - 4x^3 = 0$

   $3y^2\frac{dy}{dx} = 4x^3 - 15x^2$

   $\frac{dy}{dx} = \frac{4x^3 - 15x^2}{3y^2}$.

9) $\frac{d}{dx}(x^{1/4}) + \frac{d}{dx}(y^{1/4}) = \frac{d}{dx}(1)$

   $\frac{1}{4}x^{-3/4} + \frac{1}{4}y^{-3/4}\frac{dy}{dx} = 0$.

9) Continued

   Solving for $\frac{dy}{dx}$ yields:

   $\frac{1}{4}x^{-3/4} + \frac{1}{4}y^{-3/4}\frac{dy}{dx} = 0$

   $\frac{1}{4}y^{-3/4}\frac{dy}{dx} = -\frac{1}{4}x^{-3/4}$

   $\frac{dy}{dx} = -\frac{\frac{1}{4}x^{-\frac{3}{4}}}{\frac{1}{4}y^{-\frac{3}{4}}} = -\frac{y^{\frac{3}{4}}}{x^{\frac{3}{4}}} = -\left(\frac{y}{x}\right)^{\frac{3}{4}}$.

11) $\frac{d}{dx}(x^2) + \frac{d}{dx}(xy) = \frac{d}{dx}(6)$

    $2x + \frac{d}{dx}(x)\bullet y + x\bullet\frac{d}{dx}(y) = 0$

    $2x + y + x\frac{dy}{dx} = 0$. Solving for $\frac{dy}{dx}$ yields:

    $2x + y + x\frac{dy}{dx} = 0$

    $x\frac{dy}{dx} = -2x - y$

    $\frac{dy}{dx} = -\frac{2x+y}{x}$.

13) $\frac{d}{dx}(x^3) - \frac{d}{dx}(y^3) + \frac{d}{dx}(12xy) = \frac{d}{dx}(0)$

    $3x^2 - 3y^2\frac{dy}{dx} + 12\frac{d}{dx}(x)\bullet y + 12x\bullet\frac{d}{dx}(y) = 0$

    $3x^2 - 3y^2\frac{dy}{dx} + 12y + 12x\frac{dy}{dx} = 0$. Solving for $\frac{dy}{dx}$ yields:

    $3x^2 - 3y^2\frac{dy}{dx} + 12y + 12x\frac{dy}{dx} = 0$

    $-3y^2\frac{dy}{dx} + 12x\frac{dy}{dx} = -3x^2 - 12y$

    $-3(y^2 - 4x)\frac{dy}{dx} = -3(x^2 + 4y)$

    $\frac{dy}{dx} = \frac{x^2 + 4y}{y^2 - 4x}$.

15) $\frac{d}{dx}[(x+y)^2] - \frac{d}{dx}(1) = \frac{d}{dx}(7x^2)$

    $2(x+y)\bullet\frac{d}{dx}(x+y) - 0 = 14x$

    $2(x+y)(1 + \frac{dy}{dx}) = 14x$

    $2x + 2y + 2(x+y)\frac{dy}{dx} = 14x$

15) Continued
Solving for $\frac{dy}{dx}$ yields:

$$2x + 2y + 2(x+y)\frac{dy}{dx} = 14x$$

$$2(x+y)\frac{dy}{dx} = 12x - 2y$$

$$\frac{dy}{dx} = \frac{2(6x-y)}{2(x+y)} = \frac{6x-y}{x+y}.$$

17) $\frac{d}{dx}[y \cdot \ln(x)] = \frac{d}{dx}(10) - \frac{d}{dx}(y)$

$\frac{d}{dx}(y) \cdot \ln(x) + y \cdot \frac{d}{dx}(\ln(x)) = 0 - \frac{dy}{dx}$

$\frac{dy}{dx}\ln(x) + \frac{y}{x} = -\frac{dy}{dx}$. Solving for $\frac{dy}{dx}$ yields:

$$\frac{dy}{dx}\ln(x) + \frac{y}{x} = -\frac{dy}{dx}$$

$$\frac{dy}{dx}\ln(x) + \frac{dy}{dx} = -\frac{y}{x}$$

$$(\ln(x) + 1)\frac{dy}{dx} = -\frac{y}{x}$$

$$\frac{dy}{dx} = -\frac{y}{x(\ln(x)+1)}.$$

19) $\frac{d}{dx}[xe^y] + \frac{d}{dx}(x^2) = \frac{d}{dx}(y^2)$

$\frac{d}{dx}(x) \cdot e^y + x\frac{d}{dx}(e^y) + 2x = 2y\frac{dy}{dx}$

$e^y + xe^y\frac{dy}{dx} + 2x = 2y\frac{dy}{dx}$. Solving for $\frac{dy}{dx}$ yields:

$$e^y + xe^y\frac{dy}{dx} + 2x = 2y\frac{dy}{dx}$$

$$xe^y\frac{dy}{dx} - 2y\frac{dy}{dx} = -(e^y + 2x)$$

$$-(2y - xe^y)\frac{dy}{dx} = -(e^y + 2x)$$

$$\frac{dy}{dx} = \frac{e^y + 2x}{2y - xe^y}.$$

21) $\frac{d}{dx}[5^y] = \frac{d}{dx}(x^3)$

$5^y \ln(5)\frac{dy}{dx} = 3x^2$. Solving for $\frac{dy}{dx}$ yields:

$$5^y \ln(5)\frac{dy}{dx} = 3x^2$$

$$\frac{dy}{dx} = \frac{3x^2}{5^y \ln(5)}.$$

23) $\frac{d}{dx}[10^{y-2}] = \frac{d}{dx}(x-3)$

$10^{y-2}\ln(10) \cdot \frac{d}{dx}(y-2) = 1$

$10^{y-2}\ln(10)\frac{dy}{dx} = 1$. Solving for $\frac{dy}{dx}$ yields:

$$10^{y-2}\ln(10)\frac{dy}{dx} = 1$$

$$\frac{dy}{dx} = \frac{1}{10^{y-2}\ln(10)}.$$

25) $\frac{d}{dx}[y^{1/3}] = \frac{d}{dx}(x^2 + 6)$

$\frac{1}{3}y^{-2/3}\frac{dy}{dx} = 2x$. Solving for $\frac{dy}{dx}$ yields:

$$\frac{1}{3}y^{-2/3}\frac{dy}{dx} = 2x$$

$$\frac{dy}{dx} = 2x \cdot 3y^{2/3} = 6x\sqrt[3]{y^2}.$$

27a) $\frac{d}{dx}(x) + \frac{d}{dx}(2y) = \frac{d}{dx}(6)$

$1 + 2\frac{dy}{dx} = 0$. Solving for $\frac{dy}{dx}$ yields:

$$1 + 2\frac{dy}{dx} = 0$$

$$2\frac{dy}{dx} = -1$$

$$\frac{dy}{dx} = -\frac{1}{2}.$$

27b) Solving for y yields:
$x + 2y = 6$
$2y = -x + 6$
$y = -\frac{1}{2}x + 3$.
Thus, $\frac{dy}{dx} = \frac{d}{dx}\left(-\frac{1}{2}x + 3\right)$
$\frac{dy}{dx} = -\frac{1}{2}.$

29a) $\frac{d}{dx}(xy) + \frac{d}{dx}(4) = \frac{d}{dx}(x^4)$

$\frac{d}{dx}(xy) + 0 = \frac{d}{dx}(x^4)$

$\frac{d}{dx}(x) \cdot y + x \cdot \frac{d}{dx}(y) = 4x^3$

$y + x\frac{dy}{dx} = 4x^3$. Solving for $\frac{dy}{dx}$ yields:

$$y + x\frac{dy}{dx} = 4x^3$$

$$x\frac{dy}{dx} = 4x^3 - y$$

$$\frac{dy}{dx} = \frac{4x^3 - y}{x}.$$

29b) Solving for y yields:
$$xy + 4 = x^4$$
$$xy = x^4 - 4$$
$$y = x^3 - \frac{4}{x}.$$
Thus, $\frac{dy}{dx} = \frac{d}{dx}(x^3 - 4x^{-1})$
$$= 3x^2 + 4x^{-2} = 3x^2 + \frac{4}{x^2}.$$
To show that the answer in part a is equivalent, substitute $y = x^3 - \frac{4}{x}$ into the answer from part a:
$$\frac{dy}{dx} = \frac{4x^3 - y}{x} = \frac{4x^3 - (x^3 - \frac{4}{x})}{x} = \frac{4x^3 - x^3 + \frac{4}{x}}{x}$$
$$= \frac{3x^3 + \frac{4}{x}}{x} = 3x^2 + \frac{4}{x^2}.$$

31a) $\frac{d}{dx}\left(\frac{x}{y}\right) - \frac{d}{dx}(x^2) = \frac{d}{dx}(1)$
$$\frac{\frac{d}{dx}[x] \cdot y - x \cdot \frac{d}{dx}[y]}{y^2} - 2x = 0$$
$$\frac{(1)y - x \cdot \frac{dy}{dx}}{y^2} - 2x = 0$$
$$\frac{y - x\frac{dy}{dx}}{y^2} - 2x = 0. \text{ Solving for } \frac{dy}{dx} \text{ yields:}$$
$$\frac{y - x\frac{dy}{dx}}{y^2} - 2x = 0$$
$$\frac{y - x\frac{dy}{dx}}{y^2} = 2x$$
$$y - x\frac{dy}{dx} = 2xy^2$$
$$-x\frac{dy}{dx} = 2xy^2 - y$$
$$\frac{dy}{dx} = \frac{-2xy^2 + y}{x}.$$

31b) Solving for y yields:
$$\frac{x}{y} - x^2 = 1$$
$$\frac{x}{y} = x^2 + 1$$
$$\frac{y}{x} = \frac{1}{x^2 + 1}$$
$$y = \frac{x}{x^2 + 1}.$$

31b) Continued

Thus, $\frac{dy}{dx} = \frac{d}{dx}\left[\frac{x}{x^2+1}\right]$
$$= \frac{\frac{d}{dx}[x] \cdot (x^2+1) - x \cdot \frac{d}{dx}[x^2+1]}{(x^2+1)^2} = \frac{(x^2+1) - x(2x)}{(x^2+1)^2}$$
$$= \frac{x^2 + 1 - 2x^2}{(x^2+1)^2} = \frac{-x^2+1}{(x^2+1)^2}.$$
To show that the answer in part a is equivalent, substitute $y = \frac{x}{x^2+1}$ into the answer from part a:
$$\frac{dy}{dx} = \frac{-2xy^2 + y}{x} = (-2xy^2 + y) \cdot \frac{1}{x}$$
$$= \left(-2x\left(\frac{x}{x^2+1}\right)^2 + \frac{x}{x^2+1}\right) \cdot \frac{1}{x}$$
$$= -2\left(\frac{x}{x^2+1}\right)^2 + \frac{1}{x^2+1}$$
$$= \frac{-2x^2}{(x^2+1)^2} + \frac{1}{x^2+1} \cdot \frac{x^2+1}{x^2+1}$$
$$= \frac{-2x^2}{(x^2+1)^2} + \frac{x^2+1}{(x^2+1)^2} = \frac{-x^2+1}{(x^2+1)^2}.$$

33) $\frac{d}{dx}(x^2) + \frac{d}{dx}(y^2) = \frac{d}{dx}(13)$
$2x + 2y\frac{dy}{dx} = 0$. Solving for $\frac{dy}{dx}$ yields:
$$2x + 2y\frac{dy}{dx} = 0$$
$$2y\frac{dy}{dx} = -2x$$
$$\frac{dy}{dx} = -\frac{x}{y}.$$
At $(3, 2)$, $\frac{dy}{dx} = -\frac{3}{2} = -1.5$.
Thus, we can find the equation of the tangent line by plugging $m = -1.5$ and $(3, 2)$ into
$$y - y_1 = m(x - x_1)$$
$$y - 2 = -1.5(x - 3)$$
$$y - 2 = -1.5x + 4.5$$
$$y = -1.5x + 6.5.$$

35) $\frac{d}{dx}(4x^2) + \frac{d}{dx}(9y^2) = \frac{d}{dx}(36)$
$8x + 18y\frac{dy}{dx} = 0$. Solving for $\frac{dy}{dx}$ yields:
$$8x + 18y\frac{dy}{dx} = 0$$
$$18y\frac{dy}{dx} = -8x$$
$$\frac{dy}{dx} = -\frac{4x}{9y}.$$

35) Continued

At (0, 2), $\frac{dy}{dx} = -\frac{4(0)}{9(2)} = 0$.

Thus, we can find the equation of the tangent line by plugging $m = 0$ and $(0, 2)$ into
$$y - y_1 = m(x - x_1)$$
$$y - 2 = 0(x - 0)$$
$$y = 2.$$

37) $\frac{d}{dx}[x \ln(y)] = \frac{d}{dx}[2x^3] - \frac{d}{dx}[2y]$

$\frac{d}{dx}[x] \cdot \ln(y) + x \frac{d}{dx}[\ln(y)] = 6x^2 - 2\frac{dy}{dx}$

$\ln(y) + \frac{x}{y}\frac{dy}{dx} = 6x^2 - 2\frac{dy}{dx}$. Solving for $\frac{dy}{dx}$ yields:

$\ln(y) + \frac{x}{y}\frac{dy}{dx} = 6x^2 - 2\frac{dy}{dx}$

$\frac{x}{y}\frac{dy}{dx} + 2\frac{dy}{dx} = 6x^2 - \ln(y)$

$(\frac{x}{y} + 2)\frac{dy}{dx} = 6x^2 - \ln(y)$

$\frac{dy}{dx} = \frac{6x^2 - \ln(y)}{\frac{x}{y} + 2} \cdot \frac{y}{y} = \frac{6x^2 y - y\ln(y)}{x + 2y}$.

At (1, 1), $\frac{dy}{dx} = \frac{6(1)^2(1) - (1)\ln(1)}{(1) + 2(1)} = \frac{6}{3} = 2$.

Thus, we can find the equation of the tangent line by plugging $m = 2$ and $(1, 1)$ into
$$y - y_1 = m(x - x_1)$$
$$y - 1 = 2(x - 1)$$
$$y - 1 = 2x - 2$$
$$y = 2x - 1.$$

39) $\frac{d}{dx}(x^2) + \frac{d}{dx}(y^2) = \frac{d}{dx}(e^y)$

$2x + 2y\frac{dy}{dx} = e^y \frac{dy}{dx}$. Solving for $\frac{dy}{dx}$ yields:

$2x + 2y\frac{dy}{dx} = e^y \frac{dy}{dx}$

$2y\frac{dy}{dx} - e^y \frac{dy}{dx} = -2x$

$(2y - e^y)\frac{dy}{dx} = -2x$

$\frac{dy}{dx} = \frac{-2x}{2y - e^y}$.

39) Continued

At (1, 0), $\frac{dy}{dx} = \frac{-2(1)}{2(0) - e^{(0)}} = \frac{-2}{-1} = 2$.

Thus, we can find the equation of the tangent line by plugging $m = 2$ and $(1, 0)$ into
$$y - y_1 = m(x - x_1)$$
$$y - 0 = 2(x - 1)$$
$$y = 2x - 2.$$

41a) $\frac{d}{dx}[p] + \frac{d}{dx}[x^2] = \frac{d}{dx}[150]$

$\frac{dp}{dx} + 2x = 0$. Solving for $\frac{dp}{dx}$ yields:

$\frac{dp}{dx} + 2x = 0$

$\frac{dp}{dx} = -2x$.

41b) $\frac{dp}{dx}\big|_{(11, 29)} = -2(11) = -22$.

At a demand level of 1100 mini picture frames and a price of $29, the price is decreasing at a rate of $22 per hundred picture frames.

43a) $\frac{d}{dx}[px] + \frac{d}{dx}[2x] = \frac{d}{dx}[1000]$

$\frac{d}{dx}[p] \cdot x + p \cdot \frac{d}{dx}[x] + 2 = 0$.

$\frac{dp}{dx}x + p + 2 = 0$. Solving for $\frac{dp}{dx}$ yields:

$\frac{dp}{dx}x + p + 2 = 0$

$\frac{dp}{dx}x = -(p + 2)$

$\frac{dp}{dx} = -\frac{p+2}{x}$.

43b) $\frac{dp}{dx}\big|_{(20, 48)} = -\frac{48+2}{20} = -2.5$.

At a demand level of 2000 tents and a price of $48, the price is decreasing at a rate of $2.50 per hundred tents.

45) $\frac{d}{dt}(2x) + \frac{d}{dt}(3y) = \frac{d}{dt}(20)$

$2\frac{dx}{dt} + 3\frac{dy}{dt} = 0$.

47) $\frac{d}{dt}(x^2) - \frac{d}{dt}(3y) = \frac{d}{dt}(1)$

$2x\frac{dx}{dt} - 3\frac{dy}{dt} = 0$.

138 Chapter 4 Additional Differentiation Techniques

49) $\frac{d}{dt}(x^2) + \frac{d}{dt}(y^2) = \frac{d}{dt}(5x)$

$2x\frac{dx}{dt} + 2y\frac{dy}{dt} = 5\frac{dx}{dt}$.

51) $\frac{d}{dt}(5xy) + \frac{d}{dt}(y^4) = \frac{d}{dt}(x)$

$5\frac{d}{dt}(x)\bullet y + 5x\bullet\frac{d}{dt}(y) + 4y^3\frac{dy}{dt} = \frac{dx}{dt}$

$5y\frac{dx}{dt} + 5x\frac{dy}{dt} + 4y^3\frac{dy}{dt} = \frac{dx}{dt}$.

53) $\frac{d}{dt}(xy) = \frac{d}{dt}(7)$

$\frac{d}{dt}(x)\bullet y + x\bullet\frac{d}{dt}(y) = 0$

$y\frac{dx}{dt} + x\frac{dy}{dt} = 0$. Substituting $x = 7$, $y = 1$,

$\frac{dx}{dt} = 2$, and solving, we get:

$(1)(2) + (7)\frac{dy}{dt} = 0$

$2 + 7\frac{dy}{dt} = 0$

$7\frac{dy}{dt} = -2$

$\frac{dy}{dt} = -\frac{2}{7}$.

55) $\frac{d}{dt}(y^2) + \frac{d}{dt}(x) = \frac{d}{dt}(3)$

$2y\frac{dy}{dt} + \frac{dx}{dt} = 0$. Substituting $x = 2$, $y = 1$,

$\frac{dy}{dt} = 2$, and solving, we get:

$2(1)(2) + \frac{dx}{dt} = 0$

$4 + \frac{dx}{dt} = 0$

$\frac{dx}{dt} = -4$.

57) $\frac{d}{dt}(x^2) + \frac{d}{dt}(y^2) = \frac{d}{dt}(25)$

$2x\frac{dx}{dt} + 2y\frac{dy}{dt} = 0$. Substituting $x = 3$,

$y = -4$, $\frac{dx}{dt} = 2$, and solving, we get:

$2(3)(2) + 2(-4)\frac{dy}{dt} = 0$

$12 - 8\frac{dy}{dt} = 0$

$-8\frac{dy}{dt} = -12$

$\frac{dy}{dt} = 1.5$.

59) We begin by implicitly differentiating with respect to time :

$\frac{d}{dt}(A) = \pi\frac{d}{dt}(r^2)$

$\frac{dA}{dt} = 2\pi r\frac{dr}{dt}$. Substituting $r = 2$ in,

$\frac{dA}{dt} = 12$ in^2 per min., and solving for $\frac{dr}{dt}$, we get:

$12 = 2\pi(2)\frac{dr}{dt}$

$12 = 4\pi\frac{dr}{dt}$

$\frac{dr}{dt} = \frac{3}{\pi} \approx 0.955$.

When the radius was two inches, the radius was increasing at a rate of about 0.955 inches per minute.

61) We begin by implicitly differentiating with respect to time:

$\frac{d}{dt}(A) = \frac{d}{dt}(x^2)$

$\frac{dA}{dt} = 2x\frac{dx}{dt}$. Substituting $x = 3$ in,

$\frac{dA}{dt} = 10$ in^2 per min., and solving for $\frac{dx}{dt}$, we get:

$10 = 2(3)\frac{dx}{dt}$

$10 = 6\frac{dx}{dt}$

$\frac{dx}{dt} = \frac{5}{3} \approx 1.67$.

When the sides were three inches, the sides were increasing at a rate of about 1.67 inches per minute.

63) We begin by implicitly differentiating with respect to time:

$\frac{d}{dt}(A) = \pi\frac{d}{dt}(r^2)$

$\frac{dA}{dt} = 2\pi r\frac{dr}{dt}$. Substituting $r = 8$ in,

$\frac{dr}{dt} = 0.02$ in per min., and solving for $\frac{dA}{dt}$, we get:

$\frac{dA}{dt} = 2\pi(8)(0.02) \approx 1.01$

When the radius was eight inches, the area was increasing at a rate of about 1.01 square inches per minute.

65) Let x be the distance between the person and the lamp. Let y be the distance between the tip of the person's shadow and the lamp.

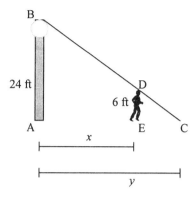

In the diagram, the two triangles, $\triangle ABC$ and $\triangle EDC$, are similar. Thus, the ratios of corresponding sides are equal. Hence, $\frac{AB}{ED} = \frac{AC}{EC}$. Substituting $AB = 24$ ft, $ED = 6$ ft, $AC = y$, $EC = y - x$, and solving for y, we get:

$$\frac{24}{6} = \frac{y}{y-x}$$
$$4 = \frac{y}{y-x}$$
$$4y - 4x = y$$
$$-4x = -3y$$
$$y = \frac{4}{3}x.$$

Now we will implicitly differentiate with respect to time and substitute $\frac{dx}{dt} = 8$ ft/s:

$$\frac{d}{dt}(y) = \frac{d}{dt}\left(\frac{4}{3}x\right)$$
$$\frac{dy}{dt} = \frac{4}{3}\frac{dx}{dt}$$
$$\frac{dy}{dt} = \frac{4}{3}(8) = \frac{32}{3} \text{ ft/s}$$

Thus, the tip of the shadow was moving at a rate of $\frac{32}{3}$ feet per second.

67a) $\frac{d}{dt}(R) = \frac{d}{dt}(250x) - \frac{d}{dt}\left(\frac{2}{5}x^2\right)$
$\frac{dR}{dt} = 250\frac{dx}{dt} - \frac{2}{5} \cdot 2x\frac{dx}{dt}$
$\frac{dR}{dt} = 250\frac{dx}{dt} - \frac{4}{5}x\frac{dx}{dt}$.

67b) $\frac{dR}{dt} = 250\frac{dx}{dt} - \frac{4}{5}x\frac{dx}{dt}$
$= 250(200) - \frac{4}{5}(100)(200) = 34000$.

When 100 suits are produced and sold per month, the revenue is increasing at a rate of \$34,000 per month.

69a) $\frac{d}{dt}(x^2) + \frac{d}{dt}(y^2) = \frac{d}{dt}(20^2)$
$2x\frac{dx}{dt} + 2y\frac{dy}{dt} = 0$.

69b) Here, $\frac{dx}{dt} = 3$ feet per second and $y = 8$ feet. We can find x by using the Pythagorean Theorem Equation:
$x^2 + (8)^2 = 20^2$
$x^2 + 64 = 400$
$x^2 = 336$
$x = \pm\sqrt{336} = \pm 4\sqrt{21} \approx \pm 18.33$.
Since x is positive, we will use
$x = 4\sqrt{21} \approx 18.33$. Hence,
$2x\frac{dx}{dt} + 2y\frac{dy}{dt} = 0$
$2(4\sqrt{21})(3) + 2(8)\frac{dy}{dt} = 0$
$24\sqrt{21} + 16\frac{dy}{dt} = 0$
$16\frac{dy}{dt} = -24\sqrt{21}$
$\frac{dy}{dt} = -1.5\sqrt{21} \approx -6.87$

When the top of the ladder is 8 feet from the ground, the ladder is sliding down the wall at a rate of about 6.87 feet per second.

71a) $\frac{d}{dt}(y) = \frac{d}{dt}(2500(1+x)^{-1})$
$\frac{dy}{dt} = -2500(1+x)^{-2}\frac{dx}{dt} = \frac{-2500}{(1+x)^2}\frac{dx}{dt}$.

71b) When $y = 100$, we can solve for x to find the PCB level:
$100 = \frac{2500}{1+x}$
$100 + 100x = 2500$
$100x = 2400$
$x = 24$ parts per million.

71b) Continued

To find $\frac{dy}{dt}$, we can substitute $x = 24$ and $\frac{dx}{dt} = 40$ ppm/year into the result found in part a:

$$\frac{dy}{dt} = \frac{-2500}{(1+x)^2} \frac{dx}{dt} = \frac{-2500}{(1+24)^2}(40) \approx -160$$

When the bass population is 100 fish, the number of bass in the pond is decreasing at a rate of 160 bass per year.

73) First, we will implicitly differentiate with respect to time:

$$\frac{d}{dt}(p) = \frac{d}{dt}(0.2x^2 + 2x)$$

$$\frac{dp}{dt} = 0.4x \frac{dx}{dt} + 2\frac{dx}{dt}$$

Substituting $x = 150$ and $\frac{dx}{dt} = 10$, we get

$\frac{dp}{dt} = 0.4(150)(10) + 2(10) = 620$. Thus, when 150 players are sold per week, the price is increasing at a rate of $620 per week.

Section 4.5 Elasticity of Demand

1) $E_a = -\dfrac{\frac{q_2-q_1}{q_1}}{\frac{p_2-p_1}{p_1}} = -\dfrac{\frac{55-50}{50}}{\frac{100-110}{110}} = -\dfrac{\frac{1}{10}}{\frac{-1}{11}} = \frac{1}{10} \div \frac{1}{11}$
$= \frac{1}{10} \cdot \frac{11}{1} = \frac{11}{10} = 1.1.$

3) $E_a = -\dfrac{\frac{q_2-q_1}{q_1}}{\frac{p_2-p_1}{p_1}} = -\dfrac{\frac{260-300}{300}}{\frac{45-40}{40}} = -\dfrac{\frac{-2}{15}}{\frac{1}{8}} = \frac{2}{15} \div \frac{1}{8}$
$= \frac{2}{15} \cdot \frac{8}{1} = \frac{16}{15} \approx 1.07.$

5) $E_a = -\dfrac{\frac{q_2-q_1}{q_1}}{\frac{p_2-p_1}{p_1}} = -\dfrac{\frac{5600-5700}{5700}}{\frac{400-390}{390}} = -\dfrac{\frac{-1}{57}}{\frac{1}{39}}$
$= \frac{1}{57} \div \frac{1}{39} = \frac{1}{57} \cdot \frac{39}{1} = \frac{13}{19} \approx 0.68.$

7a) $d(5) = 20 - 2(5) = 20 - 10 = 10.$
When the price is set at $5, the number of mouse pads sold is 1000 mouse pads.
$d(6) = 20 - 2(6) = 20 - 12 = 8.$
When the price is set at $6, the number of mouse pads sold is 800 mouse pads.

7b) $E_a = -\dfrac{\frac{q_2-q_1}{q_1}}{\frac{p_2-p_1}{p_1}} = -\dfrac{\frac{8-10}{10}}{\frac{6-5}{5}} = -\dfrac{\frac{-1}{5}}{\frac{1}{5}} = \frac{1}{5} \div \frac{1}{5}$
$= 1.$ This means that a change in price will cause a relatively equal change in demand.

9) Here, $q_1 = 320$, $p_1 = \$8.50$, $q_2 = 410$, and $p_2 = \$7.25$. Thus,
$E_a = -\dfrac{\frac{q_2-q_1}{q_1}}{\frac{p_2-p_1}{p_1}} = -\dfrac{\frac{410-320}{320}}{\frac{7.25-8.50}{8.50}} = -\dfrac{\frac{9}{32}}{\frac{-5}{34}}$
$= \frac{9}{32} \div \frac{5}{34} = \frac{9}{32} \cdot \frac{34}{5} = \frac{153}{80} = 1.9125.$
This means that a change in price will cause a relatively large change in demand.

11) Here, $q_1 = 52$, $p_1 = \$149$, $q_2 = 45$, and $p_2 = \$159$. Thus,
$E_a = -\dfrac{\frac{q_2-q_1}{q_1}}{\frac{p_2-p_1}{p_1}} = -\dfrac{\frac{45-52}{52}}{\frac{159-149}{149}} = -\dfrac{\frac{-7}{52}}{\frac{10}{149}}$
$= \frac{7}{52} \div \frac{10}{149} = \frac{7}{52} \cdot \frac{149}{10} = \frac{1043}{520} \approx 2.01.$
This means that a change in price will cause a relatively large change in demand.

13) Solving the equation for x yields:
$p = 600 - 100x$
$p - 600 = -100x$
$-0.01p + 6 = x$
Thus, $x = d(p) = 6 - 0.01p.$

15) Solving the equation for x yields:
$p = \dfrac{300}{x^2} + 10$
$p - 10 = \dfrac{300}{x^2}$
$\dfrac{1}{p-10} = \dfrac{x^2}{300}$
$\dfrac{300}{p-10} = x^2$
$x = \pm\sqrt{\dfrac{300}{p-10}}$
We will only be using the positive root.
Thus, $x = d(p) = \sqrt{\dfrac{300}{p-10}}.$

17) Solving the equation for x yields:
$p = 12 - 0.04x$
$p - 12 = -0.04x$
$-25p + 300 = x$
Thus, $x = d(p) = 300 - 25p.$

19) Solving the equation for x yields:
$$p = \sqrt{300 - x^2}$$
$$p^2 = 300 - x^2$$
$$p^2 - 300 = -x^2$$
$$300 - p^2 = x^2$$
$$x = \pm\sqrt{300 - p^2}$$
We will only be using the positive root.
Thus, $x = d(p) = \sqrt{300 - p^2}$.

21) Solving the equation for x yields:
$$p = 100e^{-0.1x}$$
$$\frac{p}{100} = e^{-0.1x}$$
$$\ln\left(\frac{p}{100}\right) = -0.1x$$
$$-10 \ln\left(\frac{p}{100}\right) = x$$
Thus, $x = d(p) = -10 \ln\left(\frac{p}{100}\right)$.

23) Since $d'(p) = \frac{d}{dp}[220 - 5p] = -5$, then
$$E(p) = \frac{-p \cdot d'(p)}{d(p)} = \frac{-p \cdot (-5)}{220 - 5p} = \frac{5p}{220 - 5p}.$$
Thus, $E(10) = \frac{5(10)}{220 - 5(10)} = \frac{50}{170} = \frac{5}{17} \approx 0.29$.
Since $\frac{5}{17} < 1$, then the demand is inelastic.

25) Since $d'(p) = \frac{d}{dp}[200 - p^2] = -2p$, then
$$E(p) = \frac{-p \cdot d'(p)}{d(p)} = \frac{-p \cdot (-2p)}{200 - p^2} = \frac{2p^2}{200 - p^2}.$$
Thus, $E(8) = \frac{2(8)^2}{200 - (8)^2} = \frac{128}{136} = \frac{16}{17} \approx 0.94$.
Since $\frac{16}{17} < 1$, then the demand is inelastic.

27) Since $d'(p) = \frac{d}{dp}[200p^{-1}] = -200p^{-2}$
$$= -\frac{200}{p^2}, \text{ then } E(p) = \frac{-p \cdot d'(p)}{d(p)}$$
$$= \frac{-p \cdot \left(\frac{-200}{p^2}\right)}{\frac{200}{p}} = \frac{\frac{200}{p}}{\frac{200}{p}} = 1.$$
Thus, $E(15) = 1$
Since $1 = 1$, then the demand is unitary.

29) Since $d'(p) = \frac{d}{dp}[(150 - 3p)^{1/2}]$
$$= \frac{1}{2}(150 - 3p)^{-1/2} \frac{d}{dp}[150 - 3p]$$
$$= \frac{1}{2}(150 - 3p)^{-1/2}(-3) = \frac{-3}{2\sqrt{150 - 3p}},$$
then $E(p) = \frac{-p \cdot d'(p)}{d(p)} = \frac{-p \cdot \frac{-3}{2\sqrt{150-3p}}}{\sqrt{150-3p}}$
$$= \frac{3p}{2\sqrt{150-3p}} \cdot \frac{1}{\sqrt{150-3p}} = \frac{3p}{2(150-3p)}.$$
Thus, $E(30) = \frac{3(30)}{2(150 - 3(30))} = \frac{90}{120} = \frac{3}{4}$.
Since $\frac{3}{4} < 1$, then the demand is inelastic.

31) Since $d'(p) = \frac{d}{dp}[100p^{-2}] = -200p^{-3}$
$$= -\frac{200}{p^3}, \text{ then } E(p) = \frac{-p \cdot d'(p)}{d(p)}$$
$$= \frac{-p \cdot \left(\frac{-200}{p^3}\right)}{\frac{100}{p^2}} = \frac{\frac{200}{p^2}}{\frac{100}{p^2}} = 2. \text{ Thus, } E(30) = 2.$$
Since $2 > 1$, then the demand is elastic.

33) Since $d'(p) = \frac{d}{dp}[4500e^{-0.02p}]$
$$= 4500e^{-0.02p} \frac{d}{dp}[-0.02p]$$
$$= 4500e^{-0.02p}(-0.02) = -90e^{-0.02p}, \text{ then}$$
$$E(p) = \frac{-p \cdot d'(p)}{d(p)} = \frac{-p \cdot -90e^{-0.02p}}{4500e^{-0.02p}} = \frac{p}{50}.$$
Thus, $E(200) = \frac{200}{50} = 4$.
Since $4 > 1$, then the demand is elastic.

35) Since $d'(p) = \frac{d}{dp}[100 \ln(1000 - 10p)]$
$$= \frac{100}{1000 - 10p} \cdot \frac{d}{dp}[1000 - 10p]$$
$$= \frac{100}{1000 - 10p} \cdot (-10) = \frac{-100}{100 - p}, \text{ then}$$
$$E(p) = \frac{-p \cdot d'(p)}{d(p)} = \frac{-p \cdot \left(\frac{-100}{100-p}\right)}{100 \ln(1000 - 10p)}$$
$$= \frac{p}{(100-p) \ln(1000 - 10p)}.$$
Thus, $E(19) = \frac{19}{(100-19) \ln(1000 - 10(19))}$
$$= \frac{19}{(81) \ln(810)} \approx 0.04.$$
Since $0.04 < 1$, then the demand is inelastic.

Section 4.5 Elasticity of Demand 143

37) Since $d'(p) = \frac{d}{dp}[100e^{-0.05p}]$
$= 100e^{-0.05p} \frac{d}{dp}[-0.05p]$
$= 100e^{-0.05p}(-0.05) = -5e^{-0.05p}$, then
$E(p) = \frac{-p \cdot d'(p)}{d(p)} = \frac{-p \cdot -5e^{-0.05p}}{100e^{-0.05p}} = \frac{p}{20}$.
Thus, $E(40) = \frac{40}{20} = 2$.
Since $2 > 1$, then the demand is elastic.

39a) Since $d'(p) = \frac{d}{dp}[3000 - 600p^{1/2}]$
$= -600 \cdot \frac{1}{2} p^{-1/2} = -300p^{-1/2} = \frac{-300}{\sqrt{p}}$, then
$E(p) = \frac{-p \cdot d'(p)}{d(p)} = \frac{-p \cdot -300p^{-1/2}}{3000 - 600p^{1/2}}$
$= \frac{p^{1/2}}{10 - 2p^{1/2}} = \frac{\sqrt{p}}{10 - 2\sqrt{p}}$.
Thus, $E(4) = \frac{\sqrt{4}}{10 - 2\sqrt{4}} = \frac{2}{6} = \frac{1}{3}$.
Since $\frac{1}{3} < 1$, then the demand is inelastic.

39b) Since the demand is inelastic, the price should be raised to increase revenue.

41a) Since $d'(p) = \frac{d}{dp}[60 - 3p] = -3$, then
$E(p) = \frac{-p \cdot d'(p)}{d(p)} = \frac{-p \cdot -3}{60 - 3p} = \frac{p}{20 - p}$.
Thus, $E(15) = \frac{15}{20 - 15} = \frac{3}{1} = 3$.
Since $3 > 1$, then the demand is elastic.

41b) Since the demand is elastic, the price should be lowered to increase revenue.

43a) Since $d'(p) = \frac{d}{dp}[50 - 2p] = -2$, then
$E(p) = \frac{-p \cdot d'(p)}{d(p)} = \frac{-p \cdot -2}{50 - 2p} = \frac{p}{25 - p}$.
Thus, $E(4) = \frac{4}{25 - 4} = \frac{4}{21} \approx 0.19$.
Since $0.19 < 1$, then the demand is inelastic. The price should be raised to increase revenue

43b) Setting $E(p)$ equal to zero and solving
yields: $\frac{p}{25 - p} = 1$
$p = 25 - p$
$2p = 25$
$p = 12.5$.
In order to maximize the revenue, the price needs to be set at $12.50 per quart.

45a) Since $d'(p) = \frac{d}{dp}[2 - \frac{p^2}{5}] = -\frac{2p}{5}$, then
$E(p) = \frac{-p \cdot d'(p)}{d(p)} = \frac{-p \cdot -\frac{2p}{5}}{2 - \frac{p^2}{5}} = \frac{2p^2}{10 - p^2}$.
Thus, $E(1.5) = \frac{2(1.5)^2}{10 - (1.5)^2} = \frac{4.5}{7.75} \approx 0.58$.
Since $0.58 < 1$, then the demand is inelastic. The price should be raised to increase revenue.

45b) Setting $E(p)$ equal to zero and solving
yields: $\frac{2p^2}{10 - p^2} = 1$
$2p^2 = 10 - p^2$
$3p^2 = 10$
$p \approx \pm 1.83$.
Since p cannot be negative, then $p \approx 1.83$. In order to maximize the revenue, the price needs to be set at $1.83 per night-light.

47a) Since $d'(p) = \frac{d}{dp}[100 \ln(150-p)]$

$= \frac{100}{150-p} \cdot \frac{d}{dp}[150-p]$

$= \frac{100}{150-p} \cdot (-1) = \frac{-100}{150-p}$, then

$E(p) = \frac{-p \cdot d'(p)}{d(p)} = \frac{-p \cdot (\frac{-100}{150-p})}{100 \ln(150-p)}$

$= \frac{p}{(150-p)\ln(150-p)}$. Thus, $E(100)$

$= \frac{100}{(150-100)\ln(150-100)} = \frac{100}{(50)\ln(50)} \approx 0.51$.

Since $0.51 < 1$, then the demand is inelastic. The price should be raised to increase revenue.

47b) Setting $E(p)$ equal to zero and solving yields: $\frac{p}{(150-p)\ln(150-p)} = 1$

$p = (150-p)\ln(150-p)$

Here, we will use a graphing calculator to get the answer by graphing each side of the equation and seeing where they intersect. Let $Y_1 = p$ and $Y_2 = (150-p)\ln(150-p)$.

The graph is displayed in viewing window [0, 150] by [0, 150]

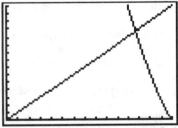

Using the intersect command, we find that:

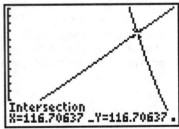

Thus, the revenue is maximized when the price is about $116.71 per T.V.

Chapter 4 Review Exercises

1) $f'(x) = \frac{d}{dx}[(x+2)^3] = 3(x+2)^2 \cdot \frac{d}{dx}(x+2)$
$= 3(x+2)^2 \cdot 1 = 3(x+2)^2.$

3) $f'(x) = \frac{d}{dx}[(8-x)^3] = 3(8-x)^2 \cdot \frac{d}{dx}(8-x)$
$= 3(8-x)^2 \cdot (-1) = -3(8-x)^2.$

5) $f'(x) = \frac{d}{dx}[(2x+5)^4] = 4(2x+5)^3 \frac{d}{dx}(2x+5)$
$= 4(2x+5)^3 \cdot 2 = 8(2x+5)^3.$

7) $g'(x) = \frac{d}{dx}[3(x^2-5x+3)^2]$
$= 3 \cdot 2(x^2-5x+3) \cdot \frac{d}{dx}(x^2-5x+3)$
$= 6(x^2-5x+3) \cdot (2x-5)$
$= 6(2x-5)(x^2-5x+3).$

9) $g'(x) = \frac{d}{dx}[(2x^2-5x+7)^{1/3}]$
$= \frac{1}{3}(2x^2-5x+7)^{-2/3} \cdot \frac{d}{dx}(2x^2-5x+7)$
$= \frac{1}{3}(2x^2-5x+7)^{-2/3}(4x-5)$
$= \frac{4x-5}{3\sqrt[3]{(2x^2-5x+7)^2}}.$

11a) $f'(x) = \frac{\frac{d}{dx}[3] \cdot (2x+9) - 3 \cdot \frac{d}{dx}[2x+9]}{(2x+9)^2}$
$= \frac{0 \cdot (2x+9) - 3 \cdot (2)}{(2x+9)^2} = \frac{-6}{(2x+9)^2}.$

11b) $f'(x) = \frac{d}{dx}[3(2x+9)^{-1}]$
$= 3(-1)(2x+9)^{-2} \cdot \frac{d}{dx}(2x+9)$
$= -3(2x+9)^{-2}(2) = -6(2x+9)^{-2}$
$= \frac{-6}{(2x+9)^2}.$

13a) $f'(x) = \frac{\frac{d}{dx}[2] \cdot (3x+5)^3 - 2 \cdot \frac{d}{dx}[(3x+5)^3]}{((3x+5)^3)^2}$
$= \frac{0 \cdot (3x+5)^3 - 2 \cdot 3(3x+5)^2 \frac{d}{dx}[3x+5]}{(3x+5)^6}$
$= \frac{-6(3x+5)^2(3)}{(3x+5)^6} = \frac{-18}{(3x+5)^4}.$

13b) $f'(x) = \frac{d}{dx}[2(3x+5)^{-3}]$
$= 2(-3)(3x+5)^{-4} \cdot \frac{d}{dx}(3x+5)$
$= -6(3x+5)^{-4}(3) = -18(3x+5)^{-4}$
$= \frac{-18}{(3x+5)^4}.$

15) $f'(x) = \frac{d}{dx}[(5x+3)^5]$
$= 5(5x+3)^4 \cdot \frac{d}{dx}(5x+3) = 5(5x+3)^4 \cdot 5$
$= 25(5x+3)^4.$ Since the slope of the tangent line at $x = -1$ is equal to $f'(-1)$
$= 25(5(-1)+3)^4 = 25(-2)^4 = 400$, we can find the equation of the tangent line by plugging $m = 400$ & $(-1, -32)$ into:
$y - y_1 = m(x - x_1)$
$y - (-32) = 400(x - (-1))$
$y + 32 = 400x + 400$
$y = 400x + 368.$

17) $f'(x) = \frac{d}{dx}[(x^2-5x+8)^4]$
$= 4(x^2-5x+8)^3 \cdot \frac{d}{dx}(x^2-5x+8)$
$= 4(x^2-5x+8)^3 (2x-5)$
$= 4(2x-5)(x^2-5x+8)^3.$ Since the slope of the tangent line at $x = 3$ is equal to $f'(3)$
$= 4(2(3)-5)((3)^2-5(3)+8)^3 = 4(1)(2)^3$
$= 32$, we can find the equation of the tangent line by plugging $m = 32$ & $(3, 16)$ into:
$y - y_1 = m(x - x_1)$
$y - 16 = 32(x - 3)$
$y - 16 = 32x - 96$
$y = 32x - 80.$

19) $g'(x) = \frac{d}{dx}[\sqrt{7x-12}] = \frac{d}{dx}[(7x-12)^{1/2}]$
$= \frac{1}{2}(7x-12)^{-1/2} \cdot \frac{d}{dx}[7x-12]$
$= \frac{1}{2}(7x-12)^{-1/2}(7) = \frac{7}{2\sqrt{7x-12}}.$

21) $f'(x) = \frac{d}{dx}[\frac{3}{\sqrt{4x+5}}] = \frac{d}{dx}[3(4x+5)^{-1/2}]$

$= -\frac{1}{2} \cdot 3(4x+5)^{-3/2} \cdot \frac{d}{dx}[4x+5]$

$= -\frac{1}{2} \cdot 3(4x+5)^{-3/2}(4) = \frac{-6}{(4x+5)\sqrt{4x+5}}$.

23) $f'(x) = \frac{d}{dx}[x](x^2+5)^3 + x\frac{d}{dx}[(x^2+5)^3]$

$= 1(x^2+5)^3 + x \cdot 3(x^2+5)^2 \frac{d}{dx}[x^2+5]$

$= (x^2+5)^3 + 3x(x^2+5)^2(2x)$

$= (x^2+5)^3 + 6x^2(x^2+5)^2$

$= (x^2+5)^2(x^2+5+6x^2) = (x^2+5)^2(7x^2+5)$.

25) $f'(x) = \dfrac{\frac{d}{dx}[3x-7] \cdot (5x-6)^{\frac{1}{2}} - (3x-7) \cdot \frac{d}{dx}[(5x-6)^{\frac{1}{2}}]}{((5x-6)^{\frac{1}{2}})^2}$

$= \dfrac{3(5x-6)^{\frac{1}{2}} - (3x-7) \cdot \frac{1}{2}(5x-6)^{-\frac{1}{2}} \frac{d}{dx}[5x-6]}{5x-6}$

$= \dfrac{3\sqrt{5x-6} - \frac{3x-7}{2\sqrt{5x-6}} \cdot (5)}{5x-6} \cdot \dfrac{2\sqrt{5x-6}}{2\sqrt{5x-6}}$

$= \dfrac{6(5x-6) - (3x-7)(5)}{2(5x-6)\sqrt{5x-6}} = \dfrac{30x - 36 - 15x + 35}{2(5x-6)\sqrt{5x-6}}$

$= \dfrac{15x-1}{2(5x-6)\sqrt{5x-6}}$.

27) $y' = \frac{d}{dx}[(3x^2 - x + 1)^{0.67}]$

$= 0.67(3x^2 - x + 1)^{-0.33} \cdot \frac{d}{dx}(3x^2 - x + 1)$

$= 0.67(3x^2 - x + 1)^{-0.33} \cdot (6x - 1)$

$= 0.67(6x-1)(3x^2 - x + 1)^{-0.33} = \dfrac{0.67(6x-1)}{(3x^2-x+1)^{0.33}}$.

29) $f'(x) = \frac{d}{dx}[(x^3 - x^2 + 5x + 1)^{-0.7}] =$

$-0.7(x^3 - x^2 + 5x + 1)^{-1.7} \frac{d}{dx}(x^3 - x^2 + 5x + 1)$

$= -0.7(x^3 - x^2 + 5x + 1)^{-1.7}(3x^2 - 2x + 5)$

$= \dfrac{-0.7(3x^2 - 2x + 5)}{(x^3 - x^2 + 5x + 1)^{1.7}}$.

31a) $MC(x) = \frac{d}{dx}[(10x-8)^{1.5} + 480]$

$= 1.5(10x-8)^{0.5} \cdot \frac{d}{dx}[10x-8] + 0$

$= 1.5(10x-8)^{0.5}(10) = 15(10x-8)^{0.5}$

$= 15\sqrt{10x-8}$.

31b) $MC(7) = 15\sqrt{10(7)-8} = 15\sqrt{62}$

≈ 118.11.

When production is at 700 windows, the cost is increasing at a rate of about $11,811 per hundred windows.

31c) $AC(x) = \dfrac{C(x)}{x} = \dfrac{(10x-8)^{1.5} + 480}{x}$

$= \dfrac{(10x-8)^{1.5}}{x} + \dfrac{480}{x}$.

31d) In general, $MAC(x) = \frac{d}{dx}[\frac{C(x)}{x}]$

$= \dfrac{\frac{d}{dx}[C(x)] \cdot x - C(x) \cdot \frac{d}{dx}[x]}{x^2} = \dfrac{MC(x) \cdot x - C(x)}{x^2}$.

Substituting $MC(x) = 15\sqrt{10x-8}$ from part a and $C(x) = (10x-8)^{1.5} + 480$, we get:

$MAC(x) = \dfrac{MC(x) \cdot x - C(x)}{x^2}$

$= \dfrac{(15\sqrt{10x-8}) \cdot x - [(10x-8)^{1.5} + 480]}{x^2}$

$= \dfrac{15x\sqrt{10x-8} - (10x-8)^{1.5} - 480}{x^2}$

$= \dfrac{15\sqrt{10x-8}}{x} - \dfrac{(10x-8)^{1.5}}{x^2} - \dfrac{480}{x^2}$.

31e) $MAC(7) = \dfrac{50(7)^2 + 40(7) - 64 - 480\sqrt{10(7)-8}}{(7)^2\sqrt{10(7)-8}}$

$= \dfrac{2450 + 280 - 64 - 480\sqrt{62}}{49\sqrt{62}} \approx \dfrac{-1113.52377953}{385.826385827}$

≈ -2.89. When production is at 700 windows, the average cost is decreasing at a rate of about $289 per hundred windows.

33) $f'(x) = \frac{d}{dx}[\ln(x^5)] = \frac{d}{dx}[5\ln(x)] = 5 \cdot \frac{1}{x} = \frac{5}{x}$.

35) $f'(x) = \frac{d}{dx}[\ln(3x^2 - 5)] = \frac{1}{3x^2-5} \frac{d}{dx}(3x^2-5)$

$= \dfrac{1}{3x^2-5}(6x) = \dfrac{6x}{3x^2-5}$.

37) $f'(x) = \frac{d}{dx}\left[\frac{x^2+5x-2}{\ln(x+4)}\right]$

$= \frac{\frac{d}{dx}[x^2+5x-2]\cdot\ln(x+4)-(x^2+5x-2)\cdot\frac{d}{dx}[\ln(x+4)]}{[\ln(x+4)]^2}$

$= \frac{(2x+5)\cdot\ln(x+4)-(x^2+5x-2)\cdot\frac{1}{x+4}\cdot\frac{d}{dx}[x+4]}{[\ln(x+4)]^2}$

$= \frac{(2x+5)\cdot\ln(x+4)-(x^2+5x-2)\cdot\frac{1}{x+4}\cdot(1)}{[\ln(x+4)]^2} \cdot \frac{x+4}{x+4}$

$= \frac{(2x+5)(x+4)\cdot\ln(x+4)-(x^2+5x-2)}{(x+4)[\ln(x+4)]^2}$

$= \frac{(2x^2+13x+20)\ln(x+4)-x^2-5x+2}{(x+4)[\ln(x+4)]^2}$ or

$= \frac{2x+5}{\ln(x+4)} - \frac{x^2+5x-2}{(x+4)[\ln(x+4)]^2}$.

39) $f'(x) = \frac{d}{dx}[4x^3 \cdot \log_2(x)]$

$= \frac{d}{dx}[4x^3]\cdot \log_2(x) + 4x^3 \cdot \frac{d}{dx}[\log_2(x)]$

$= 12x^2 \cdot \log_2(x) + 4x^3 \cdot \frac{1}{x\ln(2)}$

$= 12x^2 \log_2(x) + \frac{4x^2}{\ln(2)}$

$= 4x^2[3\log_2(x) + \frac{1}{\ln(2)}]$.

41) $g'(x) = \frac{d}{dx}\left[\log_4\left(\frac{x^2+5}{2x-3}\right)\right]$

$= \frac{d}{dx}[\log_4(x^2+5) - \log_4(2x-3)]$

$= \frac{1}{(x^2+5)\ln(4)}\frac{d}{dx}(x^2+5) - \frac{1}{(2x-3)\ln(4)}\frac{d}{dx}(2x-3)$

$= \frac{1}{(x^2+5)\ln(4)}(2x) - \frac{1}{(2x-3)\ln(4)}(2)$

$= \frac{1}{\ln(4)}\left[\frac{2x}{x^2+5} - \frac{2}{2x-3}\right]$.

43) $f'(x) = \frac{d}{dx}[\sqrt{e^x-5}] = \frac{d}{dx}[(e^x-5)^{\frac{1}{2}}]$

$= \frac{1}{2}(e^x-5)^{-\frac{1}{2}}\frac{d}{dx}[e^x-5]$

$= \frac{1}{2}(e^x-5)^{-\frac{1}{2}}(e^x) = \frac{e^x}{2\sqrt{e^x-5}}$.

45) $f'(x) = \frac{d}{dx}[e^{2x-13}] = e^{2x-13}\cdot\frac{d}{dx}[2x-13]$

$= e^{2x-13}(2) = 2e^{2x-13}$.

47) $f'(x) = \frac{d}{dx}\left[\frac{e^{2x}}{\ln(x-4)}\right]$

$= \frac{\frac{d}{dx}[e^{2x}]\cdot\ln(x-4)-(e^{2x})\cdot\frac{d}{dx}[\ln(x-4)]}{[\ln(x-4)]^2}$

$= \frac{e^{2x}\frac{d}{dx}[2x]\cdot\ln(x-4)-e^{2x}\frac{1}{x-4}\frac{d}{dx}(x-4)}{[\ln(x-4)]^2}$

$= \frac{e^{2x}[2]\cdot\ln(x-4)-e^{2x}\frac{1}{x-4}(1)}{[\ln(x-4)]^2}\cdot\frac{x-4}{x-4}$

$= \frac{2(x-4)e^{2x}\ln(x-4)-e^{2x}}{(x-4)[\ln(x-4)]^2}$ or

$= \frac{2e^{2x}}{\ln(x-4)} - \frac{e^{2x}}{(x-4)[\ln(x-4)]^2}$.

49a) $f'(x) = \frac{d}{dx}[\ln(x^4)] = \frac{d}{dx}[4\ln(x)] = \frac{4}{x}$.

The slope of the tangent line at $x = e^4$ is equal to $f'(e^4) = \frac{4}{e^4} = 4e^{-4}$. We can find the equation of the tangent line by plugging $m = 4e^{-4}$ & $(e^4, 16)$ into:

$y - y_1 = m(x - x_1)$
$y - 16 = 4e^{-4}(x - e^4)$
$y - 16 = 4e^{-4}x - 4$
$y = \frac{4}{e^4}x + 12 \approx 0.0733x + 12$.

49b) The graph is displayed in viewing window [0, 100] by [–4, 20]

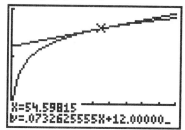

We see that the graph supports the answer in part a.

51a) $f'(x) = \frac{d}{dx}[e^x - e^{-x}] = e^x - e^{-x}\frac{d}{dx}[-x]$
$= e^x - e^{-x}(-1) = e^x + e^{-x}$.
The slope of the tangent line at $x = 0$ is equal to $f'(0) = e^x + e^{-x} = e^0 + e^0 = 2$. We can find the equation of the tangent line by plugging $m = 2$ & $(0, 0)$ into:
$y - y_1 = m(x - x_1)$
$y - 0 = 2(x - 0)$
$y = 2x$.

51b) The graph is displayed in viewing window $[-2, 2]$ by $[-8, 8]$

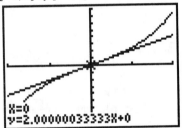

We see that the graph supports the answer in part a.

53) $g'(x) = \frac{d}{dx}[\frac{3^x}{15^x}] = \frac{d}{dx}[(\frac{1}{5})^x] = \frac{d}{dx}[5^{-x}]$
$= 5^{-x}\ln(5)\frac{d}{dx}(-x) = 5^{-x}\ln(5)(-1)$
$= -5^{-x}\ln(5)$.

55) $g'(x) = \frac{d}{dx}[\sqrt[3]{10^x}] = \frac{d}{dx}[10^{x/3}]$
$= 10^{x/3}\ln(10)\frac{d}{dx}[\frac{x}{3}] = 10^{x/3}\ln(10)[\frac{1}{3}]$
$= \frac{1}{3}\sqrt[3]{10^x}\ln(10)$.

57) $f'(x) = \frac{d}{dx}[0.4^{\sqrt{x}}] = 0.4^{\sqrt{x}}\ln(0.4)\cdot\frac{d}{dx}[x^{1/2}]$
$= 0.4^{\sqrt{x}}\ln(0.4)\cdot\frac{1}{2}[x^{-1/2}] = \frac{0.4^{\sqrt{x}}\ln(0.4)}{2\sqrt{x}}$.

59) $f'(x) = \frac{d}{dx}[\log_7(3x) - \log_3(7x)]$
$= \frac{d}{dx}[\log_7(3) + \log_7(x) - \log_3(7) - \log_3(x)]$
$= 0 + \frac{1}{x\ln(7)} - 0 - \frac{1}{x\ln(3)} = \frac{1}{x\ln(7)} - \frac{1}{x\ln(3)}$.

61a) The function $2149.6\cdot(1.036)^x$ is increasing on its domain, i.e., as $x \to \infty, f(x) \to \infty$. Hence, the function is an exponential growth function.

61b) The graph is displayed in viewing window $[1, 26]$ by $[2000, 6000]$

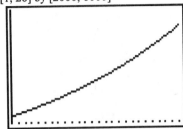

61c) $f'(x) = \frac{d}{dx}[2149.6\cdot(1.036)^x]$
$= 2149.6(1.036)^x\ln(1.036)$.

61d) $f'(7) = 2149.6(1.036)^{(7)}\ln(1.036) \approx 97.38$
In 1976, the total number of FM radio stations was growing at a rate of about 97.38 stations per year.
$f'(12) = 2149.6(1.036)^{(12)}\ln(1.036) \approx 116.22$.
In 1981, the total number of FM radio stations was growing at a rate of about 116.22 stations per year.
Thus, the rate of growth in 1981 was almost 1.2 times the rate of growth in 1976.

63) $\frac{d}{dx}(3x) - \frac{d}{dx}(5y) = \frac{d}{dx}(7)$
$3 - 5\frac{dy}{dx} = 0$. Solving for $\frac{dy}{dx}$ yields:
$3 - 5\frac{dy}{dx} = 0$
$-5\frac{dy}{dx} = -3$
$\frac{dy}{dx} = \frac{3}{5}$.

65) $\frac{d}{dx}(5x^6) + \frac{d}{dx}(y^3) = \frac{d}{dx}(x^2)$

$30x^5 + 3y^2 \frac{dy}{dx} = 2x$. Solving for $\frac{dy}{dx}$ yields:

$30x^5 + 3y^2 \frac{dy}{dx} = 2x$

$3y^2 \frac{dy}{dx} = 2x - 30x^5$

$\frac{dy}{dx} = \frac{2x - 30x^5}{3y^2}$.

67) $\frac{d}{dx}(y^3) - \frac{d}{dx}(x^2 y) = \frac{d}{dx}(5)$

$3y^2 \frac{dy}{dx} - \frac{d}{dx}(x^2) \cdot y - x^2 \cdot \frac{d}{dx}(y) = 0$

$3y^2 \frac{dy}{dx} - 2xy - x^2 \frac{dy}{dx} = 0$. Solving for $\frac{dy}{dx}$ yields:

$3y^2 \frac{dy}{dx} - 2xy - x^2 \frac{dy}{dx} = 0$

$3y^2 \frac{dy}{dx} - x^2 \frac{dy}{dx} = 2xy$

$(3y^2 - x^2) \frac{dy}{dx} = 2xy$

$\frac{dy}{dx} = \frac{2xy}{3y^2 - x^2}$.

69) $\frac{d}{dx}[x \ln(y)] = \frac{d}{dx}(y + 4)$

$\frac{d}{dx}(x) \cdot \ln(y) + x \cdot \frac{d}{dx}[\ln(y)] = \frac{dy}{dx}$

$1 \cdot \ln(y) + x \cdot \frac{1}{y} \cdot \frac{dy}{dx} = \frac{dy}{dx}$

$\ln(y) + \frac{x}{y} \frac{dy}{dx} = \frac{dy}{dx}$ (now, multiply by y)

$y \ln(y) + x \frac{dy}{dx} = y \frac{dy}{dx}$. Solving for $\frac{dy}{dx}$ yields:

$y \ln(y) = y \frac{dy}{dx} - x \frac{dy}{dx}$

$y \ln(y) = (y - x) \frac{dy}{dx}$

$\frac{dy}{dx} = \frac{y \ln(y)}{y - x}$.

71a) $\frac{d}{dx}[2x + 3y] = \frac{d}{dx}[6]$

$2 + 3 \frac{dy}{dx} = 0$. Solving for $\frac{dy}{dx}$ yields:

$3 \frac{dy}{dx} = -2$

$\frac{dy}{dx} = -\frac{2}{3}$.

71b) Solving for y yields:
$2x + 3y = 6$
$3y = -2x + 6$
$y = -\frac{2}{3}x + 2$.

Thus, $\frac{dy}{dx}$ is:

$\frac{dy}{dx} = \frac{d}{dx}[-\frac{2}{3}x + 2] = -\frac{2}{3}$.

73a) $\frac{d}{dx}(x^3 y) - \frac{d}{dx}(y) = \frac{d}{dx}(2x) + \frac{d}{dx}(2)$

$\frac{d}{dx}(x^3) \cdot y + x^3 \frac{d}{dx}(y) - \frac{dy}{dx} = 2$

$3x^2 y + x^3 \frac{dy}{dx} - \frac{dy}{dx} = 2$. Solving for $\frac{dy}{dx}$ yields:

$3x^2 y + x^3 \frac{dy}{dx} - \frac{dy}{dx} = 2$

$x^3 \frac{dy}{dx} - \frac{dy}{dx} = 2 - 3x^2 y$

$(x^3 - 1) \frac{dy}{dx} = 2 - 3x^2 y$

$\frac{dy}{dx} = \frac{2 - 3x^2 y}{x^3 - 1}$.

73b) Solving for y yields:
$x^3 y - y = 2x + 2$
$(x^3 - 1)y = 2x + 2$
$y = \frac{2x + 2}{x^3 - 1}$.

Thus, $\frac{dy}{dx}$ is:

$\frac{dy}{dx} = \frac{\frac{d}{dx}(2x+2) \cdot (x^3 - 1) - (2x + 2) \cdot \frac{d}{dx}(x^3 - 1)}{(x^3 - 1)^2}$

$= \frac{2 \cdot (x^3 - 1) - (2x + 2) \cdot (3x^2)}{(x^3 - 1)^2} = \frac{2x^3 - 2 - 6x^3 - 6x^2}{(x^3 - 1)^2}$

$= \frac{-4x^3 - 6x^2 - 2}{(x^3 - 1)^2}$.

To show that this is the same as in part a, substitute $y = \frac{2x + 2}{x^3 - 1}$ into the answer from part a and simplify:

$\frac{dy}{dx} = \frac{2 - 3x^2 y}{x^3 - 1} = \frac{2 - 3x^2 \frac{2x+2}{x^3 - 1}}{x^3 - 1} \cdot \frac{x^3 - 1}{x^3 - 1}$

$= \frac{2(x^3 - 1) - 3x^2(2x + 2)}{(x^3 - 1)^2} = \frac{2x^3 - 2 - 6x^3 - 6x^2}{(x^3 - 1)^2}$

$= \frac{-4x^3 - 6x^2 - 2}{(x^3 - 1)^2}$ which is the same as above.

150 Chapter 4 Additional Differentiation Techniques

75) $\frac{d}{dx}(x^2) - \frac{d}{dx}(y^2) = \frac{d}{dx}(9)$

$2x - 2y\frac{dy}{dx} = 0$. Solving for $\frac{dy}{dx}$ yields:

$2x - 2y\frac{dy}{dx} = 0$

$-2y\frac{dy}{dx} = -2x$

$\frac{dy}{dx} = \frac{x}{y}$.

At (5, 4), $\frac{dy}{dx} = \frac{5}{4} = 1.25$.

Thus, we can find the equation of the tangent line by plugging $m = 1.25$ and (5, 4) into

$y - y_1 = m(x - x_1)$
$y - 4 = 1.25(x - 5)$
$y - 4 = 1.25x - 6.25$
$y = 1.25x - 2.25$.

77) $\frac{d}{dx}[2x \ln(y)] = \frac{d}{dx}[4x^2] - \frac{d}{dx}[4y]$

$\frac{d}{dx}[2x] \cdot \ln(y) + 2x\frac{d}{dx}[\ln(y)] = 8x - 4\frac{dy}{dx}$

$2 \ln(y) + \frac{2x}{y}\frac{dy}{dx} = 8x - 4\frac{dy}{dx}$. Solving for $\frac{dy}{dx}$

yields:

$2 \ln(y) + \frac{2x}{y}\frac{dy}{dx} = 8x - 4\frac{dy}{dx}$

$\frac{2x}{y}\frac{dy}{dx} + 4\frac{dy}{dx} = 8x - 2\ln(y)$

$(\frac{2x}{y} + 4)\frac{dy}{dx} = 8x - 2\ln(y)$

$\frac{dy}{dx} = \frac{8x - 2\ln(y)}{\frac{2x}{y} + 4} \cdot \frac{y}{y} = \frac{4xy - y\ln(y)}{x + 2y}$.

At (1, 1), $\frac{dy}{dx} = \frac{4(1)(1) - (1)\ln(1)}{(1) + 2(1)} = \frac{4}{3}$.

Thus, we can find the equation of the tangent line by plugging $m = \frac{4}{3}$ and (1, 1) into

$y - y_1 = m(x - x_1)$
$y - 1 = \frac{4}{3}(x - 1)$
$y - 1 = \frac{4}{3}x - \frac{4}{3}$
$y = \frac{4}{3}x - \frac{1}{3}$.

79a) $\frac{d}{dx}[px] + \frac{d}{dx}[10x] = \frac{d}{dx}[600]$

$\frac{d}{dx}[p] \cdot x + p \cdot \frac{d}{dx}[x] + 10 = 0$

$\frac{dp}{dx} \cdot x + p + 10 = 0$. Solving for $\frac{dp}{dx}$ yields:

$\frac{dp}{dx} \cdot x + p + 10 = 0$

$\frac{dp}{dx} \cdot x = -(p + 10)$

$\frac{dp}{dx} = -\frac{p + 10}{x}$.

79b) $\frac{dp}{dx}\big|_{(15,30)} = -\frac{30 + 10}{15} = -\frac{40}{15} = -2\frac{2}{3}$.

When the demand is 15,000 lamps, the price per lamp is decreasing at a rate of about $2.67 per thousand lamps.

81) $\frac{d}{dt}[x^3] + \frac{d}{dt}[y^3] = \frac{d}{dt}[6x]$

$3x^2 \frac{dx}{dt} + 3y^2 \frac{dy}{dt} = 6\frac{dx}{dt}$.

83) $\frac{d}{dt}[xy] - \frac{d}{dt}[y^3] = \frac{d}{dt}[2x]$

$\frac{d}{dt}[x] \cdot y + x \cdot \frac{d}{dt}[y] - 3y^2\frac{dy}{dt} = 2\frac{dx}{dt}$

$y\frac{dx}{dt} + x\frac{dy}{dt} - 3y^2\frac{dy}{dt} = 2\frac{dx}{dt}$.

85) To find the value of y, we substitute $x = 1$ into the equation and solve:
$(1)y + 4y - 3(1) = 7$
$5y = 10$
$y = 2$.

Now differentiating with respect to t, we get:

$\frac{d}{dt}[xy] + \frac{d}{dt}[4y] - \frac{d}{dt}[3x] = \frac{d}{dt}[7]$

$\frac{d}{dt}[x] \cdot y + x \cdot \frac{d}{dt}[y] + 4\frac{dy}{dt} - 3\frac{dx}{dt} = 0$

$y\frac{dx}{dt} + x\frac{dy}{dt} + 4\frac{dy}{dt} - 3\frac{dx}{dt} = 0$. Solving for

$\frac{dx}{dt}$ yields:

$y\frac{dx}{dt} + x\frac{dy}{dt} + 4\frac{dy}{dt} - 3\frac{dx}{dt} = 0$

$y\frac{dx}{dt} - 3\frac{dx}{dt} = -(x\frac{dy}{dt} + 4\frac{dy}{dt})$

$(y - 3)\frac{dx}{dt} = -(x + 4)\frac{dy}{dt}$

$\frac{dx}{dt} = -\frac{x + 4}{y - 3} \cdot \frac{dy}{dt}$.

85) Continued

Replacing $x = 1$, $y = 2$, and $\frac{dy}{dt} = 4$, we find that:
$\frac{dx}{dt} = -\frac{1+4}{2-3} \bullet (4) = 5 \bullet 4 = 20$.

87) $\frac{d}{dt}[x^2] + \frac{d}{dt}[2xy] + \frac{d}{dt}[y^2] = \frac{d}{dt}[25]$

$2x\frac{dx}{dt} + 2\frac{d}{dt}[x]\bullet y + 2x\bullet\frac{d}{dt}[y] + 2y\frac{dy}{dt} = 0$

$2x\frac{dx}{dt} + 2y\frac{dx}{dt} + 2x\frac{dy}{dt} + 2y\frac{dy}{dt} = 0$. Solving for $\frac{dy}{dt}$, we get:

$2x\frac{dx}{dt} + 2y\frac{dx}{dt} + 2x\frac{dy}{dt} + 2y\frac{dy}{dt} = 0$

$2x\frac{dy}{dt} + 2y\frac{dy}{dt} = -(2x\frac{dx}{dt} + 2y\frac{dx}{dt})$

$(2x + 2y)\frac{dy}{dt} = -(2x + 2y)\frac{dx}{dt}$

$\frac{dy}{dt} = -\frac{dx}{dt}$.

Since $\frac{dx}{dt} = 11$, then $\frac{dy}{dt} = -11$.

89a) We begin by implicitly differentiating with respect to time :
$\frac{d}{dt}(S) = 4\pi \frac{d}{dt}(r^2)$

$\frac{dS}{dt} = 8\pi r \frac{dr}{dt}$.

89b) Substituting $r = 8$ in and $\frac{dr}{dt} = 0.7$ cm per sec., we get:
$\frac{dS}{dt} = 8\pi (8)(0.7)$.

$\frac{dS}{dt} = 44.8\pi \approx 140.74$ square centimeters

When the radius is eight inches, the surface area is increasing at a rate of about 140.74 square centimeters per second.

91) $E_a = -\frac{\frac{q_2 - q_1}{q_1}}{\frac{p_2 - p_1}{p_1}} = -\frac{\frac{70-80}{80}}{\frac{55-50}{50}} = -\frac{\frac{-1}{8}}{\frac{1}{10}} = \frac{1}{8} \div \frac{1}{10}$

$= \frac{1}{8} \bullet \frac{10}{1} = \frac{5}{4} = 1.25$.

93) $E_a = -\frac{\frac{q_2 - q_1}{q_1}}{\frac{p_2 - p_1}{p_1}} = -\frac{\frac{73-75}{75}}{\frac{15250-15000}{15000}} = -\frac{\frac{-2}{75}}{\frac{1}{60}}$

$= \frac{2}{75} \div \frac{1}{60} = \frac{2}{75} \bullet \frac{60}{1} = \frac{8}{5} = 1.6$.

95) Here, $q_1 = 45$, $p_1 = \$6.95$, $q_2 = 58$, and $p_2 = \$5.50$. Thus,

$E_a = -\frac{\frac{q_2 - q_1}{q_1}}{\frac{p_2 - p_1}{p_1}} = -\frac{\frac{58-45}{45}}{\frac{5.50-6.95}{6.95}} = -\frac{\frac{13}{45}}{\frac{-1.45}{6.95}}$

$= \frac{13}{45} \div \frac{29}{139} = \frac{13}{45} \bullet \frac{139}{29} = \frac{1807}{1305} \approx 1.38$

This means that a change in price will cause a relatively large change in demand.

97) Solving for x yields:
$p = \frac{12}{x} + 5$

$p - 5 = \frac{12}{x}$

$\frac{x}{12} = \frac{1}{p-5}$

$x = \frac{12}{p-5}$.

99) Solving for x yields:
$p = 800e^{-0.02x}$

$\frac{p}{800} = e^{-0.02x}$

$\ln\left(\frac{p}{800}\right) = \ln(e^{-0.02x})$

$\ln(p) - \ln(800) = -0.02x$

$x = \frac{\ln(p) - \ln(800)}{-0.02}$

$x = \frac{\ln(200) - \ln(p)}{0.02}$.

101a) Since $d'(p) = \frac{d}{dp}[400 - p^2] = -2p$, then

$E(p) = \frac{-p \bullet d'(p)}{d(p)} = \frac{-p \bullet (-2p)}{400 - p^2} = \frac{2p^2}{400 - p^2}$.

101b) $E(12) = \frac{2(12)^2}{400 - (12)^2} = \frac{288}{256} = \frac{9}{8}$.

Since $\frac{9}{8} > 1$, then the demand is elastic.

103a) Since $d'(p) = \frac{d}{dp}[18 \ln(120 - 3p)]$

$= \frac{18}{120-3p} \cdot \frac{d}{dp}[120 - 3p]$

$= \frac{18}{120-3p} \cdot (-3) = \frac{-54}{120-3p}$, then

$E(p) = \frac{-p \cdot d'(p)}{d(p)} = \frac{-p \cdot (\frac{-54}{120-3p})}{18\ln(120-3p)}$

$= \frac{3p}{(120-3p)\ln(120-3p)}$.

103b) $E(5) = \frac{3(5)}{(120-3(5))\ln(120-3(5))}$

$= \frac{15}{(105)\ln(105)} \approx 0.03$

Since $0.03 < 1$, then the demand is inelastic.

105a) Since $d'(p) = \frac{d}{dp}[35 \ln(150 - 5p)]$

$= \frac{35}{150-5p} \cdot \frac{d}{dp}[150 - 5p]$

$= \frac{35}{150-5p} \cdot (-5) = \frac{-175}{150-5p}$, then

$E(p) = \frac{-p \cdot d'(p)}{d(p)} = \frac{-p \cdot (\frac{-175}{150-5p})}{35\ln(150-5p)}$

$= \frac{5p}{(150-5p)\ln(150-5p)}$.

Thus, $E(25) = \frac{5(25)}{(150-5(25))\ln(150-5(25))}$

$= \frac{125}{(25)\ln(25)} \approx 1.55$

Since $1.55 > 1$, then the demand is elastic. The prices should be lowered to increase the revenue.

105b) Setting $E(p)$ equal to one and solving yields: $\frac{5p}{(150-5p)\ln(150-5p)} = 1$

$5p = (150 - 5p) \ln(150 - 5p)$
$p = (30 - p) \ln(150 - 5p)$

Here, we will use a graphing calculator to get the answer by graphing each side of the equation and seeing where they intersect. Let $Y_1 = p$ and $Y_2 = (30 - p) \ln(150 - 5p)$.
The graph is displayed in viewing window [0, 30] by [0, 30]

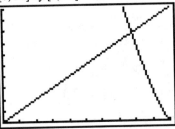

Using the intersect command, we find that

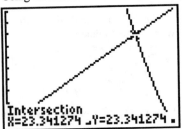

the graphs intersect at $p \approx 23.34$. Hence, the revenue is maximized at $p = \$23.34$.

CHAPTER 5 FURTHER APPLICATIONS OF THE DERIVATIVE

Section 5.1 First Derivatives and Graphs

1) The function is increasing on $(-1, 2)$, decreasing on $(-\infty, -1) \cup (2, \infty)$, and is constant at $x = -1$ and at $x = 2$.

3) The function is increasing on $(-\infty, -1)$, decreasing on $(1, \infty)$, and is constant on $[-1, 1]$.

5) The function is increasing on $(-2, 0) \cup (1, \infty)$, decreasing on $(-\infty, -2) \cup (0, 1)$, and is constant at $x = -2$, at $x = 0$, and at $x = 2$.

7a) The derivative is positive on $(-\infty, -2) \cup (0, 2)$.

7b) The derivative is negative on $(-2, 0) \cup (2, \infty)$.

7c) There are no values for which the derivative is undefined and the derivative is zero at $x = -2$, $x = 0$, and $x = 2$.

9a) The derivative is positive on $(-\infty, -2) \cup (0, 2) \cup (2, \infty)$.

9b) The derivative is negative on $(-2, 0)$.

9c) The derivative is undefined at $x = -2$ and $x = 2$, and the derivative is zero at $x = 0$.

11a) The derivative is positive on $(-\infty, 2) \cup (2, \infty)$.

11b) There are no values for which the derivative is negative.

11c) The derivative is undefined at $x = 2$, and There are no values for which the derivative is zero.

13) Since f is a polynomial, the domain is all real numbers. Computing the derivative, we get: $f'(x) = \frac{d}{dx}[2x - 3] = 2$. Since the derivative is a constant, it cannot be equal to zero or undefined. Hence, f has no critical values.

15) Since g is a polynomial, the domain is all real numbers. Computing the derivative, we get:
$g'(x) = \frac{d}{dx}[x^2 - 5x + 6] = 2x - 5$. Since the derivative is a polynomial, there are no values for which the derivative is undefined. Setting the derivative equal to zero and solving yields:
$2x - 5 = 0$
$2x = 5$
$x = 2.5$.
Thus, the only critical value of g is $x = 2.5$.

17) Since f is a polynomial, the domain is all real numbers. Computing the derivative, we get:
$f'(x) = \frac{d}{dx}[\frac{1}{3}x^3 + x^2 - 15x + 3]$
$= x^2 + 2x - 15$. Since the derivative is a polynomial, there are no values for which the derivative is undefined. Setting the derivative equal to zero and solving yields:
$x^2 + 2x - 15 = 0$
$(x + 5)(x - 3) = 0$
$x = -5$ and $x = 3$
Thus, the critical values of f are $x = -5$ and $x = 3$.

19) The domain of y is all x values such that $2x + 1 \geq 0$ or $x \geq -0.5$. Computing the derivative, we get:
$y' = \frac{d}{dx}[(2x+1)^{1/2}] = \frac{1}{2}(2x+1)^{-1/2}\frac{d}{dx}[2x+1]$
$= \frac{1}{2}\frac{1}{(2x+1)^{\frac{1}{2}}}(2) = \frac{1}{\sqrt{2x+1}}$.

19) Continued
The derivative is undefined when $2x + 1 \leq 0$ or $x \leq -0.5$. However, only $x = -0.5$ is in the domain of y. Setting the derivative equal to zero and solving yields:
$$\frac{1}{\sqrt{2x+1}} = 0$$
$$\sqrt{2x+1} \cdot \frac{1}{\sqrt{2x+1}} = 0 \cdot \sqrt{2x+1}$$
$1 = 0$, no solution.
Thus, the only critical value of y is $x = -0.5$.

21) Since f is an exponential function, its domain is all real numbers. Computing the derivative, we get:
$f'(x) = \frac{d}{dx}[e^x] = e^x$. Since the derivative is an exponential function, there are no values for which the derivative is undefined. Setting the derivative equal to zero and solving yields:
$$e^x = 0$$
$\ln(e^x) = \ln(0)$, but $\ln(0)$ is undefined. Hence, this equation has no solution. Thus, f has no critical points.

23) Since the domain of $\ln(x)$ is $x > 0$, then the domain of y is $x > 0$. Computing the derivative, we get:
$y' = \frac{d}{dx}[x + \ln(x)] = 1 + \frac{1}{x}$. Even though y' is undefined at $x = 0$, this is not a critical value since $x = 0$ is not in the domain of f. Setting the derivative equal to zero and solving yields:
$$1 + \frac{1}{x} = 0$$
$$\frac{1}{x} = -1$$
$$x = -1$$
But $x = -1$ is not in the domain of f. Thus f has no critical values.

25) Since f is an odd root, then the domain is all real numbers. Computing the derivative, we get:
$f'(x) = \frac{d}{dx}[(x+1)^{1/3}] = \frac{1}{3}(x+1)^{-2/3} \frac{d}{dx}(x+1)$
$= \frac{1}{3(x+1)^{\frac{2}{3}}}(1) = \frac{1}{3\sqrt[3]{(x+1)^2}}$.

25) Continued
The derivative is undefined at $x = -1$. Setting the derivative equal to zero and solving yields:
$$\frac{1}{3\sqrt[3]{(x+1)^2}} = 0$$
$$3\sqrt[3]{(x+1)^2} \cdot \frac{1}{3\sqrt[3]{(x+1)^2}} = 0 \cdot 3\sqrt[3]{(x+1)^2}$$
$1 = 0$, no solution.
Thus, the only critical value of f is $x = -1$.

27) Since f is a rational function, the domain of f is all real numbers except when $2x - 3 = 0$ or $x = 1.5$. Computing the derivative, we get:
$f'(x) = \frac{d}{dx}\left[\frac{x-2}{2x-3}\right]$
$= \frac{\frac{d}{dx}[x-2] \cdot (2x-3) - (x-2)\frac{d}{dx}[2x-3]}{(2x-3)^2}$
$= \frac{(1)(2x-3) - (x-2)(2)}{(2x-3)^2} = \frac{2x-3-2x+4}{(2x-3)^2}$
$= \frac{1}{(2x-3)^2}$. Even though the derivative is undefined at $x = 1.5$, this is not a critical value since $x = 1.5$ is not in the domain of f. Setting the derivative equal to zero and solving yields:
$$\frac{1}{(2x-3)^2} = 0$$
$$(2x-3)^2 \cdot \frac{1}{(2x-3)^2} = 0 \cdot (2x-3)^2$$
$1 = 0$, no solution.
Thus, f has no critical values.

29a) Recall that the domain of g was all real numbers, $g'(x) = 2x - 5$ and $x = 2.5$ was our critical value from #15. We begin by placing the critical values and any values that make the original function undefined on a sign diagram:

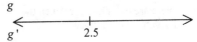

We select the test number 0 in $(-\infty, 2.5)$ and the test number 3 in $(2.5, \infty)$. Thus, $g'(0) = 2(0) - 5 = -5$ and $g'(3) = 2(3) - 5 = 6 - 5 = 1$.

Section 5.1 First Derivatives and Graphs   155

29a) Continued
Now, we can complete the sign diagram:

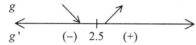

Thus, g is increasing on $(2.5, \infty)$ and decreasing on $(-\infty, 2.5)$.

29b) Since $g(2.5) = (2.5)^2 - 5(2.5) + 6 = -0.25$, then by the sign chart in part a, the function has a relative minimum at $(2.5, -0.25)$.

29c) The graph is displayed in viewing window $[-2, 8]$ by $[-2, 10]$

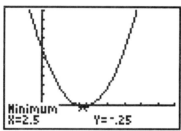

Note: The textbook writes the viewing window on the sides of the graph. The graph verifies our results.

31a) Recall that the domain of f was all real numbers, $f'(x) = x^2 + 2x - 15$ and $x = -5$ and $x = 3$ were our critical values from #17. We begin by placing the critical values and any values that make the original function undefined on a sign diagram:

We select the test number -6 in $(-\infty, -5)$, the test number 0 in $(-5, 3)$, and the test number 4 in $(3, \infty)$. Thus, $f'(-6) = 9$, $f'(0) = -15$ and $f'(4) = 9$.
Now, we can complete the sign diagram:

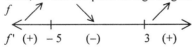

31a) Continued
Thus, f is increasing on $(-\infty, -5) \cup (3, \infty)$ and decreasing on $(-5, 3)$.

31b) Since $f(-5) = \frac{184}{3}$ and $f(3) = -24$, then by the sign chart in part a, the function has a relative minimum at $(3, -24)$ and a relative maximum at $(-5, \frac{184}{3})$.

31c) The graphs are displayed in viewing window $[-10, 10]$ by $[-50, 70]$

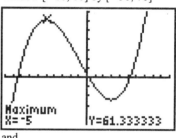

and

This verifies our results.

33a) Recall that the domain of y is all x values such that $x \geq -0.5$, $y' = \frac{1}{\sqrt{2x+1}}$ and $x = -0.5$ was our critical value. We begin by placing the critical values and any values that give us division by zero in the original function on a sign diagram:

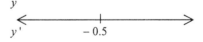

We select the test number -1 in $(-\infty, -0.5)$ and the test number 0 in $(-0.5, \infty)$. Thus, at $x = -1$, $y' =$ undefined and at $x = 0$, $y' = 1$.

33a) Continued
Now, we can complete the sign diagram:

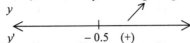

Thus, y is increasing on $(-0.5, \infty)$ and decreasing nowhere.

33b) Since there is no open interval of the domain containing -0.5, then there is no relative minimum.

33c) The graph is displayed in viewing window

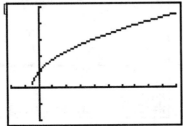

This verifies our results.

35a) Recall that the domain of y is $x > 0$, $y' = 1 + \frac{1}{x}$, and there were no critical values. Using $x = 2$ as our test value, $y' = 1.5$. Thus, y is increasing on $(0, \infty)$.

35b) Since y is not defined at $x = 0$, there are no relative extrema.

35c) The graph is displayed in viewing window $[0, -5]$ by $[-5, 5]$

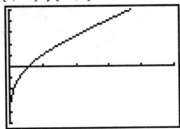

This verifies our results.

37a) Recall that the domain is all real numbers, $f'(x) = \frac{1}{3\sqrt[3]{(x+1)^2}}$, and $x = -1$ is our only critical value. We begin by placing the critical values and any values that make the original function undefined on a sign diagram:

We select the test number -2 in $(-\infty, -1)$ and the test number 0 in $(-1, \infty)$. Thus, $f'(-2) = \frac{1}{3}$ and $f'(0) = \frac{1}{3}$.

Now, we can complete the sign diagram:

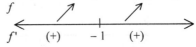

Thus, f is increasing on $(-\infty, -1) \cup (-1, \infty)$.

37b) Since f' has the same sign on both sides of $x = -1$, there are no relative extrema.

37c) The graph is displayed in viewing window $[-5, 5]$ by $[-2, 2]$

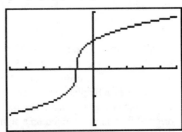

This verifies our results.

39) Since the derivative is a polynomial function, there are no values for which the derivative is undefined. Setting the derivative equal to zero and solving yields:
$$-2x + 5 = 0$$
$$-2x = -5$$
$$x = 2.5$$

39) Continued
We now place the critical values on a sign diagram:

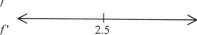

We select the test number 0 in $(-\infty, 2.5)$ and the test number 3 in $(2.5, \infty)$. Thus, $f'(0) = 5$ and $f'(3) = -1$.
Now, we can complete the sign diagram:

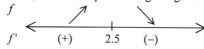

Thus, f has a relative maximum at $x = 2.5$.

41) Since the derivative is a polynomial function, there are no values for which the derivative is undefined. Setting the derivative equal to zero and solving yields:
$$3x(x + 1) = 0$$
$$3x = 0 \text{ or } x + 1 = 0$$
$$x = 0 \text{ or } x = -1$$
We now place the critical values on a sign diagram:

We select the test number -2 in $(-\infty, -1)$, the test number -0.5 in $(-1, 0)$, and the test number 1 in $(0, \infty)$. Thus, $f'(-2) = 6$, $f'(-0.5) = -0.75$ and $f'(1) = 6$.
Now, we can complete the sign diagram:

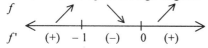

Thus, f has a relative maximum at $x = -1$ and a relative minimum at $x = 0$.

43) Since the derivative is a polynomial function, there are no values for which the derivative is undefined. Setting the derivative equal to zero and solving yields:
$$2(x + 1)^2(x - 1)(x + 3)^3 = 0$$
$$x + 1 = 0 \text{ or } x - 1 = 0 \text{ or } x + 3 = 0$$
$$x = -1 \text{ or } x = 1 \text{ or } x = -3$$

43) Continued
We now place the critical values on a sign diagram:

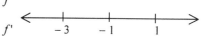

We select the test number -4 in $(-\infty, -3)$, the test number -2 in $(-3, -1)$, the test number 0 in $(-1, 1)$, and the test number 2 in $(1, \infty)$. Thus, $f'(-4) = 90$, $f'(-2) = -6$, $f'(0) = -54$, and $f'(2) = 2250$.
Now, we can complete the sign diagram:

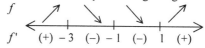

Thus, f has a relative maximum at $x = -3$ and a relative minimum at $x = 1$.

45) At the points $(0, 3)$ and $(1, 2)$, the derivative is -1. This means that the function is decreasing at those values. At the points $(3, 2)$, $(4, 3)$, and $(5, 4)$, the derivative is 1. This means that the function is increasing at those values. Since the derivative is undefined at the point $(2, 1)$, this suggests that $(2, 1)$ is a relative minimum. The graph is displayed in viewing window $[-0.1, 5.1]$ by $[-0.1, 5.1]$

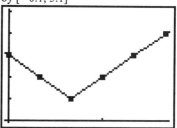

158 Chapter 5 Further Applications of the Derivative

47) Since the derivative is positive at all the points except (0, 0), the functions is increasing and (0, 0) is a possible relative minimum. The derivative is getting smaller as x gets larger suggesting that the function is "bending down." The graph is displayed in viewing window $[-0.1, 5.1]$ by $[-0.1, 2.3]$

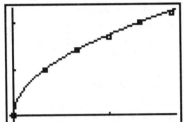

49a) The derivative is positive for $x = 1, 2, 3$, and 4 so the function is increasing at those values and the derivative is negative for the other values of x so the function is decreasing at those values. There is a relative maximum between $x = 4$ and $x = 5$. The graph is displayed in viewing window $[0.9, 10.1]$ by $[132, 146]$

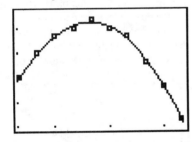

49b) The units of $f'(x)$ are the average number of patient visits per week per year.

51a) Since P is a polynomial, the domain is $[0, \infty)$ since one cannot produce a negative number of items. Computing the derivative, we get:
$P'(q) = \frac{d}{dq}[1000q - q^2] = 1000 - 2q$.

Since the derivative is a polynomial, there are no values for which the derivative is undefined. Setting the derivative equal to zero and solving yields:
$1000 - 2q = 0$
$1000 = 2q$
$q = 500$

Thus, the only critical value of P is $q = 500$. We place the critical values and any values that make the original function undefined on a sign diagram:

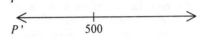

We select the test number 0 in $[0, 500)$ and the test number 501 in $(500, \infty)$. Thus, $P'(0) = 1000$ and $P'(501) = -2$ Now, we can complete the sign diagram:

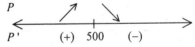

Thus, P is increasing on $[0, 500)$ and decreasing on $(500, \infty)$.

51b) Since $P(500) = 1000(500) - (500)^2$ $= 250,000$, then by the sign chart in part a, the function has a relative maximum at $(500, 250,000)$. Thus, there is a maximum profit of $250,000 when 500 golf clubs are produced and sold.

53a) Computing the derivative, we get:
$f'(x) = \frac{d}{dx}[-0.41x^2 + 13.53x + 330.69] = -0.82x + 13.53$. Since the derivative is a polynomial, there are no values for which the derivative is undefined.

53a) Continued
Setting the derivative equal to zero and solving yields:
- 0.82x + 13.53 = 0
- 0.82x = - 13.53
- x = 16.5

Thus, the only critical value of f is x = 16.5. We place the critical values and any values that make the original function undefined on a sign diagram:

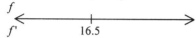

We select the test number 2 in [1, 16.5] and the test number 17 in (16.5, 31]. Thus, $f'(2) = 11.89$ and $f'(17) = -0.41$. Now, we can complete the sign diagram:

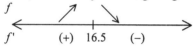

Thus, f is increasing on [1, 16.5) and decreasing on (16.5, 31].

53b) Since $f(16.5)$
$= -0.41(16.5)^2 + 13.53(16.5) + 330.69$
$= 442.3125$, then by the sign chart in part a, the function has a relative maximum at (16.5, 442.3125). Thus, half way between 1975 and 1976, the U.S. water consumption was at a maximum of about 442.31 gallons per day per capita.

55) Computing the derivative, we get:
$f'(t) = \frac{d}{dt}[0.69t^2 - 9.42t + 62.82]$
$= 1.38t - 9.42$. Since the derivative is a polynomial, there are no values for which the derivative is undefined. Setting the derivative equal to zero and solving yields:
1.38t - 9.42 = 0
1.38t = 9.42
$t \approx 6.83$

But $t \approx 6.83$ is not in the domain of f. Thus, there are no critical values of f.

55) Continued
This means that either the function in increasing on its domain or the function is decreasing on its domain. Picking 3 as our test value in the domain of f, we find that $f'(3) = 1.38(3) - 9.42 = -5.28$. Hence, f is decreasing on its domain. Thus, between 1990 and 1995, the U.S. total petroleum production was decreasing.

57a) $p(55) = 15.22e^{-0.015(55)} \approx 6.67$. When the demand is 55 pizzas, the price is $6.67.

57b) $R(x) = x \bullet p(x) = x(15.22e^{-0.015x})$
$= 15.22xe^{-0.015x}$.

57c) Since $R(x)$ is the product of an exponential function and a polynomial and the demand cannot be a negative number, then the domain of R is $[0, \infty)$. Computing the derivative, we get:
$R'(x) = \frac{d}{dx}[15.22xe^{-0.015x}]$
$= 15.22e^{-0.015x} \frac{d}{dx}(x) + 15.22x \frac{d}{dx}(e^{-0.015x})$
$= 15.22e^{-0.015x} - 0.2283xe^{-0.015x}$

Since the derivative is the product of an exponential function and a polynomial plus an exponential function, there are no values for which the derivative is undefined. Setting the derivative equal to zero and solving yields:
$15.22e^{-0.015x} - 0.2283xe^{-0.015x} = 0$
$e^{-0.015x}(15.22 - 0.2283x) = 0$
$e^{-0.015x} = 0$ or $15.22 - 0.2283x = 0$
But, $e^{-0.015x} \neq 0$ for all x.
15.22 - 0.2283x = 0
- 0.2283x = - 15.22
$x \approx 66.6667 \approx 66.67$

Thus, the only critical value of R is $x \approx 66.67$. We place the critical values and any values that make the original function undefined on a sign diagram:

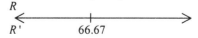

57c) Continued

We select the test number 2 in [0, 66.67) and the test number 70 in (66.67, ∞). So, $R'(2) = 15.22e^{-0.015(2)} - 0.2283(2)e^{-0.015(2)}$
$\approx 14.77 - 0.44 = 14.33$ and $R'(70) = 15.22e^{-0.015(70)} - 0.2283(70)e^{-0.015(70)}$
$\approx 5.33 - 5.59 = -0.26$

Now, we can complete the sign diagram:

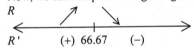

Thus, R is increasing on [0, 66.67) and decreasing on (66.67, ∞).

57d) Rounding the critical value to the nearest whole number, we get $x \approx 66.67 \approx 67$. Since $R(67) = 15.22(67)e^{-0.015(67)} \approx 373.27$, then by the sign chart in part c, the function has a relative maximum at (67, 373.27). Thus, when the demand is 67 pizzas, the maximum revenue of $373.27 is obtained.

59a) Computing the derivative, we get:

$p'(t) = \frac{d}{dt}\left[\frac{230t}{t^2 + 6t + 9}\right]$

$= \frac{(t^2 + 6t + 9)\frac{d}{dt}(230t) - 230t\frac{d}{dt}(t^2 + 6t + 9)}{(t^2 + 6t + 9)^2}$

$= \frac{(t^2 + 6t + 9)(230) - 230t(2t + 6)}{(t^2 + 6t + 9)^2}$

$= \frac{230t^2 + 1380t + 2070 - 460t^2 - 1380t}{(t^2 + 6t + 9)^2} = \frac{-230t^2 + 2070}{(t^2 + 6t + 9)^2}$

The derivative is undefined when the denominator is zero. Setting the denominator equal to zero and solving yields:
$(t^2 + 6t + 9)^2 = 0$
$(t^2 + 6t + 9) = 0$
$(t + 3)^2 = 0$
$t + 3 = 0$
$t = -3$ which is not in the domain of p so it is not a critical value.

59a) Continued

Setting the derivative equal to zero and solving yields:
$\frac{-230t^2 + 2070}{(t^2 + 6t + 9)^2} = 0$
$-230t^2 + 2070 = 0 \bullet (t^2 + 6t + 9)^2 = 0$
$-230t^2 = -2070$
$t^2 = 9$
$t = 3$ or $t = -3$

But $t = -3$ is not in the domain of p. Thus, $t = 3$ is the only critical value of p.

59b) We place the critical values and any values that make the original function undefined on a sign diagram:

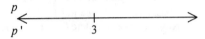

We select the test number 1 in [0, 3) and the test number 4 in (3, 20]. Thus, $p'(1)$
$= \frac{-230(1)^2 + 2070}{((1)^2 + 6(1) + 9)^2} = \frac{1840}{256} = 7.1875$ and $p'(4)$
$= \frac{-230(4)^2 + 2070}{((4)^2 + 6(4) + 9)^2} = \frac{-1610}{2401} \approx -0.67055$.

Now, we can complete the sign diagram:

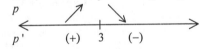

Thus, p is increasing on [0, 3) and decreasing on (3, 20].

59c) Since $p(3) = \frac{230(3)}{(3)^2 + 6(3) + 9} = \frac{690}{36} \approx 19.17$, then by the sign chart in part b, the function has a relative maximum at (3, 19.17). Three hours after the drug has been given, the concentration achieves a maximum value of about 19.17%.

61a) $AC(x) = \frac{C(x)}{x} = \frac{50 + x + \frac{x^2}{40}}{x} = \frac{50}{x} + 1 + \frac{x}{40}$.

The domain of $AC(x)$ is $(0, 100]$ since $AC(x)$ is undefined at $x = 0$.
$AC'(x) = \frac{d}{dx}\left[\frac{50}{x} + 1 + \frac{x}{40}\right] = -\frac{50}{x^2} + \frac{1}{40}$.

Even though $AC'(x)$ is undefined at $x = 0$, $x = 0$ is not a critical value since 0 is not in the domain of $AC(x)$. Setting $AC'(x) = 0$ and solving yields:

$-\frac{50}{x^2} + \frac{1}{40} = 0$

$\frac{50}{x^2} = \frac{1}{40}$

$x^2 = 2000$

$x \approx \pm 44.72$. But, -44.72 is not in the domain of $AC(x)$. So, $x \approx 44.72$ is the only critical value.

61b) We place the critical values and any values that make the original function undefined on a sign diagram:

AC
<----+---->
AC' 44.72

We select the test number 1 in $(0, 44.72)$ and the test number 50 in $(44.72, 100]$. Thus, $AC'(1) = -\frac{50}{1^2} + \frac{1}{40} = -49.975$ and $AC'(50) = -\frac{50}{50^2} + \frac{1}{40} = 0.005$.

Now, we can complete the sign diagram:

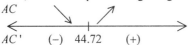

AC'   (−)   44.72   (+)

Thus, p is decreasing on $(0, 44.72)$ and increasing on $(44.72, 100]$.

61c) Rounding the critical value to the nearest whole number, we get $x \approx 44.72 \approx 45$. Since $AC(45) = \frac{50}{45} + 1 + \frac{45}{40} \approx 3.24$, then by the sign chart in part b, the function has a relative minimum at $(45, 3.24)$. This means that the minimum average cost of \$3.24 per mug is achieved when 45 mugs are produced per day.

63a) Since $P(x) = R(x) - C(x)$, then $P'(x) = \frac{d}{dx}[R(x) - C(x)] = R'(x) - C'(x)$. The profit will be maximized when $P'(x) = 0$. Thus, $R'(x) - C'(x) = 0$ which means that $R'(x) = C'(x)$. This means that the profit is maximized when the marginal revenue is equal to the marginal cost.

63b) For all production levels on (a, b), if $R'(x) > C'(x)$, then $R'(x) - C'(x) > 0$. But, $P'(x) = R'(x) - C'(x) > 0$. This means that the profit function is increasing when $R'(x) > C'(x)$.

63c) For all production levels on (a, b), if $R'(x) < C'(x)$, then $R'(x) - C'(x) < 0$. But, $P'(x) = R'(x) - C'(x) < 0$. This means that the profit function is decreasing when $R'(x) < C'(x)$.

162 Chapter 5 Further Applications of the Derivative

Section 5.2 Second Derivatives and Graphs

1) $f'(x) = \frac{d}{dx}[-4x^5 - 6x^3 + 7x]$
$= -20x^4 - 18x^2 + 7.$
$f''(x) = \frac{d}{dx}[-20x^4 - 18x^2 + 7]$
$= -80x^3 - 36x.$
$f'''(x) = \frac{d}{dx}[-80x^3 - 36x]$
$= -240x^2 - 36.$

3) $f'(x) = \frac{d}{dx}[7x^3 - 3x^2 + 4x + 5]$
$= 21x^2 - 6x + 4.$
$f''(x) = \frac{d}{dx}[21x^2 - 6x + 4]$
$= 42x - 6.$
$f'''(x) = \frac{d}{dx}[42x - 6]$
$= 42.$

5) $f'(x) = \frac{d}{dx}[e^x] = e^x.$
$f''(x) = \frac{d}{dx}[e^x] = e^x.$
$f'''(x) = \frac{d}{dx}[e^x] = e^x.$

7) $f'(x) = \frac{d}{dx}[x^{1/2}] = \frac{1}{2}x^{-1/2} = \frac{1}{2\sqrt{x}}.$
$f''(x) = \frac{d}{dx}[\frac{1}{2}x^{-1/2}] = \frac{1}{2} \cdot \frac{-1}{2}x^{-3/2}$
$= \frac{-1}{4}x^{-3/2} = -\frac{1}{4x\sqrt{x}}.$
$f'''(x) = \frac{d}{dx}[\frac{-1}{4}x^{-3/2}] = \frac{-1}{4} \cdot \frac{-3}{2}x^{-5/2}$
$= \frac{3}{8}x^{-5/2} = \frac{3}{8x^2\sqrt{x}}.$

9) $f'(x) = \frac{d}{dx}[\ln(x)] = \frac{1}{x}.$
$f''(x) = \frac{d}{dx}[\frac{1}{x}] = \frac{d}{dx}[x^{-1}] = -x^{-2} = -\frac{1}{x^2}.$
$f'''(x) = \frac{d}{dx}[-\frac{1}{x^2}] = \frac{d}{dx}[-x^{-2}] = 2x^{-3} = \frac{2}{x^3}.$

11a) Since f is "bending up" on $(1, \infty)$ and "bending down" on $(-\infty, 1)$, then f is concave up on $(1, \infty)$ and concave down on $(-\infty, 1)$.

11b) Since f is concave up on $(1, \infty)$, then f' is increasing on $(1, \infty)$. Since f is concave down on $(-\infty, 1)$, then f' is decreasing on $(-\infty, 1)$.

13a) Since f is "bending up" on $(-\infty, \infty)$ and "bending down" nowhere, then f is concave up on $(-\infty, \infty)$ and concave down nowhere.

13b) Since f is concave up on $(-\infty, \infty)$, then f' is increasing on $(-\infty, \infty)$. Since f is concave down nowhere, then f' is decreasing nowhere.

15a) Since f is "bending up" on $(-\infty, -1) \cup (1, \infty)$ and "bending down" on $(-1, 1)$, then f is concave up on $(-\infty, -1) \cup (1, \infty)$ and concave down on $(-1, 1)$.

15b) Since f is concave up on $(-\infty, -1) \cup (1, \infty)$, then f' is increasing on $(-\infty, -1) \cup (1, \infty)$. Since f is concave down on $(-1, 1)$, then f' is decreasing on $(-1, 1)$.

17) Since y is a polynomial, the domain is all real numbers. Computing the first and second derivatives, we get:
$y' = \frac{d}{dx}[x^3 + 6x^2 + 18x - 5] = 3x^2 + 12x + 18$
and $y'' = \frac{d}{dx}[3x^2 + 12x + 18] = 6x + 12.$
Since the second derivative is a polynomial, there are no values for which the second derivative is undefined. Setting the second derivative equal to zero and solving yields:
$6x + 12 = 0$
$6x = -12$
$x = -2.$

Section 5.2 Second Derivatives and Graphs 163

17) Continued
Thus, we have a possible inflection point at $x = -2$. We place this value and any values that make the original function undefined on a sign diagram:

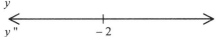

We select the test number -3 in $(-\infty, -2)$ and the test number 0 in $(-2, \infty)$. Thus, $y''|_{x=-3} = -6$ and $y''|_{x=0} = 12$.
Now, we can complete the sign diagram:

y   ∩   ∪
←—————————+—————————→
y "   (−)   −2   (+)

Hence, y is concave up on $(-2, \infty)$ and concave down on $(-\infty, -2)$. Since $y|_{x=-2} = -25$ and the concavity changes sign, then $(-2, -25)$ is an inflection point.

19) Since $g(x)$ is a polynomial, the domain is all real numbers. Computing the first and second derivatives, we get:
$g'(x) = \frac{d}{dx}[-12x^3 + 6x^2 - 24x - 11]$
$= -36x^2 + 12x - 24$
and $g''(x) = \frac{d}{dx}[-36x^2 + 12x - 24]$
$= -72x + 12$.

Since the second derivative is a polynomial, there are no values for which the second derivative is undefined. Setting the second derivative equal to zero and solving yields:
$-72x + 12 = 0$
$-72x = -12$
$x = \frac{1}{6}$.

Thus, we have a possible inflection point at $x = \frac{1}{6}$. We place this value and any values that make the original function undefined on a sign diagram:

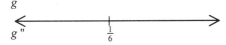

19) Continued
We select the test number 0 in $(-\infty, \frac{1}{6})$ and the test number 1 in $(\frac{1}{6}, \infty)$. Thus, $g''(0) = 12$ and $g''(1) = -60$.
Now, we can complete the sign diagram:

g   ∪   ∩
←—————————+—————————→
g "   (+) $\frac{1}{6}$ (−)

Hence, g is concave up on $(-\infty, \frac{1}{6})$ and concave down on $(\frac{1}{6}, \infty)$. Since $g(\frac{1}{6}) = -\frac{134}{9}$ and the concavity changes sign, then $(\frac{1}{6}, -\frac{134}{9})$ is an inflection point.

21) Since f is a polynomial, the domain is all real numbers. Computing the first and second derivatives, we get:
$f'(x) = \frac{d}{dx}[2x^2 + 3x + 1] = 4x + 3$
and $f''(x) = \frac{d}{dx}[4x + 3] = 4$.

Since the second derivative is a polynomial, there are no values for which the second derivative is undefined. Setting the second derivative equal to zero and solving yields:
$4 = 0$, no solution
Thus, there are no inflection points.
Since $f''(x) = 4 > 0$ for all the values in the domain of f, then f is concave up on $(-\infty, \infty)$ and concave down nowhere.

23) Since f is a polynomial, the domain is all real numbers. Computing the first and second derivatives, we get:
$f'(x) = \frac{d}{dx}[-3x^2 + 3x - 2] = -6x + 3$
and $f''(x) = \frac{d}{dx}[-6x + 3] = -6$.

Since the second derivative is a polynomial, there are no values for which the second derivative is undefined. Setting the second derivative equal to zero and solving yields:
$-6 = 0$ no solution
Thus, there are no inflection points.
Since $f''(x) = -6 < 0$ for all the values in the domain of f, then f is concave up nowhere and concave down on $(-\infty, \infty)$.

25) Since f is a rational function, the domain is all real numbers except $x = 0$. Computing the first and second derivatives, we get:
$f'(x) = \frac{d}{dx}[x + \frac{2}{x}] = \frac{d}{dx}[x + 2x^{-1}] = 1 - 2x^{-2}$
$= 1 - \frac{2}{x^2}$
and $f''(x) = \frac{d}{dx}[1 - 2x^{-2}] = 4x^{-3} = \frac{4}{x^3}$.

Since the second derivative is a rational function, it is undefined at $x = 0$, but $x = 0$ is not in the domain of f. Setting the second derivative equal to zero and solving yields:
$\frac{4}{x^3} = 0$
$4 = 0 \bullet x^3 = 0$, no solution
Thus, there are no inflection points.
We place any values that make the original function undefined on a sign diagram:
f
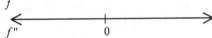
f'' 0
We select the test number -1 in $(-\infty, 0)$ and the test number 1 in $(0, \infty)$. Thus,
$f''(-1) = -4$ and $f''(1) = 4$.
Now, we can complete the sign diagram:
f ∩   ∪
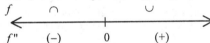
f'' $(-)$ 0 $(+)$
Hence, f is concave up on $(0, \infty)$ and concave down on $(-\infty, 0)$, but there are no inflection points.

27) Since $f(x)$ is an odd root function, the domain is all real numbers. Computing the first and second derivatives, we get:
$f'(x) = \frac{d}{dx}[x^{2/3}] = \frac{2}{3}x^{-1/3} = \frac{2}{3\sqrt[3]{x}}$
and $f''(x) = \frac{d}{dx}[\frac{2}{3}x^{-1/3}] = \frac{2}{3} \bullet \frac{-1}{3}x^{-4/3}$
$= \frac{-2}{9x\sqrt[3]{x}}$.

The second derivative is undefined at $x = 0$ which is in the domain of f. Thus, f has a possible inflection point at $x = 0$. Setting the second derivative equal to zero and solving yields: $\frac{-2}{9x\sqrt[3]{x}} = 0$
$-2 = 0 \bullet 9x\sqrt[3]{x} = 0$, no solution

27) Continued
Thus, we have a possible inflection point at $x = 0$. We place this value and any values that make the original function undefined on a sign diagram:
f
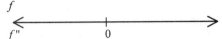
f'' 0
We select the test number -1 in $(-\infty, 0)$ and the test number 1 in $(0, \infty)$. Thus,
$f''(-1) = -\frac{2}{9}$ and $f''(1) = -\frac{2}{9}$.
Now, we can complete the sign diagram:
f ∩   ∩
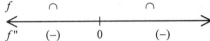
f'' $(-)$ 0 $(-)$
Hence, f is concave up nowhere and concave down on $(-\infty, 0) \cup (0, \infty)$. Since the concavity does not change sign, then there are no inflection points.

29) Since y is an exponential function, the domain is all real numbers. Computing the first and second derivatives, we get:
$y' = \frac{d}{dx}[e^{2x}] = e^{2x}\frac{d}{dx}(2x) = 2e^{2x}$
and $y'' = \frac{d}{dx}[2e^{2x}] = 2e^{2x}\frac{d}{dx}(2x) = 4e^{2x}$.

Since the second derivative is an exponential function, there are no values for which the second derivative is undefined. Setting the second derivative equal to zero and solving yields:
$4e^{2x} = 0$, no solution
Thus, there are no inflection points.
Since $y'' = 4e^{2x} > 0$ for all the values in the domain of y, then y is concave up on $(-\infty, \infty)$ and concave down nowhere.

31) Since y is an exponential function, the domain is all real numbers. Computing the first and second derivatives, we get:
$y' = \frac{d}{dx}[e^{-2x}] = e^{-2x}\frac{d}{dx}(-2x) = -2e^{-2x}$
and $y'' = \frac{d}{dx}[-2e^{-2x}] = -2e^{-2x}\frac{d}{dx}(-2x)$
$= 4e^{-2x}$.
Since the second derivative is an exponential function, there are no values for which the second derivative is undefined.

31) Continued
Setting the second derivative equal to zero and solving yields:
$4e^{-2x} = 0$, no solution
Thus, there are no inflection points.
Since $y'' = 4e^{-2x} > 0$ for all the values in the domain of y, then y is concave up on $(-\infty, \infty)$ and concave down nowhere.

33) Since g is a logarithmic function, the domain is all values of $x > -1$. Computing the first and second derivatives, we get:
$g'(x) = \frac{d}{dx}[\ln(x+1)] = \frac{1}{x+1} = (x+1)^{-1}$
and $g''(x) = \frac{d}{dx}[(x+1)^{-1}] = -1(x+1)^{-2}$
$= \frac{-1}{(x+1)^2}$
The second derivative is undefined at $x = -1$, but $x = -1$ is not in the domain of g. Setting the second derivative equal to zero and solving yields:
$\frac{-1}{(x+1)^2} = 0$
$-1 = 0 \bullet (x+1)^2 = 0$, no solution
Thus, there are no inflection points.
Since $g''(x) = \frac{-1}{(x+1)^2} < 0$ for all the values in the domain of g, then g is concave up nowhere and concave down on $(-1, \infty)$.

35a) Since f is a polynomial, the domain of f is all real numbers. Computing the derivative, we get:
$f'(x) = \frac{d}{dx}[2x^2 + 3x + 1] = 4x + 3$.
Since the derivative is a polynomial, there are no values for which the derivative is undefined. Setting $f'(x) = 0$ and solving yields:
$4x + 3 = 0$
$4x = -3$
$x = -0.75$
So, $x = -0.75$ is the critical value.

35a) Continued
We place the critical values and any values that make the original function undefined on a sign diagram:

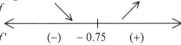

We select the test number -1 in $(-\infty, -0.75)$ and the test number 0 in $(-0.75, \infty)$. Thus $f'(-1) = -1$ and $f'(0) = 3$. Now, we can complete the sign diagram:

Thus, f is increasing on $(-0.75, \infty)$ and decreasing on $(-\infty, -0.75)$.

35b) Since $f(-0.75) = -0.125$, then by the sign chart in part a, the function has a relative minimum at $(-0.75, -0.125)$.

35c) Computing the second derivatives, we get:
$f''(x) = \frac{d}{dx}[4x + 3] = 4$.
Since the second derivative is a polynomial, there are no values for which the second derivative is undefined. Setting the second derivative equal to zero and solving yields:
$4 = 0$, no solution
Thus, there are no inflection points.
Since $f''(x) = 4 > 0$ for all the values in the domain of f, then f is concave up on $(-\infty, \infty)$ and concave down nowhere.

35d) As stated in part c, there are no inflection points.

37a) Since y is a polynomial, the domain of y is all real numbers. Computing the derivative, we get:
$y' = \frac{d}{dx}[x^3 + 6x^2 - 15x - 5]$
$= 3x^2 + 12x - 15$.
Since the derivative is a polynomial, there are no values for which the derivative is undefined.

37a) Continued
Setting $y' = 0$ and solving yields:
$3x^2 + 12x - 15 = 0$
$3(x + 5)(x - 1) = 0$
$x = -5$ or $x = 1$
So, $x = -5$ and $x = 1$ are the critical values. We place the critical values and any values that make the original function undefined on a sign diagram:

We select the test number -6 in $(-\infty, -5)$, the test number 0 in $(-5, 1)$, and the test number 2 in $(1, \infty)$. Thus, $y'|_{x=-6} = 21$, $y'|_{x=0} = -15$, and $y'|_{x=2} = 21$. Now, we can complete the sign diagram:

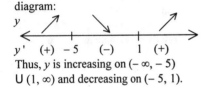

Thus, y is increasing on $(-\infty, -5) \cup (1, \infty)$ and decreasing on $(-5, 1)$.

37b) Since $y|_{x=-5} = 95$, then by the sign chart in part a, the function has a relative maximum at $(-5, 95)$. Since $y|_{x=1} = -13$, then by the sign chart in part a, the function has a relative minimum at $(1, -13)$.

37c) Computing the second derivative, we get:
$y'' = \frac{d}{dx}[3x^2 + 12x - 15] = 6x + 12$.
Since the second derivative is a polynomial, there are no values for which the second derivative is undefined. Setting the second derivative equal to zero and solving yields:
$6x + 12 = 0$
$x = -2$.
Thus, we have a possible inflection point at $x = -2$.

37c) Continued
We place this value and any values that make the original function undefined on a sign diagram:

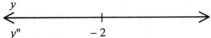

We select the test number -3 in $(-\infty, -2)$ and the test number 0 in $(-2, \infty)$. Thus, $y''|_{x=-3} = -6$ and $y''|_{x=0} = 12$. Now, we can complete the sign diagram:

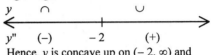

Hence, y is concave up on $(-2, \infty)$ and concave down on $(-\infty, -2)$.

37d) Since $y|_{x=-2} = 41$ and the concavity changes sign, then $(-2, 41)$ is an inflection point.

39a) Since y is a polynomial, the domain of y is all real numbers. Computing the derivative, we get:
$y' = \frac{d}{dx}[-3x^3 + 5x^2 + 2x - 9]$
$= -9x^2 + 10x + 2$.
Since the derivative is a polynomial, there are no values for which the derivative is undefined. Setting $y' = 0$ and solving yields:
$-9x^2 + 10x + 2 = 0$
Using the quadratic formula, we get:
$x = \frac{-10 \pm \sqrt{(10)^2 - 4(-9)(2)}}{2(-9)} = \frac{5 \pm \sqrt{43}}{9}$
≈ -0.17 or 1.28
So, $x \approx -0.17$ and $x \approx 1.28$ are the critical values. We place the critical values and any values that make the original function undefined on a sign diagram:

39a) Continued
We select the test number −1 in (−∞, −0.17), the test number 0 in (−0.17, 1.28), and the test number 2 in (1.28, ∞). Thus, $y'|_{x=-1} = -17$, $y'|_{x=0} = 2$, and $y'|_{x=2} = -14$. Now, we can complete the sign diagram:

f ↘ ↗ ↘
f' (−) −0.17 (+) 1.28 (−)

Thus, y is increasing on (−0.17, 1.28) and decreasing on (−∞, −0.17) ∪ (1.28, ∞).

39b) Since $y|_{x=-0.17} = -9.18$, then by the sign chart in part a, the function has a relative minimum at (−0.17, −9.18). Since $y|_{x=1.28} = -4.54$, then by the sign chart in part a, the function has a relative maximum at (1.28, −4.54).

39c) Computing the second derivative, we get:
$y'' = \frac{d}{dx}[-9x^2 + 10x + 2]$
$= -18x + 10$.
Since the second derivative is a polynomial, there are no values for which the second derivative is undefined. Setting the second derivative equal to zero and solving yields:
$-18x + 10 = 0$
$-18x = -10$
$x = \frac{5}{9}$.
Thus, we have a possible inflection point at $x = \frac{5}{9}$. We place this value and any values that make the original function undefined on a sign diagram:

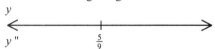

We select the test number 0 in (−∞, $\frac{5}{9}$) and the test number 1 in ($\frac{5}{9}$, ∞). Thus, $y''|_{x=0} = 10$ and $y''|_{x=1} = -8$.

39c) Continued
Now, we can complete the sign diagram:

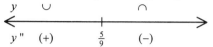

Hence, y is concave up on (−∞, $\frac{5}{9}$) and concave down on ($\frac{5}{9}$, ∞).

39d) Since $y|_{x=\frac{5}{9}} = -\frac{1667}{243}$ and the concavity changes sign, then ($\frac{5}{9}$, $-\frac{1667}{243}$) is an inflection point.

41a) Since f is a rational function, the domain is all real numbers except $x = 0$. Computing the first derivative, we get:
$f'(x) = \frac{d}{dx}[x + \frac{5}{x}] = \frac{d}{dx}[x + 5x^{-1}]$
$= 1 - 5x^{-2} = 1 - \frac{5}{x^2}$
Since the derivative is a rational function, it is undefined at $x = 0$, but $x = 0$ is not in the domain of f. Setting the derivative equal to zero and solving yields:
$1 - \frac{5}{x^2} = 0$
$1 \cdot x^2 - \frac{5}{x^2} \cdot x^2 = 0 \cdot x^2$
$x^2 - 5 = 0$
$x^2 = 5$
$x = \pm\sqrt{5} \approx \pm 2.24$
So, $x \approx -2.24$ and $x \approx 2.24$ are the critical values. We place the critical values and any values that make the original function undefined on a sign diagram:

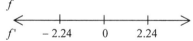

We select the test number −3 in (−∞, −2.24), the test number −1 in (−2.24, 0), the test number 1 in (0, 2.24), and the test number 3 in (2.24, ∞). Thus, $f'(-3) \approx 0.44$, $f(-1) = -4$, $f(1) = -4$ and $f(3) \approx 0.44$.

41a) Continued
Now, we can complete the sign diagram:

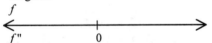

f' (+) -2.24 (−) 0 (−) 2.24 (+)
Thus, f is increasing on $(-\infty, -2.24) \cup (2.24, \infty)$ and decreasing on $(-2.24, 0) \cup (0, 2.24)$.

41b) Since $f(-2.24) \approx -4.47$, then by the sign chart in part a, the function has a relative maximum at $(-2.24, -4.47)$. Since $f(2.24) \approx 4.47$, then by the sign chart in part a, the function has a relative minimum at $(2.24, 4.47)$.

41c) Computing the second derivative, we get:
$f''(x) = \frac{d}{dx}[1 - 5x^{-2}] = 10x^{-3} = \frac{10}{x^3}$.

Since the second derivative is a rational function, it is undefined at $x = 0$, but $x = 0$ is not in the domain of f. Setting the second derivative equal to zero and solving yields:
$\frac{10}{x^3} = 0$
$10 = 0 \cdot x^3 = 0$, no solution
Thus, there are no inflection points. We replace any values that make the original function undefined on a sign diagram:

f

f'' 0

We select the test number -1 in $(-\infty, 0)$ and the test number 1 in $(0, \infty)$. Thus, $f''(-1) = -10$ and $f''(1) = 10$.
Now, we can complete the sign diagram:

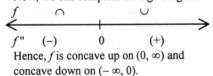

f'' (−)   0   (+)
Hence, f is concave up on $(0, \infty)$ and concave down on $(-\infty, 0)$.

41d) As stated in part c, there are no inflection points.

43a) Since $f(x)$ is a odd root function, the domain is all real numbers. Computing the first derivative, we get:
$f'(x) = \frac{d}{dx}[(x-1)^{2/3}] = \frac{2}{3}(x-1)^{-1/3}$
$= \frac{2}{3\sqrt[3]{x-1}}$.

The derivative is undefined at $x = 1$ which is in the domain of f. Thus $x = 1$ is a critical value of f. Setting the derivative equal to zero and solving yields:
$\frac{2}{3\sqrt[3]{x-1}} = 0$
$2 = 0 \cdot 3\sqrt[3]{x-1} = 0$, no solution
Thus, $x = 1$ is the only critical point. We place the critical values and any values that make the original function undefined on a sign diagram:

f

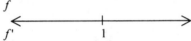

f' 1

We select the test number 0 in $(-\infty, 1)$ and the test number 2 in $(1, \infty)$. Thus, $f'(0) = -\frac{2}{3}$ and $f'(2) = \frac{2}{3}$. Now, we can complete the sign diagram:

f

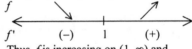

f' (−)   1   (+)
Thus, f is increasing on $(1, \infty)$ and decreasing on $(-\infty, 1)$.

43b) Since $f(1) = 0$, then by the sign chart in part a, the function has a relative minimum at $(1, 0)$.

43c) Computing the second derivative, we get:
$f''(x) = \frac{d}{dx}[\frac{2}{3}(x-1)^{-1/3}]$
$= \frac{2}{3} \cdot \frac{-1}{3}(x-1)^{-4/3} = \frac{-2}{9(x-1)\sqrt[3]{x-1}}$.

The second derivative is undefined at $x = 1$ which is in the domain of f. Thus, f has a possible inflection point at $x = 1$.
Setting the second derivative equal to zero and solving yields: $\frac{-2}{9(x-1)\sqrt[3]{x-1}} = 0$
$-2 = 0 \cdot 9(x-1)\sqrt[3]{x-1} = 0$, no solution

43c) Continued

Thus, we have a possible inflection point at $x = 1$. We place this value and any values that make the original function undefined on a sign diagram:

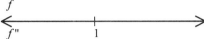

We select the test number 0 in $(-\infty, 1)$ and the test number 2 in $(1, \infty)$. Thus, $f''(0) = -\frac{2}{9}$ and $f''(2) = -\frac{2}{9}$.

Now, we can complete the sign diagram:

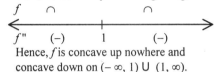

Hence, f is concave up nowhere and concave down on $(-\infty, 1) \cup (1, \infty)$.

43d) Since the concavity does not change sign, then there are no inflection points.

45a) Since y is an exponential function, the domain is all real numbers. Computing the first derivative, we get:
$$y' = \tfrac{d}{dx}[e^{3x}] = e^{3x}\tfrac{d}{dx}(3x) = 3e^{3x}.$$
Since the derivative is an exponential function, there are no values for which the derivative is undefined. Setting the derivative equal to zero and solving yields:
$3e^{3x} = 0$, no solution
Thus, there are no critical points. Since $y' = 3e^{3x} > 0$ for all the values in the domain of y, then y is increasing on $(-\infty, \infty)$ and decreasing nowhere.

45b) As stated in part a, there are no critical points. Thus, there are no relative extrema.

45c) Computing the second derivative, we get:
$$y'' = \tfrac{d}{dx}[3e^{3x}] = 3e^{3x}\tfrac{d}{dx}(3x) = 9e^{3x}.$$
Since the second derivative is an exponential function, there are no values for which the second derivative is undefined. Setting the second derivative equal to zero and solving yields:
$9e^{3x} = 0$, no solution
Thus, there are no inflection points. Since $y'' = 9e^{3x} > 0$ for all the values in the domain of y, then y is concave up on $(-\infty, \infty)$ and concave down nowhere.

45d) As stated in part c, there are no inflection points.

47a) Since g is a logarithmic function, the domain is all values of $x > -1$. Computing the first derivative, we get:
$$g'(x) = \tfrac{d}{dx}[\ln(x+1)] = \tfrac{1}{x+1} = (x+1)^{-1}$$
The derivative is undefined at $x = -1$, but $x = -1$ is not in the domain of g. Setting the derivative equal to zero and solving yields:
$\tfrac{1}{x+1} = 0$
$1 = 0 \bullet (x+1) = 0$, no solution
Thus, there are no critical points. Since $g'(x) = \tfrac{1}{x+1} > 0$ for all the values in the domain of g, then g is increasing on $(-1, \infty)$ and decreasing nowhere.

47b) As stated in part a, there are no critical points. Thus, there are no relative extrema.

47c) Computing the second derivative, we get:
$$g''(x) = \tfrac{d}{dx}[(x+1)^{-1}] = -1(x+1)^{-2}$$
$$= \tfrac{-1}{(x+1)^2}$$
The second derivative is undefined at $x = -1$, but $x = -1$ is not in the domain of g.

170 Chapter 5 Further Applications of the Derivative

47c) Continued
Setting the second derivative equal to zero and solving yields:
$$\frac{-1}{(x+1)^2} = 0$$
$-1 = 0 \bullet (x+1)^2 = 0$, no solution
Thus, there are no inflection points.
Since $g''(x) = \frac{-1}{(x+1)^2} < 0$ for all the values in the domain of g, then g is concave up nowhere and concave down on $(-1, \infty)$.

47d) As stated in part c, there are no inflection points.

49) We begin by constructing the sign diagrams for the first and second derivatives:

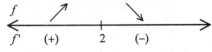

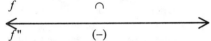

The information tells us that we have a relative maximum at $(2, 5)$ since the range is less than or equal to 5. There are no inflection points. f is increasing on $(-\infty, 2)$ and decreasing on $(2, \infty)$ and f is concave down on $(-\infty, \infty)$. Thus, we can sketch the graph:

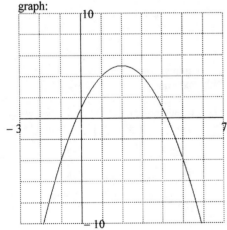

51) We begin by constructing the sign diagrams for the first and second derivatives:

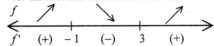

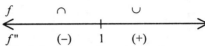

The information tells us that we have a relative maximum at $x = -1$, a relative minimum at $x = 3$ and an inflection point at $x = 1$. f is increasing on $(-\infty, -1) \cup (3, \infty)$ and decreasing on $(-1, 3)$ and f is concave up on $(1, \infty)$ and concave down on $(-\infty, 1)$. Thus, we can sketch the graph:

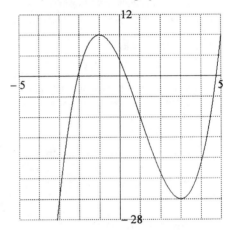

53) Computing the first and second derivatives, we get:
$f'(x) = \frac{d}{dx}[ax^2 + bx + c] = 2ax + b$ and
$f''(x) = \frac{d}{dx}[2ax + b] = 2a$.

If $a > 0$, then $f''(x) = 2a > 0$ for all values of x in the domain of f and hence f is concave up for all x. If $a < 0$, then $f''(x) = 2a < 0$ for all values of x in the domain of f and hence f is concave down for all x.

55a) $C(5) = (5)^3 - 6(5)^2 + 13(5) + 10 = 50$.
Thus, the cost to produce five items is $5,000.
$C'(x) = \frac{d}{dx}[x^3 - 6x^2 + 13x + 10] = 3x^2 - 12x + 13$.
Hence, $C'(5) = 3(5)^2 - 12(5) + 13 = 28$.
When 5 items are produced, the cost to produce the sixth item is about $2,800.

55b) Since $MC(x) = C'(x)$, then from part a, $MC(x) = 3x^2 - 12x + 13$. Since MC is a polynomial, the domain of MC is $[0, \infty)$ since a negative number of items cannot be produced. Computing the derivative, we get:
$MC'(x) = \frac{d}{dx}[3x^2 - 12x + 13] = 6x - 12$.

Since the derivative is a polynomial, there are no values for which the derivative is undefined. Setting $MC'(x) = 0$ and solving yields:
$6x - 12 = 0$
$6x = 12$
$x = 2$
So, $x = 2$ is the critical value.
We place the critical values and any values that make the original function undefined on a sign diagram:
MC

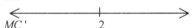

We select the test number 0 in $[0, 2)$ and the test number 3 in $(2, \infty)$. Thus $MC'(0) = -12$ and $MC'(3) = 6$. Now, we can complete the sign diagram:
MC

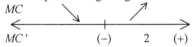

Thus, MC is increasing on $(2, \infty)$ and decreasing on $[0, 2)$. Since $MC(2) = 3(2)^2 - 12(2) + 13 = 1$, by the sign diagram above, MC has a relative minimum of 1 at $x = 2$. When 1 item is produced, the cost to produce the next item is minimized at $100.

55c) $C''(x) = \frac{d}{dx}[3x^2 - 12x + 13] = 6x - 12$.
Since the second derivative is a polynomial, there are no values for which the derivative is undefined. Setting $C''(x) = 0$ and solving yields:
$6x - 12 = 0$
$6x = 12$
$x = 2$
So, there is a possible inflection point at $x = 2$. We place this value and any values that make the original function undefined on a sign diagram:
C

We select the test number 0 in $[0, 2)$ and the test number 3 in $(2, \infty)$. Thus $C''(0) = -12$ and $C''(3) = 6$. Now, we can complete the sign diagram:
C

C" $(-)$ 2 $(+)$

Hence, since $C(2) = 20$ and the concavity changes, then $(2, 20)$ is an inflection point of C.

57a) $P(35) = -2.3(35)^3 + 445(35)^2 - 1500(35) - 200 = 393,812.5$.
Thus, the profit from producing and selling 3,500 refrigerators is $393,812.50.
$P'(x) = \frac{d}{dx}[-2.3x^3 + 445x^2 - 1500x - 200]$
$= -6.9x^2 + 890x - 1500$.
Hence, $P'(35) = -6.9(35)^2 + 890(35) - 1500 = 21,197.50$.
When 3500 refrigerators are produced and sold, the profit gained from producing and selling the next 100 refrigerators is $21,197.50.

57b) Since $MP(x) = P'(x)$, then from part a, $MP(x) = -6.9x^2 + 890x - 1500$. Since MP is a polynomial, the domain of P is $[0, \infty)$ since a negative numbers of items cannot be produced.

172 Chapter 5 Further Applications of the Derivative

57b) Continued
Computing the derivative, we get:
$MP'(x) = \frac{d}{dx}[-6.9x^2 + 890x - 1500]$
$= -13.8x + 890$. Since the derivative is a polynomial, there are no values for which the derivative is undefined. Setting $MP'(x) = 0$ and solving yields:
$-13.8x + 890 = 0$
$-13.8x = -890$
$x \approx 64.49$
So, $x \approx 64.49$ is the critical value. We place the critical values and any values that make the original function undefined on a sign diagram:
MP

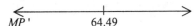
MP' 64.49

We select the test number 0 in [0, 64.49) and the test number 70 in (64.49, ∞). Thus $MP'(0) = 890$ and $MP'(70) = -76$. Now, we can complete the sign diagram:
MP

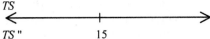
MP' (+) 64.49 (−)

Thus, MP is increasing on $(-\infty, 64.49)$ and decreasing on $(64.49, \infty)$. Since $MP(64.49) = -6.9(64.49)^2 + 890(64.49) - 1500 \approx 27{,}199.28$, by the sign diagram above, MP has a relative maximum of $\$27{,}199.28$ for the next 100 refrigerators produced and sold at $x = 6{,}449$ refrigerators.

59) Since $P'(x) = MP(x)$, then $\frac{d}{dx}[P'(x)] = \frac{d}{dx}[MP(x)]$. Thus, $P''(x) = MP'(x)$. This means that $P''(x)$ and $MP'(x)$ are going to have the same x-values since you are working with the same expression. So, when $MP'(x) > 0$, MP is increasing, then $P''(x) > 0$, P is concave up, and when $MP'(x) < 0$, MP is decreasing, then $P''(x) < 0$, P is concave down.

61) Since TS is a polynomial, the domain is all real numbers. Computing the first and second derivatives, we get:
$TS'(x) = \frac{d}{dx}[-2x^3 + 90x^2 - 1200x + 10{,}000]$
$= -6x^2 + 180x - 1200$
and $TS''(x) = \frac{d}{dx}[-6x^2 + 180x - 1200]$
$= -12x + 180$.
Since the second derivative is a polynomial, there are no values for which the second derivative is undefined. Setting the second derivative equal to zero and solving yields:
$-12x + 180 = 0$
$-12x = -180$
$x = 15$.
Thus, we have a possible inflection point at $x = 15$. We place this value and any values that make the original function undefined on a sign diagram:
TS

TS" 15

We select the test number 12 in [10, 15) and the test number 18 in (15, 25]. Thus, $TS''(12) = 36$ and $TS''(18) = -36$. Now, we can complete the sign diagram:
TS ∪   ∩

TS" (+) 15 (−)
Hence, TS is concave up on [10, 15) and concave down on (15, 25]. Since $TS(15) = 5500$ and the concavity changes sign, then (15, 5500) is an inflection point. This is the point that the greatest growth in sales is occurring due to advertising.

63a) $b(5) = 200(5)^{1/4} \approx 299.070$. In 1989, the number of births to women age 35 to 39 years old was about 299,070 births.
$b'(x) = \frac{d}{dx}[200x^{1/4}] = 200 \cdot \frac{1}{4} x^{-3/4}$
$= 50x^{-3/4} = \frac{50}{\sqrt[4]{x^3}}$.
Hence, $b'(5) = \frac{50}{\sqrt[4]{(5)^3}} \approx 14.953$.
In 1989, the number of births to women age 35 to 39 was increasing at a rate of about 14,953 births per year.

Section 5.2 Second Derivatives and Graphs   173

63b) The derivative is undefined for all $x \leq 0$, but these values are not in the domain of b. Setting the derivative equal to zero and solving yields:

$$\frac{50}{\sqrt[4]{x^3}} = 0$$

$50 = 0 \bullet \sqrt[4]{x^3} = 0$, no solution. So, there are no critical points in the domain of b. Thus, for all x in $[1, 10]$, $b'(x) = \frac{50}{\sqrt[4]{x^3}} > 0$. Hence, b is increasing on $[1, 10]$.

63c) Computing the second derivative, we get:
$b''(x) = \frac{d}{dx}[50x^{-3/4}] = 50\bullet(-\frac{3}{4})x^{-7/4}$
$= \frac{-75}{2x\sqrt[4]{x^3}}$.

The second derivative is undefined for all $x \leq 0$, but these values are not in the domain of b. Setting the second derivative equal to zero and solving yields:

$$\frac{-75}{2x\sqrt[4]{x^3}} = 0$$

$-75 = 0 \bullet 2x\sqrt[4]{x^3} = 0$, no solution. So, there are no inflection points in the domain of b. Thus, for all x in $[1, 10]$, $b''(x) = \frac{-75}{2x\sqrt[4]{x^3}} < 0$. Hence, b is concave down on $[1, 10]$.

63d) From 1985 to 1994, the number of births to women 35 to 39 years old was <u>increasing</u>, and it was increasing at a <u>slower</u> rate.

65a) $f(50) = 0.278(1.026)^{50} \approx 1.00$. When the arterial pressure is 50 mm Hg, the blood flow is about 1 mL/min.
$f'(x) = \frac{d}{dx}[0.278(1.026)^x]$
$= 0.278(1.026)^x \ln(1.026)$.
Hence, $f'(50) = 0.278(1.026)^{(50)} \ln(1.026) \approx 0.026$. When the arterial pressure is 50 mm Hg, the blood flow is increasing at a rate of about 0.026 mL/min per mm Hg.

65b) $f'(x) = 0.278(1.026)^x \ln(1.026)$ is an exponential function so it is defined for all values of x in $[20, 120]$. Since $f'(x) = 0.278(1.026)^x \ln(1.026) > 0$ for all x in $[20, 120]$, f is increasing on $[20, 120]$. Computing the second derivative, we get:
$f''(x) = \frac{d}{dx}[0.278(1.026)^x \ln(1.026)]$
$= 0.278(1.026)^x \ln(1.026) \bullet \ln(1.026)$.
But $f''(x)$ is an exponential function so it's defined for all values of x in $[20, 120]$ and $f''(x) > 0$ for all values of x in $[20, 120]$. Thus, f is concave up on $[20, 120]$. This means that the blood flow is increasing and it is increasing at a faster rate.

67a) $CI(20) = \frac{7.644}{\sqrt[4]{20}} \approx 3.61$. When a person is 20 years old, the cardiac output is about 3.61 $\frac{\text{liters per minute}}{\text{square meters}}$.
$CI'(x) = \frac{d}{dx}[7.644x^{-1/4}]$
$= 7.644(-0.25)x^{-5/4} = \frac{-1.911}{x\sqrt[4]{x}}$.
Hence, $CI'(20) = \frac{-1.911}{20\sqrt[4]{20}} \approx -0.045$.

When a person is 20 years old, the cardiac output is decreasing by about 0.045 $\frac{\text{liters per minute}}{\text{square meters}}$ per year.

67b) The derivative is undefined for all $x \leq 0$, but these values are not in the domain of CI. Setting the derivative equal to zero and solving yields:

$$\frac{-1.911}{x\sqrt[4]{x}} = 0$$

$-1.911 = 0 \bullet x\sqrt[4]{x} = 0$, no solution. So, there are no critical points in the domain of CI. Thus, for all x in $[10, 80]$, $CI'(x) = \frac{-1.911}{x\sqrt[4]{x}} < 0$. Hence, CI is increasing nowhere and decreasing on $[10, 80]$.

174　Chapter 5　Further Applications of the Derivative

67c)　Computing the second derivative, we get:
$CI''(x) = \frac{d}{dx}[-1.911x^{-5/4}]$
$= -1.911(-\frac{5}{4})x^{-9/4} = \frac{9.555}{4x^2 \sqrt[4]{x}}$.

The second derivative is undefined for all $x \leq 0$, but these values are not in the domain of CI. Setting the second derivative equal to zero and solving yields:
$\frac{9.555}{4x^2 \sqrt[4]{x}} = 0$
$9.555 = 0 \bullet 4x^2 \sqrt[4]{x} = 0$, no solution

Thus, there are no inflection points in the domain of CI. Thus, for all x in $[10, 80]$, $CI''(x) = \frac{9.555}{4x^2 \sqrt[4]{x}} > 0$. Hence, CI is concave up on $[10, 80]$. Since CI is decreasing and concave up, as a person ages, the cardiac output decreases and the cardiac output decreases at a slow rate.

69a)　$f'(x) = \frac{d}{dx}[10.28x^2 + 113.98x + 749.6]$
$= 20.56x + 113.98$.
Since $f'(x)$ is a polynomial, it is defined for all values of x in $[1, 13]$. Setting the derivative equal to zero and solving yields:
$20.56x + 113.98 = 0$
$20.56x = -113.98$
$x \approx -5.54$ which is not in the domain of f. Thus, there are no critical points. For all values of x in $[1, 13]$, $f'(x) = 20.56x + 113.98 > 0$. Hence, f is increasing on $[1, 13]$.

69b)　Computing the second derivative, we get:
$f''(x) = \frac{d}{dx}[20.56x + 113.98] = 20.56$.
Hence, $f''(x) = 20.56 > 0$ for all values for x. Since $f''(x) = (f'(x))'$, then the function f' is increasing on $[1, 13]$. This means that the rate of change of the U.S. federal debt was increasing from 1980 to 1992.

Section 5.3 Graphical Analysis and Curve Sketching

1) f is positive on $(-5, -1) \cup (\frac{10}{3}, 7)$ and negative on $(-\infty, -5) \cup (-1, \frac{10}{3}) \cup (7, \infty)$.

3) f' is positive on $(-\infty, -3) \cup (1, 5)$ and f' is negative on $(-3, 1) \cup (5, \infty)$.

5) f is concave up on $(-1, 4)$ and concave down on $(-\infty, -1) \cup (4, \infty)$.

7) f is positive on $(-\infty, -5) \cup (-\frac{5}{3}, \infty)$ and negative on $(-5, -\frac{5}{3})$.

9) f' is positive on $(-3, 1) \cup (5, \infty)$ and f' is negative on $(-\infty, -3) \cup (1, 5)$.

11) f is concave up on $(-\infty, -1) \cup (3, \infty)$ and concave down on $(-1, 3)$.

13) Since f is a polynomial, it is defined for all real numbers. Computing the derivative, we get: $f'(x) = \frac{d}{dx}[3x^2 - 2x - 3] = 6x - 2$.
Since f' is a polynomial, it is defined for all real numbers. Setting $f'(x) = 0$ and solving yields:
$6x - 2 = 0$
$6x = 2$
$x = \frac{1}{3}$, but $f(\frac{1}{3}) = -\frac{10}{3}$.
So, the critical value is $x = \frac{1}{3}$. Computing the second derivative, we get: $f''(x) = \frac{d}{dx}[6x - 2]$
$= 6$. Since $f''(\frac{1}{3}) = 6 > 0$, then f has a relative minimum of $-\frac{10}{3}$ at $x = \frac{1}{3}$.

15) Since y is a polynomial, it is defined for all real numbers. Computing the derivative, we get: $y' = \frac{d}{dx}[x^3 - 2x^2 - 13x - 10]$
$= 3x^2 - 4x - 13$.
Since y' is a polynomial, it is defined for all real numbers. Setting $y' = 0$ and solving yields:
$3x^2 - 4x - 13 = 0$.

15) Continued
Using the quadratic formula, we get:
$x = \frac{4 \pm \sqrt{(-4)^2 - 4(3)(-13)}}{2(3)} = \frac{2 \pm \sqrt{43}}{3}$
$x \approx -1.52$ or 2.85, but $y|_{x=-1.52} \approx 1.63$ and $y|_{x=2.85} \approx -40.15$.
So, the critical values are $x \approx -1.52$ or 2.85. Computing the second derivative, we get:
$y'' = \frac{d}{dx}[3x^2 - 4x - 13] = 6x - 4$. Since $y''|_{x=-1.52} = -5.12 < 0$, then y has a relative maximum of 1.63 at $x \approx -1.52$. Since $y''|_{x=2.85} = 13.1 > 0$, then y has a relative minimum of -40.15 at $x \approx 2.85$.

17) Since y is a polynomial, it is defined for all real numbers. Computing the derivative, we get: $y' = \frac{d}{dx}[\frac{1}{3}x^3 + \frac{5}{2}x^2 + 6x - 2]$
$= x^2 + 5x + 6$.
Since y' is a polynomial, it is defined for all real numbers. Setting $y' = 0$ and solving yields:
$x^2 + 5x + 6 = 0$.
$(x + 2)(x + 3) = 0$
$(x + 2) = 0$ or $(x + 3) = 0$
$x = -2$ or -3, but $y|_{x=-3} = -6.5$
and $y|_{x=-2} = \frac{-20}{3}$.
So, the critical values are $x = -3$ and -2. Computing the second derivative, we get:
$y'' = \frac{d}{dx}[x^2 + 5x + 6] = 2x + 5$. Since $y''|_{x=-3} = -1 < 0$, then y has a relative maximum of -6.5 at $x = -3$. Since $y''|_{x=-2} = 1 > 0$, then y has a relative minimum of $\frac{-20}{3}$ at $x = -2$.

19) Since f is a polynomial, it is defined for all real numbers. Computing the derivative, we get: $f'(x) = \frac{d}{dx}[x^3 + \frac{3}{2}x^2 - 6x - 3]$
$= 3x^2 + 3x - 6$.
Since f' is a polynomial, it is defined for all real numbers. Setting $f'(x) = 0$ and solving yields:
$3x^2 + 3x - 6 = 0$

19) Continued
$3(x^2 + x - 2) = 0$
$3(x + 2)(x - 1) = 0$
$(x + 2) = 0$ or $(x - 1) = 0$
$x = -2$ or 1, but $f(-2) = 7$
and $f(1) = -6.5$.
So, the critical values are $x = -2$ and 1.
Computing the second derivative, we get:
$f''(x) = \frac{d}{dx}[3x^2 + 3x - 6] = 6x - 3$. Since
$f''(-2) = -15 < 0$, then f has a relative maximum of 7 at $x = -2$. Since
$f''(1) = 3 > 0$, then f has a relative minimum of -6.5 at $x = 1$.

21) Since y is a polynomial, it is defined for all real numbers. Computing the derivative, we get: $y' = \frac{d}{dx}[x^4 + x^3 - 7x^2 - x + 6]$
$= 4x^3 + 3x^2 - 14x - 1$.
Since y' is a polynomial, it is defined for all real numbers. Setting $y' = 0$ and solving yields:
$4x^3 + 3x^2 - 14x - 1 = 0$.
In order to solve this equation, we will graph y' and utilize the ZERO command on our calculator. The graphs are displayed in viewing window $[-5, 5]$ by $[-20, 20]$.

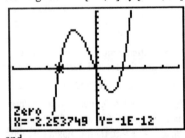

and

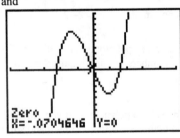

21) Continued

So, $x \approx -2.25, -0.07$, or 1.57. But $y|_{x=-2.25} \approx -12.95, y|_{x=-0.07} \approx 6.04$, and $y|_{x=1.57} \approx -2.88$. So, the critical values are $x = -2.25, -0.07$, and 1.57. Computing the second derivative, we get:
$y'' = \frac{d}{dx}[4x^3 + 3x^2 - 14x - 1]$
$= 12x^2 + 6x - 14$.
Since $y''|_{x=-2.25} \approx 33.43 > 0$, then f has a relative minimum of -12.95 at $x = -2.25$.
Since $y''|_{x=-0.07} \approx -14.36 < 0$, then f has a relative maximum of 6.04 at $x = -0.07$.
Since $y''|_{x=1.57} \approx 25.18 > 0$, then f has a relative minimum of -2.88 at $x = 1.57$.

23) Since g is a polynomial, it is defined for all real numbers. Computing the derivative, we get: $g'(x) = \frac{d}{dx}[3x^4 - 24x^2 + 16]$
$= 12x^3 - 48x$.
Since g' is a polynomial, it is defined for all real numbers. Setting $g'(x) = 0$ and solving yields:
$12x^3 - 48x = 0$
$12x(x^2 - 4) = 0$
$12x(x + 2)(x - 2) = 0$
$12x = 0, (x + 2) = 0$ or $(x - 2) = 0$
$x = -2, 0$, or 2, but $g(-2) = -32, g(0) = 16$ and $g(2) = -32$.
So, the critical values are $x = -2, 0$, and 2.
Computing the second derivative, we get:
$g''(x) = \frac{d}{dx}[12x^3 - 48x] = 36x^2 - 48$. Since
$g''(-2) = 96 > 0$, then g has a relative minimum of -32 at $x = -2$. Since
$g''(0) = -48 > 0$, then g has a relative maximum of 16 at $x = 0$. Since $g''(2) = 96 > 0$, then g has a relative minimum of -32 at $x = 2$.

25) Since f is a rational function and $x^2 + 1 \neq 0$, it is defined for all real numbers. Computing the derivative, we get:
$$f'(x) = \tfrac{d}{dx}[(x^2+1)^{-1}] = -(x^2+1)^{-2}(2x)$$
$$= \frac{-2x}{(x^2+1)^2}.$$
Since f' is a rational function and $x^2 + 1 \neq 0$, it is defined for all real numbers. Setting $f'(x) = 0$ and solving yields:
$$\frac{-2x}{(x^2+1)^2} = 0$$
$$(x^2+1)^2 \cdot \frac{-2x}{(x^2+1)^2} = 0 \cdot (x^2+1)^2$$
$$-2x = 0$$
$x = 0$, but $f(0) = 1$.
So, the critical value is $x = 0$. Computing the second derivative, we get:
$$f''(x) = \tfrac{d}{dx}\left[\frac{-2x}{(x^2+1)^2}\right]$$
$$= \frac{\tfrac{d}{dx}[-2x]\cdot(x^2+1)^2 - (-2x)\cdot\tfrac{d}{dx}[(x^2+1)^2]}{(x^2+1)^4}$$
$$= \frac{-2(x^2+1)^2 + 2x(2)(x^2+1)(2x)}{(x^2+1)^4}$$
$$= \frac{-2x^4 - 4x^2 - 2 + 8x^4 + 8x^2}{(x^2+1)^4} = \frac{6x^4 + 4x^2 - 2}{(x^2+1)^4}$$
$$= \frac{2(x^2+1)(3x^2-1)}{(x^2+1)^4} = \frac{2(3x^2-1)}{(x^2+1)^3}. \text{ Since}$$
$f''(0) = -2 < 0$, then f has a relative maximum of 1 at $x = 0$ and no relative minimum.

27) Since y is a polynomial, it is defined for all real numbers. Computing the derivative, we get: $y' = \tfrac{d}{dx}[3x^6 + 9x^4 - 5]$
$= 18x^5 + 36x^3$.
Since y' is a polynomial, it is defined for all real numbers. Setting $y' = 0$ and solving yields:
$$18x^5 + 36x^3 = 0$$
$$18x^3(x^2 + 2) = 0$$
$$18x^3 = 0 \text{ or } (x^2 + 2) = 0 \text{ (no solution)}$$
$x = 0$, but $y|_{x=0} = -5$.
So, the critical value is $x = 0$.

27) Continued
Computing the second derivative, we get: $y'' = \tfrac{d}{dx}[18x^5 + 36x^3] = 90x^4 + 108x^2$. Since $y''|_{x=0} = 0$, the second derivative test fails, so we need to use the first derivative test. We place the critical values and any values that make the original function undefined on a sign diagram:

y
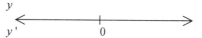
y' 0

We select the test number -1 in $(-\infty, 0)$ and the test number 1 in $(0, \infty)$. Thus, $y'|_{x=-1} = 18(-1)^5 + 36(-1)^3 = -54$ and $y'|_{x=1} = 18(1)^5 + 36(1)^3 = 54$. Now, we can complete the sign diagram:

y
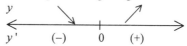
y' $(-)$ 0 $(+)$

Thus, y is increasing on $(0, \infty)$ and decreasing on $(-\infty, 0)$. This means that y has a relative minimum of -5 at $x = 0$.

29) Since y is a polynomial, it is defined for all real numbers. Computing the derivative, we get: $y' = \tfrac{d}{dx}[x^3 + 3x^2 - x - 3]$
$= 3x^2 + 6x - 1$.
Since y' is a polynomial, it is defined for all real numbers. Setting $y' = 0$ and solving yields:
$$3x^2 + 6x - 1 = 0.$$
Using the quadratic formula, we get:
$$x = \frac{-6 \pm \sqrt{(6)^2 - 4(3)(-1)}}{2(3)} = \frac{-3 \pm 2\sqrt{3}}{3}$$
$x \approx -2.15$ or 0.15, but $y|_{x=-2.15} \approx 3.08$ and $y|_{x=0.15} \approx -3.08$.
So, the critical values are $x \approx -2.15$ and 0.15. We place the critical values and any values that make the original function undefined on a sign diagram:

y

y' -2.15 0.15

29) Continued

We select the test number -3 in $(-\infty, -2.15)$, the test number 0 in $(-2.15, 0.15)$ and the test number 1 in $(0.15, \infty)$. Thus,
$y'|_{x=-3} = 3(-3)^2 + 6(-3) - 1 = 8$,
$y'|_{x=0} = 3(0)^2 + 6(0) - 1 = -1$
and $y'|_{x=1} = 3(1)^2 + 6(1) - 1 = 8$. Now, we can complete the sign diagram:

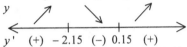

y' $(+)$ -2.15 $(-)$ 0.15 $(+)$

Thus, y is increasing on $(\infty, -2.15) \cup (0.15, \infty)$ and decreasing on $(-2.15, 0.15)$. Also, y has a relative maximum of 3.08 at $x = -2.15$ and a relative minimum of -3.08 at $x = 0.15$.

Computing the second derivative, we get:
$y'' = \frac{d}{dx}[3x^2 + 6x - 1] = 6x + 6$.

Since the second derivative is a polynomial, there are no values for which the second derivative is undefined. Setting the second derivative equal to zero and solving yields:
$6x + 6 = 0$
$6x = -6$
$x = -1$.

Thus, we have a possible inflection point at $x = -1$. We place this value and any values that make the original function undefined on a sign diagram:

y

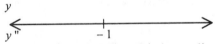

We select the test number -2 in $(-\infty, -1)$ and the test number 0 in $(-1, \infty)$. Thus,
$y''|_{x=-2} = -6$ and $y''|_{x=0} = 6$.
Now, we can complete the sign diagram:

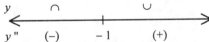

y'' $(-)$ -1 $(+)$

Hence, y is concave up on $(-1, \infty)$ and concave down on $(-\infty, -1)$. Since $y|_{x=-1} = 0$ and the concavity changes sign, then $(-1, 0)$ is an inflection point.

29) Continued

We can now sketch the graph:

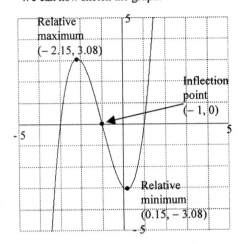

31)
Since y is a polynomial, it is defined for all real numbers. Computing the derivative, we get: $y' = \frac{d}{dx}[\frac{1}{3}x^3 - \frac{1}{2}x^2 - 6x + 2]$
$= x^2 - x - 6$.

Since y' is a polynomial, it is defined for all real numbers. Setting $y' = 0$ and solving yields:
$x^2 - x - 6 = 0$.
$(x + 2)(x - 3) = 0$
$(x + 2) = 0$ or $(x - 3) = 0$
$x = -2$ or 3, but $y|_{x=-2} = \frac{28}{3}$
and $y|_{x=3} = -11.5$.

So, the critical values are $x = -2$ and 3. We place the critical values and any values that make the original function undefined on a sign diagram:

y

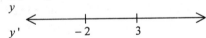

We select the test number -3 in $(-\infty, -2)$, the test number 0 in $(-2, 3)$ and the test number 4 in $(3, \infty)$. Thus,
$y'|_{x=-3} = (-3)^2 - (-3) - 6 = 6$,
$y'|_{x=0} = (0)^2 - (0) - 6 = -6$
and $y'|_{x=4} = (4)^2 - (4) - 6 = 6$.

31) Continued

Now, we can complete the sign diagram:

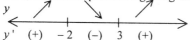

Thus, y is increasing on $(\infty, -2) \cup (3, \infty)$ and decreasing on $(-2, 3)$.

Also, y has a relative maximum of $\frac{28}{3}$ at $x = -2$ and a relative minimum of -11.5 at $x = 3$.

Computing the second derivative, we get:
$y'' = \frac{d}{dx}[x^2 - x - 6] = 2x - 1$.

Since the second derivative is a polynomial, there are no values for which the second derivative is undefined. Setting the second derivative equal to zero and solving yields:
$2x - 1 = 0$
$2x = 1$
$x = 0.5$.

Thus, we have a possible inflection point at $x = 0.5$. We place this value and any values that make the original function undefined on a sign diagram:

We select the test number 0 in $(-\infty, 0.5)$ and the test number 1 in $(0.5, \infty)$. Thus, $y''|_{x=0} = -1$ and $y''|_{x=1} = 1$.

Now, we can complete the sign diagram:

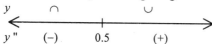

Hence, y is concave up on $(0.5, \infty)$ and concave down on $(-\infty, 0.5)$. Since $y|_{x=0.5} = -\frac{13}{12}$ and the concavity changes sign, then $(0.5, -\frac{13}{12})$ is an inflection point.

31) Continued

Now, we can sketch the graph.

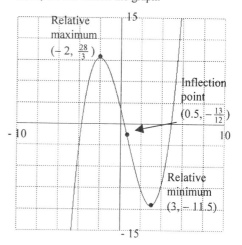

33)

Since f is a polynomial, it is defined for all real numbers. Computing the derivative, we get: $f'(x) = \frac{d}{dx}[x^3 - 6x^2 + 5]$
$= 3x^2 - 12x$.

Since f' is a polynomial, it is defined for all real numbers. Setting $f'(x) = 0$ and solving yields:
$3x^2 - 12x = 0$.
$3x(x - 4) = 0$
$3x = 0$ or $x - 4 = 0$
$x = 0$ or 4, but $f(0) = 5$ and $f(4) = -27$.

So, the critical values are $x = 0$ and 4. We place the critical values and any values that make the original function undefined on a sign diagram:

We select the test number -1 in $(-\infty, 0)$, the test number 2 in $(0, 4)$ and the test number 5 in $(4, \infty)$. Thus,
$f'(-1) = 3(-1)^2 - 12(-1) = 15$,
$f'(2) = 3(2)^2 - 12(2) = -12$, and
$f'(5) = 3(5)^2 - 12(5) = 15$.

33) Continued
Now, we can complete the sign diagram:

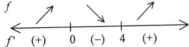

Thus, f is increasing on $(\infty, 0) \cup (4, \infty)$ and decreasing on $(0, 4)$.
Also, f has a relative maximum of 5 at $x = 0$ and a relative minimum of -27 at $x = 4$.
Computing the second derivative, we get:
$$f''(x) = \tfrac{d}{dx}[3x^2 - 12x] = 6x - 12.$$
Since the second derivative is a polynomial, there are no values for which the second derivative is undefined. Setting the second derivative equal to zero and solving yields:
$$6x - 12 = 0$$
$$6x = 12$$
$$x = 2.$$
Thus, we have a possible inflection point at $x = 2$. We place this value and any values that make the original function undefined on a sign diagram:

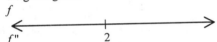

We select the test number 0 in $(-\infty, 2)$ and the test number 4 in $(2, \infty)$. Thus, $f''(0) = -12$ and $f''(4) = 12$.
Now, we can complete the sign diagram:

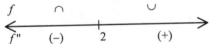

Hence, f is concave up on $(2, \infty)$ and concave down on $(-\infty, 2)$. Since $f(2) = -11$ and the concavity changes sign, then $(2, -11)$ is an inflection point.

33) Continued
We can now sketch the graph.

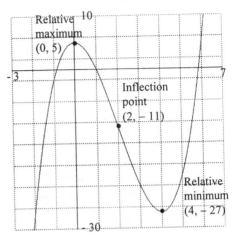

35) Since y is a polynomial, it is defined for all real numbers. Computing the derivative, we get: $y' = \tfrac{d}{dx}[x^4 - 3x^3 - 8x^2 + 12x + 16]$
$= 4x^3 - 9x^2 - 16x + 12$.
Since y' is a polynomial, it is defined for all real numbers. Setting $y' = 0$ and solving yields:
$$4x^3 - 9x^2 - 16x + 12 = 0.$$
In order to solve this equation, we will graph y' and utilize the ZERO command on our calculator. The graphs are displayed in viewing window $[-5, 5]$ by $[-25, 25]$.

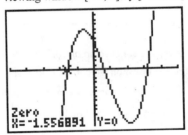

35) Continued
and

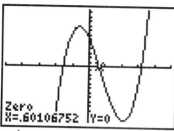

and

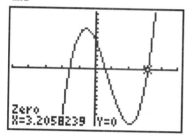

So, $x \approx -1.56, 0.60,$ or 3.21. But $y|_{x=-1.56} \approx -4.88, y|_{x=0.60} \approx 19.80$, and $y|_{x=3.21} \approx -20.97$. So, the critical values are $x = -1.56, 0.60,$ and 3.21.
We place the critical values and any values that make the original function undefined on a sign diagram:

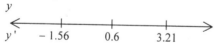

We select the test number -2 in $(-\infty, -1.56)$, the test number 0 in $(-1.56, 0.6)$, the test number 1 in $(0.6, 3.21)$, and the test number 4 in $(3.21, \infty)$. Thus,
$y'|_{x=-2} = 4(-2)^3 - 9(-2)^2 - 16(-2) + 12$
$= -24,$
$y'|_{x=0} = 4(0)^3 - 9(0)^2 - 16(0) + 12 = 12,$
$y'|_{x=1} = 4(1)^3 - 9(1)^2 - 16(1) + 12 = -9,$
and
$y'|_{x=4} = 4(4)^3 - 9(4)^2 - 16(4) + 12 = 60.$
Now, we can complete the sign diagram:

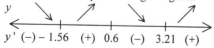

35) Continued
Thus, y is increasing on $(-1.56, 0.60) \cup (3.21, \infty)$ and decreasing on $(-\infty, -1.56) \cup (0.60, 3.21)$. Also, y has a relative minimum of -4.88 at $x = -1.56$, has a relative maximum of 19.8 at $x = 0.6$ and a relative minimum of -20.97 at $x = 3.21$.
Computing the second derivative, we get:
$y'' = \frac{d}{dx}[4x^3 - 9x^2 - 16x] = 12x^2 - 18x - 16.$
Since the second derivative is a polynomial, there are no values for which the second derivative is undefined. Setting the second derivative equal to zero and solving yields:
$12x^2 - 18x - 16 = 0$
$2(6x^2 - 9x - 8) = 0$
Using the quadratic formula, we get:
$x = \frac{9 \pm \sqrt{(-9)^2 - 4(6)(-8)}}{2(6)} = \frac{9 \pm \sqrt{273}}{12}$
$x \approx 2.13$ or $x \approx -0.63$
Thus, we have two possible inflection points at $x \approx 2.13$ and $x \approx -0.63$. We place these values and any values that make the original function undefined on a sign diagram:

y
$y'' \quad \overset{\longleftarrow}{\underset{-0.63 \quad 2.13}{\quad + \quad + \quad}} \longrightarrow$

We select the test number -1 in $(-\infty, -0.63)$, the test number 0 in $(-0.63, 2.13)$ and the test number 3 in $(2.13, \infty)$. Thus, $y''|_{x=-1} = 14, y''|_{x=0} = -16,$ and $y''|_{x=3} = 38.$ Now, we can complete the sign diagram:

$y \qquad \cup \qquad \cap \qquad \cup$
$y'' \quad \overset{\longleftarrow}{\underset{-0.63 \quad 2.13}{\quad + \quad + \quad}} \longrightarrow$

Hence, y is concave up on $(-\infty, -0.63) \cup (2.13, \infty)$ and concave down on $(-0.63, 2.13)$. Since $y|_{x=-0.63} = 6.17$ and $y|_{x=2.13} = -3.14$ and the concavity changes sign, then $(-0.63, 6.17)$ and $(2.13, -3.14)$ are inflection points.

35) Continued
Now, we can sketch the graph.

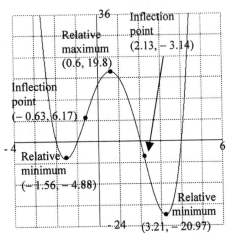

37) Since f is a polynomial, it is defined for all real numbers. Computing the derivative, we get: $f'(x) = \frac{d}{dx}[x^4 - 2x^2 + 1]$
$= 4x^3 - 4x$.
Since f' is a polynomial, it is defined for all real numbers. Setting $f'(x) = 0$ and solving yields:
$4x^3 - 4x = 0$.
$4x(x^2 - 1) = 0$
$4x(x - 1)(x + 1) = 0$
$x = -1, 0,$ or 1, but $f(-1) = 0, f(0) = 1,$ and $f(1) = 0$.
So, the critical values are $x = -1, 0,$ and 1. We place the critical values and any values that make the original function undefined on a sign diagram:
f

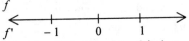

We select the test number -2 in $(-\infty, -1)$, the test number -0.5 in $(-1, 0)$, the test number 0.5 in $(0, 1)$, and the test number 2 in $(1, \infty)$. Thus,
$f'(-2) = 4(-2)^3 - 4(-2) = -24$,
$f'(-0.5) = 4(-0.5)^3 - 4(-0.5) = 1.5$
$f'(0.5) = 4(0.5)^3 - 4(0.5) = -1.5$ and
$f'(2) = 4(2)^3 - 4(2) = 24$.

37) Continued
Now, we can complete the sign diagram:
f

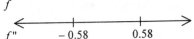

Thus, f is increasing on $(-1, 0) \cup (1, \infty)$ and decreasing on $(-\infty, -1) \cup (0, 1)$.
Also, f has a relative minimum of 0 at $x = -1$, a relative maximum of 1 at $x = 0$, and a relative minimum of 0 at $x = 1$.
Computing the second derivative, we get:
$f'' = \frac{d}{dx}[4x^3 - 4x] = 12x^2 - 4$.
Since the second derivative is a polynomial, there are no values for which the second derivative is undefined. Setting the second derivative equal to zero and solving yields:
$12x^2 - 4 = 0$
$12x^2 = 4$
$x^2 = \frac{1}{3}$.
$x = \pm \frac{1}{\sqrt{3}} \approx \pm 0.58$

Thus, we have two possible inflection points at $x = \pm 0.58$. We place these values and any values that make the original function undefined on a sign diagram:
f

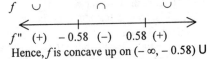

We select the test number -1 in $(-\infty, -0.58)$, the test number 0 in $(-0.58, 0.58)$, and the test number 1 in $(0.58, \infty)$. Thus,
$f''(-1) = 8, f''(0) = -4,$ and $f''(1) = 8$.
Now, we can complete the sign diagram:

Hence, f is concave up on $(-\infty, -0.58) \cup (0.58, \infty)$ and concave down on $(-0.58, 0.58)$. Since $f(-0.58) \approx 0.44$ and $f(0.58) \approx 0.44$ and the concavity changes sign, then $(-0.58, 0.44)$ and $(0.58, 0.44)$ are inflection points.

37) Continued
We can now sketch the graph.

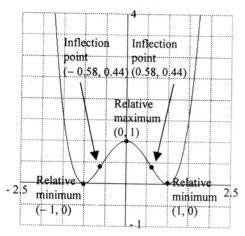

39) Since y is a rational function, it is undefined when $x = -2$. Thus, the domain of y is all real numbers except $x = -2$. Computing the derivative, we get: $y' = \frac{d}{dx}\left[\frac{x-1}{x+2}\right]$

$= \frac{\frac{d}{dx}(x-1)\cdot(x+2)-(x-1)\cdot\frac{d}{dx}(x+2)}{(x+2)^2}$

$= \frac{1\cdot(x+2)-(x-1)\cdot 1}{(x+2)^2} = \frac{x+2-x+1}{(x+2)^2} = \frac{3}{(x+2)^2}$.

Since y' is a rational function, it is undefined when $x = -2$. But $x = -2$ is not in the domain of y. Setting $y' = 0$ and solving yields:

$\frac{3}{(x+2)^2} = 0$

$(x+2)^2 \frac{3}{(x+2)^2} = 0\bullet(x+2)^2$

$3 = 0$, no solution

So, there are no critical values. We place the critical values and any values that make the original function undefined on a sign diagram:

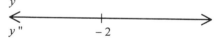

We select the test number -3 in $(-\infty, -2)$ and the test number 0 in $(-2, \infty)$.

39) Continued
Thus,

$y'|_{x=-3} = \frac{3}{(-3+2)^2} = 3$,

$y'|_{x=0} = \frac{3}{(0+2)^2} = 0.75$

Now, we can complete the sign diagram:

```
      y
    ←——————————+——————————→
    y'    (+)  -2   (+)
```

Thus, y is increasing on $(\infty, -2) \cup (-2, \infty)$ and decreasing nowhere.
Also, y has neither a relative nor a relative minimum. Computing the second derivative, we get: $y'' = \frac{d}{dx}\left[\frac{3}{(x+2)^2}\right] = \frac{d}{dx}[3(x+2)^{-2}]$

$= -6(x+2)^{-3}(1) = \frac{-6}{(x+2)^3}$.

Since the second derivative is a rational function, it is undefined when $x = -2$. But $x = -2$ is not in the domain of y. Setting the second derivative equal to zero and solving yields:

$\frac{-6}{(x+2)^3} = 0$

$(x+2)^3 \frac{-6}{(x+2)^3} = 0\bullet(x+2)^3$

$-6 = 0$, no solution.

Thus, we have no possible inflection points. We place the values that make the original function undefined on a sign diagram:

```
      y
    ←——————————+——————————→
    y''            -2
```

We select the test number -3 in $(-\infty, -2)$ and the test number 0 in $(-2, \infty)$. Thus,

$y''|_{x=-3} = 6$ and $y''|_{x=0} = -0.75$.

Now, we can complete the sign diagram:

```
      y        ∪         ∩
    ←——————————+——————————→
    y''   (−)  -2   (+)
```

Hence, y is concave up on $(-\infty, -2)$ and concave down on $(-2, \infty)$.

39) Continued

Since $\lim_{x \to -2^+} \frac{x-1}{x+2} = -\infty$ and $\lim_{x \to -2^-} \frac{x-1}{x+2} = \infty$, then $x = -2$ is a vertical asymptote. Since

$$\lim_{x \to \infty} \frac{x-1}{x+2} = \lim_{x \to \infty} \frac{\frac{x}{x} - \frac{1}{x}}{\frac{x}{x} + \frac{2}{x}} = \lim_{x \to \infty} \frac{1 - \frac{1}{x}}{1 + \frac{2}{x}} = \frac{1-0}{1+0} = 1$$

and similarly, $\lim_{x \to -\infty} \frac{x-1}{x+2}$, then $y = 1$ is a horizontal asymptote.

We can sketch the graph.

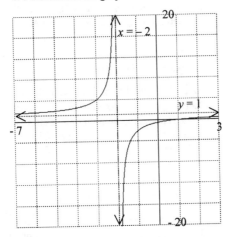

41) Since f is a rational function and $x^2 + 1 \neq 0$, it is defined for all real numbers. Computing the derivative, we get:

$$f'(x) = \frac{d}{dx}[(x^2+1)^{-1}] = -(x^2+1)^{-2}(2x)$$

$$= \frac{-2x}{(x^2+1)^2}$$

Since f' is a rational function and $x^2 + 1 \neq 0$, it is defined for all real numbers. Setting $f'(x) = 0$ and solving yields:

$$\frac{-2x}{(x^2+1)^2} = 0$$

$$(x^2+1)^2 \cdot \frac{-2x}{(x^2+1)^2} = 0 \cdot (x^2+1)^2$$

$$-2x = 0$$

$$x = 0, \text{ but } f(0) = 1.$$

So, the critical value is $x = 0$.

41) Continued

We place the critical values and any values that make the original function undefined on a sign diagram:

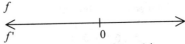

We select the test number -1 in $(-\infty, 0)$, and the test number 1 in $(0, \infty)$. Thus,

$f'(-1) = \frac{-2(-1)}{((-1)^2+1)^2} = \frac{1}{2}$ and

$f'(1) = \frac{-2(1)}{((1)^2+1)^2} = -\frac{1}{2}$. Now, we can complete the sign diagram:

```
f          ↗        ↘
─────────────────────────→
f'      (+)    0    (−)
```

Thus, f is increasing on $(-\infty, 0)$ and decreasing on $(0, \infty)$.

Also, f has a relative maximum of 1 at $x = 0$ and no relative minimum. Computing the second derivative, we get:

$$f''(x) = \frac{d}{dx}\left[\frac{-2x}{(x^2+1)^2}\right]$$

$$= \frac{\frac{d}{dx}[-2x] \cdot (x^2+1)^2 - (-2x) \cdot \frac{d}{dx}[(x^2+1)^2]}{(x^2+1)^4}$$

$$= \frac{-2(x^2+1)^2 + 2x(2)(x^2+1)(2x)}{(x^2+1)^4}$$

$$= \frac{-2x^4 - 4x^2 - 2 + 8x^4 + 8x^2}{(x^2+1)^4} = \frac{6x^4 + 4x^2 - 2}{(x^2+1)^4}$$

$$= \frac{2(x^2+1)(3x^2-1)}{(x^2+1)^4} = \frac{2(3x^2-1)}{(x^2+1)^3}.$$ Since the second derivative is a rational function and $x^2 + 1 \neq 0$, there are no values for which the second derivative is undefined. Setting the second derivative equal to zero and solving yields:

$$\frac{2(3x^2-1)}{(x^2+1)^3} = 0$$

$$(x+1)^3 \cdot \frac{2(3x^2-1)}{(x^2+1)^3} = 0 \cdot (x+1)^3$$

$$2(3x^2-1) = 0$$

$$3x^2 = 1$$

$$x^2 = \frac{1}{3}$$

$$x = \pm\frac{1}{\sqrt{3}} \approx \pm 0.58$$

41) Continued

Thus, we have two possible inflection points at $x = \pm 0.58$. We place these values and any values that make the original function undefined on a sign diagram:

We select the test number -1 in $(-\infty, -0.58)$ and the test number 0 in $(-0.58, 0.58)$ and the test number 1 in $(0.58, \infty)$. Thus, $f''(-1) = 0.5$, $f''(0) = -2$, and $f''(1) = 0.5$. Now, we can complete the sign diagram:

f $\qquad \cup \qquad \cap \qquad \cup$
$\xleftarrow{\qquad\qquad\qquad\qquad\qquad\qquad}\rightarrow$
f'' $(+)$ -0.58 $(-)$ 0.58 $(+)$

Hence, f is concave up on $(-\infty, -0.58) \cup (0.58, \infty)$ and concave down on $(-0.58, 0.58)$. Since $f(-0.58) = 0.75$ and $f(0.58) = 0.75$ and the concavity changes sign, then $(-0.58, 0.75)$ and $(0.58, 0.75)$ are inflection points. Since $\lim_{x \to \infty} \frac{1}{x^2+2}$

$= \lim_{x \to \infty} \frac{\frac{1}{x^2}}{1+\frac{2}{x^2}} = \frac{0}{0+2} = 0$ and similarly, $\lim_{x \to -\infty} \frac{1}{x^2+2} = 0$, then $y = 0$ is a horizontal asymptote.

We can now sketch the graph:

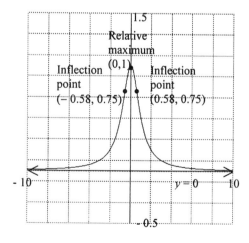

43)

Since y is a rational function, it is undefined at $x = 0$ so the domain is all real numbers except 0. Computing the derivative, we get:
$y' = \frac{d}{dx}[0.2x + 40 + 20x^{-1}] = 0.2 - 20x^{-2}$
$= 0.2 - \frac{20}{x^2}$.

Since y' is a rational function, it is undefined at $x = 0$, but 0 is not in the domain of y. Setting $y' = 0$ and solving yields:

$0.2 - \frac{20}{x^2} = 0$ (multiply by x^2)
$0.2x^2 - 20 = 0$
$0.2x^2 = 20$
$x^2 = 100$
$x = \pm 10$, but $y|_{x=-10} = 36$ and $y|_{x=10} = 44$. So, the critical values are $x = -10$ and 10. We place the critical values and any values that make the original function undefined on a sign diagram:

y
$\xleftarrow{\qquad\qquad\qquad\qquad\qquad\qquad}\rightarrow$
y' $\quad -10 \quad 0 \quad 10$

We select the test number -20 in $(-\infty, -10)$, the test number -5 in $(-10, 0)$, the test number 5 in $(0, 10)$ and the test number 20 in $(10, \infty)$. Thus,

$y'|_{x=-20} = 0.2 - \frac{20}{(-20)^2} = 0.15$,

$y'|_{x=-5} = 0.2 - \frac{20}{(-5)^2} = -0.6$,

$y'|_{x=5} = 0.2 - \frac{20}{(5)^2} = -0.6$, and

$y'|_{x=20} = 0.2 - \frac{20}{(20)^2} = 0.15$. Now, we can complete the sign diagram:

y $\quad \nearrow \quad \searrow \quad \searrow \quad \nearrow$
$\xleftarrow{\qquad\qquad\qquad\qquad\qquad\qquad}\rightarrow$
y' $(+)$ -10 $(-)$ 0 $(-)$ 10 $(+)$

Thus, y is increasing on $(-\infty, -10) \cup (10, \infty)$ and decreasing on $(-10, 0) \cup (0, 10)$. Also, y has a relative maximum of 36 at $x = -10$ and a relative minimum of 44 at $x = 10$. Computing the second derivative, we get: $y'' = \frac{d}{dx}[0.2 - 20x^{-2}] = 40x^{-3} = \frac{40}{x^3}$.

186 Chapter 5 Further Applications of the Derivative

43)   Continued
Since the second derivative is a rational function, it is undefined at $x = 0$, but 0 is not in the domain of y. Setting the second derivative equal to zero and solving yields:
$$\frac{40}{x^3} = 0$$
$$\frac{40}{x^3} \bullet x^3 = 0 \bullet x^3$$
$$40 = 0, \text{ no solution}$$
Thus, we have no possible inflection points. We place any values that make the original function undefined on a sign diagram:

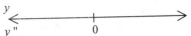
y'' 0

We select the test number -1 in $(-\infty, 0)$ and the test number 1 in $(0, \infty)$. Thus,
$y''\big|_{x=-1} = -40$ and $y''\big|_{x=1} = 40$.
Now, we can complete the sign diagram:

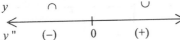
y'' $(-)$ 0 $(+)$

Hence, y is concave up on $(0, \infty)$ and concave down on $(-\infty, 0)$.
Since $\lim\limits_{x \to 0^+} 0.2x + 40 + \frac{20}{x} = \infty$
and $\lim\limits_{x \to 0^-} 0.2x + 40 + \frac{20}{x} = -\infty$, then $x = 0$ is a vertical asymptote. We can now sketch the graph.

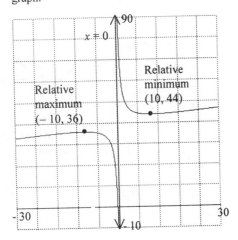

45) Since f is a logarithmic function, it is defined for all real numbers greater than zero. Computing the derivative, we get:
$f'(x) = \frac{d}{dx}[x - \ln(x)] = 1 - \frac{1}{x}$.
Since f' is a rational function, it is undefined at $x = 0$, but 0 is not in the domain of f.
Setting $f'(x) = 0$ and solving yields:
$1 - \frac{1}{x} = 0$ (multiply by x)
$x - 1 = 0$
$x = 1$, but $f(1) = 1$
So, the critical value is $x = 1$. We place the critical values and any values that make the original function undefined on a sign diagram:

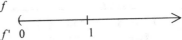
f' 0   1

We select the test number 0.5 in $(0, 1)$ and the test number 2 in $(1, \infty)$. Thus,
$f'(0.5) = 1 - \frac{1}{0.5} = -1$ & $f'(2) = 1 - \frac{1}{2} = 0.5$.
Now, we can complete the sign diagram:

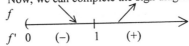
f' 0 $(-)$ 1 $(+)$

Thus, f is increasing on $(1, \infty)$ and decreasing on $(0, 1)$.
Also, f has a relative minimum of 1 at $x = 1$ and no relative maximum.
Computing the second derivative, we get:
$f''(x) = \frac{d}{dx}[1 - \frac{1}{x}] = \frac{d}{dx}[1 - x^{-1}] = x^{-2} = \frac{1}{x^2}$.
Since the second is a rational function, it is undefined at $x = 0$, but 0 is not in the domain of f. Setting the second derivative equal to zero and solving yields:
$\frac{1}{x^2} = 0$ (multiply by x^2)
$1 = 0$, no solution.
Thus, we have no possible inflection points. We place any values that make the original function undefined on a sign diagram:

f'' 0

We select the test number 1 in $(0, \infty)$. Thus,
$f''(1) = \frac{1}{(1)^2} = 1$

45) Continued
Now, we can complete the sign diagram:

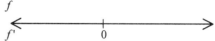

Hence, f is concave up on $(0, \infty)$ and concave down nowhere.
Since $\lim\limits_{x \to 0^+} x - \ln(x) = \infty$, then $x = 0$ is a vertical asymptote. We can now sketch the graph.

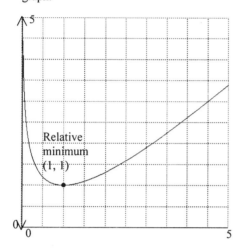

47) Since f is a radical function with an odd index, it is defined for all real numbers. Computing the derivative, we get:
$f'(x) = \frac{d}{dx}[\sqrt[3]{x^2}] = \frac{d}{dx}[x^{2/3}] = \frac{2}{3}x^{-1/3}$
$= \frac{2}{3\sqrt[3]{x}}$.

Since $f'(x)$ is undefined at $x = 0$ and $x = 0$ is in the domain of f, then $x = 0$ is a critical value and $f(0) = 0$. Setting $f'(x) = 0$ and solving yields:
$\frac{2}{3\sqrt[3]{x}} = 0$ (multiply by the denominator)
$2 = 0$, no solution
So, the only critical value is $x = 0$.

47) Continued
We place the critical values and any values that make the original function undefined on a sign diagram:

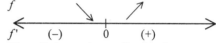

We select the test number -1 in $(-\infty, 0)$, and the test number 1 in $(0, \infty)$. Thus,
$f'(-1) = \frac{2}{3\sqrt[3]{-1}} = -\frac{2}{3}$ and $f'(1) = \frac{2}{3\sqrt[3]{1}} = \frac{2}{3}$.

Now, we can complete the sign diagram:

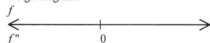

Thus, f is increasing on $(0, \infty)$ and decreasing on $(-\infty, 0)$. Also, f has a relative minimum of 0 at $x = 0$ and no relative maximum.
Computing the second derivative, we get:
$f''(x) = \frac{d}{dx}[\frac{2}{3}x^{-1/3}] = \frac{2}{3} \cdot \frac{-1}{3}x^{-4/3}$
$= -\frac{2}{9x\sqrt[3]{x}}$.

Since $f''(x)$ is undefined at $x = 0$ and $x = 0$ is in the domain of f, then there is a possible inflection point at $x = 0$. Setting the second derivative equal to zero and solving yields:
$\frac{-2}{9x\sqrt[3]{x}} = 0$ (multiply by the denominator)
$-2 = 0$, no solution
Thus, we have a possible inflection point at $x = 0$. We place this value and any values that make the original function undefined on a sign diagram:

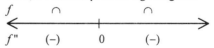

We select the test number -1 in $(-\infty, 0)$ and the test number 1 in $(0, \infty)$. Thus,
$f''(-1) = -\frac{2}{9}$ and $f''(1) = -\frac{2}{9}$.

Now, we can complete the sign diagram:

188 Chapter 5 Further Applications of the Derivative

47) Continued
Hence, f is concave up nowhere and concave down on $(-\infty, 0) \cup (0, \infty)$. Since the concavity does not changes sign, then there are no inflection points.
We can now sketch the graph.

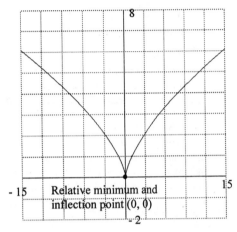

49a) Since $f''(x) < 0$ on $(-\infty, 0)$, then f' is decreasing on $(-\infty, 0)$ and since $f''(x) > 0$ on $(0, \infty)$, then f' is increasing on $(0, \infty)$. Using the first derivative test, f' is concave up on $(-\infty, \infty)$. Also, f' has a relative minimum at $x = 0$.
One possible graph would be:

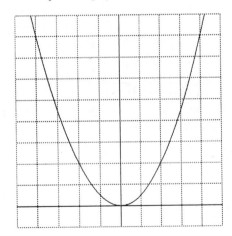

49b) Since $f''(x) < 0$ on $(-\infty, 0)$, then f is concave down on $(-\infty, 0)$ and since $f''(x) > 0$ on $(0, \infty)$, then f is concave up on $(0, \infty)$. Also, f has an inflection point at $x = 0$. One possible graph would be:

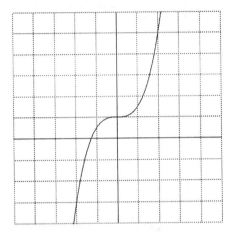

51a) Since $f''(x) < 0$ on $(-2, 2)$, then f' is decreasing on $(-2, 2)$ and since $f''(x) > 0$ on $(-\infty, -2) \cup (2, \infty)$, then f' is increasing on $(-\infty, -2) \cup (2, \infty)$. Using the first derivative test, f' is concave up on $(0, \infty)$ and concave down on $(-\infty, 0)$ meaning that there is an inflection point at $x = 0$. Also, f' has a relative maximum at $x = -2$ and a relative minimum at $x = 2$.
One possible graph would be:

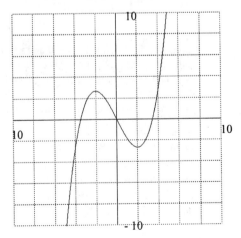

51b) Since $f''(x) < 0$ on $(-2, 2)$, then f is concave down on $(-2, 2)$ and since $f''(x) > 0$ on $(-\infty, -2) \cup (2, \infty)$, then f is concave up on $(-\infty, -2) \cup (2, \infty)$. Also, f has inflection points at $x = -2$ and $x = 2$. One possible graph would be:

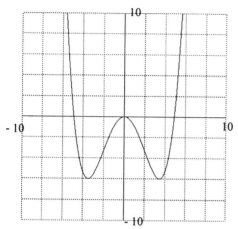

53a) Since $R(x)$ is a polynomial, its domain is all real numbers such that $0 \leq x \leq 50$. Computing the derivative, we get:
$R'(x) = \frac{d}{dx}[216x - 0.08x^3] = 216 - 0.24x^2$.

Since $R'(x)$ is a polynomial, there are no values of x that make it undefined. Setting the derivative equal to zero and solving yields:
$216 - 0.24x^2 = 0$
$216 = 0.24x^2$
$x^2 = 900$
$x = \pm 30$, but $x = -30$ is not in the domain of R, so $x = 30$ is the only critical value. We begin by placing the critical values and any values that make the original function undefined on a sign diagram:

R

R' |———————————|—————|
    0   30   50

We select the test number 10 in $[0, 30)$ and the test number 40 in $(30, 50]$. Thus, $R'(10) = 216 - 0.24(10)^2 = 192$ and $R'(40) = 216 - 0.24(40)^2 = -168$.

53a) Now, we can complete the sign diagram:

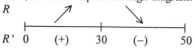

Thus, R is increasing on $[0, 30]$ and decreasing on $(30, 50]$.

53b) Computing the second derivative, we get:
$R''(x) = \frac{d}{dx}[216 - 0.24x^2] = -0.48x$.

Since the second derivative is a polynomial, there are no values for which the second derivative is undefined. Setting the second derivative equal to zero and solving yields:
$-0.48x = 0$
$x = 0$

Thus, we have a possible inflection point at $x = 0$. We place this value and any values that make the original function undefined on a sign diagram:

R

R" |———————————————————|
    0   50

We select the test number 10 in $(0, 50]$. Thus,
$R''(10) = -4.8$.

Now, we can complete the sign diagram:

Hence, R is concave down on $(0, 50]$. Since the function is not defined for values of x less than zero, then there is no inflection point at $x = 0$.

55a)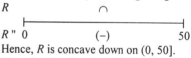

$AP(x) = \frac{P(x)}{x} = \frac{-0.023x^3 + 4.45x^2 - 15x - 2}{x}$
$= -0.023x^2 + 4.45x - 15 - \frac{2}{x}$.

190 Chapter 5 Further Applications of the Derivative

55b) Since AP is a rational function, it is defined for all real numbers except $x = 0$. But, since $x = 0$ is not in the domain of AP, then AP is defined for all real numbers in its domain. Computing the derivative, we get: $AP'(x)$
$= \frac{d}{dx}[-0.023x^2 + 4.45x - 15 - 2x^{-1}]$
$= -0.046x + 4.45 + 2x^{-2}$
$= -0.046x + 4.45 + \frac{2}{x^2}$.

Since $AP'(x)$ is a rational function, it is defined for all real numbers except $x = 0$. But $x = 0$ is not in the domain of AP. Setting $AP'(x) = 0$ and solving yields:
$-0.046x + 4.45 + \frac{2}{x^2} = 0$ (multiply by x^2)
$-0.046x^3 + 4.45x^2 + 2 = 0$
In order to solve this equation, we will graph AP' and utilize the ZERO command on our calculator. The graph is displayed in viewing window [0, 130] by [−8000, 8000]

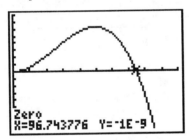

Thus, $x \approx 96.74$ is the only critical value and $AP(96.74) \approx 200.22$. We place the critical values and any values that make the original function undefined on a sign diagram:
AP

```
    (————————+————————————)
AP'  0        96.74       130
```
We select the test number 10 in $(0, 96.74)$, and the test number 100 in $(96.74, 130]$.

55b) Continued
Thus,
$AP'(10) = -0.046(10) + 4.45 + \frac{2}{(10)^2}$
$= 4.01$ and
$AP'(100) = -0.046(100) + 4.45 + \frac{2}{(100)^2}$
$= -0.1498$
Now, we can complete the sign diagram:

AP ↗   ↘
```
    (————————+————————————)
AP'  0  (+)  96.74  (−)   130
```
Thus, AP is increasing on $(0, 96.74)$ and decreasing on $(96.74, 130)$. Also, AP has a relative maximum of 200.22 at $x = 96.74$ and no relative minimum.
Computing the second derivative, we get:
$AP''(x) = \frac{d}{dx}[-0.046x + 4.45 + 2x^{-2}]$
$= -0.046 - 4x^{-3} = -0.046 - \frac{4}{x^3}$
Since the second derivative is a rational function, it is defined for all real numbers except $x = 0$. But $x = 0$ is not in the domain of AP. Setting $AP''(x) = 0$ and solving yields:
$-0.046 - \frac{4}{x^3} = 0$ (multiply by x^3)
$-0.046x^3 - 4 = 0$
$-0.046x^3 = 4$
$x^3 \approx -\frac{2000}{23}$

$x \approx -4.43$ which is not in the domain of AP. Thus, we have no possible inflection points. We place any values that make the original function undefined on a sign diagram:
AP
```
    (—————————————————————)
AP"  0                    130
```
We select the test number 10 in $(0, 130]$. Thus,
$AP''(10) = -0.046 - \frac{4}{(10)^3} = -0.05$.

Now, we can complete the sign diagram:
AP ⌒
```
    (—————————————————————)
AP"  0         (−)        130
```
Hence, AP is concave up nowhere and concave down on $(0, 130)$.

55b) Continued
We can now sketch the graph.

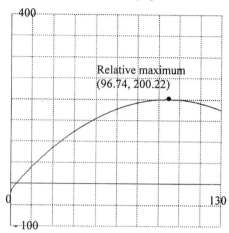

57) The maximum average profit occurs when the marginal profit function is equal to the average profit function.

59a) $C'(x) = \frac{d}{dx}[150 + 3x + \frac{x^2}{15}]$
$= 3 + \frac{2x}{15}$

59b) The graph is displayed in viewing window [0, 200] by [0, 50].

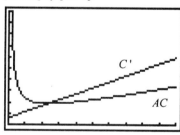

59c) Using the INTERSECT command on the calculator, we find:

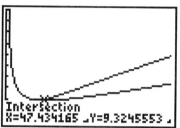

The two curves cross at (47.43, 9.32).

59d) The x-coordinate of the result from part c is the same as the x-coordinate of the relative minimum for AC in #58, part b.

61a) $AC(t) = \frac{C(t)}{t} = \frac{100t^2 + 400t + 11000}{t}$
$= 100t + 400 + \frac{11000}{t}$.

61b) The average cost per year is a minimum when $C'(t) = AC(t)$ (see exercises #58 through 60). Computing $C'(t)$, we get:
$C'(t) = \frac{d}{dt}[100t^2 + 400t + 11,000]$
$= 200t + 400$. Setting $C'(t) = AC(t)$ and solving yields:
$200t + 400 = 100t + 400 + \frac{11000}{t}$
$100t = \frac{11000}{t}$ (multiply by t)
$100t^2 = 11000$
$t^2 = 110$
$t \approx \pm 10.5$, but -10.5 is not in the domain of $AC(t)$, thus $t \approx 10.5$. Hence, the average cost per year is a minimum at $t \approx 10.5$ years.

61c) $AC(10.5) = 100(10.5) + 400 + \frac{11000}{10.5} \approx$ $\$2,497.62$. The minimum average cost per year of the car is about $\$2,497.62$.

192　Chapter 5　Further Applications of the Derivative

63a) Since the domain of f is $40 \leq x \leq 90$, then f is defined for all real numbers in [40, 90]. Computing $f'(x)$, we get:
$f'(x) = \frac{d}{dx}[-10,822 + 3800 \ln(x)]$
$= \frac{3800}{x}$. For $40 \leq x \leq 90$, $f'(x) = \frac{3800}{x} > 0$, thus f is increasing on [40, 90].

63b) Computing $f''(x)$, we get:
$f''(x) = \frac{d}{dx}[\frac{3800}{x}] = -\frac{3800}{x^2}$. For $40 \leq x \leq 90$, $f''(x) = -\frac{3800}{x^2} < 0$, thus f is concave down on [40, 90].

65a) $f(20) = 12 + 16.694 \ln(20 + 1) \approx 62.83\%$. When the wind velocity is 20 miles per hour, the percent total heat loss by convection is about 62.83%.
$f'(x) = \frac{d}{dx}[12 + 16.694 \ln(x+1)]$
$= \frac{16.694}{x+1}$. Thus, $f'(20) = \frac{16.694}{20+1} \approx 0.79$.
When the wind velocity is 20 miles per hour, the percent total heat loss by convection is increasing at a rate of about 0.79% per mile per hour.

65b) Since $f'(x) = \frac{16.694}{x+1}$, then for $0 \leq x \leq 60$, $f'(x) > 0$. Thus, f is increasing on [0, 60].

65c) Computing $f''(x)$, we get:
$f''(x) = \frac{d}{dx}[\frac{16.694}{x+1}] = -\frac{16.694}{(x+1)^2}$. For $0 \leq x \leq 60$, $f''(x) = -\frac{16.694}{(x+1)^2} < 0$. Thus, f is concave down on [0, 60].

65d) Using the information above and the fact there are no extrema or inflection points since both f' and f'' are defined for all x in [0, 60] and both $f'(x) \neq 0$ and $f''(x) \neq 0$ in [0, 60], we can sketch a graph of f:

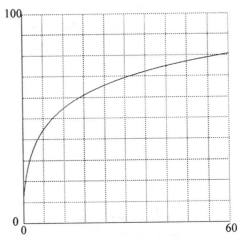

67a) $f(4) = -1.32(4)^3 + 19.66(4)^2 - 72.76(4) + 1441.47 = 1380.51$.
In 1992, the total carbon content of carbon dioxide emissions released by the United States was 1,380.51 million metric tons.
$f'(t) = \frac{d}{dt}[-1.32t^3 + 19.66t^2 - 72.76t + 1441.47]$
$= -3.96t^2 + 39.32t - 72.76$.
Thus, $f'(4) = -3.96(4)^2 + 39.32(4) - 72.76 = 21.16$. In 1992, the total carbon content of carbon dioxide emissions released by the United States was increasing at a rate of 21.16 million metric tons per year.

67b) Since f is a polynomial, it is defined for all t in [1, 7]. Since $f'(t) = -3.96t^2 + 39.32t - 72.76$ is a polynomial, it is defined for all t in [1, 7]. Setting $f'(x) = 0$ and solving yields:
$-3.96t^2 + 39.32t - 72.76 = 0$

67b) Continued
Using the quadratic formula, we get:
$$t = \frac{-39.32 \pm \sqrt{(39.32)^2 - 4(-3.96)(-72.76)}}{2(-3.96)}$$
≈ 7.47 or 2.46, but 7.47 is not in $[1, 7]$. So, the critical value is $t \approx 2.46$ and $f(2.46) \approx 1361.80$. We place the critical value and any values that make the original function undefined on a sign diagram:

f

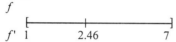

f' 1   2.46   7

We select the test number 2 in $[1, 2.46]$, and the test number 4 in $[2.46, 7]$. Thus,
$f'(2) = -3.96(2)^2 + 39.32(2) - 72.76$
≈ -9.96 and
$f'(4) = -3.96(4)^2 + 39.32(4) - 72.76$
≈ 21.16. Now, we can complete the sign diagram:

f

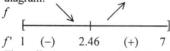

f' 1   (−)   2.46   (+)   7

Thus, f is decreasing on $[1, 2.46)$ and increasing on $(2.46, 7]$.
Also, f has a relative minimum of 1361.80 at $t = 2.46$ and no relative maximum.

67c) Computing the second derivative, we get:
$f''(t) = \frac{d}{dx}[-3.96t^2 + 39.32t - 72.76]$
$= -7.92t + 39.32$.
Since the second derivative is a polynomial, there are no values for which the second derivative is undefined. Setting the second derivative equal to zero and solving yields:
$-7.92t + 39.32 = 0$
$-7.92t = -39.32$
$t \approx 4.96$
Thus, we have a possible inflection point at $t = 4.96$.

67c) Continued
We place this value and any values that make the original function undefined on a sign diagram:

f

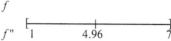

f'' 1   4.96   7

We select the test number 3 in $[1, 4.96)$ and the test number 6 in $[4.96, 7]$. Thus, $f''(3) = 15.56$ and $f''(6) = -8.2$.
Now, we can complete the sign diagram:

f

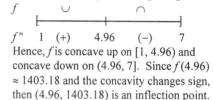

f'' 1   (+)   4.96   (−)   7

Hence, f is concave up on $[1, 4.96)$ and concave down on $(4.96, 7]$. Since $f(4.96) \approx 1403.18$ and the concavity changes sign, then $(4.96, 1403.18)$ is an inflection point.

67d) We can now sketch the graph:

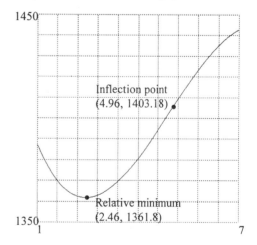

194 Chapter 5 Further Applications of the Derivative

Section 5.4 Optimizing Functions on a Closed Interval

1) Since f is a polynomial, it is continuous on the stated interval. By the Extreme Value Theorem, it has an absolute maximum and minimum.
Computing the derivative, we get:
$f'(x) = \frac{d}{dx}[x^2 - 2x - 7] = 2x - 2$.
Since $f'(x)$ is defined for all x, the only critical values are when $f'(x) = 0$. Solving yields:
$2x - 2 = 0$
$x = 1$.
Thus, $x = 1$ is the only critical value. Evaluating f at $x = 1$ and the endpoints $x = -2$ and $x = 1$, we get:
$f(1) = (1)^2 - 2(1) - 7 = -8$ and
$f(-2) = (-2)^2 - 2(-2) - 7 = 1$.
Thus, there is an absolute maximum of 1 at $x = -2$ and an absolute minimum of -8 at $x = 1$.

3) Since f is a polynomial, it is continuous on the stated interval. By the Extreme Value Theorem, it has an absolute maximum and minimum.
Computing the derivative, we get:
$f'(x) = \frac{d}{dx}[x^3 - 2x^2 - 5x + 6] = 3x^2 - 4x - 5$.
Since $f'(x)$ is defined for all x, the only critical values are when $f'(x) = 0$. Solving yields:
$3x^2 - 4x - 5 = 0$
Using the quadratic formula, we get:
$x = \frac{4 \pm \sqrt{(-4)^2 - 4(3)(-5)}}{2(3)} = \frac{2 \pm \sqrt{19}}{3}$
$x \approx 2.12$ or $x \approx -0.79$, but $x \approx 2.12$ is not in $[-2, 2]$. Thus, $x \approx -0.79$ is the only critical value. Evaluating f at $x \approx -0.79$ and the endpoints $x = -2$ and $x = 2$, we get:
$f(-0.79) = (-0.79)^3 - 2(-0.79)^2$
$\qquad - 5(-0.79) + 6 \approx 8.21$
$f(-2) = (-2)^3 - 2(-2)^2 - 5(-2) + 6 = 0$
and $f(2) = (2)^3 - 2(2)^2 - 5(2) + 6 = -4$.
Thus, there is an absolute maximum of 8.21 at $x = -0.79$ and an absolute minimum of -4 at $x = 2$.

5) Since f is a polynomial, it is continuous on the stated interval. By the Extreme Value Theorem, it has an absolute maximum and minimum.
Computing the derivative, we get:
$f'(x) = \frac{d}{dx}[x^3 - 3x^2] = 3x^2 - 6x$.
Since $f'(x)$ is defined for all x, the only critical values are when $f'(x) = 0$. Solving yields:
$3x^2 - 6x = 0$
$3x(x - 2) = 0$
$x = 0$ and $x = 2$
Thus, $x = 0$ and $x = 2$ are the critical values. Evaluating f at $x = 0$, $x = 2$, and the endpoints $x = -1$ and $x = 3$, we get:
$f(0) = (0)^3 - 3(0)^2 = 0$
$f(2) = (2)^3 - 3(2)^2 = -4$
$f(-1) = (-1)^3 - 3(-1)^2 = -4$ and
$f(3) = (3)^3 - 3(3)^2 = 0$.
Thus, there is an absolute maximum of 0 at $x = 0$ and at $x = 3$ and an absolute minimum of -4 at $x = -1$ and at $x = 2$.

7) Since f is a polynomial, it is continuous on the stated interval. By the Extreme Value Theorem, it has an absolute maximum and minimum.
Computing the derivative, we get:
$f'(x) = \frac{d}{dx}[x^4 - 15x^2 - 10x + 24]$
$\qquad = 4x^3 - 30x - 10$.
Since $f'(x)$ is defined for all x, the only critical values are when $f'(x) = 0$. Solving yields:
$4x^3 - 30x - 10 = 0$
To solve this, we will graph $Y_1 = f'(x)$ and use the zero command.

Section 5.4 Optimizing Functions on a Closed Interval 195

7) Continued
The graphs are displayed in viewing window [– 3, 3] by [– 70, 50]

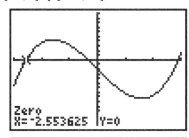

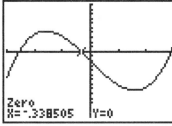

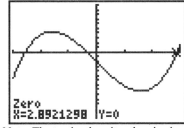

Note: The textbook writes the viewing window on the sides of the graph.
Thus, $x \approx -2.55$, $x \approx -0.34$, and $x \approx 2.89$ are the critical values. Evaluating f at these values and the endpoints $x = -3$ and $x = 3$, we get:
$f(-2.55) = (-2.55)^4 - 15(-2.55)^2$
$\quad - 10(-2.55) + 24 \approx -5.76$
$f(-0.34) = (-0.34)^4 - 15(-0.34)^2$
$\quad - 10(-0.34) + 24 \approx 25.68$
$f(2.89) = (2.89)^4 - 15(2.89)^2$
$\quad - 10(2.89) + 24 \approx -60.42$
$f(-3) = (-3)^4 - 15(-3)^2 - 10(-3) + 24 = 0$
$f(3) = (3)^4 - 15(3)^2 - 10(3) + 24 = -60$.
Thus, there is an absolute maximum of 25.68 at $x = -0.34$ and an absolute minimum of -60.42 at $x = 2.89$.

9) Since h is a polynomial, it is continuous on the stated interval. By the Extreme Value Theorem, it has an absolute maximum and minimum.
Computing the derivative, we get:
$h'(x) = \frac{d}{dx}[x^4 - x^3 + 5] = 4x^3 - 3x^2$.
Since $h'(x)$ is defined for all x, the only critical values are when $h'(x) = 0$. Solving yields:
$\quad 4x^3 - 3x^2 = 0$
$\quad x^2(4x - 3) = 0$
$\quad x = 0$ and $x = 0.75$
Thus, $x = 0$ and $x = 0.75$ are the critical values. Evaluating h at $x = 0$ and $x = 0.75$ and the endpoints $x = -2$ and $x = 2$, we get:
$h(0) = (0)^4 - (0)^3 + 5 = 5$
$h(0.75) = (0.75)^4 - (0.75)^3 + 5 \approx 4.89$
$h(-2) = (-2)^4 - (-2)^3 + 5 = 29$ and
$h(2) = (2)^4 - (2)^3 + 5 = 13$.
Thus, there is an absolute maximum of 29 at $x = -2$ and an absolute minimum of 4.89 at $x = 0.75$.

11) Since y is a polynomial, it is continuous on the stated interval. By the Extreme Value Theorem, it has an absolute maximum and minimum.
Computing the derivative, we get:
$y' = \frac{d}{dx}[(2x^2 - 1)^4] = 4(2x^2 - 1)^3 (4x)$
$\quad = 16x(2x^2 - 1)^3$.
Since y' is defined for all x, the only critical values are when $y' = 0$. Solving yields:
$\quad 16x(2x^2 - 1)^3 = 0$
$\quad 16x = 0$ or $2x^2 - 1 = 0$
$\quad x = 0$ or $x^2 = 0.5$
$\quad x = 0$ or $x \approx \pm 0.71$, but $x \approx -0.71$ is not in [0, 2]. Thus, $x = 0$ and $x \approx 0.71$ are the critical values. Evaluating y at $x = 0$ and $x \approx 0.71$ and the endpoints $x = 0$ and $x = 2$, we get:
$y\big|_{x=0} = (2(0)^2 - 1)^4 = 1$
$y\big|_{x=0.71} = (2(0.71)^2 - 1)^4 = 0$ and
$y\big|_{x=2} = (2(2)^2 - 1)^4 = 2401$.
There is an absolute maximum of 2401 at $x = 2$ & an absolute minimum of 0 at $x = 0.71$.

13) Since y is a radical with an odd root, it is continuous on the stated interval. By the Extreme Value Theorem, it has an absolute maximum and minimum.
Computing the derivative, we get:
$y' = \frac{d}{dx}[x^{1/3}] = \frac{1}{3}x^{-2/3} = \frac{1}{3\sqrt[3]{x^2}}$. Since y' is undefined at $x = 0$ and $x = 0$ is in the domain of y, then $x = 0$ is a critical value of y.
Setting $y' = 0$ and solving yields:
$\frac{1}{3\sqrt[3]{x^2}} = 0$ (multiply by $3\sqrt[3]{x^2}$)
$1 = 0$, no solution
Thus, $x = 0$ is the only critical value.
Evaluating y at $x = 0$ and the endpoints $x = -1$ and $x = 1$, we get:
$y\big|_{x=0} = \sqrt[3]{0} = 0$,
$y\big|_{x=-1} = \sqrt[3]{-1} = -1$,
and $y\big|_{x=1} = \sqrt[3]{1} = 1$.
Thus, there is an absolute maximum of 1 at $x = 1$ and an absolute minimum of -1 at $x = -1$.

15) Since f is a rational function, the only values where it is not continuous are the values that make it undefined. The function f is undefined at $x = 2$, but 2 is not in $[0, 1]$. Thus, f is continuous on the interval. By the Extreme Value Theorem, it has an absolute maximum and minimum.
Computing the derivative, we get:
$f'(x) = \frac{d}{dx}[(x-2)^{-1}] = -1(x-2)^{-2} = \frac{-1}{(x-2)^2}$.
$f'(x)$ is undefined when $x = 2$, but 2 is not in $[0, 1]$. Thus, the only critical values are when $f'(x) = 0$. Solving yields:
$\frac{-1}{(x-2)^2} = 0$ (multiply by $(x-2)^2$)
$-1 = 0$, no solution.
Thus, there are no critical values.

15) Continued
Evaluating f at the endpoints $x = 0$ and $x = 1$, we get:
$f(0) = \frac{1}{(0-2)} = -0.5$
$f(1) = \frac{1}{(1-2)} = -1$
Thus, there is an absolute maximum of -0.5 at $x = 0$ and an absolute minimum of -1 at $x = 1$.

17) Since y is a rational function, the only values where it is not continuous are the values that make it undefined. But $x^2 + 1 \neq 0$, for all x. Thus, y is continuous on the interval. By the Extreme Value Theorem, it has an absolute maximum and minimum.
Computing the derivative, we get:
$y' = \frac{d}{dx}[(x^2+1)^{-1}] = -1(x^2+1)^{-2}(2x)$
$= \frac{-2x}{(x^2+1)^2}$.
y' is defined for all x since $x^2 + 1 \neq 0$. Thus, the only critical values are when $y' = 0$. Solving yields:
$\frac{-2x}{(x^2+1)^2} = 0$ (multiply by $(x^2+1)^2$)
$-2x = 0$
$x = 0$, but $x = 0$ is not in $[1, 4]$.
Thus, there are no critical values.
Evaluating y at the endpoints $x = 1$ and $x = 4$, we get:
$y\big|_{x=1} = \frac{1}{(1)^2+1} = \frac{1}{2}$ and
$y\big|_{x=4} = \frac{1}{(4)^2+1} = \frac{1}{17}$.
Thus, there is an absolute maximum of $\frac{1}{2}$ at $x = 1$ and an absolute minimum of $\frac{1}{17}$ at $x = 4$.

19) Since P is a polynomial, it is continuous on the stated interval. By the Extreme Value Theorem, it has an absolute maximum and minimum.
Computing the derivative, we get:
$P'(x) = \frac{d}{dx}[-x^2 + 92x - 180] = -2x + 92$.

19) Continued
Since $P'(x)$ is defined for all x, the only critical values are when $P'(x) = 0$. Solving yields:
$$-2x + 92 = 0$$
$$x = 46.$$
Thus, $x = 46$ is the only critical value. Evaluating P at $x = 46$ and the endpoints $x = 40$ and $x = 60$, we get:
$P(46) = -(46)^2 + 92(46) - 180 = 1936$
$P(40) = -(40)^2 + 92(40) - 180 = 1900$ and
$P(60) = -(60)^2 + 92(60) - 180 = 1740.$
Thus, there is an absolute maximum of $1,936 at $x = $46. So, the owner should charge $46 per day to maximize the profit.

21) Since HB is a polynomial, it is continuous on the stated interval. By the Extreme Value Theorem, it has an absolute maximum and minimum. Computing the derivative, we get:
$$HB'(x) = \frac{d}{dx}[15x^2 - 210x + 750]$$
$$= 30x - 210.$$
Since $HB'(x)$ is defined for all x, the only critical values are when $HB'(x) = 0$. Solving yields:
$$30x - 210 = 0$$
$$x = 7.$$
Thus, $x = 7$ is the only critical value. Evaluating HB at $x = 7$ and the endpoints $x = 0$ and $x = 14$, we get:
$HB(7) = 15(7)^2 - 210(7) + 750 = 15$
$HB(0) = 15(0)^2 - 210(0) + 750 = 750$ and
$HB(14) = 15(14)^2 - 210(14) + 750 = 750.$
Thus, there is an absolute minimum of 15 bacteria per cubic centimeter at $x = 7$ days.
a) Seven days after treatment, the concentration was minimized.
b) The minimum concentration was of 15 bacteria per cubic centimeter.

23) Since y is a polynomial, it is continuous on the stated interval. By the Extreme Value Theorem, it has an absolute maximum and minimum.

23) Continued
Computing the derivative, we get:
$$y' = \frac{d}{dx}[-0.037x^3 + 0.605x^2 - 2.776x + 9.333]$$
$$= -0.111x^2 + 1.21x - 2.776.$$
Since y' is defined for all x, the only critical values are when $y' = 0$. Solving yields:
$$-0.111x^2 + 1.21x - 2.776 = 0$$
Using the quadratic formula, we get
$$x = \frac{-1.21 \pm \sqrt{(1.21)^2 - 4(-0.111)(-2.776)}}{2(-0.111)}$$
$$= \frac{1.21 \pm \sqrt{0.231556}}{0.222}$$
$x \approx 7.62$ or $x \approx 3.28$. Thus, $x \approx 7.62$ and $x \approx 3.28$ are the critical values. Evaluating y at $x \approx 7.62$ and $x \approx 3.28$ and the endpoints $x = 1$ and $x = 10$, we get:
$y|_{x=3.28} = -0.037(3.28)^3 + 0.605(3.28)^2$
$\qquad - 2.776(3.28) + 9.333 \approx 5.43$
$y|_{x=7.62} = -0.037(7.62)^3 + 0.605(7.62)^2$
$\qquad - 2.776(7.62) + 9.333 \approx 6.94$
$y|_{x=1} = -0.037(1)^3 + 0.605(1)^2$
$\qquad - 2.776(1) + 9.333 = 7.125$ and
$y|_{x=10} = (-0.037(3.28)^3 + 0.605(3.28)^2$
$\qquad - 2.776(3.28) + 9.333 = 5.073.$
There is an absolute maximum of 7.125 at $x = 1$ and an absolute minimum of 5.073 at $x = 10$. Thus, in 1986, the unemployment rate was at its highest (7.125%) and in 1995, the unemployment rate was at its lowest (5.073%).

25) Since f is a polynomial, it is continuous on the stated interval. By the Extreme Value Theorem, it has an absolute maximum and minimum.
Computing the derivative, we get:
$$f'(x) = \frac{d}{dx}[-4.21x^2 + 10.6x + 8613.57]$$
$$= -8.42x + 10.6.$$
Since $f'(x)$ is defined for all x, the only critical values are when $f'(x) = 0$. Solving yields:
$$-8.42x + 10.6 = 0$$
$$x \approx 1.26$$

25) Continued
Thus, $x \approx 1.26$ is the only critical value.
Evaluating f at $x \approx 1.26$ and the endpoints $x = 1$ and $x = 22$, we get:
$f(1.26) = -4.21(1.26)^2 + 10.6(1.26) + 8613.57 \approx 8620$
$f(1) = -4.21(1)^2 + 10.6(1) + 8613.57 \approx 8620$ and
$f(22) = -4.21(22)^2 + 10.6(22) + 8613.57 \approx 6809$.
Thus, in 1974, the number of bowling alleys reached its peak at 8,620 alleys and in 1995, the number of bowling alleys hit a low of 6,809 alleys.

27) Since f is an exponential function, it is continuous on the stated interval. By the Extreme Value Theorem, it has an absolute maximum and minimum.
Computing the derivative, we get:
$f'(x) = \frac{d}{dx}[10{,}560e^{-0.012x}]$
$= 10{,}560e^{-0.012x}(-0.012) = -126.72e^{-0.012x}$.
Since $f'(x)$ is defined for all x, the only critical values are when $f'(x) = 0$. Solving yields:
$-126.72e^{-0.012x} = 0$
which has no solution since it is an exponential decay function.
Thus, there are no critical values.
Evaluating f at the endpoints $x = 1$ and $x = 36$, we get:
$f(1) = 10{,}560e^{-0.012(1)} \approx 10{,}434$
$f(36) = 10{,}560e^{-0.012(36)} \approx 6{,}856$
Thus, in 1960, the number of Catholic elementary schools was at its peak of 10,434 schools and in 1995, the number of Catholic elementary schools hit a low of 6,856 schools.

29a) $y_3 = 10{,}560e^{-0.012x} + 2410e^{-0.019x}$ represents the total number of Catholic elementary and secondary schools.

29b) Since y_3 is the sum of two continuous functions, it is continuous on the stated interval. By the Extreme Value Theorem, it has an absolute maximum and minimum. Computing the derivative, we get:
$y_3' = \frac{d}{dx}[10{,}560e^{-0.012x} + 2410e^{-0.019x}]$
$= -126.72e^{-0.012x} - 45.79e^{-0.019x}$.
Since $f'(x)$ is defined for all x, the only critical values are when $f'(x) = 0$. Solving yields:
$-126.72e^{-0.012x} - 45.79e^{-0.019x} = 0$
$126.72e^{-0.012x} = -45.79e^{-0.019x}$
Divide both sides by $126.72e^{-0.019x}$:
$e^{0.007x} \approx -0.3613478535$
which has no solution since it is an exponential growth function.
Thus, there are no critical values.
Evaluating y_3 at the endpoints $x = 1$ and $x = 36$, we get:
$y_3|_{x=1} = 10{,}560e^{-0.012(1)} + 2410e^{-0.019(1)}$
$\approx 10{,}434 + 2{,}365 = 12{,}799$
$y_3|_{x=36} = 10{,}560e^{-0.012(36)} + 2410e^{-0.019(36)}$
$\approx 6{,}856 + 1{,}216$
Thus, in 1960, the number of Catholic elementary and secondary schools was at its peak of 12,799 schools and in 1995, the number of Catholic elementary and secondary schools hit a low of 8,072 schools.

31) Since f is a polynomial, it is continuous on the stated interval. By the Extreme Value Theorem, it has an absolute maximum and minimum.
Computing the derivative, we get:
$f'(x) = \frac{d}{dx}[0.02x^4 - 0.31x^3 + 1.6x^2 - 3.09x + 3.51]$
$= 0.08x^3 - 0.93x^2 + 3.2x - 3.09$
Since $f'(x)$ is defined for all x, the only critical values are when $f'(x) = 0$. Solving yields:
$0.08x^3 - 0.93x^2 + 3.2x - 3.09 = 0$
To solve this, we will graph $Y_1 = f'(x)$ and use the zero command.

31) Continued
The graphs are displayed in viewing window [1, 7] by [– 2, 2]

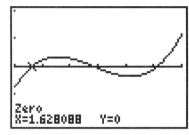

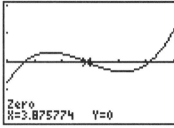

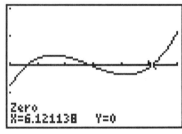

Thus, $x \approx 1.63$, $x \approx 3.88$, and $x \approx 6.12$ are the critical values.
Evaluating f at $x \approx 1.63$, $x \approx 3.88$, $x \approx 6.12$ and the endpoints $x = 1$ and $x = 7$, we get:
$f(1.63) = 0.02(1.63)^4 - 0.31(1.63)^3$
$\qquad + 1.6(1.63)^2 - 3.09(1.63) + 3.51$
$\qquad \approx 1.52.$
$f(3.88) = 0.02(3.88)^4 - 0.31(3.88)^3$
$\qquad + 1.6(3.88)^2 - 3.09(3.88) + 3.51$
$\qquad \approx 2.03.$
$f(6.12) = 0.02(6.12)^4 - 0.31(6.12)^3$
$\qquad + 1.6(6.12)^2 - 3.09(6.12) + 3.51$
$\qquad \approx 1.52.$
$f(1) = 0.02(1)^4 - 0.31(1)^3 + 1.6(1)^2$
$\qquad - 3.09(1) + 3.51 = 1.73$ and

31) Continued
$f(7) = 0.02(7)^4 - 0.31(7)^3 + 1.6(7)^2$
$\qquad - 3.09(7) + 3.51 = 1.97$
Thus, in 1993, natural gas prices hit a peak of $2.03 per 1000 cubic feet and in 1991 and 1995, natural gas prices hit a low of $1.52 per 1000 cubic feet.

33) The area of the rectangular enclosure is $A = x \cdot y = 10{,}000$. Solving for y yields: $y = \frac{10000}{x}$ The perimeter, P, of the fence needed is $P = x + x + y + y = 2x + 2y$. Substituting in for y in $2x + 2y$, we get:
$$P(x) = 2x + 2\left(\frac{10000}{x}\right) = 2x + \frac{20000}{x}$$
The domain is [50, 200]. Since P is a rational function, the only values where it is not continuous are the values that make it undefined. But $x = 0$ is not in [50, 200]. Thus, P is continuous on the interval. By the Extreme Value Theorem, it has an absolute maximum and minimum.
Computing the derivative, we get:
$P'(x) = \frac{d}{dx}[2x + 20{,}000x^{-1}] = 2 - 20{,}000x^{-2}$
$= 2 - \frac{20000}{x^2}$.
$P'(x)$ is undefined at $x = 0$, but $x = 0$ is not in [50, 200]. So, the only critical values are when $P'(x) = 0$. Solving yields:
$\qquad 2 - \frac{20000}{x^2} = 0$ (multiply by x^2)
$\qquad 2x^2 - 20{,}000 = 0$
$\qquad x^2 = 10{,}000$
$\qquad x = \pm 100$, but -100 is not in [50, 200].
Thus, $x = 100$ is the only critical value.
Evaluating P at $x = 100$ and the endpoints $x = 50$ and $x = 200$, we get:
$P(100) = 2(100) + \frac{20000}{100} = 400$
$P(50) = 2(50) + \frac{20000}{50} = 500$ and
$P(200) = 2(200) + \frac{20000}{200} = = 500.$
If $x = 100$, then $y = \frac{10000}{100} = 100$.
Thus, the dimensions need to be 100 feet by 100 feet to minimize the amount of fencing needed at 400 feet.

35) Since $R(x) = x \bullet p(x) = x(2.75 - 0.01x) =$
$2.75x - 0.01x^2$ & $C(x) = 5 + 0.5x + 0.003x^2$,
then $P(x) = R(x) - C(x) =$
$2.75x - 0.01x^2 - (5 + 0.5x + 0.003x^2)$
$= -0.013x^2 + 2.25x - 5$.
Since P is a polynomial, it is continuous on the stated interval. By the Extreme Value Theorem, it has an absolute maximum and minimum.
Computing the derivative, we get:
$P'(x) = \frac{d}{dx}[-0.013x^2 + 2.25x - 5]$
$= -0.026x + 2.25$.
Since $P'(x)$ is defined for all x, the only critical values are when $P'(x) = 0$. Solving yields:
$\quad -0.026x + 2.25 = 0$
$\quad x \approx 86.54$.
Thus, $x = 86.54$ is the only critical value. Evaluating P at $x = 86.54$ and the endpoints $x = 0$ and $x = 275$, we get:
$P(86.54) = -0.013(86.54)^2 + 2.25(86.54)$
$\qquad - 5 \approx 92.36$
$P(0) = -0.013(0)^2 + 2.25(0) - 5 = -5$ and
$P(275) = -0.013(275)^2 + 2.25(275) - 5$
≈ -369.38

a) Thus, the company must produce and sell 8,654 copies of the magazine in order to maximize their profit.

b) $p(86.54) = 2.75 - 0.01(86.54) \approx 1.88$. The price of the magazine should be $1.88 per issue.

37) Since C is a rational function, the only values where it is not continuous are the values that make it undefined. Setting the denominator equal to zero and solving yields: $\quad t^2 + 4t + 5 = 0$
Using the quadratic formula, we get:
$t = \frac{-4 \pm \sqrt{(4)^2 - 4(1)(5)}}{2(1)} = \frac{-4 \pm \sqrt{-4}}{2}$
which is undefined in the real numbers. Hence, $t^2 + 4t + 5 \neq 0$ for all t. Thus, C is continuous on the interval. By the Extreme Value Theorem, it has an absolute maximum and minimum.

37) Continued
Computing the derivative, we get:
$C'(t) = \frac{d}{dt}[\frac{5.3t}{t^2 + 4t + 5}]$
$= \frac{(t^2 + 4t + 5)(5.3) - (2t + 4)(5.3t)}{(t^2 + 4t + 5)^2} = \frac{-5.3t^2 + 26.5}{(t^2 + 4t + 5)^2}$.
$C'(t)$ is undefined when $t^2 + 4t + 5 = 0$, but $t^2 + 4t + 5 \neq 0$ for all t. Thus, the only critical values are when $C'(t) = 0$. Solving yields:
$\frac{-5.3t^2 + 26.5}{(t^2 + 4t + 5)^2} = 0$
(multiply by $(t^2 + 4t + 5)^2$)
$-5.3t^2 + 26.5 = 0$
$t^2 = 5$
$t = \pm\sqrt{5} \approx \pm 2.24$, -2.24 is not in $[0, 8]$. Thus, $t \approx 2.24$ is the only critical value. Evaluating C at $t \approx 2.24$ and at the endpoints $t = 0$ and $t = 8$, we get:
$C(2.24) = \frac{5.3(2.24)}{(2.24)^2 + 4(2.24) + 5} \approx 0.626$
$C(0) = \frac{5.3(0)}{(0)^2 + 4(0) + 5} = 0$ and
$C(8) = \frac{5.3(8)}{(8)^2 + 4(8) + 5} \approx 0.420$

a) After about 2.24 hours, the concentration is at a maximum.

b) The maximum concentration is 0.626 milligrams per cubic centimeter.

39a) $R(x) = x \bullet p(x)$
$= x(0.14x^2 - 15.52x + 561.21)$
$= 0.14x^3 - 15.52x^2 + 561.21x$

39b) Since R is a polynomial, it is continuous on the stated interval. By the Extreme Value Theorem, it has an absolute maximum and minimum. Computing the derivative, we get:
$R'(x) = \frac{d}{dx}[0.14x^3 - 15.52x^2 + 561.21x]$
$= 0.42x^2 - 31.04x + 561.21$.

39b) Continued
Since $R'(x)$ is defined for all x, the only critical values are when $R'(x) = 0$. Solving yields:
$0.42x^2 - 31.04x + 561.21 = 0$
Using the quadratic formula, we get:
$x = \frac{31.04 \pm \sqrt{(-31.04)^2 - 4(0.42)(561.21)}}{2(0.42)}$
$= \frac{31.04 \pm \sqrt{20.6488}}{0.84}$
$x \approx 31.54$ or $x \approx 42.36$.
Thus, $x \approx 31.54$ and $x \approx 42.36$ are the critical values. Evaluating R at $x \approx 31.54$, $x \approx 42.36$ and the endpoints $x = 5$ and $x = 55$, we get:
$R(31.54) = -0.14(31.54)^3 - 15.52(31.54)^2 + 561.21(31.54) \approx 6654.22$
$R(42.36) = -0.14(42.36)^3 - 15.52(42.36)^2 + 561.21(42.36) \approx 6565.57$
$R(5) = -0.14(5)^3 - 15.52(5)^2 + 561.21(5) = 2435.55$ and
$R(55) = -0.14(55)^3 - 15.52(55)^2 + 561.21(55) = 7211.05$
The maximum revenue is $7,211.05.

39c) From above, the revenue is maximized when 55 units are demanded.

39d) $p(55) = 0.14(55)^2 - 15.52(55) + 561.21 = \$131.11$ per unit. So, the item needs to be priced at $131.11 per unit in order to maximize the revenue.

41a) $y' = \frac{d}{dx}[-0.0007x^4 + 0.045x^3 - 0.868x^2 + 6.347x + 60.574]$
$= -0.0028x^3 + 0.135x^2 - 1.736x + 6.347$
Since the derivative is an polynomial function, there are no values for which the derivative is undefined. Setting the derivative equal to zero and solving yields:
$-0.0028x^3 + 0.135x^2 - 1.736x + 6.347 = 0$
To solve this, we will graph $Y_1 = f'(x)$ and use the zero command.

41a) Continued
The graph is displayed in viewing window [1, 25] by [−2, 5]

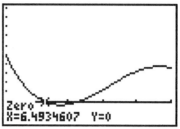

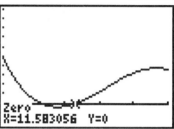

The critical values $x \approx 6.49$ and $x \approx 11.58$.
$y\big|_{x=6.49} = 76.27$ and $y\big|_{x=11.58} = 74.97$
We now place the critical values on a sign diagram:

We select the test number 6 in [1, 6.49), the test number 7 in (6.49, 11.58), and the test number 12 in (11.58, 25].
Thus, $y'\big|_{x=6} = 0.1922$,
$y'\big|_{x=7} = -0.1434$, and
$y'\big|_{x=12} = 0.1286$.
Now, we can complete the sign diagram:

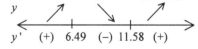

Rounding to the nearest year, in 1975, the total energy consumption hit a peak of 76.27 quadrillion BTUs whereas in 1981, the total energy consumption hit a low of 74.97 quadrillion BTUs.

41b) Since y is a polynomial, it is continuous on the stated interval. By the Extreme Value Theorem, it has an absolute maximum and minimum. Using the results from part a, the critical values are $x \approx 6.49$ and $x \approx 11.58$. Evaluating y at $x \approx 6.49$, $x \approx 11.58$ and the endpoints $x = 1$ and $x = 25$, we get:

$y\big|_{x=6.49} = 76.27$

$y\big|_{x=11.58} = 74.97$

$y\big|_{x=1} = 66.10$ and

$y\big|_{x=25} = 106.44$

In 1970, the total energy consumption was at its lowest value of 66.10 quadrillion BTUs and in 1994, the total energy consumption was at its highest value of 106.44 quadrillion BTUs.

43) Since f is a polynomial, it is continuous on the stated interval. By the Extreme Value Theorem, it has an absolute maximum and minimum.
Computing the derivative, we get:
$f'(x) = \frac{d}{dx}[-0.41x^2 + 13.53x + 330.69]$
$= -0.82x + 13.53$.
Since $f'(x)$ is defined for all x, the only critical values are when $f'(x) = 0$. Solving yields:
$-0.82x + 13.53 = 0$
$x = 16.5$
Thus, $x = 16.5$ is the only critical value. Evaluating f at $x = 16.5$ and the endpoints $x = 1$ and $x = 31$, we get:
$f(16.5) = -0.41(16.5)^2 + 13.53(16.5)$
$\qquad + 330.69 \approx 442.31$
$f(1) = -0.41(1)^2 + 13.53(1) + 330.69$
$\qquad = 343.81$
$f(31) = -0.41(31)^2 + 13.53(31) + 330.69$
$\qquad = 356.11$
Rounding off to nearest year, in 1960, water consumption was at its lowest rate of 343.81 gallons per day per capita while in 1976, water consumption was at its highest rate of 442.31 gallons per day per capita.

45) Since v is a polynomial, it is continuous on the stated interval. By the Extreme Value Theorem, it has an absolute maximum and minimum.
Computing the derivative, we get:
$v'(h) = \frac{d}{dh}[0.0015h^2 - 0.019h + 1.563]$
$= 0.003h - 0.019$.
Since $v'(h)$ is defined for all h, the only critical values are when $v'(h) = 0$. Solving yields:
$0.003h - 0.019 = 0$
$h \approx 6.33$, but $h \approx 6.33$ is not in $[10, 80]$
Thus, there are no critical values. Evaluating v at the endpoints $h = 10$ and $h = 80$, we get:
$v(10) = 0.0015(10)^2 - 0.019(10) + 1.563$
$\qquad = 1.523$
$v(80) = 0.0015(80)^2 - 0.019(80) + 1.563$
$\qquad = 9.643$
The maximum viscosity of 9.643 occurs at a hematocrit of 80 and the minimum viscosity of 1.523 occurs at a hematocrit of 10.

47a) $C(x) = $ (cost per light)$\bullet x + $ fixed costs
$= 60x + 1000$.

47b) We need to first find $p(x)$. If $x = 100$ lamps are sold, the price $p(x) = \$100$. If the price is decreased by \$5, the number of lamps sold increases by 7. In other words, if $x = 107$ lamps ($100 + 7$) are sold, then the price $p(x) = \$95$ (\$100 − \$5). Thus, $(100, 100)$ and $(107, 95)$ are two points that lie on the line $p(x)$. We can calculate the slope:
$m = \frac{p(107) - p(100)}{107 - 100} = \frac{95 - 100}{7} = -\frac{5}{7}$.
Using the point-slope formula, we can find the equation of $p(x)$:
$p(x) - p(100) = -\frac{5}{7}(x - 100)$
$p(x) - 100 = -\frac{5}{7}x + \frac{500}{7}$
$p(x) = -\frac{5}{7}x + \frac{1200}{7}$
Now, $R(x) = x \bullet p(x) = x(-\frac{5}{7}x + \frac{1200}{7})$
$= -\frac{5}{7}x^2 + \frac{1200}{7}x$.

47c) $P(x) = R(x) - C(x)$
$= -\frac{5}{7}x^2 + \frac{1200}{7}x - (60x + 1000)$
$= -\frac{5}{7}x^2 + \frac{780}{7}x - 1000.$

Since P is a polynomial, it is continuous on the stated interval. By the Extreme Value Theorem, it has an absolute maximum and minimum. Computing the derivative, we get:
$P'(x) = \frac{d}{dx}[-\frac{5}{7}x^2 + \frac{780}{7}x - 1000]$
$= -\frac{10}{7}x + \frac{780}{7}.$

Since $P'(x)$ is defined for all x, the only critical values are when $P'(x) = 0$. Solving yields:
$-\frac{10}{7}x + \frac{780}{7} = 0$
$x = 78.$

Thus, $x = 78$ is the only critical value. Evaluating P at $x = 78$ and the endpoints $x = 50$ and $x = 200$, we get:
$P(78) = -\frac{5}{7}(78)^2 + \frac{780}{7}(78) - 1000$
≈ 3345.71
$P(50) = -\frac{5}{7}(50)^2 + \frac{780}{7}(50) - 1000$
≈ 2785.71 and
$P(200) = -\frac{5}{7}(200)^2 + \frac{780}{7}(200) - 1000$
$\approx -7285.71.$

Thus, there is an absolute maximum of $3345.71 at $x = 78$. So, 78 lamps need to be produced and sold to maximize the profit.

49a) We need to first find $p(x)$. If $x = 3000$ pets are sold, the price $p(x) = \$10$. If the price is decreased by \$0.17, the number of pets sold increases by 180. In other words, if $x = 3180$ pets $(3000 + 180)$ are sold, then the price $p(x) = \$9.83$ ($\$10 - \$0.17$). Thus, $(3000, 10)$ and $(3180, 9.83)$ are two points that lie on the line $p(x)$. We can calculate the slope: $m = \frac{p(3180) - p(3000)}{3180 - 3000} = \frac{10 - 9.83}{180} = -\frac{0.17}{180} = -\frac{17}{18000}.$

49a) Continued
Using the point-slope formula, we can find the equation of $p(x)$:
$p(x) - p(3000) = -\frac{17}{18000}(x - 3000)$
$p(x) - 10 = -\frac{17}{18000}x + \frac{17}{6}$
$p(x) = -\frac{17}{18000}x + \frac{77}{6}$
Now, $R(x) = x \bullet p(x) = x(-\frac{17}{18000}x + \frac{77}{6})$
$= -\frac{17}{18000}x^2 + \frac{77}{6}x$

49b) Since R is a polynomial, it is continuous on the stated interval. By the Extreme Value Theorem, it has an absolute maximum and minimum. Computing the derivative, we get:
$R'(x) = \frac{d}{dx}[-\frac{17}{18000}x^2 + \frac{77}{6}x]$
$= -\frac{17}{9000}x + \frac{77}{6}$

Since $R'(x)$ is defined for all x, the only critical values are when $R'(x) = 0$. Solving yields:
$-\frac{17}{9000}x + \frac{77}{6} = 0$ (multiply by 9000)
$-17x + 115500 = 0$
$x \approx 6794.12$

Thus, $x \approx 6794.12$ is the only critical value. Evaluating R at $x \approx 6794.12$ and the endpoints $x = 3000$ and $x = 10{,}000$, we get:
$R(6794.12) = -\frac{17}{18000}(6794.12)^2$
$+ \frac{77}{6}(6794.12) \approx 43{,}595.59$
$R(3000) = -\frac{17}{18000}(3000)^2$
$+ \frac{77}{6}(3000) = 30{,}000$ and
$R(10{,}000) = -\frac{17}{18000}(10{,}000)^2$
$+ \frac{77}{6}(10{,}000) \approx 33{,}888.89$

Thus, there is an absolute maximum of about \$43,596 at $x \approx 6794$. So, 6794 pets need to be produced and sold to maximize the revenue.

Section 5.5 The Second Derivative Test and Optimization

1) Since f is a rational function, the only values where it is not continuous are the values that make it undefined. But $x = 0$ is not in $(0, 20)$. Thus, f is continuous on the interval. Computing the derivative, we get:
$f'(x) = \frac{d}{dx}[2x + 6x^{-1}] = 2 - 6x^{-2} = 2 - \frac{6}{x^2}$.
$f'(x)$ is undefined when $x = 0$, but 0 is not in $(0, 20)$. Thus, the only critical values are when $f'(x) = 0$. Solving yields:
$2 - \frac{6}{x^2} = 0$ (multiply by x^2)
$2x^2 - 6 = 0$
$x^2 = 3$
$x = \pm \sqrt{3}$, but $-\sqrt{3}$ is not in $(0, 20)$.
Thus, $x = \sqrt{3}$ is the only critical value and $f(\sqrt{3}) = \sqrt{48} = 4\sqrt{3}$. Computing the second derivative, we get:
$f''(x) = \frac{d}{dx}[2 - 6x^{-2}] = 12x^{-3} = \frac{12}{x^3}$.
Hence, $f''(\sqrt{3}) = \frac{12}{(\sqrt{3})^3} > 0$. So, f has an absolute minimum at $(\sqrt{3}, 4\sqrt{3})$
$\approx (1.73, 6.93)$.

3) Since f is a rational function, the only values where it is not continuous are the values that make it undefined. But $x = 0$ is not in $(0, 20)$. Thus, f is continuous on the interval. Computing the derivative, we get:
$f'(x) = \frac{d}{dx}[4x - 3 + 2x^{-1}] = 4 - 2x^{-2}$
$= 4 - \frac{2}{x^2}$.
$f'(x)$ is undefined when $x = 0$, but 0 is not in $(0, 20)$. Thus, the only critical values are when $f'(x) = 0$. Solving yields:
$4 - \frac{2}{x^2} = 0$ (multiply by x^2)
$4x^2 - 2 = 0$
$x^2 = 0.5$
$x = \pm \sqrt{0.5}$, but $-\sqrt{0.5}$ is not in $(0, 20)$.

3) Continued
Thus, $x = \sqrt{0.5} = \frac{\sqrt{2}}{2}$ is the only critical value and $f(\frac{\sqrt{2}}{2}) = 4\sqrt{2} - 3$.
Computing the second derivative, we get:
$f''(x) = \frac{d}{dx}[4 - 2x^{-2}] = 4x^{-3} = \frac{4}{x^3}$.
Hence, $f''(\frac{\sqrt{2}}{2}) = \frac{4}{\left(\frac{\sqrt{2}}{2}\right)^3} > 0$. So, f has an absolute minimum at $(\frac{\sqrt{2}}{2}, 4\sqrt{2} - 3) \approx$
$(0.71, 2.66)$.

5) Since y is a rational function, the only values where it is not continuous are the values that make it undefined. But $x = 0$ is not in $(0, 5)$. Thus, y is continuous on the interval. Computing the derivative, we get:
$y' = \frac{d}{dx}[3x^2 - 2 + 6x^{-2}] = 6x - 12x^{-3}$
$= 6x - \frac{12}{x^3}$.
y' is undefined when $x = 0$, but 0 is not in $(0, 5)$. Thus, the only critical values are when $y' = 0$. Solving yields:
$6x - \frac{12}{x^3} = 0$ (multiply by x^3)
$6x^4 - 12 = 0$
$x^4 = 2$
$x = \pm \sqrt[4]{2}$, but $-\sqrt[4]{2}$ is not in $(0, 5)$.
Thus, $x = \sqrt[4]{2}$ is the only critical value and $y\big|_{x = \sqrt[4]{2}} = 6\sqrt{2} - 2$. Computing the second derivative, we get:
$y'' = \frac{d}{dx}[6x - 12x^{-3}] = 6 + 36x^{-4} = 6 + \frac{36}{x^4}$.
Hence, $y''\big|_{x = \sqrt[4]{2}} = 6 + \frac{36}{(\sqrt[4]{2})^4} = 24 > 0$. So, y has an absolute minimum at
$(\sqrt[4]{2}, 6\sqrt{2} - 2) \approx (1.19, 6.49)$.

7) Since g is a rational function, the only values where it is not continuous are the values that make it undefined. But $x = 0$ is not in $(0, 10)$. Thus, g is continuous on the interval. Computing the derivative, we get:
$g'(x) = \frac{d}{dx}[3x^2 - 1 + 2x^{-2}] = 6x - 4x^{-3}$
$= 6x - \frac{4}{x^3}$.

$g'(x)$ is undefined when $x = 0$, but 0 is not in $(0, 10)$. Thus, the only critical values are when $g'(x) = 0$. Solving yields:
$6x - \frac{4}{x^3} = 0$ (multiply by x^3)
$6x^4 - 4 = 0$
$x^4 = \frac{2}{3}$
$x = \pm \sqrt[4]{\frac{2}{3}}$, but $-\sqrt[4]{\frac{2}{3}}$ is not in $(0, 10)$.

Thus, $x = \sqrt[4]{\frac{2}{3}}$ is the only critical value and $g(\sqrt[4]{\frac{2}{3}}) = 2\sqrt{6} - 1$. Computing the second derivative, we get:
$g''(x) = \frac{d}{dx}[6x - 4x^{-3}] = 6 + 12x^{-4} = 6 + \frac{12}{x^4}$.

Hence, $g''(\sqrt[4]{\frac{2}{3}}) = 6 + \frac{12}{(\sqrt[4]{\frac{2}{3}})^4} = 24 > 0$. So, g has an absolute minimum at
$(\sqrt[4]{\frac{2}{3}}, 2\sqrt{6} - 1) \approx (0.90, 3.90)$.

9) Since f is a rational function, the only values where it is not continuous are the values that make it undefined. But $x = 0$ is not in $(0, 20)$. Thus, f is continuous on the interval. Computing the derivative, we get:
$f'(x) = \frac{d}{dx}[-3x - 2x^{-1}] = -3 + 2x^{-2}$
$= -3 + \frac{2}{x^2}$.

$f'(x)$ is undefined when $x = 0$, but 0 is not in $(0, 20)$. Thus, the only critical values are when $f'(x) = 0$. Solving yields:
$-3 + \frac{2}{x^2} = 0$ (multiply by x^2)
$-3x^2 + 2 = 0$
$x^2 = \frac{2}{3}$
$x = \pm \sqrt{\frac{2}{3}}$, but $-\sqrt{\frac{2}{3}}$ is not in $(0, 20)$.

9) Continued
Thus, $x = \sqrt{\frac{2}{3}} = \frac{\sqrt{6}}{3}$ is the only critical value and $f(\frac{\sqrt{6}}{3}) = -2\sqrt{6}$. Computing the second derivative, we get:
$f''(x) = \frac{d}{dx}[-3 + 2x^{-2}] = -4x^{-3} = -\frac{4}{x^3}$.

Hence, $f''(\frac{\sqrt{6}}{3}) = -\frac{4}{(\frac{\sqrt{6}}{3})^3} < 0$. So, f has an absolute maximum at $(\frac{\sqrt{6}}{3}, -2\sqrt{6}) \approx$
$(0.82, -4.90)$.

11) Since y is a rational function, the only values where it is not continuous are the values that make it undefined. But $x = 0$ is not in $(0, 10)$. Thus, y is continuous on the interval. Computing the derivative, we get:
$y' = \frac{d}{dx}[-2x - 3x^{-1}] = -2 + 3x^{-2}$
$= -2 + \frac{3}{x^2}$.

y' is undefined when $x = 0$, but 0 is not in $(0, 10)$. Thus, the only critical values are when $y' = 0$. Solving yields:
$-2 + \frac{3}{x^2} = 0$ (multiply by x^2)
$-2x^2 + 3 = 0$
$x^2 = 1.5$
$x = \pm \sqrt{1.5}$, but $-\sqrt{1.5}$ is not in
$(0, 10)$. Thus, $x = \sqrt{1.5} = \frac{\sqrt{6}}{2}$ is the only critical value and $y\big|_{x = \frac{\sqrt{6}}{2}} = -2\sqrt{6}$.
Computing the second derivative, we get:
$y'' = \frac{d}{dx}[-2 + 3x^{-2}] = -6x^{-3} = -\frac{6}{x^3}$.

Hence, $y''\big|_{x = \frac{\sqrt{6}}{2}} = -\frac{6}{(\frac{\sqrt{6}}{2})^3} < 0$. So, y has an absolute maximum at $(\frac{\sqrt{6}}{2}, -2\sqrt{6})$
$\approx (1.22, -4.90)$.

13) Since g is a rational function, the only values where it is not continuous are the values that make it undefined. But $x = 0$ is not in $(0, 7)$. Thus, g is continuous on the interval. Computing the derivative, we get:
$g'(x) = \frac{d}{dx}[2 - 3x^2 - 2x^{-2}] = -6x + 4x^{-3}$
$= -6x + \frac{4}{x^3}$.

$g'(x)$ is undefined when $x = 0$, but 0 is not in $(0, 7)$. Thus, the only critical values are when $g'(x) = 0$. Solving yields:
$-6x + \frac{4}{x^3} = 0$ (multiply by x^3)
$-6x^4 + 4 = 0$
$x^4 = \frac{2}{3}$
$x = \pm \sqrt[4]{\frac{2}{3}}$, but $-\sqrt[4]{\frac{2}{3}}$ is not in $(0, 7)$.

Thus, $x = \sqrt[4]{\frac{2}{3}}$ is the only critical value and
$g(0.90) = -2\sqrt{6} + 2 \approx -2.90$.
Computing the second derivative, we get:
$g''(x) = \frac{d}{dx}[-6x + 4x^{-3}] = -6 - 12x^{-4}$
$= -6 - \frac{12}{x^4}$. Hence, $g''(\sqrt[4]{\frac{2}{3}})$
$= -6 - \frac{12}{(\sqrt[4]{\frac{2}{3}})^4} = -24 < 0$. So, g has an

absolute maximum at $(\sqrt[4]{\frac{2}{3}}, -2\sqrt{6} + 2) \approx (0.90, -2.90)$.

15) Since y is a rational function, the only values where it is not continuous are the values that make it undefined. But $x = 0$ is not in $(0, 10)$. Thus, y is continuous on the interval. Computing the derivative, we get:
$y' = \frac{d}{dx}[3 - 2x - 5x^{-1}] = -2 + 5x^{-2}$
$= -2 + \frac{5}{x^2}$.

y' is undefined when $x = 0$, but 0 is not in $(0, 10)$. Thus, the only critical values are when $y' = 0$. Solving yields:
$-2 + \frac{5}{x^2} = 0$ (multiply by x^2)
$-2x^2 + 5 = 0$
$x^2 = 2.5$
$x = \pm\sqrt{2.5}$, but $-\sqrt{2.5}$ is not in $(0, 10)$.

15) Continued
Thus, $x = \sqrt{2.5} = \frac{\sqrt{10}}{2}$ is the only critical
value and $y\big|_{x = \frac{\sqrt{10}}{2}} = -2\sqrt{10} + 3$
Computing the second derivative, we get:
$y'' = \frac{d}{dx}[-2 + 5x^{-2}] = -10x^{-3} = -\frac{10}{x^3}$.
Hence, $y''\big|_{x = \frac{\sqrt{10}}{2}} = -\frac{10}{(\frac{\sqrt{10}}{2})^3} < 0$. So, y has
an absolute maximum at
$(\frac{\sqrt{10}}{2}, -2\sqrt{10} + 3) \approx (1.58, -3.32)$.

17) The area of the rectangular lots is
$A = x \cdot y$. Since the lots are adjacent and a canal is to be used as one side of the fencing, the perimeter, P, of the fence needed is
$P = x + y + y + y = 400$ or $x + 3y = 400$.
Solving for x yields: $x = 400 - 3y$.
Substituting in for x in $x \cdot y$, we get:
$A(y) = (400 - 3y)y = 400y - 3y^2$.
The domain is $(0, \infty)$. Since A is a polynomial, it is continuous on the interval. Computing the derivative, we get:
$A'(y) = \frac{d}{dy}[400y - 3y^2] = 400 - 6y$.

$A'(y)$ is defined for all values in the domain so the only critical values are when $A'(y) = 0$. Solving yields:
$400 - 6y = 0$
$y = \frac{200}{3}$
But $x = 400 - 3(\frac{200}{3}) = 200$ and
$A(\frac{200}{3}) = 400(\frac{200}{3}) - 3(\frac{200}{3})^2 = \frac{40000}{3}$.
Computing the second derivative, we get:
$A''(y) = \frac{d}{dy}[400 - 6y] = -6 < 0$ for all y.
Hence, A has an absolute maximum area of $13,333\frac{1}{3}$ square feet when $x = 200$ ft and $y = 66\frac{2}{3}$ feet.

19) The perimeter of the outer boundary is $x + x + y + x + x + y = 4x + 2y$. The cost of the outer boundary then is $8.50(4x + 2y) = 34x + 17y$. The cost for the inner partition is $3.20y$, thus the total cost C is
$$34x + 17y + 3.2y = 34x + 20.2y$$
The area of one of the rectangular plots is $A = x \cdot y = 1500$. Solving for y yields: $y = \frac{1500}{x}$. Substituting in for y in $34x + 20.2y$, we get:
$$C(x) = 34x + 20.2\left(\frac{1500}{x}\right) = 34x + \frac{30300}{x}.$$
The domain is $(0, \infty)$. Since C is a rational function and $x = 0$ is not in $(0, \infty)$, it is continuous on the interval. Computing the derivative, we get:
$$C'(x) = \frac{d}{dx}[34x + 30{,}300x^{-1}]$$
$$= 34 - 30{,}300x^{-2} = 34 - \frac{30300}{x^2}.$$
$C'(x)$ is defined for all values in the domain so the only critical values are when $C'(x) = 0$. Solving yields:
$$34 - \frac{30300}{x^2} = 0 \text{ (multiply by } x^2)$$
$$34x^2 - 30{,}300 = 0$$
$$x^2 \approx 891.18$$
$x \approx \pm 29.85$, but $x \approx -29.85$ is not in $(0, \infty)$, so $x \approx 29.85$ is the only critical value. But $y = \frac{1500}{29.85} \approx 50.25$ and
$$C(29.85) = 34(29.85) + \frac{30300}{29.85} = 2029.98.$$
Computing the second derivative, we get:
$$C''(x) = \frac{d}{dx}[34 - 30{,}300x^{-2}] = 60{,}600x^{-3}$$
$$= \frac{60600}{x^3}. \text{ Thus, } C''(29.85) = \frac{60600}{(29.85)^3} \approx 2.28.$$
Hence, C has an absolute minimum cost of $2029.98 when $x = 29.85$ ft and $y = 50.25$ feet.

21) We first make a drawing of the situation:

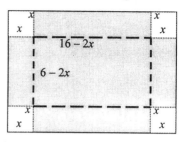

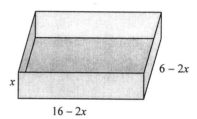

The volume of the box is $V(x) = lwh$
$= (16 - 2x)(6 - 2x)x = 4x^3 - 44x^2 + 96x$
The domain of the V is $(0, 3)$ since if $x = 0$, no cut would be made and if $x = 3$, the width $6 - 2x$ would be zero. In either case, we could not make a box. Since V is a polynomial, it is continuous on the interval. Computing the derivative, we get:
$$V'(x) = \frac{d}{dx}[4x^3 - 44x^2 + 96x]$$
$$= 12x^2 - 88x + 96.$$
$V'(x)$ is defined for all values in the domain so the only critical values are when $V'(x) = 0$. Solving yields:
$$12x^2 - 88x + 96 = 0$$
$$4(3x^2 - 22x + 24) = 0$$
$$4(3x - 4)(x - 6) = 0$$
$$x = \tfrac{4}{3} \text{ or } 6, \text{ but } 6 \text{ is not in } (0, 3)$$
which means that $x = \tfrac{4}{3}$ is the only critical value. But $l = 16 - 2(\tfrac{4}{3}) = \tfrac{40}{3}$, $w = 6 - 2(\tfrac{4}{3}) = \tfrac{10}{3}$, and $V(\tfrac{4}{3}) = 4(\tfrac{4}{3})^3 - 44(\tfrac{4}{3})^2 + 96(\tfrac{4}{3}) = \tfrac{1600}{27}$.

21) Continued
Computing the second derivative, we get:
$V''(x) = \frac{d}{dx}[12x^2 - 88x + 96] = 24x - 88$.
Hence, $V''(\frac{4}{3}) = 24(\frac{4}{3}) - 88$
$= -56 < 0$. Thus, V has an absolute maximum of $\frac{1600}{27}$ cubic inches when the $l = \frac{40}{3}$ in, $w = \frac{10}{3}$ in, and $h = \frac{4}{3}$ in.

23) The surface area of the base is x^2 so the cost for the base is $0.3x^2$. The surface area for the four sides is $4xy$, so the cost for the sides is $0.2 \bullet 4xy = 0.8xy$. The total cost for the box is $0.3x^2 + 0.8xy$. The volume of the box is $x^2 y = 6000$. Solving for y yields: $y = \frac{6000}{x^2}$.
Substituting in for y in $0.3x^2 + 0.8xy$ yields:
$C(x) = 0.3x^2 + 0.8x(\frac{6000}{x^2}) = 0.3x^2 + \frac{4800}{x}$.

The domain of C is $(0, \infty)$. Since C is a rational function and $x = 0$ is not in $(0, \infty)$, it is continuous on the interval. Computing the derivative, we get:
$C'(x) = \frac{d}{dx}[0.3x^2 + 4800x^{-1}]$
$= 0.6x - 4800x^{-2} = 0.6x - \frac{4800}{x^2}$.

$C'(x)$ is defined for all values in the domain so the only critical values are when $C'(x) = 0$.
Solving yields:
$0.6x - \frac{4800}{x^2} = 0$ (multiply by x^2)
$0.6x^3 - 4800 = 0$
$x^3 = 8000$
$x = 20$, (note: there are two complex solutions which we will ignore). So $x = 20$ is the only critical value.
But $y = \frac{6000}{20^2} = 15$ and
$C(20) = 0.3(20)^2 + \frac{4800}{20} = 360$.
Computing the second derivative, we get:
$C''(x) = \frac{d}{dx}[0.6x - 4800x^{-2}]$
$= 0.6 + 9600x^{-3} = 0.6 + \frac{9600}{x^3}$.

23) Continued
Thus,
$C''(20) = 0.6 + \frac{9600}{(20)^3} = 1.8 > 0$. Hence, C has an absolute minimum cost of $360 when $x = 20$ in and $y = 15$ in.

25a) $AC(x) = \frac{C(x)}{x} = \frac{5 + 0.5x + 0.003x^2}{x}$
$= 0.003x + 0.5 + \frac{5}{x}$. The domain is $x > 0$.

25b) Since AC is a rational function and $x = 0$ is not in $(0, \infty)$, it is continuous on the interval. Computing the derivative, we get:
$AC'(x) = \frac{d}{dx}[0.003x + 0.5 + 5x^{-1}]$
$= 0.003 - 5x^{-2} = 0.003 - \frac{5}{x^2}$.

$AC'(x)$ is defined for all values in the domain so the only critical values are when $AC'(x) = 0$. Solving yields:
$0.003 - \frac{5}{x^2} = 0$ (multiply by x^2)
$0.003x^2 - 5 = 0$
$x^2 = \frac{5000}{3}$
$x \approx \pm 40.82$, but -40.82 is not in $(0, \infty)$, so $x \approx 40.82$ is the only critical value. But,
$AC(40.82) = 0.003(40.82) + 0.5 + \frac{5}{40.82}$
≈ 0.745.
Computing the second derivative, we get:
$AC''(x) = \frac{d}{dx}[0.003 - 5x^{-2}] = 10x^{-3} = \frac{10}{x^3}$.
Thus, $AC''(40.82) = \frac{10}{(40.82)^3} \approx 0.0001 > 0$.

Hence, AC has an absolute minimum cost of $74.50 per 100 magazines when 4,082 magazines are produced.

27) $AC(t) = \frac{C(t)}{t} = \frac{100t^2 + 600t + 15000}{t}$

$= 100t + 600 + \frac{15000}{t}$. The domain is $t > 0$.

Since AC is a rational function and $t = 0$ is not in $(0, \infty)$, it is continuous on the interval. Computing the derivative, we get:

$AC'(t) = \frac{d}{dt}[100t + 600 + 15,000t^{-1}]$

$= 100 - 15,000t^{-2} = 100 - \frac{15000}{t^2}$.

$AC'(t)$ is defined for all values in the domain so the only critical values are when $AC'(t) = 0$. Solving yields:

$100 - \frac{15000}{t^2} = 0$ (multiply by t^2)

$100t^2 - 15,000 = 0$

$t^2 = 150$

$t \approx \pm 12.25$, but -12.25 is not in $(0, \infty)$, so $t \approx 12.25$ is the only critical value. But,

$AC(12.25) = 100(12.25) + 600 + \frac{15000}{(12.25)}$

≈ 3049.49.

Computing the second derivative, we get:

$AC''(t) = \frac{d}{dt}[100 - 15,000t^{-2}] = 30,000t^{-3}$

$= \frac{30000}{t^3}$.

Thus, $AC''(12.25) = \frac{30000}{(12.25)^3} \approx 16.32 > 0$.

Hence, AC has an absolute minimum cost of $3049.49 per year 12.25 years after the car is bought.

29) The total cost per day is equal to the number of rooms occupied times $4.50. So, $C(x) = 4.5x$.

$P(x) = R(x) - C(x)$

$= -0.75x^2 + 200x - 4.5x = -0.75x^2 + 195.5x$.

The domain of P is $(0, 200]$. Since P is a polynomial, it is continuous on the interval. Computing the derivative, we get:

$P'(x) = \frac{d}{dx}[-0.75x^2 + 195.5x]$

$= -1.5x + 195.5$

$P'(x)$ is defined for all values in the domain so the only critical values are when $P'(x) = 0$. Solving yields:

$-1.5x + 195.5 = 0$

$x \approx 130$

So $x \approx 130$ is the only critical value.

29) Continued

But, $P(130) = -0.75(130)^2 + 195.5(130)$

$= 12,740$ and

$p(130) = -0.75(130) + 200 = 102.50$

Computing the second derivative, we get:

$P''(x) = \frac{d}{dx}[-1.5x + 195.5] = -1.5 < 0$

for all x. Hence, P has an absolute maximum profit of $12,740 when the rate is $102.50 per room.

31) We need to first find the yield per tree, $y(x)$. If $x = 40$ trees per acre, the yield per tree $y(x) = 160$ pounds. The number of trees per acre is increased by one, the yield per tree decreases by 2 pounds. In other words, if $x = 41$ trees per acre $(40 + 1)$, then the yield $y(x) = 158\ (160 - 2)$. Thus, $(40, 160)$ and $(41, 158)$ are two points that lie on the line $y(x)$. We can calculate the slope:

$m = \frac{y(41) - y(40)}{41 - 40} = \frac{158 - 160}{1} = -2$.

Using the point-slope formula, we can find the equation of $y(x)$:

$y(x) - y(40) = -2(x - 40)$

$y(x) - 160 = -2x + 80$

$y(x) = -2x + 240$

Now, the total yield $Y(x) = x \bullet y(x)$

$= x(-2x + 240)$

$= -2x^2 + 240x$.

The domain of Y is $[40, 120]$. Since Y is a polynomial, it is continuous on the interval. Computing the derivative, we get:

$Y'(x) = \frac{d}{dx}[-2x^2 + 240x] = -4x + 240$

$Y'(x)$ is defined for all values in the domain so the only critical values are when $Y'(x) = 0$. Solving yields:

$-4x + 240 = 0$

$x = 60$

So $x = 60$ is the only critical value. But,

$Y(60) = -2(60)^2 + 240(60) = 7,200$

Computing the second derivative, we get:

$Y''(x) = \frac{d}{dx}[-4x + 240] = -4 < 0$

for all x. Hence, Y has an absolute maximum yield of 7,200 pounds per acre per year when 60 trees are planted per acre.

33) We need to first find the yield per tree, $y(x)$. If $x = 30$ trees per acre, the yield per tree $y(x) = 50$ pounds. The number of trees per acre is increased by one, the yield per tree decreases by 1.25 pounds. In other words, if $x = 31$ trees per acre $(30 + 1)$, then the yield $y(x) = 48.75$ $(50 - 1.25)$. Thus, $(30, 50)$ and $(31, 48.75)$ are two points that lie on the line $y(x)$. We can calculate the slope:
$$m = \frac{y(31) - y(30)}{31 - 30} = \frac{48.75 - 50}{1} = -1.25.$$
Using the point-slope formula, we can find the equation of $y(x)$:
$y(x) - y(30) = -1.25(x - 30)$
$y(x) - 50 = -1.25x + 37.5$
$y(x) = -1.25x + 87.5$
Now, the total yield $Y(x) = x \bullet y(x)$
$= x(-1.25x + 87.5)$
$= -1.25x^2 + 87.5x.$
The domain of Y is $[30, 70)$. Since Y is a polynomial, it is continuous on the interval. Computing the derivative, we get:
$Y'(x) = \frac{d}{dx}[-1.25x^2 + 87.5x] = -2.5x + 87.5$
$Y'(x)$ is defined for all values in the domain so the only critical values are when $Y'(x) = 0$. Solving yields:
$-2.5x + 87.5 = 0$
$x = 35$
So $x = 35$ is the only critical value. But, $Y(35) = -1.25(35)^2 + 87.5(35) = 1531.25$. Computing the second derivative, we get:
$Y''(x) = \frac{d}{dx}[-2.5x + 87.5] = -2.5 < 0$
for all x. Hence, Y has an absolute maximum yield of 1531.25 pounds per acre when 35 trees are planted per acre.

35) The yearly harvest $h(x) = f(x) - x$
$= 1.75x - 0.003x^2 - x = 0.75x - 0.003x^2$.
The domain of h is $(0, \infty)$. Since h is a polynomial, it is continuous on the interval. Computing the derivative, we get:
$h'(x) = \frac{d}{dx}[-0.003x^2 + 0.75x]$
$= -0.006x + 0.75$
$h'(x)$ is defined for all values in the domain so the only critical values are when $h'(x) = 0$.

35) Continued
Solving yields:
$-0.006x + 0.75 = 0$
$x = 125$
So $x = 125$ is the only critical value. But,
$h(125) = -0.003(125)^2 + 0.75(125)$
$= 46.875$
Computing the second derivative, we get:
$h''(x) = \frac{d}{dx}[-0.006x + 0.75] = -0.006 < 0$
for all x. The population should grow to 12,500 deer. It will then sustain a maximal yearly harvest of 4687.5 deer.

37) The yearly harvest $h(x) = f(x) - x$
$= 2.2x - 0.01x^2 - x = 1.2x - 0.01x^2$. The domain of h is $(0, \infty)$. Since h is a polynomial, it is continuous on the interval. Computing the derivative, we get:
$h'(x) = \frac{d}{dx}[-0.01x^2 + 1.2x] = -0.02x + 1.2$
$h'(x)$ is defined for all values in the domain so the only critical values are when $h'(x) = 0$. Solving yields:
$-0.02x + 1.2 = 0$
$x = 60$
So $x = 60$ is the only critical value. But,
$h(60) = -0.01(60)^2 + 1.2(60) = 36$
Computing the second derivative, we get:
$h''(x) = \frac{d}{dx}[-0.02x + 1.2] = -0.02 < 0$
for all x. The population should grow to 60,000 pheasants. It will then sustain a maximal yearly harvest of 36,000 pheasants.

39) $IC(x) =$ storage costs + reorder costs
$=$ (storage per item)$\bullet \frac{x}{2}$
$+$ (cost per item$\bullet x$ + fixed cost)$\bullet \frac{\text{yearly sale}}{x}$
$IC(x) = 300 \bullet \frac{x}{2} + (600x + 600) \bullet \frac{200}{x}$
$= 150x + 120,000 + \frac{120000}{x}$. The domain is $(0, 200]$. Since IC is a rational function, the only values where it is not continuous are the values that make it undefined. But $x = 0$ is not in $(0, 200]$. Thus, IC is continuous on the interval.

39) Continued

Computing the derivative, we get:

$IC'(x) = \frac{d}{dx}[150x + 120{,}000 + 120{,}000x^{-1}]$

$= 150 - 120{,}000x^{-2} = 150 - \frac{120000}{x^2}$.

$IC'(x)$ is undefined when $x = 0$, but 0 is not in $(0, 200]$. Thus, the only critical values are when $IC'(x) = 0$. Solving yields:

$150 - \frac{120000}{x^2} = 0$ (multiply by x^2)

$150x^2 - 120{,}000 = 0$

$x^2 = 800$

$x = \pm 28.28$, but -28.28 is not in $(0, 200]$. Rounding off to the nearest lot size that produces an integer value for the number orders placed, we get $x = 25$ is the only critical value.

$IC(25) = 150(25) + 120{,}000 + \frac{120000}{25}$

$= 128550$ and the number of orders is $\frac{\text{yearly sale}}{x} = \frac{200}{25} = 8$. Computing the second derivative, we get:

$IC''(x) = \frac{d}{dx}[150 - 120{,}000x^{-2}]$

$= 240{,}000x^{-3} = \frac{240000}{x^3}$.

Hence, $IC''(25) = \frac{240000}{(25)^3} = 15.36 > 0$. So, IC has an absolute minimum cost of $128{,}550 when each lot size has 25 desks and 8 orders are placed per year.

41) $IC(x)$ = storage costs + reorder costs

= (storage per item)$\bullet \frac{x}{2}$

+ (cost per item$\bullet x$ + fixed cost)$\bullet \frac{\text{yearly sale}}{x}$

$IC(x) = 0.5 \bullet \frac{x}{2} + (26x + 30) \bullet \frac{480}{x}$

$= 0.25x + 12{,}480 + \frac{14400}{x}$. The domain is $(0, 480]$. Since IC is a rational function, the only values where it is not continuous are the values that make it undefined. But $x = 0$ is not in $(0, 480]$. Thus, IC is continuous on the interval. Computing the derivative, we get:

$IC'(x) = \frac{d}{dx}[0.25x + 12{,}480 + 14{,}400x^{-1}]$

$= 0.25 - 14{,}400x^{-2} = 0.25 - \frac{14400}{x^2}$.

41) Continued

$IC'(x)$ is undefined when $x = 0$, but 0 is not in $(0, 480]$. Thus, the only critical values are when $IC'(x) = 0$. Solving yields:

$0.25 - \frac{14400}{x^2} = 0$ (multiply by x^2)

$0.25x^2 - 14{,}400 = 0$

$x^2 = 57{,}600$

$x = \pm 240$, but -240 is not in $(0, 480]$. Rounding off to the nearest lot size that produces an integer value for the number orders placed, we get $x = 240$ is the only critical value.

$IC(240) = 0.25(240) + 12{,}480 + \frac{14400}{240}$

$= 12{,}600$ and the number of orders is $\frac{\text{yearly sale}}{x} = \frac{480}{240} = 2$. Computing the second derivative, we get:

$IC''(x) = \frac{d}{dx}[0.25 - 14{,}400x^{-2}]$

$= 28{,}800x^{-3} = \frac{28{,}800}{x^3}$.

Hence, $IC''(240) = \frac{28{,}800}{(240)^3} \approx 0.002 > 0$. So, IC has an absolute minimum cost of $12{,}600 when each lot size has 240 cases and 2 orders are placed per year.

43) $IC(x)$ = storage costs + reorder costs

= (storage per item)$\bullet \frac{x}{2}$

+ (cost per item$\bullet x$ + fixed cost)$\bullet \frac{\text{yearly sale}}{x}$

$IC(x) = 9 \bullet \frac{x}{2} + (120x + 54) \bullet \frac{2400}{x}$

$= 4.5x + 288{,}000 + \frac{129600}{x}$. The domain is $(0, 2400]$. Since IC is a rational function, the only values where it is not continuous are the values that make it undefined. But $x = 0$ is not in $(0, 2400]$. Thus, IC is continuous on the interval. Computing the derivative, we get:

$IC'(x) = \frac{d}{dx}[4.5x + 288{,}000 + 129{,}600x^{-1}]$

$= 4.5 - 129{,}600x^{-2} = 4.5 - \frac{129{,}600}{x^2}$.

$IC'(x)$ is undefined when $x = 0$, but 0 is not in $(0, 2400]$. Thus, the only critical values are when $IC'(x) = 0$.

43) Continued
Solving yields:
$4.5 - \frac{129{,}600}{x^2} = 0$ (multiply by x^2)
$4.5x^2 - 129{,}600 = 0$
$x^2 = 28{,}800$
$x = \pm 169.71$, but -169.71 is not in $(0, 2400]$. Rounding off to the nearest lot size that produces an integer value for the number orders placed, we get $x = 160$ is the only critical value.

$IC(160) = 4.5(160) + 288{,}000 + \frac{129{,}600}{160}$

= 289,530 and the number of orders is $\frac{\text{yearly sale}}{x} = \frac{2400}{160} = 15$. Computing the second derivative, we get:

$IC''(x) = \frac{d}{dx}[4.5 - 129{,}600x^{-2}]$

$= 259{,}200x^{-3} = \frac{259200}{x^3}$.

Hence, $IC''(160) = \frac{259200}{(160)^3} \approx 0.063 > 0$. So, IC has an absolute minimum cost of $289,530 when each lot size has 160 ovens and 15 orders are placed per year.

45) $IC(x)$ = storage costs + reorder costs
= (storage per item)$\cdot \frac{x}{2}$
+ (cost per item$\cdot x$ + fixed cost)$\cdot \frac{\text{yearly sale}}{x}$

$IC(x) = 1500 \cdot \frac{x}{2} + (8500x + 750) \cdot \frac{300}{x}$

$= 750x + 2{,}550{,}000 + \frac{225000}{x}$. The domain is $(0, 300]$. Since IC is a rational function, the only values where it is not continuous are the values that make it undefined. But $x = 0$ is not in $(0, 300]$. Thus, IC is continuous on the interval. Computing the derivative, we get:

$IC'(x) = \frac{d}{dx}[750x + 2{,}550{,}000 + 225{,}000x^{-1}]$

$= 750 - 225{,}000x^{-2} = 750 - \frac{225000}{x^2}$.

$IC'(x)$ is undefined when $x = 0$, but 0 is not in $(0, 300]$. Thus, the only critical values are when $IC'(x) = 0$.

45) Continued
Solving yields:
$750 - \frac{225000}{x^2} = 0$ (multiply by x^2)
$750x^2 - 225{,}000 = 0$
$x^2 = 300$
$x = \pm 17.32$, but -17.32 is not in $(0, 300]$. Rounding off to the nearest lot size that produces an integer value for the number orders placed, we get $x = 15$ is the only critical value.

$IC(15) = 750(15) + 2{,}550{,}000 + \frac{225000}{15}$

= 2,576,250 and the number of orders is $\frac{\text{yearly sale}}{x} = \frac{300}{15} = 20$. Computing the second derivative, we get:

$IC''(x) = \frac{d}{dx}[750 - 225{,}000x^{-2}]$

$= 450{,}000x^{-3} = \frac{450000}{x^3}$.

Hence, $IC''(15) = \frac{450000}{(15)^3} \approx 133.33 > 0$. So, IC has an absolute minimum cost of $2,576,250 when each lot size has 15 cars and 20 orders are placed per year.

47) $PC(x)$ = storage costs + reorder costs
= (storage per item)$\cdot \frac{x}{2}$
+ (cost per item$\cdot x$ + fixed cost)$\cdot \frac{\text{yearly sale}}{x}$

$PC(x) = 3 \cdot \frac{x}{2} + (17x + 1800) \cdot \frac{5000}{x}$

$= 1.5x + 85{,}000 + \frac{9000000}{x}$. The domain is $(0, 5000]$. Since PC is a rational function, the only values where it is not continuous are the values that make it undefined. But $x = 0$ is not in $(0, 5000]$. Thus, PC is continuous on the interval. Computing the derivative, we get:

$PC'(x) = \frac{d}{dx}[1.5x + 85{,}000 + 9{,}000{,}000x^{-1}]$

$= 1.5 - 9{,}000{,}000x^{-2} = 1.5 - \frac{9000000}{x^2}$.

$PC'(x)$ is undefined when $x = 0$, but 0 is not in $(0, 5000]$. Thus, the only critical values are when $PC'(x) = 0$.

Section 5.5 The Second Derivative Test and Optimization 213

47)   Continued
Solving yields:
$1.5 - \frac{9000000}{x^2} = 0$ (multiply by x^2)
$1.5x^2 - 9{,}000{,}000 = 0$
$x^2 = 6{,}000{,}000$
$x = \pm 2449.49$, but -2449.49 is not in $(0, 5000]$. Rounding off to the nearest lot size that produces an integer value for the number orders placed, we get $x = 2500$ is the only critical value.
$PC(2500) = 1.5(2500) + 85{,}000 + \frac{9000000}{2500}$
$= 92{,}350$ and the number of orders is
$\frac{\text{yearly sale}}{x} = \frac{5000}{2500} = 2$. Computing the second derivative, we get:
$PC''(x) = \frac{d}{dx}[1.5 - 9{,}000{,}000x^{-2}]$
$= 18{,}000{,}000x^{-3} = \frac{18000000}{x^3}$.
Hence, $PC''(2500) = \frac{18000000}{(2500)^3} \approx 0.0012 > 0$.

So, PC has an absolute minimum cost of $\$92{,}350$ when each lot size has 2500 books and 2 printings are done per year.

49) $PC(x) =$ storage costs + reorder costs
$=$ (storage per item)$\bullet \frac{x}{2}$
$+$ (cost per item$\bullet x$ + fixed cost)$\bullet \frac{\text{yearly sale}}{x}$
$PC(x) = 2 \bullet \frac{x}{2} + (3x + 500) \bullet \frac{50000}{x}$
$= x + 150{,}000 + \frac{25000000}{x}$. The domain is $(0, 50{,}000]$. Since PC is a rational function, the only values where it is not continuous are the values that make it undefined. But $x = 0$ is not in $(0, 50{,}000]$. Thus, PC is continuous on the interval. Computing the derivative, we get:
$PC'(x) = \frac{d}{dx}[x + 150{,}000 + 25{,}000{,}000x^{-1}]$
$= 1 - 25{,}000{,}000x^{-2} = 1 - \frac{25000000}{x^2}$.
$PC'(x)$ is undefined when $x = 0$, but 0 is not in $(0, 50{,}000]$. Thus, the only critical values are when $PC'(x) = 0$.

49)   Continued
Solving yields:
$1 - \frac{25000000}{x^2} = 0$ (multiply by x^2)
$x^2 - 25{,}000.000 = 0$
$x^2 = 25{,}000{,}000$
$x = \pm 5000$, but -5000 is not in $(0, 50{,}000]$. Rounding off to the nearest lot size that produces an integer value for the number orders placed, we get $x = 5000$ is the only critical value.
$PC(5000) = (5000) + 150{,}000 + \frac{25000000}{5000}$
$= 160{,}000$ and the number of orders is
$\frac{\text{yearly sale}}{x} = \frac{50000}{5000} = 10$. Computing the second derivative, we get:
$PC''(x) = \frac{d}{dx}[1 - 25{,}000{,}000x^{-2}]$
$= 50{,}000{,}000x^{-3} = \frac{50000000}{x^3}$.
Hence, $PC''(5000) = \frac{50000000}{(5000)^3} \approx 0.0004 > 0$.

So, PC has an absolute minimum cost of $\$160{,}000$ when each lot size has 5000 toys and 10 production runs are done per year.

Chapter 5 Review Exercises

1a) y is increasing on $(-\infty, -2) \cup (3, \infty)$.

1b) y is decreasing on $(-2, 3)$.

1c) y is constant at $x = -2$ and $x = 3$.

3a) y is increasing on $(-1, 1)$.

3b) y is decreasing on $(1, \infty)$.

3c) y is constant on $(-\infty, -1)$ and at $x = 1$.

5a) Since $y' > 0$ on $(-\infty, -3) \cup (1, \infty)$, then y is increasing on $(-\infty, -3) \cup (1, \infty)$.

5b) Since $y' < 0$ on $(-3, 1)$, then y is decreasing on $(-3, 1)$.

5c) y is constant at $x = -3$ and $x = 1$.

7a) Since $y' > 0$ on $(-4, -1) \cup (3, \infty)$, then y is increasing on $(-4, -1) \cup (3, \infty)$.

7b) Since $y' < 0$ on $(-\infty, -4) \cup (-1, 3)$, then y is decreasing on $(-\infty, -4) \cup (-1, 3)$.

7c) y is constant at $x = -4$, $x = 1$, and $x = 3$.

9a) Since f is a polynomial, the domain is all real numbers. Computing the derivative, we get:
$f'(x) = \frac{d}{dx}[\frac{1}{2}x - 4] = \frac{1}{2}$. Since the derivative is a polynomial, there are no values for which the derivative is undefined. Setting the derivative equal to zero and solving yields:
$\frac{1}{2} = 0$
No solution
Thus, f has no critical values.

9b) Since $f'(x) = \frac{1}{2} > 0$ for all x, the f is increasing on $(-\infty, \infty)$ and decreasing nowhere.

9c) Since f has no critical values, then it has no relative extrema.

9d) The graph is displayed in viewing window $[-10, 10]$ by $[-10, 10]$

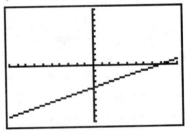

Note: The textbook writes the viewing window on the sides of the graph. The graph verifies our results.

11a) Since y is a polynomial, the domain is all real numbers. Computing the derivative, we get:
$y' = \frac{d}{dx}[x^3 + 3x^2 - 24x + 20]$
$= 3x^2 + 6x - 24$. Since the derivative is a polynomial, there are no values for which the derivative is undefined. Setting the derivative equal to zero and solving yields:
$3x^2 + 6x - 24 = 0$
$3(x^2 + 2x - 8) = 0$
$3(x + 4)(x - 2) = 0$
$(x + 4) = 0$ or $(x - 2) = 0$
$x = -4$ and $x = 2$.
Thus, the critical values of y are $x = -4$ and $x = 2$.

11b) We begin by placing the critical values and any values that make the original function undefined on a sign diagram:

We select the test number -5 in $(-\infty, -4)$, the test number 0 in $(-4, 2)$, and the test number 3 in $(2, \infty)$. Thus,
$y'|_{x=-5} = 3(-5)^2 + 6(-5) - 24 = 21$,
$y'|_{x=0} = 3(0)^2 + 6(0) - 24 = -24$,
and $y'|_{x=3} = 3(3)^2 + 6(3) - 24 = 21$.
Now, we can complete the sign diagram:

y ↗ ↘ ↗
y' (+) −4 (−) 2 (+)

Thus, y is increasing on $(-\infty, -4) \cup (2, \infty)$ and decreasing on $(-4, 2)$.

11c) $y|_{x=-4} = (-4)^3 + 3(-4)^2 - 24(-4) + 20 = 100$. Thus, by the sign chart in part b, the function has a relative maximum at $(-4, 100)$.
$y|_{x=2} = (2)^3 + 3(2)^2 - 24(2) + 20 = -8$.
Thus, by the sign chart in part b, the function has a relative minimum at $(2, -8)$.

11d) The graphs are displayed in viewing window $[-10, 10]$ by $[-40, 120]$

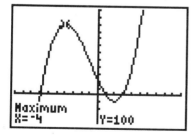

11d) Continued and

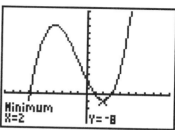

The graphs verify our results.

13a) Since g is a polynomial, the domain is all real numbers. Computing the derivative, we get:
$g'(x) = \frac{d}{dx}[(3x-2)^4] = 4(3x-2)^3(3)$
$= 12(3x-2)^3$. Since the derivative is a polynomial, there are no values for which the derivative is undefined. Setting the derivative equal to zero and solving yields:
$12(3x-2)^3 = 0$
$(3x-2) = 0$
$x = \frac{2}{3}$.

Thus, the only critical value of g is $x = \frac{2}{3}$.

13b) We begin by placing the critical values and any values that make the original function undefined on a sign diagram:

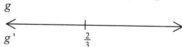

We select the test number 0 in $(-\infty, \frac{2}{3})$, and the test number 1 in $(\frac{2}{3}, \infty)$. Thus,
$g'(0) = 12(3(0) - 2)^3 = -96$ and
$g'(1) = 12(3(1) - 2)^3 = 12$.
Now, we can complete the sign diagram:

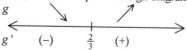

Thus, g is increasing on $(\frac{2}{3}, \infty)$ and decreasing on $(-\infty, \frac{2}{3})$.

216 Chapter 5 Further Applications of the Derivative

13c) $g(\frac{2}{3}) = (3(\frac{2}{3}) - 2)^4 = 0$. Thus, by the sign chart in part b, the function has a relative minimum at $(\frac{2}{3}, 0)$.

13d) The graph is displayed in viewing window $[-1, 2]$ by $[-2, 10]$

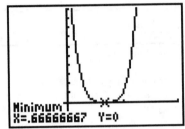

The graphs verify our results.

15) Since the derivative is an polynomial function, there are no values for which the derivative is undefined. Setting the derivative equal to zero and solving yields:
$$-3x + 8 = 0$$
$$-3x = -8$$
$$x = \tfrac{8}{3}$$

We now place the critical values on a sign diagram:

We select the test number 0 in $(-\infty, \tfrac{8}{3})$ and the test number 3 in $(\tfrac{8}{3}, \infty)$. Thus, $f'(0) = 8$ and $f'(3) = -1$. Now, we can complete the sign diagram:

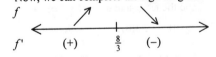

Thus, f has a relative maximum at $x = \tfrac{8}{3}$.

17) Since the derivative is a polynomial function, there are no values for which the derivative is undefined. Setting the derivative equal to zero and solving yields:
$$x^2 - 9 = 0$$
$$(x - 3)(x + 3) = 0$$
$$x - 3 = 0 \text{ or } x + 3 = 0$$
$$x = 3 \text{ or } x = -3$$

We now place the critical values on a sign diagram:

We select the test number -4 in $(-\infty, -3)$, the test number 0 in $(-3, 3)$, and the test number 4 in $(3, \infty)$. Thus, $f'(-4) = 7$, $f'(0) = -9$, and $f'(4) = 7$.
Now, we can complete the sign diagram:

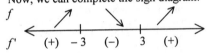

Thus, f has a relative maximum at $x = -3$ and a relative minimum at $x = 3$.

19) At $x = 0$ and 1, the derivative is negative so the function is decreasing at those values. At $x = 3, 4$, and 5, the derivative is positive so the function is increasing at those values. At $x = 2$, the derivative is zero so we have a relative minimum at $x = 2$. The graph is displayed in viewing window $[-0.1, 5.1]$ by $[-5, 8]$

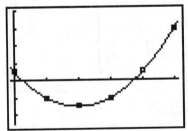

21a) $p(30) = 4.5e^{-0.03(30)} \approx 1.83$. The price is $1.83 per container.

21b) $R(x) = x \cdot p(x) = x(4.5e^{-0.03x}) = 4.5xe^{-0.03x}$

21c) Since R is a exponential function times a polynomial, its domain is $[0, \infty)$.
Computing the derivative, we get:
$R'(x) = \frac{d}{dx}[4.5xe^{-0.03x}]$
$= (4.5)e^{-0.03x} + (4.5x)e^{-0.03x}(-0.03)$
$= 4.5e^{-0.03x}[1 - 0.03x]$
Since $R'(x)$ is defined for all x, the only critical values are when $R'(x) = 0$. Solving yields:
$4.5e^{-0.03x}[1 - 0.03x] = 0$
But, $4.5e^{-0.03x} = 0$ has no solution since it is an exponential decay equation.
$1 - 0.03x = 0$
$x \approx 33.333$, but $R(33.333) \approx 55.182$.
We now place the critical values on a sign diagram:
R

R'   33.333

We select the test number 0 in $(-\infty, 33.333)$ and the test number 40 in $(33.333, \infty)$. Thus, $R'(0) = 4.5$ and $R'(40) \approx -0.271$.
Now, we can complete the sign diagram:
R
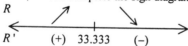
R'   (+)   33.333   (−)

Thus, R is increasing on $[0, 33.333)$ and decreasing on $(33.333, \infty)$.

21d) From the sign chart in part c, R has a relative maximum of 55.18 at $x = 33.33$. This means that the total revenue peaks at $55,180 when 33,330 containers are produced and sold.

21e) The graph is displayed in viewing window [0, 100] by [0, 100]

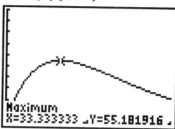

23a) $AC(x) = \frac{C(x)}{x} = \frac{800 + 15x + \frac{x^2}{18}}{x}$
$= \frac{800}{x} + 15 + \frac{x}{18}$. The domain is (0, 500).

23b) $AC(x)$ is undefined at $x = 0$, but $x = 0$ is not in (0, 500). Computing the derivative, we get:
$AC'(x) = \frac{d}{dx}[800x^{-1} + 15 + \frac{x}{18}]$
$= -800x^{-2} + \frac{1}{18} = \frac{-800}{x^2} + \frac{1}{18}$.
$AC'(x)$ is undefined at $x = 0$ but $x = 0$ is not in (0, 500). Thus, the only critical values are when $AC'(x) = 0$. Solving yields:
$\frac{-800}{x^2} + \frac{1}{18} = 0$ (multiply by $18x^2$)
$-14,400 + x^2 = 0$
$x^2 = 14,400$
$x = \pm 120$, but $x = -120$ is not in (0, 500). Thus, $x = 120$ is the only critical value and $AC(120) \approx 28.33$.

23c) We now place the critical values on a sign diagram:
AC

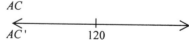
AC'   120
We select the test number 100 in (0, 120) and the test number 140 in (120, 500). Thus, $AC'(100) \approx -0.024$ and $AC'(140) \approx 0.015$.

23c) Continued
Now, we can complete the sign diagram:

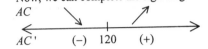

Thus, AC is decreasing on $(0, 120)$ and increasing on $(120, 500)$.

23d) From parts b and c, AC has a relative minimum of 28.33 at $x = 120$. In other words, the average cost is minimized at \$28.33 per phone when 120 phones are produced per day.

25) $f'(x) = \frac{d}{dx}[x^3 - 7x^2 + 5] = 3x^2 - 14x$.

$f''(x) = \frac{d}{dx}[3x^2 - 14x] = 6x - 14$

$f'''(x) = \frac{d}{dx}[6x - 14] = 6$.

27) $y' = \frac{d}{dx}[(x-5)^{1/3}] = \frac{1}{3}(x-5)^{-2/3}$ (1)

$= \frac{1}{3\sqrt[3]{(x-5)^2}}$.

$y'' = \frac{d}{dx}[\frac{1}{3}(x-5)^{-2/3}] = \frac{1}{3} \cdot \frac{-2}{3}(x-5)^{-5/3}$

$= \frac{-2}{9\sqrt[3]{(x-5)^5}}$

$y''' = \frac{d}{dx}[\frac{-2}{9}(x-5)^{-5/3}]$

$= \frac{-2}{9} \cdot \frac{-5}{3}(x-5)^{-8/3} = \frac{10}{27\sqrt[3]{(x-5)^8}}$.

29a) Since f is a polynomial, the domain is all real numbers. Computing the first and second derivatives, we get:

$f'(x) = \frac{d}{dx}[x^3 - x^2 + 5x - 7] = 3x^2 - 2x + 5$

and $f''(x) = \frac{d}{dx}[3x^2 - 2x + 5] = 6x - 2$.

Since the second derivative is a polynomial, there are no values for which the second derivative is undefined. Setting the second derivative equal to zero and solving yields:
$6x - 2 = 0$
$6x = 2$
$x = \frac{1}{3}$.

29a) Continued
Thus, we have a possible inflection point at $x = \frac{1}{3}$. We place this value and any values that make the original function undefined on a sign diagram:

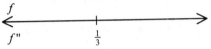

We select the test number 0 in $(-\infty, \frac{1}{3})$ and the test number 1 in $(\frac{1}{3}, \infty)$. Thus, $f''(0) = -2$ and $f''(1) = 4$.
Now, we can complete the sign diagram:

Hence, f is concave up on $(\frac{1}{3}, \infty)$ and concave down on $(-\infty, \frac{1}{3})$.

29b) Since $f(\frac{1}{3}) \approx -5.41$ and the concavity changes sign, then $(\frac{1}{3}, -5.41)$ is an inflection point.

31a) Since y is a square root function, the domain is $[0, \infty)$. Computing the first and second derivatives, we get:

$y' = \frac{d}{dx}[12 - x^{1/2}] = -0.5x^{-1/2} = \frac{-1}{2\sqrt{x}}$

and $y'' = \frac{d}{dx}[-0.5x^{-1/2}] = 0.25x^{-3/2}$

$= \frac{1}{4\sqrt{x^3}}$.

The second derivative is undefined for $x \le 0$, but only $x = 0$ is in $[0, \infty)$. Setting the second derivative equal to zero and solving yields:

$\frac{1}{4\sqrt{x^3}} = 0$ (multiply by $\sqrt{x^3}$)

$1 = 0$
No solution
Thus, there is a possible inflection point at $x = 0$ and $y\big|_{x=0} = 12$.

31a) Continued
We place this value and any values that make the original function undefined on a sign diagram:

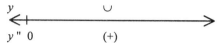
y″   0

We select the test number 1 in [0, ∞).
Thus, $y''\big|_{x=1} = 0.25$.
Now, we can complete the sign diagram:

y   ∪
←———|——————————→
y″   0   (+)

Hence, y is concave up on [0, ∞) and concave down nowhere.

31b) From part a, there are no inflection points since the concavity does not change sign.

33a) Since y is a rational function, it is undefined at $x = 0$. Computing the first and second derivatives, we get:
$y' = \frac{d}{dx}[3x - 5x^{-1}] = 3 + 5x^{-2} = 3 + \frac{5}{x^2}$
and $y'' = \frac{d}{dx}[3 + 5x^{-2}] = -10x^{-3} = \frac{-10}{x^3}$.

The second derivative is undefined at $x = 0$, but $x = 0$ is not in the domain of y. Setting the second derivative equal to zero and solving yields:
$\frac{-10}{x^3} = 0$ (multiply by x^3)
$-10 = 0$
No solution

Thus, there are no possible inflection points. We place this value and any values that make the original function undefined on a sign diagram:

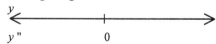
y″   0

We select the test number −1 in (−∞, 0) and the test number 1 in (0, ∞). Thus,
$y''\big|_{x=-1} = 10$ and $y''\big|_{x=1} = -10$.

33a) Continued
Now, we can complete the sign diagram:

y   ∪   ∩
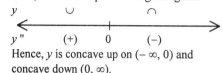
y″   (+)   0   (−)

Hence, y is concave up on (−∞, 0) and concave down (0, ∞).

33b) From part a, there are no inflection points since y is not defined at $x = 0$.

35a) Since y is a exponential function, it is defined for all real numbers. Computing the first and second derivatives, we get:
$y' = \frac{d}{dx}[e^{5x}] = 5e^{5x}$.
and $y'' = \frac{d}{dx}[5e^{5x}] = 25e^{5x}$.

The second derivative is defined for all x. Setting the second derivative equal to zero and solving yields:
$25e^{5x} = 0$

No solution since it is an exponential growth function. Thus, there are no possible inflection points. We place this value and any values that make the original function undefined on a sign diagram:

y

y″

We select the test number 0 in (−∞, ∞).
Thus, $y''\big|_{x=0} = 25$.
Now, we can complete the sign diagram:

y   ∪
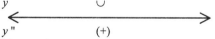
y″   (+)

Hence, y is concave up on (−∞, ∞) and concave down nowhere.

35b) From part a, there are no inflection points since concavity does not change sign.

37a) Since f is a polynomial, the domain of f is all real numbers. Computing the derivative, we get:
$f'(x) = \frac{d}{dx}[2x^3 + 3x^2 - 36x]$
$= 6x^2 + 6x - 36.$
Since the derivative is a polynomial, there are no values for which the derivative is undefined. Setting $f'(x) = 0$ and solving yields:
$6x^2 + 6x - 36 = 0$
$6(x + 3)(x - 2) = 0$
$x = -3$ or $x = 2$
So, $x = -3$ and $x = 2$ are the critical values. We place the critical values and any values that make the original function undefined on a sign diagram:

We select the test number -4 in $(-\infty, -3)$, the test number 0 in $(-3, 2)$, and the test number 3 in $(2, \infty)$. Thus, $f'(-4) = 36, f'(0) = -36,$ and $f'(3) = 36$. Now, we can complete the sign diagram:

f
$\quad\quad\nearrow\quad\quad\searrow\quad\quad\nearrow$
$f' \ (+)\ -3\ (-)\ \ 2\ (+)$

Thus, f is increasing on $(-\infty, -3)$ ∪ $(2, \infty)$ and decreasing on $(-3, 2)$.

37b) Since $f(-3) = 81$, then by the sign chart in part a, the function has a relative maximum at $(-3, 81)$. Since $f(2) = -44$, then by the sign chart in part a, the function has a relative minimum at $(2, -44)$.

37c) Computing the second derivative, we get:
$f''(x) = \frac{d}{dx}[6x^2 + 6x - 36] = 12x + 6.$
Since the second derivative is a polynomial, there are no values for which the second derivative is undefined. Setting the $f''(x) = 0$ and solving yields:
$12x + 6 = 0$
$x = -0.5$. Thus, we have a possible inflection point at $x = -0.5$.

37c) Continued
We place this value and any values that make the original function undefined on a sign diagram:

f
$f'' \quad\quad\quad -0.5$

We select the test number -1 in $(-\infty, -0.5)$ and the test number 0 in $(-0.5, \infty)$. Thus, $f''(-1) = -6$ and $f''(0) = 6$. Now, we can complete the sign diagram:

$f \quad\quad \cap \quad\quad \cup$
$f'' \ (-) \quad -0.5 \quad (+)$

Hence, f is concave up on $(-0.5, \infty)$ and concave down on $(-\infty, -.05)$.

37d) Since $f(-0.5) = 18.5$ and the concavity changes sign, then $(-0.5, 18.5)$ is an inflection point.

39a) Since y is a logarithmic function, the domain is all values of $x < 2$. Computing the first derivative, we get:
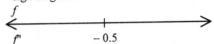
The derivative is undefined at $x = 2$, but $x = 2$ is not in the domain of y. Setting the derivative equal to zero and solving yields:
$\frac{-1}{2-x} = 0$
$-1 = 0\bullet(2 - x) = 0$, no solution
Thus, there are no critical points.
Since $y' = \frac{-1}{2-x} < 0$ for all the values in the domain of y, then y is decreasing on $(-\infty, 2)$ and increasing nowhere.

39b) As stated in part a, there are no critical points. Thus, there are no relative extrema.

39c) Computing the second derivative, we get:
$y'' = \frac{d}{dx}[-(2-x)^{-1}] = -(2-x)^{-2}$
$= \frac{-1}{(2-x)^2}$

The second derivative is undefined at $x = 2$, but $x = 2$ is not in the domain of y. Setting the second derivative equal to zero and solving yields:
$\frac{-1}{(2-x)^2} = 0$
$-1 = 0 \bullet (2-x)^2 = 0$, no solution
Thus, there are no inflection points.
Since $y'' = \frac{-1}{(2-x)^2} < 0$ for all the values in the domain of y, then y is concave up nowhere and concave down on $(-\infty, 2)$.

39d) As stated in part c, there are no inflection points.

41a) Note that the domain of P is $[0, \infty)$. $P(4) = -0.5(4)^3 + 3(4)^2 + 68(4) - 133 = 155$. When 400 cookies are baked and sold per day, the daily profit is $155.
$P'(x) = \frac{d}{dx}[-0.5x^3 + 3x^2 + 68x - 133]$
$= -1.5x^2 + 6x + 68$.
Hence, $P'(4) = 68$. When 400 cookies are baked and sold per day, the profit is increasing at a rate of $68 per 100 cookies.

41b) The derivative is defined for all real numbers. Setting the derivative equal to zero and solving yields:
$-1.5x^2 + 6x + 68 = 0$
Using the quadratic formula, we get:
$x = \frac{-6 \pm \sqrt{(6)^2 - 4(-1.5)(68)}}{2(-1.5)} = \frac{6 \pm 2\sqrt{111}}{3}$
$x \approx 9$ or $x \approx -5$, but $x = -5$ is not in $[0, \infty)$. So, $x \approx 9$ is the only critical value in the domain of P. We place this value and any values that make the original function undefined on a sign diagram:
P
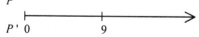
P' 0   9

41b) Continued
We select the test number 1 in $[0, 9)$ and test number 10 in $(9, \infty)$. Thus, $P'(1) = 72.5$ and $P'(10) = -22$
Now, we can complete the sign diagram:
P
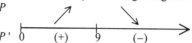
P' 0   (+)   9   (−)
Hence, P is increasing on $[0, 9)$ and decreasing on $(9, \infty)$. Since $P(9) = 357.5$ and by the sign chart above, the profit is maximized at $357.50 when 900 cookies are baked and sold.

41c) Since $MP(x) = P'(x)$, then $MP(x) = -1.5x^2 + 6x + 68$. Computing the second derivative, we get:
$MP'(x) = \frac{d}{dx}[-1.5x^2 + 6x + 68] = -3x + 6$.
The second derivative is defined for all x. Setting the second derivative equal to zero and solving yields:
$-3x + 6 = 0$
$x = 2$ and $MP(2) = 74$.
Thus, $x = 2$ is the only critical value of $MP(x)$. We place this value and any values that make the original function undefined on a sign diagram:
MP
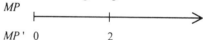
MP' 0   2
We select the test number 1 in $[0, 2)$ and test number 3 in $(2, \infty)$. Thus, $MP'(1) = 3$ and $MP'(3) = -3$
Now, we can complete the sign diagram:
MP
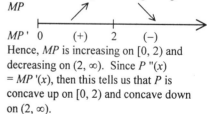
MP' 0   (+)   2   (−)
Hence, MP is increasing on $[0, 2)$ and decreasing on $(2, \infty)$. Since $P''(x) = MP'(x)$, then this tells us that P is concave up on $[0, 2)$ and concave down on $(2, \infty)$.

41d) Since $P(2) = 11$, using the results from part c, we have a point of inflection at $(2, 11)$.

43a) The graph is displayed in viewing window [0, 100] by [0, 1500]

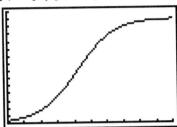

43b) $p(84) = \dfrac{1400}{1+43e^{-0.09(84)}} \approx 1369$. Eighty-four years after the goats were introduced, there were about 1,369 goats

$p'(t) = \dfrac{d}{dt}[1400(1+43e^{-0.09t})^{-1}] =$
$-1400(1+43e^{-0.09t})^{-2}(43e^{-0.09t})(-0.09)$
$= \dfrac{5418e^{-0.09t}}{(1+43e^{-0.09t})^2}$.

Hence, $p'(84) \approx 2.70$. Eighty-four years after the goats were introduced, the goat population was increasing at a rate of about 2.7 goats per year.

43c) Let $p'(t) = \dfrac{r(t)}{s(t)} = \dfrac{5418e^{-0.09t}}{(1+43e^{-0.09t})^2}$.

$p''(t) = \dfrac{d}{dt}\left[\dfrac{r(t)}{s(t)}\right] = \dfrac{r'(t)s(t) - r(t)s'(t)}{[s(t)]^2}$

but $r'(t)s(t) - r(t)s'(t)$
$= 5418e^{-0.09t}(-0.09)(1+43e^{-0.09t})^2$
$- 5418e^{-0.09t}\, 2(1+43e^{-0.09t})(43e^{-0.09t})$
$\bullet(-0.09)$
$= 5418e^{-0.09t}(-0.09)(1+43e^{-0.09t})$
$\bullet(1+43e^{-0.09t} - 2(43e^{-0.09t})) =$
$-487.62e^{-0.09t}(1+43e^{-0.09t})(1-43e^{-0.09t})$

Thus, $p''(t) =$
$\dfrac{-487.62e^{-0.09t}(1+43e^{-0.09t})(1-43e^{-0.09t})}{(1+43e^{-0.09t})^4}$

$= \dfrac{-487.62e^{-0.09t}(1-43e^{-0.09t})}{(1+43e^{-0.09t})^3}$.

43d) We will use the graphing calculator to determine when $p''(t)$ is positive, negative, and zero. The graph is displayed in viewing window [0, 100] by [−2, 2]

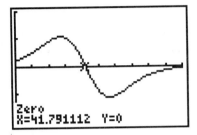

Since $p''(t) > 0$ on [0, 41.79), then p is concave up on [0, 41.79). Since $p''(t) < 0$ on (41.79, 100], then p is concave down on (41.79, 100].

43e) The goat population was growing the fastest when $p''(x) = 0$. Using the results from part d and rounding off to the nearest year, the population was growing the fastest 42 years after it was introduced.

45) f is positive on (1, 5) and f is negative on (−5, 1).

47) f' is positive on (−5, −3) ∪ (−1, 3) and f' is negative on (−3, −1) ∪ (3, 5).

49) f is concave up on (−2, 1) and concave down on (−5, −2) ∪ (1, 5).

51) The concavity changes sign at $x = -2$ and $x = 1$, thus the points (−2, −3) and (1, 0) are inflection points.

53) Since y is a polynomial, it is defined for all real numbers. Computing the derivative, we get: $y' = \dfrac{d}{dx}[x^3 - 5x^2 - 8x]$
$= 3x^2 - 10x - 8$.
Since y' is a polynomial, it is defined for all real numbers. Setting $y' = 0$ and solving yields:
$3x^2 - 10x - 8 = 0$.

53) Continued
$(3x + 2)(x - 4) = 0$
$3x + 2 = 0$ or $x - 4 = 0$
$x = -\frac{2}{3}$ or 4, but $y|_{x=-2/3} \approx 2.815$
and $y|_{x=4} \approx -48$.
So, the critical values are $x = -\frac{2}{3}$ or 4.
Computing the second derivative, we get:
$y'' = \frac{d}{dx}[3x^2 - 10x - 8] = 6x - 10$. Since
$y''|_{x=-2/3} = -14 < 0$, then y has a relative maximum of 2.815 at $x = -\frac{2}{3}$. Since
$y''|_{x=4} = 14 > 0$, then y has a relative minimum of -48 at $x = 4$.

55) Since g is a rational function and $x^2 + 9 \neq 0$, it is defined for all real numbers. Computing the derivative, we get:
$g'(x) = \frac{d}{dx}[(x^2 + 9)^{-1}] = -(x^2 + 9)^{-2}(2x)$
$= \frac{-2x}{(x^2+9)^2}$.
Since g' is a rational function and $x^2 + 9 \neq 0$, it is defined for all real numbers. Setting $g'(x) = 0$ and solving yields:
$\frac{-2x}{(x^2+9)^2} = 0$
$(x^2+9)^2 \cdot \frac{-2x}{(x^2+9)^2} = 0 \cdot (x^2+9)^2$
$-2x = 0$
$x = 0$, but $g(0) = \frac{1}{9}$. So, the critical value is $x = 0$. Computing the second derivative, we get: $g''(x) = \frac{d}{dx}\left[\frac{-2x}{(x^2+9)^2}\right]$
$= \frac{\frac{d}{dx}[-2x]\cdot(x^2+9)^2 - (-2x)\cdot\frac{d}{dx}[(x^2+9)^2]}{(x^2+9)^4}$
$= \frac{-2(x^2+9)^2 + 2x(2)(x^2+9)(2x)}{(x^2+9)^4}$
$= \frac{-2x^4 - 36x^2 - 162 + 8x^4 + 72x^2}{(x^2+9)^4} = \frac{6x^4 + 36x^2 - 162}{(x^2+9)^4}$
$= \frac{6(x^2+9)(x^2-3)}{(x^2+9)^4} = \frac{6(x^2-3)}{(x^2+9)^3}$. Since
$g''(0) = -\frac{2}{81} < 0$, then g has a relative maximum of $\frac{1}{9}$ at $x = 0$ and no relative minimum.

57) Since f is a polynomial, it is defined for all real numbers. Computing the derivative, we get: $f'(x) = \frac{d}{dx}[2x^3 - \frac{3}{2}x^2 - 9x + 5]$
$= 6x^2 - 3x - 9$.
Since $f'(x)$ is a polynomial, it is defined for all real numbers. Setting $f'(x) = 0$ and solving yields:
$6x^2 - 3x - 9 = 0$.
$3(2x^2 - x - 3) = 0$
$3(2x - 3)(x + 1) = 0$
$x = 1.5$ or -1, but $f(-1) = 10.5$ and $f(1.5) = -5.125$. So, the critical values are $x = -1$ and 1.5. We place the critical values and any values that make the original function undefined on a sign diagram:

We select the test number -2 in $(-\infty, -1)$, the test number 0 in $(-1, 1.5)$ and the test number 2 in $(1.5, \infty)$. Thus,
$f'(-2) = 6(-2)^2 - 3(-2) - 9 = 21$,
$f'(0) = 6(0)^2 - 3(0) - 9 = -9$,
and $f'(2) = 6(2)^2 - 3(2) - 9 = 9$. Now, we can complete the sign diagram:

f ↗ ↘ ↗
f' (+) -1 (−) 1.5 (+)

Thus, f is increasing on $(-\infty, -1) \cup (1.5, \infty)$ and decreasing on $(-1, 1.5)$. Also, f has a relative maximum of 10.5 at $x = -1$ and a relative minimum of -5.125 at $x = 1.5$. Computing the second derivative, we get:
$f''(x) = \frac{d}{dx}[6x^2 - 3x - 9] = 12x - 3$.
Since the second derivative is a polynomial, there are no values for which the second derivative is undefined. Setting the second derivative equal to zero and solving yields:
$12x - 3 = 0$
$12x = 3$
$x = 0.25$.
Thus, we have a possible inflection point at $x = 0.25$

57) Continued
We place this value and any values that make the original function undefined on a sign diagram:

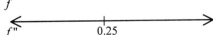

We select the test number 0 in $(-\infty, 0.25)$ and the test number 1 in $(0.25, \infty)$. Thus, $f''(0) = -3$ and $f''(1) = 9$.
Now, we can complete the sign diagram:

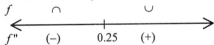

Hence, f is concave up on $(0.25, \infty)$ and concave down on $(-\infty, 0.25)$. Since $f(0.25) = 2.6875$ and the concavity changes sign, then $(0.25, 2.6875)$ is an inflection point. We can now sketch the graph.

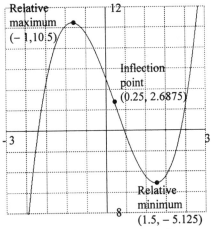

59) Since y is a rational function and $x^2 - 4 = 0$ when $x = 2$ and $x = -2$, then y is defined for all real numbers except -2 and 2.
Computing the derivative, we get:

$$y' = \frac{d}{dx}\left[\frac{1}{x^2-4}\right] = \frac{\frac{d}{dx}[1]\cdot(x^2-4) - 1\cdot\frac{d}{dx}[x^2-4]}{(x^2-4)^2}$$

$$= \frac{0-2x}{(x^2-4)^2} = \frac{-2x}{(x^2-4)^2}.$$

y' is a rational function and y' is undefined at $x = \pm 2$, but ± 2 is not in the domain of y.

59) Continued
Setting $y' = 0$ and solving yields:

$$\frac{-2x}{(x^2-4)^2} = 0$$

$$(x^2-4)^2 \cdot \frac{-2x}{(x^2-4)^2} = 0 \cdot (x^2-4)^2$$

$$-2x = 0$$

$$x = 0, \text{ but } y\big|_{x=0} = -0.25$$

So, $x = 0$ is the only critical value. We place this value and any values that make the original function undefined on a sign diagram:

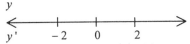

We select the test number -3 in $(-\infty, -2)$, the test number -1 in $(-2, 0)$ the test number 1 in $(0, 2)$, and the test number 3 in $(2, \infty)$. Thus,

$$y'\big|_{x=-3} = \frac{-2(-3)}{((-3)^2-4)^2} = 0.24,$$

$$y'\big|_{x=-1} = \frac{-2(-1)}{((-1)^2-4)^2} = \frac{2}{9},$$

$$y'\big|_{x=1} = \frac{-2(1)}{((1)^2-4)^2} = -\frac{2}{9}, \text{ and}$$

$$y'\big|_{x=3} = \frac{-2(3)}{((3)^2-4)^2} = -0.24, \text{ Now, we can}$$

complete the sign diagram:

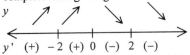

Thus, y is increasing on $(-\infty, -2) \cup (-2, 0)$ and decreasing on $(0, 2) \cup (2, \infty)$.
Also, y has a relative maximum of -0.25 at $x = 0$ and no relative minimum. Computing the second derivative, we get:

$$y'' = \frac{d}{dx}\left[\frac{-2x}{(x^2-4)^2}\right]$$

$$= \frac{\frac{d}{dx}[-2x]\cdot(x^2-4)^2 - (-2x)\cdot\frac{d}{dx}[(x^2-4)^2]}{(x^2-4)^4}$$

$$= \frac{-2(x^2-4)^2 + 2x(2)(x^2-4)(2x)}{(x^2-4)^4}$$

$$= \frac{-2x^4 + 16x^2 - 32 + 8x^4 - 32x^2}{(x^2-4)^4} = \frac{6x^4 - 16x^2 - 32}{(x^2-4)^4}$$

$$= \frac{2(x^2-4)(3x^2+4)}{(x^2-4)^4} = \frac{2(3x^2+4)}{(x^2-4)^3}$$

59) Continued

Since the second derivative is a rational function and y'' is undefined at $x = \pm 2$, but ± 2 is not in the domain of y.
Setting the second derivative equal to zero and solving yields:

$$\frac{2(3x^2+4)}{(x^2-4)^3} = 0$$

$$(x^2-4)^3 \cdot \frac{2(3x^2+4)}{(x^2-4)^3} = 0 \cdot (x^2-4)^3$$

$$2(3x^2+4) = 0$$
$$3x^2 + 4 = 0$$
$$x^2 = -\frac{4}{3}, \text{ no solution}$$

Thus, there are no inflection points. We place any values that make the original function undefined on a sign diagram:

y
← ——+——+——→
y''   −2   2

We select the test number -3 in $(-\infty, -2)$, the test number 0 in $(-2, 2)$, and the test number 3 in $(2, \infty)$. Thus,
$y''|_{x=-3} = 0.496$, $y''|_{x=0} = -0.125$,
and $y''|_{x=3} = 0.496$.
Now, we can complete the sign diagram:

y   ∪   ∩   ∪
← ——+——+——→
y''   −2   2

Hence, f is concave up on $(-\infty, -2) \cup (2, \infty)$ and concave down on $(-2, 2)$.

Since $\lim\limits_{x \to \infty} \frac{1}{x^2-4} = \lim\limits_{x \to \infty} \frac{\frac{1}{x^2}}{1-\frac{4}{x^2}} = \frac{0}{1-0} = 0$ and

similarly, $\lim\limits_{x \to -\infty} \frac{1}{x^2-4} = 0$, then $y = 0$ is a

horizontal asymptote. Also, observe that there is a vertical asymptote of $x = -2$ since
$\lim\limits_{x \to -2^-} f(x) = \infty$ and $\lim\limits_{x \to -2^+} f(x) = -\infty$.

There is also a vertical asymptote of $x = 2$ since $\lim\limits_{x \to 2^-} f(x) = -\infty$ and $\lim\limits_{x \to 2^+} f(x) = \infty$.

59) Continued

We can now sketch the graph.

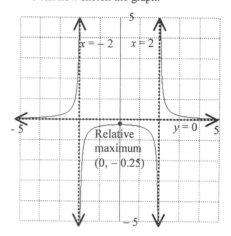

Relative maximum $(0, -0.25)$

61) Since g is an exponential function, it is defined for all real numbers. Computing the derivative, we get:

$g'(x) = \frac{d}{dx}[e^x - 4x] = e^x - 4$.

Since g' is an exponential function, it is defined for all real numbers. Setting $g'(x) = 0$ and solving yields:

$$e^x - 4 = 0$$
$$e^x = 4$$

$x = \ln(4) \approx 1.39$, but $g(1.39) = -1.55$.
So, the critical value is $x = 1.39$. We place the critical values and any values that make the original function undefined on a sign diagram:

g
← ——+——→
g'   1.39

We select the test number 0 in $(-\infty, 1.39)$ and the test number 2 in $(1.39, \infty)$. Thus, $g'(0) = e^0 - 4 = -3$ & $g'(2) = e^2 - 4 \approx 3.39$.

226   Chapter 5   Further Applications of the Derivative

61) Continued
Now, we can complete the sign diagram:

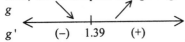

Thus, g is increasing on $(1.39, \infty)$ and decreasing on $(-\infty, 1.39)$.
Also, g has a relative minimum of -1.55 at $x = 1.39$ and no relative maximum.
Computing the second derivative, we get:
$g''(x) = \frac{d}{dx}[e^x - 4] = e^x$.
Since the second is an exponential function, it is defined for all real numbers. Setting the second derivative equal to zero and solving yields:
$e^x = 0$, no solution.
Thus, we have no possible inflection points. We place any values that make the original function undefined on a sign diagram:

We select the test number 0 in $(-\infty, \infty)$.
Thus, $g''(0) = e^0 = 1$
Now, we can complete the sign diagram:

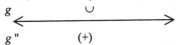

Hence, g is concave up on $(-\infty, \infty)$ and concave down nowhere. We can now sketch the graph.

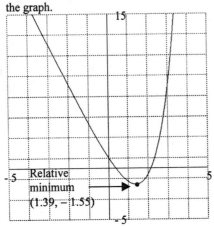

63) Since $f''(x) > 0$ on $(0, \infty)$, then f' is increasing on $(0, \infty)$ and since $f''(x) < 0$ on $(-\infty, 0)$, then f' is decreasing on $(-\infty, 0)$. Also, f' has a relative minimum at $x = 0$.
One possible graph would be:

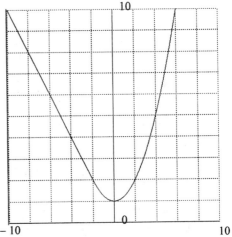

Since $f''(x) > 0$ on $(0, \infty,)$, then f is concave up on $(0, \infty)$ and since $f''(x) < 0$ on $(-\infty, 0)$, then f is concave down on $(-\infty, 0)$. Also, f has an inflection point at $x = 0$. One possible graph would be:

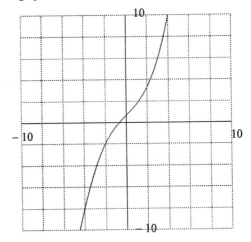

65a) $AC(x) = \frac{C(x)}{x} = \frac{15x + 200}{x}$
$= 15 + \frac{200}{x}$. The domain is $(0, 26]$ since AC is undefined at $x = 0$.

65b) The average cost per year is a minimum when $C'(x) = AC(x)$ (see exercises #58 through 60 on page 316). Computing $C'(x)$, we get:
$C'(x) = \frac{d}{dx}[15x + 200]$
$= 15$. Setting $C'(x) = AC(x)$ and solving yields:
$15 = 15 + \frac{200}{x}$
$0 = \frac{200}{x}$ (multiply by x)
$0 = 200$, no solution.
But $AC'(x) = \frac{d}{dx}[15 + 200x^{-1}] = \frac{-200}{x^2} < 0$ for all x in $(0, 26]$. Thus, AC is decreasing on $(0, 26]$ which means that the minimum value of AC will occur at $x = 26$. Hence, the average cost per sculpture is a minimum at $x = 26$ sculptures.

65c) $AP(x) = \frac{P(x)}{x} = \frac{-0.12x^3 + 70x - 200}{x}$
$= -0.12x^2 + 70 - \frac{200}{x}$. The domain is $(0, 26]$ since AP is undefined at $x = 0$.

65d) The average profit per year is a maximum when $P'(x) = AP(x)$ (see exercises #55 through 57 on pages 315 & 316). Computing $P'(x)$, we get:
$P'(x) = \frac{d}{dx}[-0.12x^3 + 70x - 200]$
$= -0.36x^2 + 70$. Setting $P'(x) = AP(x)$ and solving yields:
$-0.36x^2 + 70 = -0.12x^2 + 70 - \frac{200}{x}$
$-0.24x^2 + 70 + \frac{200}{x} = 0$ (multiply by x)
$-0.24x^3 + 70x + 200 = 0$
To solve this, we will graph $y_1 = -0.24x^3 + 70x + 200$ on a graphing calculator and use the zero command.

65d) Continued
The graph is displayed in viewing window $[0, 26]$ by $[-1000, 1000]$

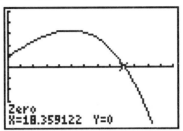

Rounding to the nearest whole number, the average profit per sculpture is a maximum at $x = 18$ sculptures.

67a) $f(9) = 15.7e^{0.058(9)} = 15.7e^{0.522} \approx 26$. In 1989, there were about 26 African American members of the U.S. House of Representatives. Computing $f'(t)$, we get:
$f'(t) = \frac{d}{dt}[15.7e^{0.058t}]$
$= 15.7e^{0.058t}(0.058) = 0.9106e^{0.058t}$.
Thus, $f'(9) = 0.9106e^{0.058(9)}$
$= 0.9106e^{0.522} \approx 1.53$. In 1989, the number of African Americans in the U.S. House of Representatives was growing at a rate of about 1.5 members per year.

67b) Since $f'(t) = 0.9106e^{0.058t}$, then for $1 \leq t \leq 15$, $f'(t) > 0$. Thus, f is increasing on $[1, 15]$.

67c) Computing $f''(t)$, we get:
$f''(t) = \frac{d}{dt}[0.9106e^{0.058t}]$
$= 0.9106e^{0.058t}(0.058) = 0.0528148e^{0.058t}$.
Since $f''(t) = 0.0528148e^{0.058t}$, then for $1 \leq t \leq 15$, $f''(t) > 0$. Thus, f is concave up on $[1, 15]$. Thus, between 1981 and 1995, the number of African Americans in the U.S. House of Representatives was increasing and it was increasing at a faster rate.

228 Chapter 5 Further Applications of the Derivative

67d) From parts b and c, f has no relative extrema or inflection points. Sketching a graph, we get:

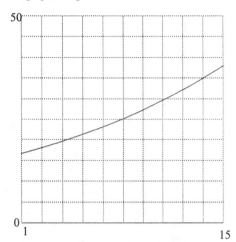

69) Since y is a polynomial, it is continuous on the stated interval. By the Extreme Value Theorem, it has an absolute maximum and minimum.
Computing the derivative, we get:
$y' = \frac{d}{dx}[x^3 + 6x^2 + 9x] = 3x^2 + 12x + 9$.
Since y' is defined for all x, the only critical values are when $y' = 0$. Solving yields:
$3x^2 + 12x + 9 = 0$
$3(x^2 + 4x + 3) = 0$
$3(x + 3)(x + 1) = 0$
$x = -3$ and $x = -1$
Thus, $x = -3$ and $x = -1$ are the critical values. Evaluating y at $x = -3$, $x = -1$, and the endpoints $x = -5$ and $x = 5$, we get:
$y\big|_{x=-3} = (-3)^3 + 6(-3)^2 + 9(-3) = 0$
$y\big|_{x=-1} = (-1)^3 + 6(-1)^2 + 9(-1) = -4$
$y\big|_{x=-5} = (-5)^3 + 6(-5)^2 + 9(-5) = -20$
and $y\big|_{x=5} = (5)^3 + 6(5)^2 + 9(5) = 320$.
Thus, there is an absolute maximum of 320 at $x = 5$ and an absolute minimum of -20 at $x = -5$.

71) Since h is a polynomial, it is continuous on the stated interval. By the Extreme Value Theorem, it has an absolute maximum and minimum.
Computing the derivative, we get:
$h'(x) = \frac{d}{dx}[\frac{1}{5}x^5 - \frac{5}{3}x^3 + 4x] = x^4 - 5x^2 + 4$.
Since $h'(x)$ is defined for all x, the only critical values are when $h'(x) = 0$. Solving yields:
$x^4 - 5x^2 + 4 = 0$
$(x^2 - 4)(x^2 - 1) = 0$
$(x - 2)(x + 2)(x - 1)(x + 1) = 0$
$x = \pm 2, \pm 1$.
Thus, $x = -2, -1, 1,$ and 2 are the critical values. Evaluating h at those value and the endpoints $x = -5$ and $x = 5$, we get:
$h(-2) = \frac{1}{5}(-2)^5 - \frac{5}{3}(-2)^3 + 4(-2) \approx -1.067$
$h(-1) = \frac{1}{5}(-1)^5 - \frac{5}{3}(-1)^3 + 4(-1) \approx -2.533$
$h(2) = \frac{1}{5}(2)^5 - \frac{5}{3}(2)^3 + 4(2) \approx 1.067$
$h(1) = \frac{1}{5}(1)^5 - \frac{5}{3}(1)^3 + 4(1) \approx 2.533$
$h(-5) = \frac{1}{5}(-5)^5 - \frac{5}{3}(-5)^3 + 4(-5)$
≈ -436.667 and
$h(5) = \frac{1}{5}(5)^5 - \frac{5}{3}(5)^3 + 4(5) \approx 436.667$.
Thus, there is an absolute maximum of 436.67 at $x = 5$ and an absolute minimum of -436.67 at $x = -5$.

73) Since f is a rational function, the only values where it is not continuous are the values that make it undefined. The function f is undefined at $x = 0$, but 0 is not in $[1, 10]$. Thus, f is continuous on the interval. By the Extreme Value Theorem, it has an absolute maximum and minimum.
Computing the derivative, we get:
$f'(x) = \frac{d}{dx}[x + 4x^{-1}] = 1 - 4x^{-2} = 1 - \frac{4}{x^2}$.
$f'(x)$ is undefined when $x = 0$, but 0 is not in $[1, 10]$. Thus, the only critical values are when $f'(x) = 0$.

73) Continued
Solving yields:
$1 - \frac{4}{x^2} = 0$ (multiply by x^2)
$x^2 - 4 = 0$
$x^2 = 4$
$x = \pm 2$, but $x = -2$ is not in [1, 10].
Thus, $x = 2$ is the only critical value.
Evaluating f at $x = 2$ and at the endpoints $x = 1$ and $x = 10$, we get:
$f(2) = (2) + \frac{4}{2} = 4$
$f(1) = (1) + \frac{4}{1} = 5$
$f(10) = (10) + \frac{4}{10} = 10.4$

Thus, there is an absolute maximum of 10.4 at $x = 10$ and an absolute minimum of 4 at $x = 2$.

75) Since y is a polynomial, it is continuous on the stated interval. By the Extreme Value Theorem, it has an absolute maximum and minimum.
Computing the derivative, we get:
$y' = \frac{d}{dx}[-0.0668x^3 + 2.04x^2 - 2.41x + 71.29]$
$= -0.2004x^2 + 4.08x - 2.41$.
Since y' is defined for all x, the only critical values are when $y' = 0$. Solving yields:
$-0.2004x^2 + 4.08x - 2.41 = 0$
Using the quadratic formula, we get:
$x = \frac{-4.08 \pm \sqrt{(4.08)^2 - 4(-0.2004)(2.41)}}{2(-0.2004)}$
$x = \frac{4.08 \pm \sqrt{17.714544}}{-0.4008}$
$x \approx 0.61$ or $x \approx 19.75$, but $x \approx 0.61$ is not in [1, 25]. Thus, $x \approx 19.75$ is the critical value. Evaluating y at $x = 19.75$, and the endpoints $x = 1$ and $x = 25$, we get:
$y|_{x=19.75} = -0.0668(19.75)^3 + 2.04(19.75)^2$
$\qquad - 2.41(19.75) + 71.29 \approx 304.81$,
$y|_{x=1} = -0.0668(1)^3 + 2.04(1)^2$
$\qquad - 2.41(1) + 71.29 \approx 70.85$, and
$y|_{x=25} = -0.0668(25)^3 + 2.04(25)^2$
$\qquad - 2.41(25) + 71.29 \approx 242.29$.
Thus, there is an absolute maximum of 304.81 at $x \approx 19.75$ and an absolute minimum of 70.85 at $x = 1$.

75) Continued
Rounding off the nearest year, in 1992, the amount of money spent by the U.S. government hit a peak of $304.81 billion and in 1973, it was at it lowest value of $70.85 billion.

77) The area of the rectangular pool is $AP = x \bullet y = 1200$. Solving for y yields: $y = \frac{1200}{x}$. The area of the garden and walkway together is
$A = (8 + x + 8)(6 + y + 6) = (x + 16)(y + 12)$
Substituting in $y = \frac{1200}{x}$, we get:
$A(x) = (x + 16)(\frac{1200}{x} + 12)$
$= 12x + 1392 + \frac{19200}{x}$. The domain is [10, 100]. Since A is a rational function, the only values where it is not continuous are the values that make it undefined. But $x = 0$ is not in [10, 100]. Thus, A is continuous on the interval. By the Extreme Value Theorem, it has an absolute maximum and minimum
Computing the derivative, we get:
$A'(x) = \frac{d}{dx}[12x + 1392 + 19{,}200x^{-1}]$
$= 12 - 19{,}200x^{-2} = 12 - \frac{19200}{x^2}$.
$A'(x)$ is undefined at $x = 0$, but $x = 0$ is not in [10, 100]. So, the only critical values are when $A'(x) = 0$. Solving yields:
$12 - \frac{19200}{x^2} = 0$ (multiply by x^2)
$12x^2 - 19{,}200 = 0$
$x^2 = 1600$
$x = \pm 40$, but -40 is not in [10, 100].
Thus, $x = 40$ is the only critical value.
Evaluating A at $x = 40$ and the endpoints $x = 10$ and $x = 100$, we get:
$A(40) = 12(40) + 1392 + \frac{19200}{40} = 2352$
$A(10) = 12(10) + 1392 + \frac{19200}{10} = 3432$ and
$A(100) = 12(100) + 1392 + \frac{19200}{100} = 2784$.
If $x = 40$, then $y = \frac{1200}{40} = 30$.
Thus, the dimensions of the pool need to be 40 feet long and 30 feet wide to minimize the total area at 2352 square feet.

79a) $R(x) = x \cdot p(x) = x(-0.002x^2 + 1.5x - 1.7)$
$= -0.002x^3 + 1.5x^2 - 1.7x$. The domain is [50, 550].

79b) Since R is a polynomial, it is continuous on the stated interval. By the Extreme Value Theorem, it has an absolute maximum and minimum.
Computing the derivative, we get:
$R'(x) = \frac{d}{dx}[-0.002x^3 + 1.5x^2 - 1.7x]$
$= -0.006x^2 + 3x - 1.7$.
Since $R'(x)$ is defined for all x, the only critical values are when $R'(x) = 0$. Solving yields:
$-0.006x^2 + 3x - 1.7 = 0$
Using the quadratic formula, we get:
$x = \frac{-3 \pm \sqrt{(3)^2 - 4(-0.006)(-1.7)}}{2(-0.006)} = \frac{3 \pm \sqrt{8.9592}}{0.012}$
$x \approx 0.57$ or $x \approx 499.43$, but 0.57 is not in [50, 550]. Thus, $x \approx 499$ is the critical value. Evaluating R at this value and the endpoints $x = 50$ and $x = 550$, we get:
$R(499) = -0.002(499)^3 + 1.5(499)^2$
$- 1.7(499) \approx 124{,}150.20$,
$R(50) = -0.002(50)^3 + 1.5(50)^2$
$- 1.7(50) = 3415$, and
$R(550) = -0.002(550)^3 + 1.5(550)^2$
$- 1.7(550) \approx 120{,}065$.
Thus, the maximum weekly revenue is $124,150.20.

79c) From part b, the maximum revenue occurs when $x = 499$ units are demanded.

79d) $p(499) = -0.002(499)^2 + 1.5(499) - 1.7$
$= \$248.80$. So the price is $248.80 per unit when the revenue is maximized.

81) Since y is a rational function, the only values where it is not continuous are the values that make it undefined. But $x = 0$ is not in (0, 10). Thus, y is continuous on the interval. Computing the derivative, we get:
$y' = \frac{d}{dx}[3x + 6x^{-1}] = 3 - 6x^{-2} = 3 - \frac{6}{x^2}$.

81) Continued
y' is undefined when $x = 0$, but 0 is not in (0, 10). Thus, the only critical values are when $y' = 0$. Solving yields:
$3 - \frac{6}{x^2} = 0$ (multiply by x^2)
$3x^2 - 6 = 0$
$x^2 = 2$
$x = \pm\sqrt{2}$, but $-\sqrt{2}$ is not in (0, 10).
Thus, $x = \sqrt{2}$ is the only critical value and $y|_{x=1.41} = 6\sqrt{2}$. Computing the second derivative, we get:
$y'' = \frac{d}{dx}[3 - 6x^{-2}] = 12x^{-3} = \frac{12}{x^3}$. Hence,
$y''|_{x=\sqrt{2}} = \frac{12}{(\sqrt{2})^3} > 0$. So, y has an absolute minimum at $(\sqrt{2}, 6\sqrt{2}) \approx (1.41, 8.49)$.

83) Since g is a rational function, the only values where it is not continuous are the values that make it undefined. But $x = 0$ is not in (0, 5). Thus, g is continuous on the interval. Computing the derivative, we get:
$g'(x) = \frac{d}{dx}[4x^2 - 2 + 9x^{-2}] = 8x - 18x^{-3}$
$= 8x - \frac{18}{x^3}$.
$g'(x)$ is undefined when $x = 0$, but 0 is not in (0, 5). Thus, the only critical values are when $g'(x) = 0$. Solving yields:
$8x - \frac{18}{x^3} = 0$ (multiply by x^3)
$8x^4 - 18 = 0$
$x^4 = \frac{9}{4}$
$x = \pm\sqrt[4]{\frac{9}{4}}$, but $-\sqrt[4]{\frac{9}{4}}$ is not in (0, 5).
Thus, $x = \sqrt[4]{\frac{9}{4}} = \frac{\sqrt{6}}{2}$ is the only critical value and $g(\frac{\sqrt{6}}{2}) = 10$. Computing the second derivative, we get:
$g''(x) = \frac{d}{dx}[8x - 18x^{-3}] = 8 + 54x^{-4} =$
$8 + \frac{54}{x^4}$. Hence, $g''(1.22) = 8 + \frac{54}{(1.22)^4} > 0$.
So, g has an absolute minimum at $(\frac{\sqrt{6}}{2}, 10)$
$\approx (1.22, 10)$.

85) Since f is a rational function, the only values where it is not continuous are the values that make it undefined. But $x = 0$ is not in $(0, 10)$. Thus, f is continuous on the interval. Computing the derivative, we get:
$f'(x) = \frac{d}{dx}[-4x - 3x^{-1}] = -4 + 3x^{-2}$
$= -4 + \frac{3}{x^2}$.
$f'(x)$ is undefined when $x = 0$, but 0 is not in $(0, 10)$. Thus, the only critical values are when $f'(x) = 0$. Solving yields:
$-4 + \frac{3}{x^2} = 0$ (multiply by x^2)
$-4x^2 + 3 = 0$
$x^2 = \frac{3}{4}$
$x = \pm\sqrt{\frac{3}{4}}$, but $-\sqrt{\frac{3}{4}}$ is not in $(0, 10)$.
Thus, $x = \sqrt{\frac{3}{4}} = \frac{\sqrt{3}}{2}$ is the only critical value and $f(\frac{\sqrt{3}}{2}) = -4\sqrt{3}$. Computing the second derivative, we get:
$f''(x) = \frac{d}{dx}[-4 + 3x^{-2}] = -6x^{-3} = -\frac{6}{x^3}$.
Hence, $f''(\frac{\sqrt{3}}{2}) = -\frac{6}{(\frac{\sqrt{3}}{2})^3} < 0$. So, f has an absolute maximum at $(\frac{\sqrt{3}}{2}, -4\sqrt{3}) \approx (0.87, -6.93)$.

87) Since f is a rational function, the only values where it is not continuous are the values that make it undefined. But $x = 0$ is not in $(0, 5)$. Thus, f is continuous on the interval. Computing the derivative, we get:
$f'(x) = \frac{d}{dx}[5 - x^2 - 9x^{-2}] = -2x + 18x^{-3}$
$= -2x + \frac{18}{x^3}$.
$f'(x)$ is undefined when $x = 0$, but 0 is not in $(0, 5)$. Thus, the only critical values are when $f'(x) = 0$. Solving yields:
$-2x + \frac{18}{x^3} = 0$ (multiply by x^3)
$-2x^4 + 18 = 0$
$x^4 = 9$
$x = \pm\sqrt{3}$, but $-\sqrt{3}$ is not in $(0, 5)$.
Thus, $x = \sqrt{3}$ is the only critical value.

87) Continued
Also, $f(\sqrt{3}) = -1$
Computing the second derivative, we get:
$f''(x) = \frac{d}{dx}[-2x + 18x^{-3}] = -2 - 54x^{-4}$
$= -2 - \frac{54}{x^4}$. Hence, $f''(\sqrt{3}) = -2 - \frac{54}{(\sqrt{3})^4}$
< 0. So, f has an absolute maximum at $(\sqrt{3}, -1) \approx (1.73, -1)$.

89) The area of the rectangular region + the semi-circle is $A = x \cdot y + \frac{\pi x^2}{8}$. Since the region is enclosed by a track, the perimeter, P, of the track is
$P = \frac{\pi x}{2} + y + y + x = 12$ or
$0.5\pi x + 2y + x = 12$. Solving for y yields: $y = 6 - 0.25\pi x - 0.5x$. Substituting in for y in $x \cdot y + \frac{\pi x^2}{8}$, we get:
$A(x) = x(6 - 0.25\pi x - 0.5x) + \frac{\pi x^2}{8} =$
$= 6x - 0.25x^2\pi - 0.5x^2 + \frac{\pi x^2}{8}$
$= 6x - 0.125x^2\pi - 0.5x^2$.
The domain is $(0, \infty)$.
Computing the derivative, we get:
$A'(x) = \frac{d}{dx}[6x - 0.125x^2\pi - 0.5x^2]$
$= 6 - 0.25\pi x - x$.
$A'(x)$ is defined for all values in the domain so the only critical values are when $A'(x) = 0$. Solving yields:
$6 - 0.25\pi x - x = 0$
$6 = (0.25\pi + 1)x$
$x = \frac{6}{0.25\pi + 1} \approx 3.36$
But $y = 6 - 0.25\pi(\frac{6}{0.25\pi+1}) - 0.5(\frac{6}{0.25\pi+1})$
≈ 1.68 and $A(\frac{6}{0.25\pi+1}) = 6(\frac{6}{0.25\pi+1})$
$- 0.125\pi(\frac{6}{0.25\pi+1})^2 - 0.5(\frac{6}{0.25\pi+1})^2 \approx 10$
Computing the second derivative, we get:
$A''(x) = \frac{d}{dx}[6 - 0.25\pi x - x.] = -0.25\pi - 1$
< 0 for all x. Hence, A has an absolute maximum area of about 10 square feet when $x \approx 3.36$ ft and $y \approx 1.68$ feet.

91) We first make a drawing of the situation:

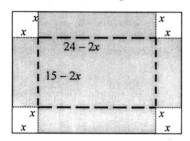

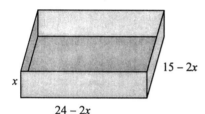

The volume of the box is $V(x) = lwh$
$= (24 - 2x)(15 - 2x)x = 4x^3 - 78x^2 + 360x$
The domain of the V is $(0, 7.5)$ since if $x = 0$, no cut would be made and if $x = 7.5$, the width $15 - 2x$ would be zero. In either case, we could not make a box. Since V is a polynomial, it is continuous on the interval. Computing the derivative, we get:
$V'(x) = \frac{d}{dx}[4x^3 - 78x^2 + 360x]$
$= 12x^2 - 156x + 360$.
$V'(x)$ is defined for all values in the domain so the only critical values are when $V'(x) = 0$. Solving yields:
$12x^2 - 156x + 360 = 0$
$12(x^2 - 13x + 30) = 0$
$12(x - 10)(x - 3) = 0$
$x = 10$ or 3, but 10 is not in $(0, 7.5)$ which means that $x = 3$ is the only critical value. But $l = 24 - 2(3) = 18$, $w = 15 - 2(3) = 9$, and $V(3) = 4(3)^3 - 78(3)^2 + 360(3) = 486$.

91) Continued
Computing the second derivative, we get:
$V''(x) = \frac{d}{dx}[12x^2 - 156x + 360] = 24x - 156$.
Hence, $V''(3) = 24(3) - 156$
$= -84 < 0$. Thus, V has an absolute maximum of 486 cubic inches when the $l = 18$ in, $w = 9$ in, and $h = 3$ in.

93) We need to first find the yield per tree, $y(x)$. If $x = 35$ trees per acre, the yield per tree $y(x) = 150$ pounds. The number of trees per acre is increased by one, the yield per tree decreases by 4 pounds. In other words, if $x = 36$ trees per acre $(35 + 1)$, then the yield $y(x) = 146 (150 - 4)$. Thus, $(35, 150)$ and $(36, 146)$ are two points that lie on the line $y(x)$. We can calculate the slope:
$m = \frac{y(36) - y(35)}{36 - 35} = \frac{146 - 150}{1} = -4$.

Using the point-slope formula, we can find the equation of $y(x)$:
$y(x) - y(35) = -4(x - 35)$
$y(x) - 150 = -4x + 140$
$y(x) = -4x + 290$
Now, the total yield $Y(x) = x \bullet y(x)$
$= x(-4x + 290)$
$= -4x^2 + 290x$.
The domain of Y is $[35, 73)$. Since Y is a polynomial, it is continuous on the interval. Computing the derivative, we get:
$Y'(x) = \frac{d}{dx}[-4x^2 + 290x] = -8x + 290$
$Y'(x)$ is defined for all values in the domain so the only critical values are when $Y'(x) = 0$. Solving yields:
$-8x + 290 = 0$
$x = 36.25 \approx 36$
So $x = 36$ is the only critical value. But, $Y(36) = -4(36)^2 + 290(36) = 5256$.
Computing the second derivative, we get:
$Y''(x) = \frac{d}{dx}[-4x + 290] = -4 < 0$
for all x. Hence, Y has an absolute maximum yield of 5256 pounds per acre when 36 trees are planted per acre.

Chapter 6.1 Exercises

1) $F'(x) = 6 = f(x)$

3) $F'(x) = 8x^7 \neq f(x)$

5) $F'(x) = 8 + \dfrac{2t}{2} = 8 + t = f(x)$

7) $F'(x) = 1 \neq f(x)$

9) $F'(x) = \dfrac{11x^{10}}{11} = x^{10} = f(x)$

11) $\int x^4 dx = \dfrac{x^5}{5} + c$

13) $\int x^{2.31} dx = \dfrac{x^{3.31}}{3.31} + c$

15) $\int \dfrac{1}{t^3} dt = \int t^{-3} dt = \dfrac{t^{-2}}{-2} + c$
$= -\dfrac{1}{2t^2} + c$

17) $\int \sqrt[4]{x^5} dx = \int x^{5/4} dx = \dfrac{4x^{9/4}}{9} + c$

19) $\int \dfrac{1}{\sqrt[3]{x}} dx = \int x^{-1/3} dx = \dfrac{3}{2} x^{2/3} + c$
$= \dfrac{3x^{2/3}}{2} + c$

21) $\int 0.4 x^6 dx = 0.4 \int x^6 dx$
$= \dfrac{0.4 x^7}{7} + c = \dfrac{2x^7}{35} + c$

23) $\int (2x+3) dx = \int 2x dx + \int 3 dx$
$= x^2 + 3x + c$

25) $\int \left(\dfrac{2}{3}x + 4\right) dx = \dfrac{2}{3} \int x dx + 4 \int dx$
$= \dfrac{2}{3}\left(\dfrac{x^2}{2}\right) + 4x + c = \dfrac{x^2}{3} + 4x + c$

27) $\int (3t^2 + 2t + 10) dt$
$= 3 \int t^2 dt + 2 \int t dt + 10 \int dt$
$= t^3 + t^2 + 10t + c$

29) $\int (1 - 2x^2 + 3x^3) dx$
$= \int dx - 2 \int x^2 dx + 3 \int x^3 dx$
$= x - \dfrac{2x^3}{3} + \dfrac{3x^4}{4} + c$

31) $\int (6.21 x^2 + 0.03x - 4.01) dx$
$= 6.21 \int x^2 dx + 0.03 \int x dx - 4.01 \int dx$
$= \dfrac{6.21 x^3}{3} + \dfrac{0.03 x^2}{2} - 4.01x + c$
$= 2.07 x^3 + 0.015 x^2 - 4.01x + c$

33) $\int \left(\dfrac{1}{x^2} - \dfrac{3}{x^3}\right) dx$
$= \int x^{-2} dx - 3 \int x^{-3} dx$
$= -\dfrac{1}{x} + \dfrac{3}{2x^2} + c$

35) $\int (3+2\sqrt{x})dx = 3\int dx + 2\int x^{1/2}dx$

$= 3x + 2\left(\dfrac{2}{3}\right)x^{3/2} + c = 3x + \dfrac{4x^{3/2}}{3} + c$

37) $\int \left(\dfrac{3}{x^3} + 2x^{3/2} - 4\right)dx$

$= 3\int x^{-3}dx + 2\int x^{3/2}dx - 4\int dx$

$= 3\left(-\dfrac{1}{2}\right)x^{-2} + 2\left(\dfrac{2}{5}\right)x^{5/2} - 4x + c$

$= \dfrac{-3}{2x^2} + \dfrac{4x^{5/2}}{5} - 4x + c$

39) $\int \dfrac{3t^3 - 2t}{6t}dt = \int \dfrac{3t^3}{6t}dt - \int \dfrac{2t}{6t}dt$

$= \dfrac{1}{2}\int t^2 dt - \dfrac{1}{3}\int dt = \dfrac{1}{2}\left(\dfrac{1}{3}\right)t^3 - \dfrac{1}{3}t + c$

$= \dfrac{t^3}{6} - \dfrac{t}{3} + c = \dfrac{t^3 - 2t}{6} + c = \dfrac{t(t^2-2)}{6} + c$

41) $\int (0.1z^{-3} + 2z^{-2} + z^3)dz$

$= 0.1\int z^{-3}dz + 2\int z^{-2}dz + \int z^3 dz$

$= 0.1\left(-\dfrac{1}{2}\right)z^{-2} + 2(-1)z^{-1} + \dfrac{1}{4}z^4 + c$

$= \dfrac{-1}{20z^2} - \dfrac{2}{z} + \dfrac{z^4}{4} + c$

43) $\int (2t^{0.13} + 5)dt = 2\int t^{0.13}dt + 5\int dt$

$= 2\left(\dfrac{1}{1.13}\right)t^{1.13} + 5t + c$

$= \dfrac{200}{113}t^{1.13} + 5t + c$

45) $f(x) = \int f'(x)dx = \int -2dx$

$= -2x + c$

$f(0) = -2(0) + c = 4 \Rightarrow c = 4$

$f(x) = -2x + 4$

47) $f(x) = \int 5x\,dx = \dfrac{5x^2}{2} + c$

$f(0) = \dfrac{5(0)^2}{2} + c = 0 \Rightarrow c = 0$

$f(x) = \dfrac{5x^2}{2} + 0 = \dfrac{5x^2}{2}$

49) $f(x) = \int (2x - 3)dx = x^2 - 3x + c$

$f(0) = 0^2 - 3(0) + c = 4 \Rightarrow c = 4$

$f(x) = x^2 - 3x + 4$

51) $f(x) = \int (500 - 0.05t)dt$

$= 500t - \dfrac{0.05t^2}{2} + c$

$= 500t - 0.025t^2 + c$

$f(0) = 0 - 0 + c = 40 \Rightarrow c = 40$

$f(t) = 500t - 0.025t^2 + 40$

53) $s(x) = \int (2x^{-2} + 3x^{-3} - 1)dx$

$= -\dfrac{2}{x} - \dfrac{3}{2x^2} - x + c$

$s(1) = \dfrac{-2}{1} - \dfrac{3}{2(1)^2} - 1 + c = 2$

$c = \dfrac{13}{2}$

$s(x) = \dfrac{-2}{x} - \dfrac{3}{2x^2} - x + \dfrac{13}{2}$

55) $y = \int y' dt = \int \dfrac{5t+2}{\sqrt[3]{t}} dt$

$= 5\int \dfrac{t}{t^{1/3}} dt + 2\int t^{-1/3} dt$

$= 5\int t^{2/3} dt + 2\int t^{-1/3} dt$

$= 5\left(\dfrac{3t^{5/3}}{5}\right) + 2\left(\dfrac{3t^{2/3}}{2}\right) + c$

$= 3t^{5/3} + 3t^{2/3} + c$

$y(0) = c = 1 \Rightarrow c = 1$

$y = 3t^{5/3} + 3t^{2/3} + 1$

57) $R(t) = \int \dfrac{1-t^4}{t^3} dt = \int t^{-3} dt - \int t\, dt$

$= -\dfrac{1}{2t^2} - \dfrac{t^2}{2} + c = -\dfrac{t^4+1}{2t^2} + c$

$R(1) = -\dfrac{(1^4+1)}{2(1)^2} + c = 4$

$= \dfrac{-2}{2} + c = 4 \Rightarrow c = 5$

$R(t) = -\dfrac{t^4+1}{2t^2} + 5$

59a) $P(q) = \int (40 - 0.05q) dq$

$= 40q - 0.025q^2 + c$

$P(0) = 0 - 0 + c = 0 \Rightarrow c = 0$

$P(q) = 40q - 0.025q^2$

59b) $P(200) = 40(200) - 0.025(200)^2$
$= 7000$
$7000 profit from selling 200 units

61a) $R(x) =$

$\int (0.000045x^2 - 0.03x + 3.75) dx$

$= 0.000015x^3 - 0.015x^2 + 3.75x + c$
$R(0) = 0 - 0 + 0 + c = 0 \Rightarrow c = 0$
$R(x) = 0.000015x^3 - 0.015x^2 + 3.75x$

61b) $p(x) = \dfrac{R(x)}{x}$

$= 0.000015x^2 - 0.015x + 3.75$

61c) $p(100) = 2.40$; When demand is 100 passengers, the price should be $2.40

63a) $AC(x) = \int \dfrac{-100}{x^2} dx$

$= -100\int x^{-2} dx$

$= -100(-1)x^{-1} + c = \dfrac{100}{x} + c$

$AC(100) = \dfrac{100}{100} + c = 2.50$

$\Rightarrow c = 1.50$

$\Rightarrow AC(x) = \dfrac{100}{x} + 1.50$

63b) $C(x) = xAC(x) = 100 + 1.50x$

63c) $C(100) = 100 + 1.50(100) = 250$
The cost of producing 100 banners is $250

65a) [0,25][-40,55]

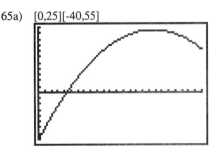

65a) continued
Arrests were decreasing where a' is negative, from 1970 until about 1974; after 1974 they began to increase.

65b) $a(t) = \int (-0.3t^2 + 10.56t - 40.31) dt$
$= -0.1t^3 + 5.28t^2 - 40.31t + c$
$a(8) = -35.76 + c = 480$
$\Rightarrow c = 515.76$
$a(t) = -0.1t^3 + 5.28t^2 - 40.31t + 515.76$

65c) [0,25],[0,1500]

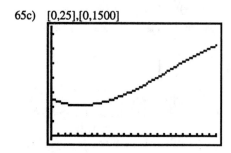

65d) $a(21) = 1071.63$; In 1991 approximately 1,071,630 arrests were made.

Chapter 6.2 Exercises

1) $\Delta x = \dfrac{2-0}{4} = \dfrac{1}{2}$

3) $\Delta x = \dfrac{4-1}{6} = \dfrac{1}{2}$

5) $\Delta x = \dfrac{5-3}{4} = \dfrac{1}{2}$

7) $\Delta x = \dfrac{2-(-1)}{6} = \dfrac{1}{2}$

9) $\Delta x = \dfrac{5-(-3)}{28} = \dfrac{2}{7}$

11a) $\Delta x = \dfrac{2-0}{4} = \dfrac{1}{2}$

$A = \dfrac{1}{2}(5) + \dfrac{1}{2}(5) + \dfrac{1}{2}(5) + \dfrac{1}{2}(5) = 10$

11b) $A = \dfrac{1}{2}(5) + \dfrac{1}{2}(5) + \dfrac{1}{2}(5) + \dfrac{1}{2}(5) = 10$

11c) $|10 - 10| = 0$

13a) $\Delta x = \dfrac{4-1}{6} = \dfrac{1}{2}$

$A = \dfrac{1}{2}(3) + \dfrac{1}{2}\left(\dfrac{9}{2}\right) + \dfrac{1}{2}(6) + \dfrac{1}{2}\left(\dfrac{15}{2}\right)$
$\quad + \dfrac{1}{2}(9) + \dfrac{1}{2}\left(\dfrac{21}{2}\right) = \dfrac{81}{4}$

13b) $A = \dfrac{1}{2}\left(\dfrac{9}{2}\right) + \dfrac{1}{2}(6) + \dfrac{1}{2}\left(\dfrac{15}{2}\right)$
$\quad + \dfrac{1}{2}(9) + \dfrac{1}{2}\left(\dfrac{21}{2}\right) + \dfrac{1}{2}(12) = \dfrac{99}{4}$

13c) $\left|\dfrac{81}{4} - \dfrac{99}{4}\right| = \dfrac{9}{2} = 4.5$

15a) $\Delta x = \dfrac{5-3}{4} = \dfrac{1}{2}$

$A = \dfrac{1}{2}(5) + \dfrac{1}{2}(6) + \dfrac{1}{2}(7) + \dfrac{1}{2}(8) = 13$

15b) $A = \dfrac{1}{2}(6) + \dfrac{1}{2}(7) + \dfrac{1}{2}(8) + \dfrac{1}{2}(9) = 15$

15c) $|13 - 15| = 2$

17a) $\Delta x = \dfrac{4-0}{4} = 1$

$A = 1(0) + (2) + 1(8) + 1(18) = 28$

17b) $A = 1(2) + 1(8) + 1(18) + 1(32) = 60$

17c) $|28 - 60| = 32$

19a) $\Delta x = \dfrac{2-(-1)}{6} = \dfrac{1}{2}$

$A = \dfrac{1}{2}(3) + \dfrac{1}{2}\left(\dfrac{15}{4}\right) + \dfrac{1}{2}(4) + \dfrac{1}{2}\left(\dfrac{15}{4}\right)$
$\quad + \dfrac{1}{2}(3) + \dfrac{1}{2}\left(\dfrac{7}{4}\right) = \dfrac{77}{8}$

19b) $A = \dfrac{1}{2}\left(\dfrac{15}{4}\right) + \dfrac{1}{2}(4) + \dfrac{1}{2}\left(\dfrac{15}{4}\right) + \dfrac{1}{2}(3)$
$\quad + \dfrac{1}{2}\left(\dfrac{7}{4}\right) + \dfrac{1}{2}(0) = \dfrac{65}{8}$

19c) $\left|\dfrac{77}{8} - \dfrac{65}{8}\right| = \dfrac{3}{2} = 1.5$

21)

	left	right	|left-right|
10	9.24	6.84	2.4
100	8.1204	7.8804	0.24
1000	8.012004	7.988004	0.024

23)

	left	right	|left-right|
10	36.11837	38.51837	2.4
100	37.21318	37.45318	0.24
1000	37.32133	37.34533	0.024

25)

	left	right	|left-right|
10	32.895	39.195	6.3
100	35.68545	36.31545	0.63
1000	35.9685	36.0315	0.063

27)

	left	right	|left-right|
10	30.24	51.04	20.8
100	38.9664	41.0464	2.08
1000	39.89606	40.10406	0.208

29)

	left	right	|left-right|
10	33.25	39	5.75
100	35.79625	36.37125	0.575
1000	36.05459	36.11209	0.0575

31a) $\int_3^9 10\,dx$

31b) Leftsum=60   Rightsum=60

33a) $\int_0^{20} \frac{1}{2} x\,dx$

33b) Leftsum=99   Rightsum=101

35a) $\int_1^4 \frac{1}{x^2}\,dx$

35b) Leftsum=0.7642101
      Rightsum=0.7360851

37a) $\int_0^3 (9 - x^2)\,dx$

37b) Leftsum=18.13455   Rightsum=17.86455

39a) $\int_{-1}^2 (x^2 + 2)\,dx$

39b) Leftsum=8.95545   Rightsum=9.04545

41a) $\int_1^2 5\,dx + \int_2^4 (x^2 + 1)\,dx$

41b) Leftsum=25.5468   Rightsum=25.7868

43) [-1,4],[-3,15]

45) [0,5],[-1,2]

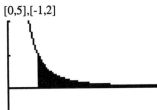

47) [0,6],[-1,3]

49) [0,3],[-1,3]

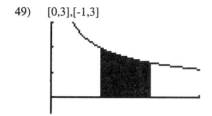

Chapter 6.3 Exercises

1) $\int_1^4 3x\,dx = \frac{3}{2}x^2 \Big|_1^4 = \frac{3}{2}(4^2 - 1^2)$
$= \frac{45}{2}$

3) $\int_0^2 5x\,dx = \frac{5}{2}x^2 \Big|_0^2 = \frac{5}{2}(2^2 - 0^2)$
$= 10$

5) $\int_3^5 (2x-1)\,dx = (x^2 - x)\Big|_3^5$
$= (5^2 - 5) - (3^2 - 3) = 14$

7) $\int_0^4 2x^2\,dx = \frac{2}{3}x^3 \Big|_0^4 = \frac{2}{3}(4^3 - 0^3)$
$= \frac{128}{3}$

9) $\int_{-1}^2 (4 - x^2)\,dx = \left(4x - \frac{1}{3}x^3\right)\Big|_{-1}^2$
$= \left[4(2) - \frac{1}{3}(2)^3\right] - \left[4(-1) - \frac{1}{3}(-1)^3\right]$
$= 9$

11) $\int_{-2}^0 3x^2\,dx = x^3 \Big|_{-2}^0 = 0^3 - (-2)^3 = 8$

13) $\int_1^4 8\sqrt{x}\,dx = 8\left(\frac{2}{3}\right)x^{3/2}\Big|_1^4 = \frac{16}{3}x^{3/2}\Big|_1^4$
$= \frac{16}{3}(4^{3/2} - 1^{3/2}) = \frac{112}{3}$

15) $\int_1^4 (x^3 - 3x)\,dx = \frac{1}{4}x^4 - \frac{3}{2}x^2 \Big|_1^4$
$= \left[\frac{(4)^4}{4} - \frac{3(4)^2}{2}\right] - \left[\frac{(1)^4}{4} - \frac{3(1)^2}{2}\right]$
$= \frac{165}{4}$

17) $\int_0^5 (0.2x^2 + 1.3x + 2.3)\,dx$
$= \frac{0.2}{3}x^3 + \frac{1.3}{2}x^2 + 2.3x \Big|_0^5$
$= \left[\frac{0.2}{3}(5)^3 + \frac{1.3}{2}(5)^2 + 2.3(5)\right] - [0]$
$= \frac{433}{12}$

19) $\int_{-2}^3 (5 + x - 6x^2)\,dx$
$= 5x + \frac{x^2}{2} + 2x^3 \Big|_{-2}^3$
$= \left[5(3) + \frac{3^2}{2} + 2(3)^3\right]$
$- \left[5(-2) + \frac{(-2)^2}{2} + 2(-2)^3\right]$
$= -\frac{85}{2}$

21) $\int_0^2 (x^4 - 2x^3)\,dx = \frac{x^5}{5} - \frac{x^4}{2} \Big|_0^2$
$= \left[\frac{2^5}{5} - \frac{2^4}{2}\right] - [0] = -\frac{8}{5}$

23) $\int_4^9 \frac{x-3}{\sqrt{x}}dx = \int_4^9 \left(\sqrt{x} - \frac{3}{\sqrt{x}}\right)dx$

$= \frac{2}{3}x^{3/2} - 6x^{1/2}\Big|_4^9$

$= \left[\frac{2}{3}(9)^{3/2} - 6(9)^{1/2}\right]$

$- \left[\frac{2}{3}(4)^{3/2} - 6(4)^{1/2}\right]$

$= \frac{20}{3}$

25a) [-10,10],[-10,10]

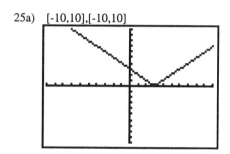

25b) $f(x) = \begin{cases} -x+3, & x < 3 \\ x-3, & x \geq 3 \end{cases}$

25c) $\int_{-2}^{3}(-x+3)dx + \int_{3}^{5}(x-3)dx$

$= \left(\frac{-x^2}{2} + 3x\right)\Big|_{-2}^{3} + \left(\frac{x^2}{2} - 3x\right)\Big|_{3}^{5}$

$= \left[\left(-\frac{9}{2} + 9\right) - (-2-6)\right]$

$+ \left[\left(\frac{25}{2} - 15\right) - \left(\frac{9}{2} - 9\right)\right]$

$= \frac{25}{2} + 2 = \frac{29}{2}$

27a) [-10,10],[-10,10]

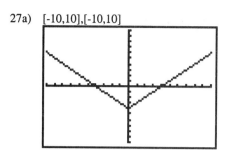

27b) $f(x) = \begin{cases} -x-4, & x \leq 0 \\ x-4, & x > 0 \end{cases}$

27c) $|x| - 4 = 0 \Rightarrow x = \pm 4$ x-Intercepts: -4, 4

27d) Net: $\int_{-2}^{0}(-x-4)dx + \int_{0}^{6}(x-4)dx$

$= \frac{-x^2}{2} - 4x\Big|_{-2}^{0} + \frac{x^2}{2} - 4x\Big|_{0}^{6}$

$= (-6) + (-6) = -12$

Gross: $\left|\int_{-2}^{0}(-x-4)dx\right| + \left|\int_{0}^{4}(x-4)dx\right|$

$+ \int_{4}^{6}(x-4)dx$

$= \left|\frac{-x^2}{2} - 4x\Big|_{-2}^{0}\right| + \left|\frac{x^2}{2} - 4x\Big|_{0}^{4}\right|$

$+ \left(\frac{x^2}{2} - 4x\right)\Big|_{4}^{6}$

$= |-6| + |-8| + 2 = 16$

29) $\int_{-3}^{0} -x\,dx + \int_{0}^{4} x\,dx = \frac{-x^2}{2}\Big|_{-3}^{0} + \frac{x^2}{2}\Big|_{0}^{4}$

$= 0 - \left[-\frac{1}{2}(9)\right] + \left[\frac{1}{2}(16) - 0\right] = \frac{25}{2}$

31) $\int_{-2}^{0}(1-x^2)dx + \int_{0}^{2}(x+1)dx$

$x - \dfrac{x^3}{3}\Big|_{-2}^{0} + \dfrac{x^2}{2} + x\Big|_{0}^{2}$

$= \left(0 + \dfrac{2}{3}\right) + (0+4) = \dfrac{14}{3}$

33) $\int_{2}^{5}(4-x^2)dx + \int_{5}^{7}6\,dx$

$= 4x - \dfrac{x^3}{3}\Big|_{2}^{5} + 6x\Big|_{5}^{7}$

$= \left(\dfrac{-65}{3}\right) - \left(\dfrac{16}{3}\right) + 42 - 30$

$= \dfrac{-45}{3} = -15$

35a) x-intercept: -2

35b) $\int_{-5}^{5}(x+2)dx = \dfrac{x^2}{2} + 2x\Big|_{-5}^{5} = 20$

35c) $\left|\int_{-5}^{-2}(x+2)dx\right| + \int_{-2}^{5}(x+2)dx$

$= \left|\dfrac{x^2}{2} + 2x\Big|_{-5}^{-2}\right| + \dfrac{x^2}{2} + 2x\Big|_{-2}^{5}$

$= \dfrac{9}{2} + \dfrac{49}{2} = 29$

37a) x-intercept: 4

37b) $\int_{0}^{9}(2\sqrt{x}-4)dx = \dfrac{4}{3}x^{3/2} - 4x\Big|_{0}^{9} = 0$

37c) $\left|\int_{0}^{4}(2\sqrt{x}-4)dx\right| + \int_{4}^{9}(2\sqrt{x}-4)dx$

$= \left|\dfrac{4}{3}x^{3/2} - 4x\Big|_{0}^{4}\right| + \dfrac{4}{3}x^{3/2} - 4x\Big|_{4}^{9}$

$= \left|\dfrac{-16}{3}\right| + \dfrac{16}{3} = \dfrac{32}{3}$

39a) x-intercepts: $-3, 3$

39b) $\int_{-1}^{4}(x^2-9)dx = \dfrac{x^3}{3} - 9x\Big|_{-1}^{4} = \dfrac{-70}{3}$

39c) $\left|\int_{-1}^{3}(x^2-9)dx\right| + \int_{3}^{4}(x^2-9)dx$

$= \left|\dfrac{x^3}{3} - 9x\Big|_{-1}^{3}\right| + \dfrac{x^3}{3} - 9x\Big|_{3}^{4}$

$= \left|\dfrac{-80}{3}\right| + \dfrac{10}{3} = 30$

41a) x-intercepts: $-1, 2$

41b) $\int_{-1}^{4}(x^2-x-2)dx = \dfrac{x^3}{3} - \dfrac{x^2}{2} - 2x\Big|_{-1}^{4}$

$= \dfrac{25}{6}$

41c) $\left|\int_{-1}^{2}(x^2-x-2)dx\right|$

$+ \int_{2}^{4}(x^2-x-2)dx$

$\left|\dfrac{x^3}{3} - \dfrac{x^2}{2} - 2x\Big|_{-1}^{2}\right| + \dfrac{x^3}{3} - \dfrac{x^2}{2} - 2x\Big|_{2}^{4}$

$= \left|\dfrac{-9}{2}\right| + \dfrac{26}{3} = \dfrac{79}{6}$

43a) x-intercepts: -2, 0, 1

43b) $\int_{-2}^{3}(x^3+x^2-2x)dx$
$= \frac{x^4}{4}+\frac{x^3}{3}-x^2\Big|_{-2}^{3} = \frac{16}{3}$

43c) $\int_{-2}^{0}(x^3+x^2-2x)dx$
$+\left|\int_{0}^{1}(x^3+x^2-2x)dx\right|$
$+\int_{1}^{3}(x^3+x^2-2x)dx$
$= \frac{8}{3}+\left|\frac{-5}{12}\right|+\frac{37}{12} = \frac{37}{6}$

45a) C'(10) = 42; The cost is increasing at a rate of $42/shoe when the production rate is 10 shoes per shift

45b) $\int_{0}^{50}(50-0.8x)dx = 50x-0.4x^2\Big|_{0}^{50}$
$=1500$; Total increase in cost of producing 0 to 50 shoes is $1500

47a) $\int_{20}^{30}(50+0.4x^3)dx = 50x+0.1x^4\Big|_{20}^{30}$
$= 65500$; Total increase in revenue of producing and selling the 20th to 30th satellite dish is $65,500

47b) [0,40],[0,20000]

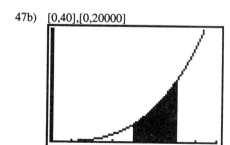

49a) $P(x) = \int(0.3x^2+0.2x)dx$
$= 0.1x^3+0.1x^2+c$
$P(20) = 0.1(20)^3+0.1(20)^2+c$
$= 704 \Rightarrow c = -136$
$P(x) = 0.1x^3+0.1x^2-136$

49b) $\int_{10}^{20}(0.3x^2+0.2x)dx$
$= 0.1x^3+0.1x^2\Big|_{10}^{20} = 730$

Total increase in profit from producing and selling 10 to 20 machines is $730

51a) $P'(6) = 6.79(6)^2-11.44(6)$
$+325.32 = 501.12$

In 1976 the rate of imports was 501,120 barrels imported each day per year

51b) $\int_{3}^{8}(6.79t^2-11.44t+325.32)dt$
$= 2.26333t^3-5.72t^2+325.32t\Big|_{3}^{8}$
$= 2409.72$; The total increase in imports of petroleum into the USA from 1973 to 1978 was 2,409,720 barrels.

53a) $m(4) = -2.08(4)-1.85 = -10.17$

In 1989, military expenditures by NATO countries was decreasing at a rate of 10.17 billion dollars per year.

53b) $\int_{4}^{8}(-2.08x-1.85)dx$
$= -1.04x^2-1.85x\Big|_{4}^{8} = -57.32$

The total decrease in military expenditures from 1989 to 1993 was $57.32 billion.

55a) $f(20) = 3.78(20)^2+25.56(20)$
$-119.48 = 1903.72$; In 1990, jet fuel consumption was increasing at a rate of 1903.72 million gallons per year.

55b) $\int_{10}^{20} (3.78x^2 + 25.56x - 119.48) dx$
$= 1.26x^3 + 12.78x^2 - 119.48x \Big|_{10}^{20}$
$= 11459.2$; 11,459,200,000 gallons

57a) $F(x) = \int f(x) dx$
$= \int (0.36x^3 - 10.26x^2 + 78.52x$
$- 123.18) dx$
$= 0.09x^4 - 3.42x^3 + 39.26x^2$
$- 123.18x + c$
$F(0) = 2961 \Rightarrow c = 2961$
$F(x) = 0.09x^4 - 3.42x^3 + 39.26x^2$
$- 123.18x + 2961$

57b) $\int_{5}^{15} f(x) dx = F(x) \Big|_{5}^{15} = 5.2$
The total increase in the number of employees from 1985 to 1995 was 5.2 thousand.

59a) $v(10) = 0.3(10) = 3 \; \dfrac{\text{feet}}{\text{second}}$

59b) $\int_{10}^{20} 0.3t \, dt = 0.15t^2 \Big|_{10}^{20} = 45$
The object moved 45 feet between 10 and 20 seconds.

61a) $v(3) = 2(3) + 10 = 16 \dfrac{\text{feet}}{\text{second}}$

61b) $\int_{3}^{5} (2t + 10) dt = t^2 + 10t \Big|_{3}^{5} = 36$
Between 3 and 5 seconds the object moved 36 feet.

Chapter 6.4 Exercises

1) $\int 3(3x+4)^2 \, dx$

$u = 3x+4$

$du = 3dx$

$\int u^2 \, du = \frac{1}{3}u^3 + c = \frac{(3x+4)^3}{3} + c$

$\frac{d}{dx}\left[\frac{(3x+4)^3}{3} + c\right] = \frac{1}{3}(3)(3x+4)^2(3)$

$= 3(3x+4)^2$

3) $\int 8x(4x^2+1)^3 \, dx$

$u = 4x^2 + 1$

$du = 8x \, dx$

$\int u^3 \, du = \frac{1}{4}u^4 + c = \frac{(4x^2+1)^4}{4} + c$

$\frac{d}{dx}\left[\frac{(4x^2+1)^4}{4}\right] = \frac{1}{4}(4)(4x^2+1)^3(8x)$

$= 8x(4x^2+1)^3$

5) $\int 2t\sqrt{t^2-1} \, dt$

$u = t^2 - 1$

$du = 2t \, dt$

$\int \sqrt{u} \, du = \frac{2}{3}u^{3/2} + c = \frac{2(t^2-1)^{3/2}}{3} + c$

$\frac{d}{dx}\left[\frac{2(t^2-1)}{3}\right] = \frac{2}{3}\left(\frac{3}{2}\right)(t^2-1)^{1/2}(2t)$

$= 2t\sqrt{t^2-1}$

7) $\int (3x^2-2)(x^3-2x) \, dx$

$u = x^3 - 2x$

$du = 3x^2 - 2$

$\int u \, du = \frac{1}{2}u^2 + c = \frac{(x^3-2x)^2}{2} + c$

$\frac{d}{dx}\left[\frac{(x^3-2x)^2}{2}\right]$

$= \frac{1}{2}(2)(x^3-2x)(3x-2)$

$= (x^3-2x)(3x-2)$

9) $\int \frac{3x^2}{\sqrt[3]{x^3-5}} \, dx$

$u = x^3 - 5$

$du = 3x^2 \, dx$

$\int \frac{1}{\sqrt[3]{u}} \, du = \frac{3}{2}u^{2/3} + c$

$= \frac{3(x^3-5)^{2/3}}{2} + c$

$\frac{d}{dx}\left[\frac{3(x^3-5)^{2/3}}{2} + c\right]$

$= \frac{3}{2}\left(\frac{2}{3}\right)(x^3-5)^{-1/3}(3x^2) = \frac{3x^2}{\sqrt[3]{x^3-5}}$

11) $\int \frac{4x}{(2x^2+3)^3} \, dx$

$u = 2x^2 + 3$

$du = 4x$

11) continued

$$\int \frac{1}{u^3} du = -\frac{1}{2} u^{-2} + c$$

$$= \frac{-1}{2(2x^2 + 3)^2} + c$$

$$\frac{d}{dx}\left[\frac{-1}{2(2x^2 + 3)^2}\right]$$

$$= -\frac{1}{2}(-2)(2x^2 + 3)^{-3}(4x)$$

$$= \frac{4x}{(2x^2 + 3)^3}$$

13) $u = 4x + 7$
$du = 4dx$
$\frac{1}{4} du = dx$

$$\frac{1}{4}\int u^4 du = \frac{1}{4}\left(\frac{1}{5}\right) u^5 + c$$

$$= \frac{(4x + 7)^5}{20} + c$$

15) $u = t^2 - 3$
$du = 2tdt$
$\frac{1}{2} du = tdt$

$$\int t(t^2 - 3)^5 dt = \frac{1}{2}\int u^5 du$$

$$= \frac{1}{2}\left(\frac{1}{6}\right) u^6 + c = \frac{(t^2 - 3)^6}{12} + c$$

17) $u = x^3 - 5$
$du = 3x^2 dx$
$\frac{1}{3} du = x^2 dx$

$$\int \frac{x^2}{\sqrt{x^3 - 5}} dx = \frac{1}{3}\int \frac{1}{\sqrt{u}} du$$

$$= \frac{1}{3}(2) u^{1/2} + c = \frac{2\sqrt{x^3 - 5}}{3} + c$$

19) $u = 10x^2 - 4$
$du = 20x dx$
$\frac{1}{4} du = 5x dx$

$$\int \frac{5x}{(10x^2 - 4)^2} dx = \frac{1}{4}\int \frac{1}{u^2} du$$

$$= \frac{-1}{4u} = \frac{-1}{4(10x^2 - 4)}$$

21) $u = 1 - x^{-1}$
$du = x^{-2} dx$

$$\int \frac{1}{x^2}\sqrt{1 - x^{-1}} dx = \int \sqrt{u} du$$

$$= \frac{2}{3} u^{3/2} + c = \frac{2}{3}(1 - x^{-1})^{3/2} + c$$

23) $u = t - 7$
$du = 1 dt$

$$\int (t - 7)^{10} dt = \int u^{10} du$$

$$= \frac{1}{11} u^{11} + c = \frac{(t - 7)^{11}}{11} + c$$

25) $u = x^2 + 2x + 5$
$du = (2x+2)dx$
$\frac{1}{2}du = (x+1)dx$

$\int \frac{x+1}{(x^2+2x+5)^5}dx = \frac{1}{2}\int \frac{1}{u^5}du$

$= \frac{1}{2}\left(\frac{-1}{4}\right)u^{-4} + c$

$= \frac{-1}{8(x^2+2x+5)^4} + c$

27) $u = x^2 + 3x - 10$
$du = (2x+3)dx$
$3du = (6x+9)dx$

$\int (6x+9)(x^2+3x-10)dx$

$= 3\int u\,du = \frac{3u^2}{2} + c$

$= \frac{3(x^2+3x-10)}{2} + c$

29) $u = x+1$
$x = u-1$

$\int x\sqrt{x+1}\,dx = \int (u-1)(u^{1/2})du$

$= \int (u^{3/2} - u^{1/2})du$

$= \frac{2u^{5/2}}{5} - \frac{2u^{3/2}}{3} + c$

$= \frac{2(x+1)^{5/2}}{5} - \frac{2(x+1)^{3/2}}{3} + c$

31) $u = t-1$
$t = u+1$

$\int t(t-1)^5 dt = \int (u-1)u^5 du$

31) continued

$= \int (u^6 + u^5)du = \frac{u^7}{7} + \frac{u^6}{6} + c$

$= \frac{(t-1)^7}{7} + \frac{(t-1)^6}{6} + c$

33) $u = x-1$
$x = u+1$
$3x = 3u+3$

$\int \frac{3x}{\sqrt{x-1}}dx = \int \frac{3u+3}{\sqrt{u}}du$

$= \int (3u^{1/2} + 3u^{-1/2})du$

$= 2u^{3/2} + 6u^{1/2} + c$

$= 2(x-1)^{3/2} + 6(x-1)^{1/2} + c$

35) $u = x^2 + 9$
$du = 2x\,dx$

$\int \frac{2x}{\sqrt{x^2+9}}dx = \int \frac{1}{\sqrt{u}}du$

$= 2u^{1/2} + c = 2\sqrt{x^2+9} + c$

$\int_0^4 \frac{2x}{\sqrt{x^2+9}}dx = 2\sqrt{x^2+9}\Big|_0^4$

$= 2\sqrt{4^2+9} - 2\sqrt{0^2+9}$

$= 10 - 6 = 4$

37) $u = x^2 - 2$
$du = 2x\,dx$
$\frac{1}{2}du = x\,dx$

$\int x(x^2-2)^3 dx = \frac{1}{2}\int u^3 du$

$= \frac{1}{2}\left(\frac{1}{4}\right)u^4 + c = \frac{(x^2-2)^4}{8} + c$

37) continued

$$\int_0^2 x(x^2-2)^3 \, dx = \left.\frac{(x^2-2)^4}{8}\right|_0^2$$

$$= \frac{(4-2)^4}{8} - \frac{(-2)^4}{8} = 0$$

39) $u = 3t - 5$
$du = 3dt$
$\frac{1}{3}du = dt$

$$\int (3t-5)^2 \, dt = \frac{1}{3}\int u^2 \, du = \frac{u^3}{9} + c$$

$$= \frac{(3t-5)^3}{9} + c$$

$$\int_0^4 (3t-5)^2 \, dt = \left.\frac{(3t-5)^3}{9}\right|_0^4 = 52$$

41) $u = 1 - x$
$du = -dx$
$-du = dx$

$$\int \sqrt{1-x} \, dx = -\int \sqrt{u} \, du$$

$$= -\frac{2}{3}u^{3/2} + c = \frac{-2(1-x)^{3/2}}{3} + c$$

$$\int_{-2}^1 \sqrt{1-x} \, dx = \left.\frac{-2(1-x)^{3/2}}{3}\right|_{-2}^1$$

$$= 0 - \left(\frac{-2}{3}\right)(3\sqrt{3}) = 2\sqrt{3}$$

43) $u = 1 + 3x^2$
$du = 6x\,dx$
$\frac{1}{6}du = x\,dx$

$$\int \frac{x}{1+3x^2} \, dx = \frac{1}{6}\int \frac{1}{u^2} \, du$$

$$= \frac{1}{6}\ln(u) + c = \frac{1}{6}\ln[1+3x^2] + c$$

$$\int_0^1 \frac{x}{1+3x^2} \, dx = \frac{1}{6}(\ln(4) - \ln(1))$$

$$= \frac{1}{6}\ln(4) = \frac{1}{3}\ln(2)$$

45) $u = t - 1$
$du = dt$

$$\int \frac{1}{\sqrt{t-1}} \, dt = \int \frac{1}{\sqrt{u}} \, du = 2u^{1/2} + c$$

$$= 2\sqrt{t-1} + c$$

$$\int_2^{10} \frac{1}{\sqrt{t-1}} \, dt = \left.2\sqrt{t-1}\right|_2^{10} = 4$$

47) $u = x^2 + 1$ if $x = 0$ then $u = 1$
$du = 2x\,dx$ if $x = 3$ then $u = 10$
$\frac{1}{2}du = x\,dx$

$$\int_0^3 \frac{x}{(x^2+1)^2} \, dx = \int_1^{10} \frac{1}{2u^2} \, du = \left.\frac{-1}{2u}\right|_1^{10}$$

$$= \frac{9}{20}$$

49) $u = x^2 + x - 1$ if $x = 1$ then $u = 1$
$du = (2x+1)dx$ if $x = 4$ then $u = 19$

$$\int_1^4 \frac{2x+1}{(x^2+x+1)^2} \, dx = \int_1^{19} \frac{1}{u^2} \, du$$

49) continued

$$= \left. \frac{-1}{u} \right|_1^{19} = \frac{18}{19}$$

51) $u = 1 + t^3$ if $t = 0$ then $u = 1$
     $du = 3t^2 dt$ if $t = 2$ then $u = 9$

$$\int_0^2 \frac{3t^2}{(1+t^3)^5} dt = \int_1^9 \frac{1}{u^5} du = \left. \frac{-1}{4u^4} \right|_1^9$$

$$= \frac{1640}{6561}$$

53) $u = x^3 + 1$ if $x = 0$ then $u = 1$
     $du = 3x^2 dx$ if $x = 2$ then $u = 9$

$$\int_0^2 3x^2 \sqrt{x^3+1}\, dx = \int_1^9 \sqrt{u}\, du$$

$$= \left. \frac{2}{3} u^{3/2} \right|_1^9 = \frac{52}{3}$$

55) $u = t^2 - 1$ if $t = -1$ then $u = 0$
     $du = 2t\, dt$ if $t = 1$ then $u = 0$

$$\frac{1}{2} du = t\, dt$$

$$\int_{-1}^1 t(t^2-1)^3 dt = \frac{1}{2}\int_0^0 u^3 du$$

$$= \left. \frac{u^4}{8} \right|_0^0 = 0$$

57) $u = 5x - 1$ if $x = 2$ then $u = 9$
     $du = 5\, dx$ if $x = 10$ then $u = 49$

$$\int_2^{10} \frac{3}{\sqrt{5x-1}} dx = \frac{3}{5}\int_9^{49} \frac{1}{\sqrt{u}} du$$

$$= \left. \frac{6}{5} u^{1/2} \right|_9^{49} = \frac{24}{5}$$

59) $u = x^2 + 1$ if $x = 0$ then $u = 1$
     $du = 2x\, dx$ if $x = 1$ then $u = 2$

59) continued

$$\int_0^1 \frac{2x}{(x^2+1)^2} dx = \int_1^2 \frac{1}{u^2} du = \left. \frac{-1}{u} \right|_1^2$$

$$= \frac{1}{2} \text{ square units}$$

61) $u = x^2 - 1$ if $x = 1$ then $u = 0$
     $du = 2x\, dx$ if $x = 3$ then $u = 8$

$$\frac{1}{2} du = x\, dx$$

$$\int_1^3 x\sqrt{x^2-1}\, dx = \frac{1}{2}\int_0^8 \sqrt{u}\, du$$

$$= \left. \frac{1}{3} u^{3/2} \right|_0^8 = \frac{16\sqrt{2}}{3}$$

63a) $MC(75) = 6$; When daily production is 75 fobs, it costs about \$6 to produce the 76th fob

63b) $u = x^2 + 10000$
     $du = 2x\, dx$
     $5du = 10x\, dx$

$$\int \frac{10x}{\sqrt{x^2+10000}} dx = 5\int \frac{1}{\sqrt{u}} du$$

$$= 10\sqrt{u} + c = 10\sqrt{x^2+10000} + c$$

$$\int_0^{75} \frac{10x}{\sqrt{x^2+10000}} dx$$

$$= \left. 10\sqrt{x^2+10000} \right|_0^{75} = 250 \text{ ; The total}$$

increase in daily costs of producing 0 to 75 fobs is \$250

250 Chapter 6 Integral Calculus

63c) $C = \int \dfrac{10x}{\sqrt{x^2 + 10000}}\,dx$

$= 10\sqrt{x^2 + 10000} + c$

$C(75) = 10\sqrt{75^2 + 10000} + c$

$C(75) = 3250 \Rightarrow c = 2000$

$C(x) = 10\sqrt{x^2 + 10000} + 2000$

63d) $C(0) = 3000$ dollars

65a) $MP(70) = \dfrac{7\sqrt{94}}{47} \approx 1.44399$

When monthly production and sales are 70 flags, the profit from the 71st flag is about \$144.40

65b) $u = 2x^2 - 400$
$du = 4x\,dx$
$\dfrac{1}{2}du = 2x\,dx$

$\int \dfrac{2x}{\sqrt{2x^2 - 400}}\,dx = \dfrac{1}{2}\int \dfrac{1}{\sqrt{u}}\,du$

$= \sqrt{u} + c = \sqrt{2x^2 - 400} + c$

$\int_{20}^{100} \dfrac{2x}{\sqrt{2x^2 - 400}}\,dx = \sqrt{2x^2 - 400}\Big|_{20}^{100}$

$= 120$; The total increase in profit of producing and selling from 20 to 100 flags each month is \$12,000

65c) $P(x) = \sqrt{2x^2 - 400} + c$

$P(20) = \sqrt{2(20)^2 - 400} + c = 0$

$c = -20$

$P(x) = \sqrt{2x^2 - 400} - 20$

67a) $u = t^2 + 9$ if t = 3 then u = 18
$du = 2t\,dt$ if t = 5 then u = 34

$\int_3^5 \dfrac{t}{\sqrt{t^2 + 9}}\,dt = \dfrac{1}{2}\int_{18}^{34} \dfrac{1}{\sqrt{u}}\,du$

$= \sqrt{u}\Big|_{18}^{34} = \sqrt{34} - \sqrt{18} \approx 1.588$ ft.

67b) $s(t) = \int \dfrac{t}{\sqrt{t^2 + 9}}\,dt = \sqrt{t^2 + 9} + c$

$s(4) = \sqrt{4^2 + 9} + c = 8 \Rightarrow c = 3$

$s(t) = \sqrt{t^2 + 9} + 3$

69a) $h'(5) \approx 33.0169$; Five years after the town was incorporated, homes were being built at a rate of about 33 homes per year

69b) $u = t^3 + 4$
$du = 3t^2\,dt$
$5du = 15t^2\,dt$

$\int \dfrac{15t^2}{\sqrt{t^3 + 4}}\,dt = 5\int \dfrac{1}{\sqrt{u}}\,du = 10\sqrt{u} + c$

$\int_0^5 \dfrac{15t^2}{\sqrt{t^3 + 4}}\,dt = 10\sqrt{t^3 + 4}\Big|_0^5$

$= 10\sqrt{129} - 20 \approx 94$; 94 homes were built during the 1st five years after the town was incorporated

69c) $h(t) = 10\sqrt{t^3 + 4} + c$

$h(0) = 10\sqrt{t^3 + 4} + c = 0 \Rightarrow c = -20$

$h(t) = 10\sqrt{t^3 + 4} - 20$

69d) $h(5) = 10\sqrt{129} - 20 \approx 94$; There were 94 homes in the town after five years

Chapter 6.5 Exercises

1) $\int \dfrac{2}{x}dx = 2\int \dfrac{1}{x}dx = 2\ln|x| + c$

3) $\int \dfrac{-2}{3t}dt = \dfrac{-2}{3}\int \dfrac{1}{t}dt = \dfrac{-2}{3}\ln|t| + c$

5) $\int \dfrac{1+e}{x}dx = \int \dfrac{1}{x}dx + e\int \dfrac{1}{x}dx$
$= \ln|x| + e\ln|x| + c = (1+e)\ln|x| + c$

7) $\int_1^5 \dfrac{3}{x}dx = 3\int_1^5 \dfrac{1}{x}dx = 3\ln|x|\Big|_1^5 = 3\ln 5$

9) $\int_2^5 \dfrac{1}{2}x^{-1}dx = \dfrac{1}{2}\int_2^5 x^{-1}dx = \dfrac{1}{2}\ln|x|\Big|_2^5$
$= \dfrac{1}{2}[\ln(5) - \ln(2)] = \dfrac{1}{2}\ln\left(\dfrac{5}{2}\right)$

11) $u = 2x + 3$
$du = 2dx$
$\dfrac{1}{2}du = dx$
$\int \dfrac{1}{2x+3}dx = \dfrac{1}{2}\int \dfrac{1}{u}du = \dfrac{1}{2}\ln|u| + c$
$= \dfrac{1}{2}\ln|2x+3| + c$

13) $u = t^2 + 1$
$du = 2tdt \Rightarrow \dfrac{1}{2}du = tdt$
$\int \dfrac{t}{t^2+1}dt = \dfrac{1}{2}\int \dfrac{1}{u}du = \dfrac{1}{2}\ln|u| + c$
$= \dfrac{1}{2}\ln(t^2+1) + c$

15) $u = x^2 - 4x + 9$
$du = (2x-4)dx \Rightarrow \dfrac{1}{2}du = (x-2)dx$
$\int \dfrac{x-2}{x^2-4x+9}dx = \dfrac{1}{2}\int \dfrac{1}{u}du$
$= \dfrac{1}{2}\ln|u| + c = \dfrac{1}{2}\ln|x^2-4x+9| + c$

17) $u = x^4 + 10$
$du = 4x^3 dx \Rightarrow \dfrac{1}{4}du = x^3 dx$
$\int \dfrac{x^3}{x^4+10}dx = \dfrac{1}{4}\int \dfrac{1}{u}du$
$= \dfrac{1}{4}\ln|u| + c = \dfrac{1}{4}\ln(x^4+10) + c$

19) $u = \ln 2x$
$du = \dfrac{1}{x}dx$
$\int \dfrac{1}{x\ln 2x}dx = \int \dfrac{1}{u}du = \ln|u| + c$
$= \ln|\ln 2x| + c$

21) $u = \sqrt{x} + 4$
$du = \dfrac{1}{2\sqrt{x}}dx \Rightarrow 4du = \dfrac{2}{\sqrt{x}}dx$
$\int \dfrac{2}{\sqrt{x}(\sqrt{x}+4)}dx = 4\int \dfrac{1}{u}du$
$= 4\ln|u| + c = 4\ln(\sqrt{x}+4) + c$
$\int_1^2 \dfrac{2}{\sqrt{x}(\sqrt{x}+4)}dx = 4\ln(\sqrt{x}+4)\Big|_1^2$

21) continued
$$= 4[\ln(\sqrt{2}+4) - \ln(\sqrt{1}+4)]$$
$$= 4\ln\frac{4+\sqrt{2}}{5}$$

23) $u = 3x - 2$ if $x = 0$ then $u = -2$
$du = 3dx$ if $x = 3$ then $u = 7$
$\frac{2}{3}du = 2dx$

$$\int_0^3 \frac{2}{3x-2}dx = \frac{2}{3}\int_{-2}^7 \frac{1}{u}du = \frac{2}{3}\ln|u|\Big|_{-2}^7$$

$$= \frac{2}{3}[\ln|7| - \ln|-2|]$$

$$= \frac{2}{3}[\ln(7) - \ln(2)] = \frac{2}{3}\ln\left(\frac{7}{2}\right)$$

25a) $s(2) = 20$; The sales rate after two days is 20 dresses per day.

25b) $\int_2^4 \frac{40}{t}dt = 40\ln|t|\Big|_2^4 = 40\ln\frac{4}{2}$
$= 40\ln 2 \approx 27.73$; About 28 dresses.

27a) $g(6) \approx 15.405$; In 1981 the foreign student enrollment was increasing at a rate of 15,405 students per year

27b) $\int_1^6 \frac{93.43}{x}dx = 92.43\ln|x|\Big|_1^6$
$= 92.43[\ln 6 - \ln 1] = 92.43\ln 6$
≈ 165.612 ; From 1976 to 1981 foreign student enrollment increased by about 165,612 students.

29a) $g(16) \approx 3.14353$; In 1996 the number of employees was increasing at a rate of about 3144 employees per year

29b) $u = x + 1$
$du = dx$
$$f(x) = \int \frac{53.44}{x+1}dx = 53.44\int \frac{1}{u}du$$
$= 53.44\ln|u| + c = 53.44\ln|x+1| + c$
$f(10) = 53.44\ln 11 + c = 209.5$
$c = 81.3565$
$f(x) = 81.3565 + 53.44\ln|x+1|$

31) $\int 2e^x dx = 2\int e^x dx = 2e^x + c$

33) $\int (e^x - 1)dx = \int e^x dx - \int dx$
$= e^x - x + c$

35) $\int (4e^t + t - 1)dt = 4e^t + \frac{t^2}{2} - t + c$

37) $\int_0^1 (e^x - 1)dx = e^x - x\Big|_0^1$
$= (e - 1) - (1) = e - 2$

39) $\int_1^2 \frac{e^x + 4}{2}dx = \frac{1}{2}\int_1^2 (e^x + 4)dx$
$= \frac{1}{2}(e^x + 4x)\Big|_1^2$
$= \frac{1}{2}[(e^2 + 8) - (e + 4)]$
$= \frac{1}{2}(e^2 - e + 4)$

41) $u = 4x + 1$
$du = 4dx \Rightarrow \frac{1}{4}du = dx$

41) continued

$$\int e^{4x+1}dx = \frac{1}{4}\int e^u du = \frac{1}{4}e^u + c$$
$$= \frac{1}{4}e^{4x+1} + c$$

43) $u = x^2 + 1$
$du = 2xdx$

$$\int 2xe^{x^2+1}dx = \int e^u du = e^u + c$$
$$= e^{x^2+1} + c$$

45) $u = e^x + 2$
$du = e^x dx$

$$\int \frac{e^x}{e^x + 2}dx = \int \frac{1}{u}du = \ln|u| + c$$
$$= \ln(e^x + 2) + c$$

47) $u = 1 - e^{-x}$
$du = e^{-x}dx$

$$\int \frac{3}{e^x(1-e^{-x})}dx = 3\int \frac{1}{u}du$$
$$= 3\ln|u| + c = 3\ln|1 - e^{-x}| + c$$

49) $u = \sqrt{x}$

$$du = \frac{1}{2\sqrt{x}}dx \Rightarrow 2du = \frac{1}{\sqrt{x}}dx$$

$$\int \frac{e^{\sqrt{x}}}{\sqrt{x}}dx = 2\int e^u du = 2e^u + c$$
$$= 2e^{\sqrt{x}} + c$$

51) $u = -4x$
$du = -4dx \Rightarrow \frac{-1}{4}du = dx$

51) continued

$$\int e^{-4x}dx = \frac{-1}{4}\int e^u du = \frac{-e^u}{4} + c$$
$$= \frac{-1}{4}e^{-4x} + c$$

$$\int_1^3 e^{-4x}dx = \frac{-1}{4}e^{-4x}\Big|_1^3$$
$$= \frac{-1}{4}\left[e^{-12} - e^{-4}\right]$$

53) $\int_0^1 \frac{1+e^x}{e^x}dx = \int_0^1 (e^{-x} + 1)dx$
$$= -e^{-x} + x\Big|_0^1$$
$$= (-e^{-1} + 1) - (-e^{-0} + 0) = 2 + e^{-1}$$

55) $u = x^2$
$du = 2xdx \Rightarrow \frac{1}{2}du = xdx$

$$\int xe^{x^2}dx = \frac{1}{2}\int e^u du = \frac{1}{2}e^u + c$$
$$= \frac{1}{2}e^{x^2} + c$$

57) $u = e^{4x} + 4$
$du = 4e^{4x}dx \Rightarrow \frac{1}{4}du = e^{4x}dx$

$$\int e^{4x}(e^{4x} + 4)dx = \frac{1}{4}\int u du$$
$$= \frac{1}{8}u^2 + c = \frac{1}{8}(e^{4x} + 4)^2 + c$$

59a) $MC(20) \approx 1.560$; At a production level of 20 faucets, the cost to produce the 21st faucet is about \$1.56

59b) $u = 0.02x$
$du = 0.02dx \Rightarrow 2du = 0.04dx$

$\int 1.50 + 0.04e^{0.02x} dx$

$= \int 1.50 dx + 2\int e^u du$

$= 1.50x + 2e^{0.02x} + c$

$\int_{100}^{150} 1.50 + 0.04e^{0.02x} dx$

$= 1.50x + 2e^{0.02x} \Big|_{100}^{150} = \$100.393$

61a) $g(11) = -0.009775$; In 1975, the number of infant deaths was decreasing at a rate of about .00978 infant deaths per 1000 live births per year

61b) $\int_1^{11} -1.38e^{-0.45x} dx$
$u = -0.45x$

$du = -0.45dx \Rightarrow \dfrac{1}{-0.45} du = dx$

$\int -1.38e^{-0.45x} dx = \dfrac{-1.38}{-0.45} \int e^u du$

$= 3.06\overline{6} e^u + c$

$\int_1^{11} -1.38e^{-0.45x} dx = 3.06\overline{6} e^{-0.45x} \Big|_1^{11}$

$= 3.06\overline{6} \left[e^{-4.95} - e^{-0.45} \right] \approx -1.93367$

From 1965 to 1975 the total decrease in infant mortality was 1.93 infant deaths per 1000 live births

63a) $g(50) = 1.97301$; In 1950 the population in the U.S. was increasing at a rate of about 1.97 million people per year

63b) $u = 0.013x$ if x=0 then u=0
$du = 0.013dx$ if x=50 then u=0.65

$\dfrac{1}{0.013} du = dx$

63b) continued

$\int_0^{50} 1.03 e^{0.013x} dx = \dfrac{1.03}{0.013} \int_0^{0.65} e^u du$

$= 79.2308 e^u \Big|_0^{0.65}$

$= 79.2308 \left[e^{0.65} - 1 \right]$

≈ 72.539 million

65) $\int 5^x dx = \int e^{x \ln 5} dx = \dfrac{e^{x \ln 5}}{\ln 5} + c$

$= \dfrac{5^x}{\ln 5} + c$

67) $\int 5\left(\dfrac{1}{2}\right)^x dx = 5\int e^{x \ln \frac{1}{2}} dx$

$= 5\left(\dfrac{1}{\ln .5}\right) e^{x \ln \frac{1}{2}} + c = \dfrac{5}{\ln .5}\left(\dfrac{1}{2}\right)^x + c$

69) $\int 7^{-x} dx = \int e^{-x \ln 7} dx$

$= \dfrac{-e^{-x \ln 7}}{\ln 7} + c = \dfrac{-7^{-x}}{\ln 7} + c$

71) $\int_1^2 10^{3x} dx = \int_1^2 e^{3x \ln 10} dx = \dfrac{e^{3x \ln 10}}{3 \ln 10}\Big|_1^2$

$= \dfrac{10^{3x}}{3 \ln 10}\Big|_1^2 = \dfrac{10^6 - 10^3}{3 \ln 10} = \dfrac{333000}{\ln 10}$

≈ 144620

73) $\int_{-1}^1 2^{3x-1} dx = \int_{-1}^1 2^{-1} 2^{3x} dx$

$= \dfrac{1}{2}\int_{-1}^1 e^{3x \ln 2} dx = \dfrac{1}{2}\left[\dfrac{1}{3 \ln 2} 2^{3x}\right]_{-1}^1$

73) continued

$$= \frac{8^x}{6\ln 2}\Big|_{-1}^{1} = \frac{21}{16\ln 2}$$

75a) $P(10) = 98.04487$; In 1990, the number of people confined in prison was increasing at a rate of about 9,804,487 inmates per year.

75b) $\int_5^{10} 0.2598(1.81)^x \, dx$

$= 0.2598 \int_5^{10} e^{x\ln 1.81} \, dx$

$= \dfrac{0.2598}{\ln 1.81}(1.81)^x \Big|_5^{10}$

$= \dfrac{0.2598}{\ln 1.81}[1.81^{10} - 1.81^5]$

≈ 156.74

The total change in the prison population from 1985 to 1990 was about 15,673,973 inmates.

Chapter 6.6 Exercises

1) $f'(x) = 5 = y'$, Yes

3) $f'(x) = 5x^4 \neq y'$, No

5) $f'(x) = \dfrac{-1}{x^2} = y'$, Yes

7) $f'(x) = 6x^2 \neq y'$, No

9) $f'(x) = \dfrac{1}{x} = y'$, Yes

11) $f'(x) = \dfrac{-12}{e^{4x}}$

$\dfrac{-12}{e^{4x}} + 4\left(\dfrac{3}{e^{4x}}\right) = 0 = y' + 4y$, Yes

13) $f'(x) = 3e^{3x}$

$3e^{3x} + 15 = 3(e^{3x} + 5)$

$\Leftrightarrow y' + 15 = 3y$, Yes

15) $dy = -2dx$

$\int dy = \int -2dx \Rightarrow y = -2x + c$

17) $dy = (1 - 3x)dx$

$\int dy = \int (1 - 3x)dx$

$\Rightarrow y = x - \dfrac{3}{2}x^2 + c$

19) $dy = (5x^3 + x - 2)dx$

$\int dy = \int (5x^3 + x - 2)dx$

$y = \dfrac{5}{4}x^4 + \dfrac{1}{2}x^2 - 2x + c$

21) $dy = (\sqrt{x} + 2)dx$

$\int dy = \int (\sqrt{x} + 2)dx$

$y = \dfrac{2}{3}x^{3/2} + 2x + c$

23) $dy = \dfrac{5}{x^2}dx$

$\int dy = \int \dfrac{5}{x^2}dx \Rightarrow y = \dfrac{-5}{x} + c$

25) $y = 2x + c$

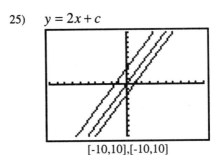

[-10,10],[-10,10]

27) $y = \dfrac{3}{2}x^2 + x + c$

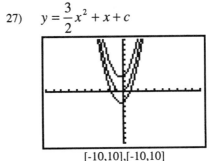

[-10,10],[-10,10]

29) $y = \dfrac{2}{3}x^{3/2} + 3x + c$

[-10,10],[-10,10]

31) $\int dy = \int 12x\,dx$
$y = 6x^2 + c$
$x = 1, y = 8 \Rightarrow c = 2$
$y = 6x^2 + 2$

33) $\int dy = \int (3x^2 + 4x)\,dx$
$y = x^3 + 2x^2 + c$
$x = 1, y = 6 \Rightarrow c = 3$
$y = x^3 + 2x^2 + 3$

35) $\int dy = \int (7 - 4t)\,dt$
$y = 7t - 2t^2 + c$
$t = 0, y = 3 \Rightarrow c = 3$
$y = 7t - 2t^2 + 3$

37) $\int dy = \int (x^3 - 2x)\,dx$
$y = \dfrac{1}{4}x^4 - x^2 + c$
$x = 0, y = -2 \Rightarrow c = -2$
$y = \dfrac{1}{4}x^4 - x^2 - 2$

39) $\int dy = \int (t^2 + 2t + 3)\,dt$
$y = \dfrac{1}{3}t^3 + t^2 + 3t + c$
$t = 0, y = 4 \Rightarrow c = 4$
$y = \dfrac{1}{3}t^3 + t^2 + 3t + 4$

41) $\int dy = \int 2x^{0.3}\,dx$
$y = \dfrac{2}{1.3}x^{1.3} + c$
$x = 0, y = 4 \Rightarrow c = 4$
$y = \dfrac{2}{1.3}x^{1.3} + 4$

43) $\int dy = \int \left(1 - \dfrac{2}{x}\right)dx$
$y = x - 2\ln|x| + c$
$x = 1, y = 6 \Rightarrow c = 5$
$y = x - 2\ln|x| + 5$

45a) $\int dC = \int (-0.02x + 6)\,dx$
$C = -0.01x^2 + 6x + c$

45c) $400 = -0.01(10^2) + 6(10) + c$
$c = 341$
$C(x) = -0.01x^2 + 6x + 341$

47a) Ten months after its introduction, the percentage of drinkers accepting the new formula is increasing at a rate of 11.1% per month

47b) $\int dA = \int (t+1.1)dt$

$A = \dfrac{1}{2}t^2 + 1.1t + c$; The percentage of drinkers accepting the new formula after t months is given by A, where c is the initial condition

47c) $t = 0, A = 5 \Rightarrow c = 5$

$A(t) = \dfrac{1}{2}t^2 + 1.1t + 5$

49a) In 1989, the number of deaths caused by heart disease was decreasing at a rate of about 1.757 death per 100,000 Americans

49b) $\int dD = \int \dfrac{-20.17}{t^{1.06}} dt$

$D = \dfrac{20.17}{0.06} t^{-0.06} + c$

$336.18 = \dfrac{20.17}{0.06} 1^{-0.06} + c$

$c = 0.013$

$D(t) = \dfrac{2017}{6} t^{-0.06} + 0.013$

49c) $D(15) = \dfrac{2017}{6}(15)^{-0.06} + 0.013$

≈ 285.76; In 1994 the number of deaths from heart disease was about 285.76 deaths per 100,000 Americans

51) $\int y\,dy = \int x\,dx$

$\dfrac{1}{2}y^2 = \dfrac{1}{2}x^2 + c \Leftrightarrow y^2 = x^2 + c$

$y = \pm\sqrt{x^2 + c}$

53) $\int \dfrac{1}{y} dy = \int 3x^2 dx$

$\ln|y| = x^3 + c \Rightarrow y = ce^{x^3}, y > 0$

55) $\dfrac{dy}{dx} = e^x e^y$

$\int e^{-y} dy = \int e^x dx$

$-e^{-y} = e^x + c \Rightarrow e^{-y} = -e^x + c$

57) $\int \dfrac{1}{y} dy = \int (2x+2)dx$

$\ln|y| = x^2 + 2x + c$

$y = ce^{x^2+2x} = ce^{x(x+2)}, y > 0$

59) $\dfrac{dy}{dx} = \sqrt{x}\sqrt{y}$

$\int \dfrac{1}{\sqrt{y}} dy = \int \sqrt{x}\,dx$

$2\sqrt{y} = \dfrac{2}{3}x^{3/2} + c \Rightarrow \sqrt{y} = \dfrac{x^{3/2}}{3} + c$

61) $\int \dfrac{1}{y} dy = \int e^x dx$

$\ln|y| = e^x + c \Rightarrow y = ce^{e^x}$

63) $\int \dfrac{1}{y} dy = \int 2x\,dx$

$\ln|y| = x^2 + c \Rightarrow y = ce^{x^2}$

$x = 0, y = 1 \Rightarrow c = 1$

$y = e^{x^2}$

65) $\int \dfrac{1}{y^4} dy = \int 2x\, dx$

$\dfrac{-1}{3y^3} = x^2 + c \Rightarrow y^{-3} = -3x^2 + c$

$x = 0, y = 1 \Rightarrow c = 1$

$\dfrac{1}{y^3} = -3x^2 + 1$

67) $\dfrac{dy}{dx} = \dfrac{e^x}{e^y} \Rightarrow \int e^y dy = \int e^x dx$

$e^y = e^x + c$

$x = 0, y = 0 \Rightarrow c = 0$

$e^y = e^x \Leftrightarrow y = x$

69) $\dfrac{dy}{dx} = x^3 y \Rightarrow \int \dfrac{1}{y} dy = \int x^3 dx$

$\ln|y| = \dfrac{1}{4} x^4 + c \Rightarrow y = c e^{x^4/4}$

$x = 3, y = 10 \Rightarrow 10 = c e^{81/4}$

$c = 10 e^{-81/4}$

$y = 10 e^{-81/4} e^{x^4/4} = 10 e^{(x^4 - 81)/4}$

71) $\int \dfrac{1}{y} dy = \int \dfrac{x}{1+x^2} dx \qquad \begin{array}{l} u = 1 + x^2 \\ du = 2x\, dx \end{array}$

$\ln|y| = \dfrac{1}{2} \int \dfrac{1}{u} du = \dfrac{1}{2} \ln|1 + x^2| + c$

$x = -1, y = 2 \Rightarrow \ln(2) = \dfrac{1}{2} \ln(2) + c$

$c = \ln(2) - \dfrac{1}{2} \ln(2) = \dfrac{\ln(2)}{2}$

$\ln|y| = \dfrac{1}{2} \ln(1 + x^2) + \dfrac{\ln(2)}{2}$

$= \dfrac{1}{2} \ln[2(1 + x^2)] = \ln\left[\sqrt{2(1 + x^2)}\right]$

$y = \sqrt{2(x^2 + 1)}$

Chapter 6.7 Exercises

1a) $P' = 0.05P, P(0) = 50000$

1b) $k = 0.05, c = 50000$
$P(t) = 50000e^{0.05t}$

1c) $P(10) \approx 82436$

3a) $P' = 0.12P$

3b) $P(t) = 5000e^{0.12t}$

3c) $P(3) = 7166.65$
$P(12) = 21103.5$
$P(24) = 89071.4$

5a) $N(0) = ae^{k(o)} = 39.9 = a$
$N(24) = 39.9e^{24k} = 60.4$

$e^{24k} = \frac{60.4}{39.9} \Rightarrow 24k = \ln\left(\frac{60.4}{39.9}\right)$

$k = \frac{1}{24}\ln\left(\frac{60.4}{39.9}\right) \approx 0.01728$

$N(t) = 39.9e^{0.01728t}$

5b) $\int_0^{20} 39.9e^{0.01728t}\, dt$

$= \frac{39.9}{0.01728}e^{0.01728t}\Big|_0^{20}$

$= 2309.03(1.41234 - 1)$

$= 953.253$ million board feet

5b) Continued

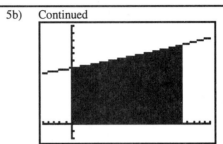

[-10,30],[-5,60]

7a) $y' = 0.034y, y(0) = 6$

7b) $y(3) = 6.6443$

$y' = \frac{d}{dt}\left[6e^{0.034t}\right] = 0.204e^{0.034t}$

$y'(3) = 0.2259$; In 1993, the percentage of wood waste generated in the United States was 6.64 and was increasing at a rate of 0.2259 percent per year

9a) $t = 14.2, y = 1500$

$1500 = 3000e^{-14.2k} \Leftrightarrow \frac{1}{2} = e^{-14.2k}$

$\ln\frac{1}{2} = -14.2k$

$k = \frac{\ln(.5)}{-14.2} \approx 0.048813$

$y = 3000e^{-0.048813t}$

9b) $2000 = 3000e^{-0.048813t}$

$\frac{2}{3} = e^{-0.048813t}$

$\ln\left(\frac{2}{3}\right) = -0.048813t$

$t = \frac{\ln(2/3)}{-0.048813} \approx 8.3065$ days

11a) 1980: $t = 0, P = 920000$
1995: $t = 15, P = 700000$
$P' = -0.01822P$
$P(0) = 920000$

11b) $920000 = ae^{-k(0)} \Rightarrow a = 920000$
$700000 = 920000e^{-k(15)}$
$\frac{35}{46} = e^{-15k} \Rightarrow \ln\left(\frac{35}{46}\right) = -15k$
$k = \frac{\ln(35/46)}{-15} \approx 0.01822$
$P(t) = 920000e^{-0.01822t}$

11c) $P(18) = 662760$

13a) $y' = -0.09y$

13b) $y(3) = 4.9925$
$y' = \frac{d}{dt}\left(6.54e^{-0.09t}\right) = -.5886e^{-0.09t}$
$y'(3) = -.4493$
In 1993 the percentage of unemployed Caucasians was 4.99, and was decreasing at a rate of about 0.45% per year

15a) $y' = (10 - y)$

15b) $f(0) = 0; L = 10$

17a) $y' = 2(8 - y)$

17b) $f(0) = 0; L = 8$

19a) $y' = 0.5(20 - y)$

19b) $f(0) = 0; L = 20$

21a) $y' = 0.1(12 - y)$

21b) $f(0) = 0; L = 12$

23a) $y' = 0.2(1.5 - y)$

23b) $f(0) = 0; L = 1.5$

25a) $y' = 0.1(15.5 - y)$

25b) $f(0) = 0; L = 15.5$

27a) $y(2) = 1353.57$
$y' = \frac{d}{dt}\left[3000(1 - e^{-0.3t})\right] = 900e^{-0.3t}$
$y'(2) = 493.93$; Two weeks after installation about 1354 employees have the system and about 494 employees are learning it per week

27b) $2500 = 3000(1 - e^{-0.3t}) \Rightarrow t = 5.97$
about 5.97 or 6 weeks

29a) $L = 300$

29b) $40 = 300(1 - e^{-k}) \Rightarrow \frac{2}{15} = 1 - e^{-k}$
$e^{-k} = \frac{13}{15} \Rightarrow -k = \ln\frac{13}{15} \Rightarrow k = \ln\frac{15}{13}$

29c) [0,26],[0,400]

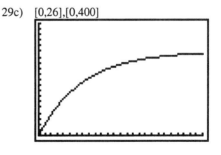

29d) No

262 Chapter 6 Integral Calculus

31a) $L = 75 \Rightarrow 15 = 75(1 - e^{-1.5k})$
$\dfrac{1}{5} = 1 - e^{-1.5k} \Rightarrow e^{-1.5k} = \dfrac{4}{5}$
$-1.5k = \ln(0.8) \Rightarrow k = 0.148762$

31b) $y = 75(1 - e^{-0.14876t})$

31c) $y' = 11.157 e^{-0.14876t}$
$y'(5) \approx 5.303;\ y'(25) \approx 0.2706$

33a) $y' = \dfrac{1}{15} y(15 - y)$

33b) $y(0) = \dfrac{15}{2};\ L = 15$

35a) $y' = \dfrac{1}{10} y(10 - y)$

35b) $y(0) = \dfrac{5}{3};\ L = 10$

37a) $y' = \dfrac{1}{30} y(90 - y)$

37b) $y(0) = 45;\ L = 90$

39a) $y' = \dfrac{1}{625}(2500 - y)$

39b) $y(0) = \dfrac{1250}{3};\ L = 2500$

41a) $y' = 0.003(4000 - y)$

41b) $y(0) = 571.429;\ L = 4000$

43a) $y(8) = 10019$; After eight completed games, 10,019 fans are expected to toss toilet paper

43b) $6000 = \dfrac{12000}{1 + 119 e^{-0.8x}} \Rightarrow x = 5.9739$
six games

45a) $L = 30000$

45b) $27300 = \dfrac{30000}{1 + 3000 e^{-6b}}$
$81900000 e^{-6b} = 27000$
$e^{-6b} = \dfrac{27000}{81900000}$
$-6b = \ln\left(\dfrac{3}{91000}\right) \Rightarrow b = 1.72$
$k = \dfrac{b}{L} = 0.000057 = 5.7 \times 10^{-5}$

47a) $y' = ky(L - y) \Rightarrow dy = ky(L - y)dt$
$\dfrac{dy}{y(L - y)} = k\,dt$

47b) $\dfrac{d}{dt}\left[\dfrac{1}{L} \ln\left(\dfrac{y}{L - y}\right)\right]$
$= \dfrac{1}{L}\left(\dfrac{1}{\frac{y}{L-y}}\right)\left(\dfrac{1(L-y) - y(-1)}{(L-y)^2}\right)$
$= \dfrac{1}{L}\left(\dfrac{L-y}{y}\right)\left(\dfrac{L}{(L-y)^2}\right) = \dfrac{1}{y(L-y)}$

47c) $\dfrac{1}{L}\ln\left(\dfrac{y}{L-y}\right) = kt + c$, solve for y

$$\dfrac{y}{L-y} = e^{kLt+cL}$$

$$y = Le^{kLt+cL} - ye^{kLt+cL}$$

$$= \dfrac{Le^{kLt+cL}}{1+e^{kLt+cL}}\left(\div \dfrac{e^{kLt+cL}}{e^{kLt+cL}}\right)$$

$$= \dfrac{L}{e^{-kLt-cL}+1} = \dfrac{L}{1+e^{-cL}e^{-kLt}}$$

$$y = \dfrac{L}{1+ae^{-kLt}}$$

47d) Differential equation: $y' = ky(L-y)$

$$y = \dfrac{L}{1+ae^{-kLt}}$$

$$y' = \dfrac{-L(-akL)e^{-kLt}}{(1+ae^{-kLt})^2}$$

$$= k\left(\dfrac{L}{1+ae^{-kLt}}\right)\left(\dfrac{Lae^{-kLt}}{1+ae^{-kLt}}\right)$$

$$= k\left(\dfrac{L}{1+ae^{-kLt}}\right)\left(L - \dfrac{L}{1+ae^{-kLt}}\right)$$

$$= ky(L-y)$$

Since $k > 0$ and $0 < y < L$, we have
$ky(L-y) = y' > 0$

47e) $\displaystyle\lim_{t\to\infty}\left(\dfrac{L}{1+ae^{-kLt}}\right) = \left(\dfrac{L}{1+0}\right) = L$

Chapter 6 Review Exercises

1) $F'(x) = 2x \neq f(x)$, no

3) $F'(t) = t^4 + t^3 + t^2 = f(x)$, yes

5) $\int x^7 dx = \dfrac{x^8}{8} + c$

7) $\int t^{-5} dt = \dfrac{-1}{4t^4} + c$

9) $\int (0.5x^7 - 6x) dx = \dfrac{1}{16}x^8 - 3x^2 + c$

11) $\int (x^2 + 0.4x - x^{5/2}) dx$
$= \dfrac{x^3}{3} + 0.2x^2 - \dfrac{2}{7}x^{7/2} + c$

13) decreasing $(-\infty, 2)$; increasing $(2, \infty)$
Concave up with a relative min at $x = 2$

15) $\int (3x+4) dx = \dfrac{3}{2}x^2 + 4x + c$
$f(0) = 11 \Rightarrow c = 11$
$f(x) = \dfrac{3}{2}x^2 + 4x + 11$

17) $\int \sqrt{x}\, dx = \dfrac{2}{3}x^{3/2} + c$
$(x, y) = (9, -5) \Rightarrow c = -23$
$y = \dfrac{2}{3}x^{3/2} - 23$

19a) $P(q) = \int (48 - 0.03q) dq$
$= 48q - 0.015q^2 + c$

19a) continued
$P(100) = 48(100) - 0.015(100)^2 + c$
$= 1600 \Rightarrow c = -3050$
$P(q) = 48q - 0.015q^2 - 3050$

19b) $48(500) - 0.015(500)^2 - 3050$
$= 17200$; $17,200 profit

21) $\Delta x = \dfrac{9 - 0}{27} = \dfrac{1}{3}$

23) $\Delta x = \dfrac{11 - 7}{32} = \dfrac{1}{8}$

25a) $f(x) = 5x, \Delta x = \dfrac{5 - 0}{5} = 1$
Leftsum: $1(0) + 1(5) + 1(10) + 1(15)$
$+ 1(20) = 50$

25b) Rightsum: $1(5) + 1(10) + 1(15) + 1(20)$
$+ 1(25) = 75$

27a) $f(x) = x^2, \Delta x = \dfrac{3 - 0}{6} = \dfrac{1}{2}$
Leftsum: $\dfrac{1}{2}(0) + \dfrac{1}{2}\left(\dfrac{1}{4}\right) + \dfrac{1}{2}(1)$
$+ \dfrac{1}{2}\left(\dfrac{9}{4}\right) + \dfrac{1}{2}(4) + \dfrac{1}{2}\left(\dfrac{25}{4}\right) = \dfrac{55}{8}$

27b) Rightsum: $\dfrac{1}{2}\left(\dfrac{1}{4}\right) + \dfrac{1}{2}(1) + \dfrac{1}{2}\left(\dfrac{9}{4}\right)$
$+ \dfrac{1}{2}(4) + \dfrac{1}{2}\left(\dfrac{25}{4}\right) + \dfrac{1}{2}(9) = \dfrac{91}{8}$

29)

	Leftsum	Rightsum	|left-right|
10	705.6	961.6	256
100	816.576	842.176	25.6
1000	828.0538	830.6138	2.56

31)

	Leftsum	Rightsum	|left-right|
10	211.8298	391.0298	179.2
100	285.5104	303.4304	17.92
1000	293.5047	295.2967	1.792

33) $\int_3^7 6x\,dx = 3x^2 \Big|_3^7$
$= 3(49) - 3(9) = 120$

35) $\int_{-2}^2 (10 - x^2)\,dx = 10x - \frac{1}{3}x^3 \Big|_{-2}^2$
$= \frac{104}{3}$

37) $\int_0^5 (2.7x^2 - 1.4x + 0.8)\,dx$
$= 0.9x^3 - 0.7x^2 + 0.8x \Big|_2^8 = 99$

39) $\int_3^9 \frac{2}{3}x\,dx = \frac{1}{3}x^2 \Big|_3^9 = 24$

41) $\int_2^8 \left(\frac{8}{x^2} + 2\right)dx = \frac{-8}{x} + 2x \Big|_2^8 = 15$

43a) $\int_0^4 \frac{x^2}{4}\,dx = \frac{1}{12}x^3 \Big|_0^4 = \frac{16}{3}$

43b) $\int_0^4 2\sqrt{x}\,dx = \frac{4}{3}x^{3/2} \Big|_0^4 = \frac{32}{3}$

45a) $[-10,10], [-10,10]$

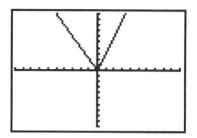

45b) $\int_{-5}^2 g(x)\,dx = \int_{-5}^0 -2x\,dx + \int_0^2 3x\,dx$
$= -x^2 \Big|_{-5}^0 + \frac{3}{2}x^2 \Big|_0^2 = 25 + 6 = 31$

47a) $[-10,10], [-10,10]$

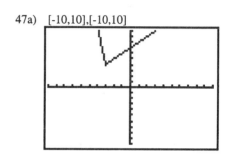

47b) $\int_{-5}^2 g(x)\,dx$
$= \int_{-5}^{-3} (x^2 - 5)\,dx + \int_{-3}^2 (x + 7)\,dx$
$= \frac{1}{3}x^3 - 5x \Big|_{-5}^{-3} + \frac{1}{2}x^2 + 7x \Big|_{-3}^2 = \frac{331}{6}$

49a) $0 = x - 3 \Rightarrow x = 3$; x-intercept: (3,0)

49b) net: $\int_0^{10} (x-3)\,dx = \frac{1}{2}x^2 - 3x \Big|_0^{10} = 20$

gross: $\left|\int_0^3 (x-3)\,dx\right| + \int_3^{10} (x-3)\,dx$
$= \frac{9}{2} + \frac{49}{2} = 29$

266 Chapter 6 Integral Calculus

51a) $(x-2)(x+1) = 0 \Rightarrow x = 2, x = -1$
x-intercepts: (-1,0), (2,0)

51b) net: $\int_{-1}^{5}(x^2 - x - 2)dx$

$= \frac{x^3}{3} - \frac{x^2}{2} - 2x \Big|_{-1}^{5} = 18$

gross: $\left|\int_{-1}^{2}(x^2 - x - 2)dx\right|$

$+ \int_{2}^{5}(x^2 - x - 2)dx = \frac{9}{2} + \frac{45}{2} = 27$

53a) $C'(160) = 20$; At a production level of 160 calculators, the cost of producing the 161st calculator is $20

53b) $\int_{0}^{200}\left(\frac{x}{20} + 12\right)dx = \frac{x^2}{40} + 12x\Big|_{0}^{200}$

$= 3400$; The total cost of producing the first 200 calculators is $3400

55a) $r(t) = 22.7739$; In 1981, U.S. government spending on defense and space-related research and development was increasing at a rate of about 22.774 billion dollars per year

55b) $\int_{5}^{11}(-0.0167t^3 + 0.448t^2 - 0.85t$

$+ 12.5)dt = -0.004175t^4$

$+ 0.14933t^3 - 0.425t^2 + 12.5t\Big|_{5}^{11}$

$= 155.779$; $155,779,000,000

57) $u = 5x - 7 \Rightarrow du = 5dx$

$\int 5(5x-7)^3 dx = \int u^3 du$

$= \frac{1}{4}u^4 + c = \frac{1}{4}(5x-7)^4 + c$

57) continued

$\frac{d}{dx}\left[\frac{1}{4}(5x-7)^4 + c\right]$

$= \frac{1}{4}(4)(5x-7)^3(5) = 5(5x-7)^3$

59) $u = x^2 - 5x + 3 \Rightarrow du = (2x-5)dx$

$\int (2x-5)(x^2 - 5x + 3)dx = \int u\,du$

$= \frac{1}{2}u^2 + c = \frac{1}{2}(x^2 - 5x + 3)^2 + c$

$\frac{d}{dx}\left[\frac{1}{2}(x^2 - 5x + 3)^2 + c\right]$

$= \frac{1}{2}(2)(x^2 - 5x + 3)(2x - 5)$

$= (2x - 5)(x^2 - 5x + 3)$

61) $u = 3x + 6$

$du = 3dx \Rightarrow \frac{1}{3}du = dx$

$\int (3x+6)^8 dx = \frac{1}{3}\int u^8 du$

$= \frac{1}{3}\left(\frac{1}{9}\right)u^9 + c = \frac{(3x+6)^9}{27} + c$

63) $u = x^2 + 4x + 6$

$du = (2x+4)dx \Rightarrow \frac{1}{2}du = (x+2)dx$

$\int \frac{x+2}{x^2+4x+6}dx = \frac{1}{2}\int \frac{1}{u}du$

$= \frac{1}{2}\ln|u| + c = \frac{1}{2}\ln|x^2 + 4x + 6| + c$

65) $u = t + 4 \Leftrightarrow t = u - 4$
$du = dt$

$\int 3t(t+4)^2 dt = \int 3(u-4)u^2 du$

$= 3\int (u^3 - 4u^2) du$

$= 3\left(\dfrac{u^4}{4} - \dfrac{4u^3}{3}\right) + c$

$= 3\left[\dfrac{(t+4)^4}{4} - \dfrac{4(t+4)^3}{3}\right] + c$

67) $u = x + 3 \Leftrightarrow x = u - 3 \Rightarrow du = dx$

$\int \dfrac{x}{\sqrt{x+3}} dx = \int \dfrac{u-3}{\sqrt{u}} du$

$= \int \left(\sqrt{u} - \dfrac{3}{\sqrt{u}}\right) du$

$= \dfrac{2}{3} u^{3/2} - 6u^{1/2} + c$

$= \dfrac{2\sqrt{(x+3)^3}}{3} - 6\sqrt{x+3} + c$

69) $u = x^2 - 64 \Rightarrow du = 2x\,dx$

$\int 2x\sqrt{x^2 - 64}\, dx = \int \sqrt{u}\, du$

$= \dfrac{2}{3} u^{3/2} + c = \dfrac{3}{2}(x^2 - 64)^{2/3} + c$

$\int_{10}^{17} 2x\sqrt{x^2 - 64}\, dx = \dfrac{3}{2}(x^2 - 64)^{2/3}\Big|_{10}^{17}$

$= 2106$

71) $u = x^3 - 4 \Rightarrow du = 3x^2 dx$

$\int \dfrac{3x^2}{\sqrt{x^3 - 4}} dx = \int \dfrac{1}{\sqrt{u}} du$

$= 2\sqrt{u} + c = 2\sqrt{x^3 - 4} + c$

71) continued

$\int_2^{11} \dfrac{3x^2}{\sqrt{x^3 - 4}} dx = 2\sqrt{x^3 - 4}\Big|_2^{11}$

$= 2(\sqrt{1327} - 2) \approx 68.856$

73) $u = x^2 - 5x + 1$ if $x = 1$ then $u = -3$
$du = (2x - 5) dx$ if $x = 4$ then $u = -3$

$\int_1^4 \dfrac{2x-5}{x^2 - 5x + 1} = \int_{-3}^{-3} \dfrac{1}{u} du = 0$

75) $u = x^3 + 17$ if $x = 4$ then $u = 81$
$du = 3x^2 dx$ if $x = 8$ then $u = 529$

$\int_4^8 3x^2 \sqrt{x^3 + 17}\, dx = \int_{81}^{529} \sqrt{u}\, du$

$= \dfrac{2}{3} u^{3/2}\Big|_{81}^{529} = \dfrac{22876}{3}$

77) $u = x^2 + 144 \Rightarrow du = 2x\,dx$

$\int \dfrac{26x}{\sqrt{x^2 + 144}} dx = 13 \int u^{-1/2} du$

$= 26\sqrt{u} + c = 26\sqrt{x^2 + 144} + c$

$\int_5^{16} \dfrac{26x}{\sqrt{x^2 + 144}} dx = 26\sqrt{x^2 + 144}\Big|_5^{16}$

$= 182$

79a) $u = t^3 - 5t \Rightarrow du = (3t^2 - 5)dt$
$x = 3 \Rightarrow u = 12, x = 6 \Rightarrow u = 186$

$\int_3^6 (t^3 - 5t)^3 (3t^2 - 5) dt = \int_{12}^{186} u^3 du$

$= \dfrac{1}{4} u^4 \Big|_{12}^{186} = 299{,}215{,}620$ meters

79b) $\int (t^3 - 5t)^3 (3t^2 - 5) dt = \int u^3 du$

$= \frac{1}{4}u^4 + c = \frac{(t^3 - 5t)^4}{4} + c$

$s(3) = \frac{(3^3 - 15)^4}{4} + c = 736$

$c = -4448$

$s(t) = \frac{1}{4}(t^3 - 5t)^4 - 4448$

81) $\int \frac{3}{4} t^{-1} dt = \frac{3}{4} \ln|t| + c$

83) $\int (5e^x - 3x) dx = 5e^x - \frac{3}{2}x^2 + c$

85) $\int_3^5 \frac{6}{x} dx = 6 \ln|x| \Big|_3^5 = 6 \ln\left(\frac{5}{3}\right)$

87) $\int_0^2 5^x dx = \int_0^2 e^{x \ln 5} dx = \frac{e^{x \ln 5}}{\ln 5} \Big|_0^2$

$= \frac{5^x}{\ln 5} \Big|_0^2 = \frac{24}{\ln 5}$

89) $u = \sqrt{x} + 4$

$du = \frac{1}{2\sqrt{x}} dx \Rightarrow 10 du = \frac{5}{\sqrt{x}} dx$

$\int \frac{5}{\sqrt{x}(\sqrt{x} + 4)} dx = 10 \int \frac{1}{u} du$

$= 10 \ln|u| + c = 10 \ln(\sqrt{x} + 4) + c$

91) $u = 5 - e^x \Rightarrow du = -e^x dx$

$\int e^x (5 - e^x)^3 dx = -\int u^3 du$

91) continued

$= \frac{-u^4}{4} + c = \frac{-(5 - e^x)^4}{4} + c$

93) $u = x^2 + 16 \Rightarrow du = 2x dx$

$\int \frac{2x}{x^2 + 16} dx = \int \frac{1}{u} du = \ln|u| + c$

$= \ln(x^2 + 16) + c$

$\int_0^3 \frac{2x}{x^2 + 16} dx = \ln(x^2 + 16) \Big|_0^3$

$= 2\ln\left(\frac{5}{4}\right)$

95) $\int_{-2}^2 2^{3x} dx = \int_{-2}^2 e^{3x \ln 2} dx = \frac{e^{3x \ln 2}}{3 \ln 2} \Big|_{-2}^2$

$= \frac{8^x}{3 \ln 2} \Big|_{-2}^2 = \frac{1365}{64 \ln 2}$

97a) $r(6) = 7659.89$; In 1980, the number of United States billed international calls was increasing at a rate of 7.65989 billion calls per year.

97b) $\int_1^{11} 63.8 e^{0.798t} dt = \frac{63.8}{0.798} e^{0.798t} \Big|_1^{11}$

$= 518688$; From 1985 to 1995, the total increase in United States billed international calls was 518.688 billion.

99) $k = 1.37, f(0) = 8.2 \Rightarrow c = 8.2$

$f(t) = 8.2 e^{1.37t}$

101) $k = -0.87, f(0) = 384 \Rightarrow c = 384$

$f(t) = 384 e^{-0.87t}$

103a) $P' = 0.08P$

103b) $P(t) = ce^{kt}$ $k = 0.08$
$P(0) = 300 \Rightarrow c = 300$
$P(t) = 300e^{0.08t}$

103c)

Number of hours, t	Number of bacteria, P
0	300
2	352.053 (352)
4	413.138 (413)
10	667.662 (668)
20	1485.91 (1486)

103d) [0,20],[0,1500]

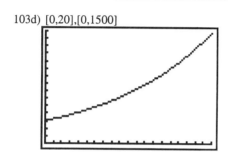

105a) $f' = -0.17f$

105b) [0,12],[0,300]

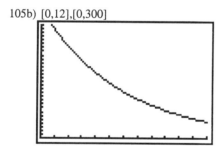

105c) $f(5) = 146.176$; In 1993, CFCs were released into the enviroment at a rate of about 146 thousand metric tons per year

105d) $f'(t) = \dfrac{d}{dt}\left[342e^{-0.17x}\right]$
$= 342(-0.17)e^{-0.17x} = -58.143^{-0.17x}$
$f'(5) = -24.849$

105d) continued
In 1993, the rate of CFCs being released was decreasing at a rate of about 24.85 thousand metric tons per year.

Chapter 7.1 Exercises

1) $\dfrac{1}{7-0}\int_0^7 (2x+5)dx = \dfrac{x^2+5x}{7}\Big|_0^7$

$= \dfrac{1}{7}[84-0] = 12$

3) $\dfrac{1}{3}\int_{-1}^2 (3x^2-2x)dx = \dfrac{1}{3}(x^3-x^2)\Big|_{-1}^2$

$= \dfrac{1}{3}[4+2] = 2$

5) $\dfrac{1}{2}\int_0^2 (x^3+2x^2-x+1)dx$

$= \dfrac{1}{2}\left(\dfrac{1}{4}x^4+\dfrac{2}{3}x^3-\dfrac{1}{2}x^2+x\right)\Big|_0^2$

$= \dfrac{1}{2}\left[\dfrac{28}{3}-0\right] = \dfrac{14}{3}$

7) $\dfrac{1}{7}\int_1^8 x^{2/3}dx = \dfrac{1}{7}\left(\dfrac{3}{5}x^{5/3}\right)\Big|_1^8$

$= \dfrac{3}{35}(32-1) = \dfrac{93}{35}$

9) $\dfrac{1}{10}\int_0^{10} 4e^{0.2x}dx = 2e^{0.2x}\Big|_0^{10}$

$= 2e^2 - 2 \approx 12.78$

11a) **RTSUM** $a=-1, b=1, n=100$

$\int_{-1}^1 \dfrac{2x+1}{x^2+1}dx \approx 1.59$

11b) $\dfrac{1}{1-(-1)}\int_{-1}^1 \dfrac{2x+1}{x^2+1}dx \approx .795$

13a) **RTSUM** $a=0, b=2, n=100$

$\int_0^2 \ln(1+x^2)dx \approx 1.45$

13b) $\dfrac{1}{2-0}(1.45) = .725$

15) $\dfrac{1}{11-1}\int_1^{11}(.13x^2-0.39x+2.79)dx$

$= \dfrac{1}{10}\left[\dfrac{0.13}{3}x^3-\dfrac{0.39}{2}x^2+2.79x\right]_1^{11}$

$= \dfrac{1}{10}[64.77-2.64] = \$6.213$ billion

17) $\dfrac{1}{10}\int_1^{11}(.11x^2-.91x+22.59)dx$

$= \dfrac{1}{10}\left(\dfrac{.11}{3}x^3-\dfrac{.91}{2}x^2+22.59x\right)\Big|_1^{11}$

$\approx \$22.007$ million

19) $\dfrac{1}{10}\int_1^{11}(-.02x^2+1.091x+3.318)dx$

$= \dfrac{1}{10}\left(-\dfrac{.02}{3}x^3+\dfrac{1.091}{2}x^2+3.318x\right)\Big|_1^{11}$

$\approx \$8.98$ billion

21a) $\dfrac{1}{3}\int_0^3 (.8x+3)dx = \dfrac{1}{3}(.4x^2+3x)\Big|_0^3$

$= \$4.2$ million

21b) $\dfrac{1}{3}\int_9^{12}(.8x+3)dx = \dfrac{1}{3}(.4x^2+3x)\Big|_9^{12}$

$= \$11.4$ million

23) $\dfrac{1}{80}\int_{120}^{200}(-.02x+8.3)dx$

$=\dfrac{8.3x-.01x^2}{80}\bigg|_{120}^{200}=\$5.1$

25) $u=2t^2+5 \qquad t=20 \Rightarrow u=805$
$du=4tdt \qquad t=0 \Rightarrow u=5$

$\dfrac{1}{20}\int_{0}^{20}\dfrac{200t}{2t^2+5}dt=\dfrac{5}{4}\int_{5}^{805}\dfrac{1}{u}du$

$=\dfrac{5}{4}\ln u\bigg|_{5}^{805}\approx 12.70\%$

27a) $\dfrac{1}{500}\int_{0}^{500}(60000+300x)dx$

$=120x+0.3x^2\bigg|_{0}^{500}=\$135000$

27b) $AC(x)=\dfrac{C(x)}{x}=\dfrac{60000}{x}+300$

$AC(500)=\$420$

27c) Part a averages the costs of producing anywhere from 0 to 500 engines, while part b is the average cost of producing each engine if 500 are produced.

29) $A(t)=300e^{.0585t}$

$\dfrac{1}{5}\int_{0}^{5}300e^{.0585t}dt=\dfrac{60e^{.0585t}}{.0585}\bigg|_{0}^{5}$

$\approx \$3484.85$

31) $A(t)=6000e^{.045t}$

$\dfrac{1}{1}\int_{0}^{1}6000e^{.045t}dt=\dfrac{6000e^{.045t}}{0.045}\bigg|_{0}^{1}$

$\approx \$6137.05$

33a) $A(t)=3500e^{.038t}$

$\dfrac{1}{1}\int_{0}^{1}3500e^{.038t}dt=\dfrac{3500e^{.038t}}{0.038}\bigg|_{0}^{1}$

$=\$3567.35$

33b) $(3567.35)(.0075)=\$26.76$

35a) $\int_{2}^{5}480e^{.06t}dt=8000e^{.06t}\bigg|_{2}^{5}$

$=\$1778.90$

35b) $A(t)=\int A'(t)=8000e^{.06t}+c$

$A(0)=8000+c=8000 \Rightarrow c=0$

$A(t)=8000e^{.06t}$

$\dfrac{1}{5}\int_{0}^{5}8000e^{.06t}dt=\dfrac{8000}{5(.06)}e^{.06t}\bigg|_{0}^{5}$

$=\$9329.57$

37a) $C(75)-C(25)=\int_{25}^{75}C'(x)dx$

$=\int\left(1+\dfrac{x}{20}\right)dx=\left(x+\dfrac{x^2}{40}\right)\bigg|_{25}^{75}\approx \$175$

37b) $C(x)=x+\dfrac{1}{40}x^2+50$

$\dfrac{1}{50}\int_{25}^{75}\left(x+\dfrac{1}{40}x^2+50\right)dx$

$=\dfrac{1}{50}\left(\dfrac{x^2}{2}+\dfrac{x^3}{120}+50x\right)\bigg|_{25}^{75}$

$\approx \$167.71$; Between the production of 25 and 75 mugs, the total average cost is $167.71

39a) [0,12],[0,600]

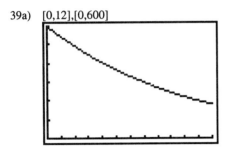

39b) $\int_0^{12} 600e^{-.1t}\,dt$

39c) $= \dfrac{600}{-.1}e^{-.1t}\Big|_0^{12} \approx 4193$; The total production in the first year is about 4,193,000 barrels

41) $\int_0^3 2\,dt = 2t\Big|_0^3 = 6$ million dollars

43) $\int_0^3 6000e^{.08t}\,dt = 75000e^{.08t}\Big|_0^3$
$\approx \$20{,}343.69$

45) $S \approx \int_0^T Pe^{r(T-t)}\,dt$

$P = 250; r = \dfrac{.065}{12} = .0054;$

$T = 25(12) = 300$

$S \approx \int_0^{300} 250e^{.0054(300-t)}\,dt$

$u = 300 - t \Rightarrow -du = dt$

$-250\int e^{.0054u}\,du = \dfrac{-250}{.0054}e^{.0054u}$

$S \approx \dfrac{-250}{.0054}e^{.0054(300-t)}\Big|_0^{300}$

$\approx \$188{,}234.72$

47) $P = 1800; r = .08; T = 43$
$u = 43 - t \Rightarrow -du = dt$

$\int 1800e^{.08(43-t)}\,dt = -1800\int e^{.08u}\,du$

$= -22500e^{.08u} = -22500e^{.08(43-t)}$

$S \approx \int_0^{43} 1800e^{.08(43-t)}\,dt$

$= -22500e^{.08(43-t)}\Big|_0^{43} \approx \$679{,}206.56$

49) $P = 2000; r = .08; T = 40$
$u = 40 - t \Rightarrow -du = dt$

$-\int 2000e^{.08u}\,du = -25000e^{.08u}$

$S \approx \int_0^{40} 2000e^{.08(40-t)}\,dt$

$= -25000e^{.08(40-t)}\Big|_0^{40} \approx \$588{,}313.25$

51a) Vicki contributed ($1800/year)(43 years)
=$77,400

51b) Jeff contributed ($2000/year)(43 years)
=$80,000

51c) Vicki's final amount surpassed Jeff's by $90,893.31. This is because Vicki's contributions had 3 years longer to earn interest.

53) $P = 75; r = .004375; T = 216$
$u = 216 - t \Rightarrow -du = dt$

$-\int 75e^{.004375u}\,du = \dfrac{-75}{.004375}e^{.004375u}$

$S \approx \int_0^{216} 75e^{.004375(216-t)}\,dt$

$= \dfrac{-75}{.004375}e^{.004375(216-t)}\Big|_0^{216}$

$\approx \$26{,}962.51$

Chapter 7.2 Exercises

1) $\int_0^1 (x+1-x^2)\,dx$

3) $\int_{-1}^{1} [x+1-(x^2-6)]\,dx$
$= \int_{-1}^{1} (x+7-x^2)\,dx$

5) $\int_a^c [g(x)-f(x)]\,dx$
$+ \int_c^b [f(x)-g(x)]\,dx$

7) $\int_0^b (d(x)-k)\,dx$

9) $\int_0^4 (3x+1-0)\,dx = \left(\dfrac{3x^2}{2}+x\right)\Big|_0^4$
$= 28$ square units

11) $\int_{-2}^{2}(2x^2+x+1-0)\,dx$
$= \left(\dfrac{2}{3}x^3+\dfrac{1}{2}x^2+x\right)\Big|_{-2}^{1} = \dfrac{15}{2}$ sq. units

13) $\int_1^2 (-x^2+x+2-0)\,dx$
$+ \int_2^4 (0-(-x^2+x+2))\,dx$
$= \left(\dfrac{-x^3}{3}+\dfrac{x^2}{2}+2x\right)\Big|_1^2$
$+ \left(\dfrac{x^3}{3}-\dfrac{x^2}{2}-2x\right)\Big|_2^4 = \dfrac{7}{6}+\dfrac{26}{3}=\dfrac{59}{6}$
square units

15) $\int_{-2}^{2}(4-x^2)\,dx = \left(4x-\dfrac{x^3}{3}\right)\Big|_{-2}^{2} = \dfrac{32}{3}$
square units

17) $\int_{-2}^{2}[(-x+4)-(-x^2+1)]\,dx$
$= \int_{-2}^{2}(x^2-x+3)\,dx$
$= \left(\dfrac{x^3}{3}-\dfrac{x^2}{2}+3x\right)\Big|_{-2}^{2} = \dfrac{52}{3}$
square units

19) $\int_{-2}^{1}[(x^2+x+1)-(x-3)]\,dx$
$= \int_{-2}^{1}(x^2+4)\,dx = \left(\dfrac{x^3}{3}+4x\right)\Big|_{-2}^{1}$
$= 15$ square units

21) $\int_1^5 [(16-\sqrt{x})-(8-\sqrt[3]{x})]\,dx$
$= \left(8x-\dfrac{2}{3}x^{3/2}+\dfrac{3}{4}x^{4/3}\right)\Big|_1^5 \approx 30.88$
square units

23) $\int_0^5 (10e^{-.08x}-5e^{-.15x})\,dx$
$= \left(-125e^{-.08x}+\dfrac{100}{3}e^{-.15x}\right)\Big|_0^5$
≈ 23.62 square units

25) $\int_2^8 \left(\dfrac{3}{x}-4x^{-2}\right)dx = \left(3\ln|x|+\dfrac{4}{x}\right)\Big|_2^8$
$= 6\ln(2)-\dfrac{3}{2} \approx 2.66$ square units

27) $\int_{-1}^{2}\left[(4-x^2)-(-x+2)\right]dx$

$=\left(2x+\frac{1}{2}x^2-\frac{1}{3}x^3\right)\Big|_{-1}^{2}=\frac{9}{2}$ sq. units

29) $\int_{-2.5}^{.5}(-4x^2-8x+5)dx$

$=\left(-\frac{4}{3}x^3-4x^2+5x\right)\Big|_{-2.5}^{.5}=18$

square units

31) $\int_{0}^{1}\left[x^3-(3x^2-2x)\right]dx$

$+\int_{1}^{2}\left[(3x^2-2x)-x^3\right]dx$

$=\left(\frac{1}{4}x^4-x^3+x^2\right)\Big|_{0}^{1}$

$+\left(x^3-x^2-\frac{1}{4}x^4\right)\Big|_{1}^{2}=\frac{1}{2}$ sq. units

33) $\int_{0}^{1}(x-x^3)dx+\int_{-1}^{0}(x^3-x)dx$

$=\left(\frac{x^2}{2}-\frac{x^4}{4}\right)\Big|_{0}^{1}+\left(\frac{x^4}{4}-\frac{x^2}{2}\right)\Big|_{-1}^{0}=\frac{1}{2}$

square units

35) $\int_{1}^{6}(2.89-.04t)dt=(2.89t-.02t^2)\Big|_{1}^{6}$

$=13.75$

$(13.75)(40)(52)=28,600$ The construction worker earned \$28,600 more than the manufacturing worker

37a) [0,17],[300,1000]

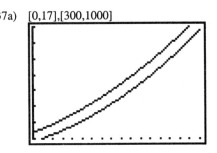

37b) $\int_{1}^{16}\left[(1.13x^2+23.51x+337.88)\right.$
$\left.-(1.28x^2+18.77x+283.09)\right]dx$
$=\int_{1}^{16}(54.79+4.74x-0.15x^2)dx$

37c) $\int_{1}^{16}(54.79+4.74x-0.15x^2)dx$

$=(54.79x+2.37x^2-0.05x^3)\Big|_{1}^{16}$

$=1221.45$ From 1980 to 1996, receipts exceeded expenditures by about 1.2 trillion dollars.

39a) $\int_{1}^{2}(4t^{18/5}-4t^{5/2})dt$

39b) $\int_{1}^{2}(4t^{18/5}-4t^{5/2})dt$

$=4\left(\frac{5}{23}t^{23/5}-\frac{2}{7}t^{7/2}\right)\Big|_{1}^{2}\approx 8.43$

Between the first and second minutes the use of the nutrient increases growth by about 8.43 mg.

39c) $\int_{0}^{2}4t^{18/5}dt-\int_{0}^{2}4t^{5/2}dt$

$=\frac{20}{23}t^{23/5}\Big|_{0}^{2}-\frac{8}{7}t^{7/2}\Big|_{0}^{2}\approx 8.16$ mg.

41) $\int_1^8 (-.07x^3 + 1.46x^2 - 6.35x + 14.38)dx$
$= -.0175x^4 + .486667x^3$
$- 3.175x^2 + 14.38x \Big|_1^8 \approx 77.66$

From 1989 to 1996, imports exceeded exports by about 77.7 billion dollars.

43) $\int_1^5 \frac{15.1}{x} dx = 15.1 \ln|x| \Big|_1^5 = 15.1 \ln 5$
≈ 24.3025; About 24,303 schools.

45a) $\int_0^{20} (-.2t^2 + 7.94t + 7.73)dt$
$= (-0.0667t^3 + 3.97t^2 + 7.73t) \Big|_0^{20}$
$= 1209.27$; If Double D cola uses the TV personality in its ads, sales are expected to increase by 1.2 million dollars, during the 1st twenty weeks.

45b) Sales increase: 1,209,270
Increase expense   1,000,000
                      209,270

Double D will still have an increase of about 2.1 thousand dollars, after paying the fee for the TV personality.

47) $\int_0^5 (12.15 e^{.04t} - .76t - 8.42)dt$
$= (303.75 e^{.04t} - 0.38t^2 - 8.42t) \Big|_0^5$
$= 15.6511$; Chocolate Time will have an increase in sales of about 15.7 million during the first five years, if it uses an ad campaign.

49a) [0,10], [0,600]

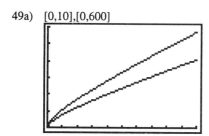

49b) $\int_4^8 (10t + 20\sqrt{t})dt = \left(5t^2 + \frac{40}{3} t^{3/2}\right) \Big|_4^8$
≈ 435.032; With the special tapes, a child can learn about 435 more new words a year between the ages of 4 and 8.

51) $\int_3^8 (2.67 e^{.09x} - .68 x^{2/3}) dx$
$= \left(\frac{89}{3} e^{.09x} - .408 x^{5/3}\right) \Big|_3^8 \approx 11.5759$

Total savings, form the third through eighth year, will be about $11.6 million.

53) $\int_{10}^{100} \left(-.2x + 40 - \frac{300}{x+1}\right) dx$

$u = x + 1 \Rightarrow du = dx; \int \frac{300}{x+1} dx$

$= 300 \int \frac{1}{u} du = 300 \ln|x+1| + c$

$\int_{10}^{100} \left(-.2x + 40 - \frac{300}{x+1}\right) dx$

$= \left(-.1x^2 + 40x - 300 \ln|x+1|\right) \Big|_{10}^{100}$

≈ 1944.83; Total profit is about $1,944.83

55) $\int_5^{40} (4 - .075x) dx = (4x - .0375x^2) \Big|_5^{40}$
$= 80.9375$; Total profit is about $80.94

57a) $\int_0^4 (-2x^2 + 11x + 13) dx$
$= \left(-\frac{2}{3} x^3 + \frac{11}{2} x^2 + 13x\right) \Big|_0^4 = \frac{292}{3}$
≈ 97 units

57b) $\int_0^4 (x^2 - 2x + 2)\,dx$

$= \left(\dfrac{1}{3}x^3 - x^2 + 2x\right)\Big|_0^4 \approx 13.3$ more units

59) $\int_3^8 \left(\dfrac{4x^2 + 16x + 9}{2x + 4} - \dfrac{2x^2 + 8x + 9}{2x + 4}\right)dx$

$= \int_3^8 \dfrac{2(x^2 + 4x)}{2(x + 2)}\,dx$

$= \int_3^8 \dfrac{(x^2 + 4x + 4) - 4}{x + 2}\,dx$

$= \int_3^8 \left(x + 2 - \dfrac{4}{x + 2}\right)dx$

$= \left(\dfrac{x^2}{2} + 2x - 4\ln|x + 2|\right)\Big|_3^8 \approx 34.727$

about 34.7 fewer feet of growth.

Chapter 7.3 Exercises

1) $\int_0^{x_m}[d(x)-d_{mp}]dx$

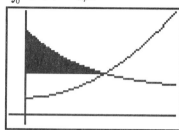

3) $(d_{mp})(x_m)$

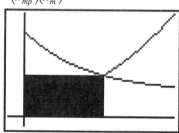

5) $\int_0^{700}[260-(.3x+50)]dx$
$=\int_0^{700}(210-.3x)dx$

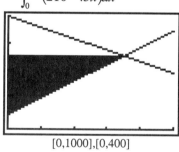

[0,1000],[0,400]

7) $s(x_m)=d(x_m)$
$.5x_m+30=-.25x_m+300$
$.75x_m=270 \Rightarrow x_m=360$
$s(x_m)=210=d_{mp}$
$\int_0^{360}(180-.5x)dx$

7) continued

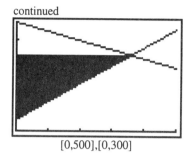

[0,500],[0,300]

9) $\int_0^{x_m}[d(x)-d_{mp}]dx$

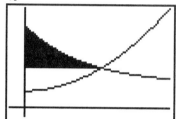

11) market price is $d(200)=2900$
consumer surplus is $\int_0^{200}(600-3x)dx$
$=\left(600x-\dfrac{3}{2}x^2\right)\Big|_0^{200}=\$60,000$

13) $d(100)=\dfrac{560}{3}$
$\int_0^{100}\left(\dfrac{100}{3}-\dfrac{1}{3}x\right)dx$
$=\left(\dfrac{100}{3}x-\dfrac{1}{6}x^2\right)\Big|_0^{100}=\$1,666.67$

15) $d(150)=510$
$\int_0^{150}(90-.6x)dx=(90x-.3x^2)\Big|_0^{150}$
$=\$6,750$

17) $d(100) = 810$

$\int_0^{100} (800 - .08x^2) dx$

$= \left(800x - \frac{2}{75} x^3 \right) \Big|_0^{100} = \$53{,}333.33$

19) $d(30) = 100$

$\int_0^{30} \left(300 - \frac{x^2}{3} \right) dx = \left(300x - \frac{x^3}{9} \right) \Big|_0^{30}$

$= \$6{,}000$

21) $d(300) = 600e^{-3}$

$\int_0^{300} (600e^{-.01x} - 600e^{-3}) dx$

$= \left(\frac{600}{-.01} e^{-.01x} - 600e^{-3} x \right) \Big|_0^{300}$

$\approx \$48{,}051.10$

23) $d(175) = 230e^{-3.5}$

$\int_0^{175} (230e^{-.02x} - 230e^{-3.5}) dx$

$= \left(\frac{230}{-.02} e^{-.02x} - 230e^{-3.5} x \right) \Big|_0^{175}$

$\approx \$9{,}937.29$

25) market price = $s(75) = 65$
producers' surplus =

$\int_0^{75} (15 - .2x) dx = (15x - .1x^2) \Big|_0^{75}$

$= \$562.50$

27) $s(100) = 95$

$\int_0^{100} (15 - .15x) dx$

$= (15x - .075x^2) \Big|_0^{100} = \$750$

29) $s(60) = 75$

$\int_0^{60} \left(20 - \frac{1}{3} x \right) dx = \left(20x - \frac{1}{6} x^2 \right) \Big|_0^{60}$

$= \$600$

31) $s(100) = 520$

$\int_0^{100} (500 - .05x^2) dx$

$= \left(500x - \frac{.05}{3} x^3 \right) \Big|_0^{100} = \$33{,}333.33$

33) $s(150) = 2260$

$\int_0^{150} (2250 - .1x^2) dx$

$= \left(2250x - \frac{1}{30} x^3 \right) \Big|_0^{150} = \$225{,}000$

35) $s(90) = 10e^{1.8}$

$\int_0^{90} (10e^{1.8} - 10e^{.02x}) dx$

$= (10e^{1.8} x - 500e^{.02x}) \Big|_0^{90} \approx \$2{,}919.86$

37) $s(110) = 13e^{1.1}$

$\int_0^{110} (13e^{1.1} - 13e^{.01x}) dx$

$= (13e^{1.1} x - 1300e^{.01x}) \Big|_0^{110} \approx \$1690.54$

39a) Equilibrium point is where $d(x) = s(x)$
$3553 - 13x = 5.70x \Rightarrow x = 190$
$d(190) = 1083$, so the E.P. is
$(190, 1083)$.

39b) $\int_0^{190}(2470-13x)dx$

$=\left(2470x-\dfrac{13}{2}x^2\right)\Big|_0^{190}=\$234{,}650$

39c) $\int_0^{190}(1083-5.7x)dx$

$=\left(1083x-2.85x^2\right)\Big|_0^{190}=\$102{,}885$

41a) $d=s\Rightarrow 398-.4x=.5x+11$
$\Rightarrow x=430$; $d(430)=226$
Equilibrium point is (430,226)

41b) $\int_0^{430}(172-.4x)dx$

$=\left(172x-.2x^2\right)\Big|_0^{430}=\$36{,}980$

41c) $\int_0^{430}(215-.5x)dx$

$=\left(215x-.25x^2\right)\Big|_0^{430}=\$46{,}225$

43a) $2743-.04x^2=.06x^2+20.5$
$\Rightarrow .1x^2=2722.5\Rightarrow x=165$
$d(165)=1654$; E.P. is (165,1654)

43b) $\int_0^{165}(1089-.04x^2)dx$

$=\left(1089x-\dfrac{.04}{3}x^3\right)\Big|_0^{165}=\$119{,}790$

43c) $\int_0^{165}(1633.5-.06x^2)dx$

$=\left(1633.5x-.02x^3\right)\Big|_0^{165}=\$179{,}685$

45a) $500-.2x^2=.2x^2+10$
$\Rightarrow .4x^2=490\Rightarrow x=35$
$d(35)=255$; (35,255)

45b) $\int_0^{35}(245-.2x^2)dx$

$=\left(245x-\dfrac{.2}{3}x^3\right)\Big|_0^{35}=\$5716.67$

45c) $\int_0^{35}(245-.2x^2)$

$=\left(245x-\dfrac{.2}{3}x^3\right)\Big|_0^{35}=\$5716.67$

47a) $600e^{-.01x}=1.125e^{.02x}$

$\dfrac{600}{1.125}=\dfrac{e^{.02x}}{e^{-.01x}}=e^{.03x}\Rightarrow x\approx 209$

$d(209)\approx 74$; (209,74)

47b) $\int_0^{209}(600e^{-.01x}-74)dx$

$=\left(\dfrac{600}{-.01}e^{-.01x}-74x\right)\Big|_0^{209}$

$=\$37{,}112.77$

47c) $\int_0^{209}(74-1.125e^{.02x})dx$

$=\left(74x-\dfrac{1.125}{.02}e^{.02x}\right)\Big|_0^{209}$

$=\$11{,}845.42$

49a) $d(x)=-1.4x+25=11\Rightarrow x=10$

49b) $\int_0^{10}(-1.4x+14)dx$

$=\left(-.7x^2+14x\right)\Big|_0^{10}=\$70$

51) $27.3e^{-0.09x} = 12.14 \Rightarrow x \approx 9$
$\int_0^9 (27.3e^{-.09x} - 12.14)dx$
$= \left(\frac{27.3}{-.09}e^{-.09x} - 12.14x\right)\Big|_0^9 \approx \$59.13$

53a) $0.24x + 3.70 = 19.30 \Rightarrow x = 65$

53b) $\int_0^{65} (15.6 - .24x)dx$
$= (15.6x - .12x^2)\Big|_0^{65} = \$507$

55) $s(x) = 6.7e^{.02x} = 19.75 \Rightarrow x \approx 54$
$\int_0^{54} (19.75 - 6.7e^{.02x})dx$
$= \left(19.75x - \frac{6.7}{.02}e^{.02x}\right)\Big|_0^{54} \approx \$415.03$

57a) $13 - .01x^2 = .1x + 1$
$\Rightarrow 0 = (x+40)(x-30) \Rightarrow x = 30$
$d(30) = 4$; E.P. is $(30,4)$

57b) $\int_0^{30} (9 - .01x^2)dx = (9x - .005x^2)\Big|_0^{30}$
$= \$180$

57c) $\int_0^{30} (3 - .1x)dx = (3x - .05x^2)\Big|_0^{30}$
$= \$45$

59a) $30 - x = \sqrt{x}$
$900 - 60x + x^2 = x$
$x = \frac{60 \pm \sqrt{(-61)^2 - 4(1)(900)}}{2(1)} = 25$
$d(25) = 5$; $(25,5)$

59b) $\int_0^{25} (25 - x)dx = \left(25x - \frac{x^2}{2}\right)\Big|_0^{25}$
$= \$312.5$

59c) $\int_0^{25} (5 - \sqrt{x})dx = \left(5x - \frac{2}{3}x^{3/2}\right)\Big|_0^{25}$
$\approx \$41.67$

61) Gini Index: $2\int_0^1 (x - (1.98x^4 - 3.31x^3$
$+ 2.61x^2 - .28x))dx = 2\Big(\frac{-1.98}{5}x^5$
$+ \frac{3.31}{4}x^4 - \frac{2.61}{3}x^3 + \frac{1.28}{2}x^2\Big)\Big|_0^1$
$= 0.4030$

63) $2\int_0^1 (x - (2.16x^4 - 3.57x^3 + 2.71x^2$
$- .3x))dx = 2\Big(\frac{-2.16}{5}x^5 + \frac{3.57}{4}x^4$
$- \frac{2.71}{3}x^3 + \frac{1.3}{2}x^2\Big)\Big|_0^1 = 0.4143$

65) $2\int_0^1 (x - (2.53x^4 - 4.22x^3 + 3.07x^2$
$- .38x))dx = \Big(\frac{-2.53}{5}x^5 + \frac{4.22}{4}x^4$
$- \frac{3.07}{3}x^3 + \frac{1.38}{2}x^2\Big)\Big|_0^1 = 0.4313$

67) The Gini index appears to be increasing over this time period, meaning inequality in income distribution became greater.

Chapter 7.4 Exercises

1) $\int 2xe^{3x}dx$

$u = 2x \qquad dv = e^{3x}dx$
$du = 2dx \qquad v = \dfrac{1}{3}e^{3x}dx$

$\int 2xe^{3x}dx = \dfrac{2}{3}xe^{3x} - \int \dfrac{2}{3}e^{3x}dx$

$= \dfrac{2}{3}xe^{3x} - \dfrac{2}{9}e^{3x} + c$

$= e^{3x}\left(\dfrac{2}{3}x - \dfrac{2}{9}\right) + c$

3) $\int xe^{4x}dx$

$u = x \qquad dv = e^{4x}dx$
$du = dx \qquad v = \dfrac{1}{4}e^{4x}$

$\int xe^{4x}dx = \dfrac{1}{4}xe^{4x} - \int \dfrac{1}{4}e^{4x}dx$

$= \dfrac{1}{4}xe^{4x} - \dfrac{1}{16}e^{4x} + c$

$= e^{4x}\left(\dfrac{1}{4}x - \dfrac{1}{16}\right) + c$

5) $\int xe^{-x}dx$

$u = x \qquad dv = e^{-x}dx$
$du = dx \qquad v = -e^{-x}$

$\int xe^{-x}dx = -xe^{-x} - \int -e^{-x}dx$

$= -xe^{-x} - e^{-x} + c = -e^{-x}(x+1) + c$

7) $\int xe^{-.03x}dx$

$u = x \qquad dv = e^{-.03x}dx$
$du = dx \qquad v = \dfrac{1}{-.03}e^{-.03x}$

$\int xe^{-.03x}dx = \dfrac{x}{-.03}e^{-.03x} - \int \dfrac{e^{-.03x}}{-.03}dx$

$= \dfrac{xe^{-.03x}}{-.03} - \dfrac{e^{-.03x}}{.0009} + c$

$= e^{-.03x}\left(\dfrac{-100}{3}x - \dfrac{10000}{9}\right) + c$

9) $\int_0^2 xe^{-x}dx$

$u = x \qquad dv = e^{-x}dx$
$du = dx \qquad v = -e^{-x}$

$\int xe^{-x}dx = -xe^{-x} - \int -e^{-x}dx$

$\int_0^2 xe^{-x} = \left(-xe^{-x} - e^{-x}\right)\Big|_0^2 \approx .594$

11) $\int_0^5 xe^{-.04x}dx$

$u = x \qquad dv = e^{-.04x}dx$
$du = dx \qquad v = -25e^{-.04x}$

$\int_0^5 xe^{-.04x}dx$

$= -25xe^{-.04x}\Big|_0^5 + 25\int_0^5 e^{-.04x}dx$

$= \left(-25e^{-.04x}(x+25)\right)\Big|_0^5 \approx 10.95$

13) $\int_1^e x\ln x\,dx$

$u = \ln x \qquad dv = xdx$
$du = x^{-1}dx \qquad v = .5x^2$

13) continued

$$\int_1^e x \ln x\, dx = .5x^2 \ln x \Big|_1^e - \int_1^e .5x\, dx$$
$$= \left(.5x^2 \ln x - .25x^2\right)\Big|_1^e \approx 2.097$$

15) $\int x^2 \ln x\, dx$

$u = \ln x \qquad dv = x^2 dx$
$du = x^{-1} dx \qquad v = \dfrac{x^3}{3}$

$$\int x^2 \ln x\, dx = \frac{1}{3}x^3 \ln x - \frac{1}{3}\int x^2 dx$$
$$= \frac{1}{3}x^3 \ln x - \frac{1}{9}x^3 + c$$
$$= \frac{1}{3}x^3\left(\ln x - \frac{1}{3}\right) + c$$

17) $\int \ln 2x\, dx$

$u = \ln 2x \qquad dv = dx$
$du = x^{-1} \qquad v = x$

$$\int \ln 2x\, dx = x \ln 2x - \int dx$$
$$= x \ln 2x - x + c = x(\ln 2x - 1) + c$$

19) $\int (x+3)e^x dx$

$u = x+3 \qquad dv = e^x dx$
$du = dx \qquad v = e^x$

$$\int (x+3)e^x dx = e^x(x+3) - \int e^x dx$$
$$= e^x(x+3) - e^x + c = e^x(x+2) + c$$

21) $\int 6te^{-.1t} dt$

$u = 6t \qquad dv = e^{-.1t} dt$
$du = 6 dt \qquad v = -10e^{-.1t}$

21) continued

$$\int 6te^{-.1t} dt = -60te^{-.1t} + 60\int e^{-.1t} dt$$
$$= -60te^{-.1t} - 600e^{-.1t} + c$$
$$= -60e^{-.1t}(t+10) + c$$

23) $\int_0^5 5te^{-.06t} dt$

$u = 5t \qquad dv = e^{-.06t} dt$
$du = 5 dt \qquad v = \dfrac{e^{-.06t}}{-.06}$

$$\int 5te^{-.06t} dt = \frac{-5}{.06}te^{-.06t} + \frac{5}{.06}\int e^{-.06t} dt$$
$$\int_0^5 5te^{-.06t} dt = \left(\frac{-5}{.06}te^{-.06t} - \frac{5}{.0036}e^{-.06t}\right)\Big|_0^5$$
$$\approx 51.300$$

25) $\int 1.2te^{-.08t} dt$

$u = 1.2t \qquad dv = e^{-.08t} dt$
$du = 1.2 dt \qquad v = -12.5e^{-.08t}$

$$\int 1.2te^{-.08t} dt = -15te^{-.08t} + 15\int e^{-.08t} dt$$
$$= -15te^{-.08t} - 187.5e^{-.08t} + c$$
$$= -15e^{-.08t}(t+12.5) + c$$

27) $\int (x+2)(x+1)^5 dx$

$u = x+2 \qquad dv = (x+1)^5 dx$
$du = dx \qquad v = \dfrac{(x+1)^6}{6}$

$$= \frac{(x+2)(x+1)^6}{6} - \frac{1}{6}\int (x+1)^6 dx$$
$$= \frac{1}{6}(x+2)(x+1)^6 - \frac{1}{42}(x+1)^7 + c$$

27) continued

$$= (x+1)^6\left(\frac{6x+13}{42}\right) + c$$

29) $\int (x-2)(x+3)^6 dx$

$u = x-2$ $dv = (x+3)^6$
$du = dx$ $v = \dfrac{(x+3)^7}{7}$

$$= \frac{(x-2)(x+3)^7}{7} - \frac{1}{7}\int (x+3)^7 dx$$

$$= \frac{1}{7}(x-2)(x+3)^7 - \frac{1}{56}(x+3)^8 + c$$

$$= (x+3)^7\left(\frac{7x-19}{56}\right) + c$$

31) $\int x^2 e^{3x} dx$

$u = x^2$ $dv = e^{3x} dx$
$du = 2x dx$ $v = \dfrac{1}{3}e^{3x} dx$

$$= \frac{x^2}{3}e^{3x} - \frac{2}{3}\int xe^{3x} dx$$

$u = x$ $dv = e^{3x} dx$
$du = dx$ $v = \dfrac{1}{3}e^{3x} dx$

$$= \frac{x^2}{3}e^{3x} - \frac{2}{3}\left[\frac{x}{3}e^{3x} - \frac{1}{3}\int e^{3x} dx\right]$$

$$= \frac{x^2}{3}e^{3x} - \frac{2x}{9}e^{3x} + \frac{2}{27}e^{3x} + c$$

$$= \frac{e^{3x}}{27}(9x^2 - 6x + 2) + c$$

33) $\int x^2 e^{-x} dx$

$u = x^2$ $dv = e^{-x} dx$
$du = 2x dx$ $v = -e^{-x}$

$$= -x^2 e^{-x} + 2\int xe^{-x} dx$$

$u = x$ $dv = e^{-x} dx$
$du = dx$ $v = -e^{-x}$

$$= -x^2 e^{-x} + 2\left[-xe^{-x} + \int e^{-x} dx\right]$$

$$= -x^2 e^{-x} - 2xe^{-x} - 2e^{-x} + c$$

$$= -e^{-x}(x^2 + 2x + 2) + c$$

35) $\int x^2 e^{5x} dx$

$u = x^2$ $dv = e^{5x} dx$
$du = 2x dx$ $v = .2e^{5x}$

$$= .2x^2 e^{5x} - .4\int xe^{5x} dx$$

$u = x$ $dv = e^{5x} dx$
$du = dx$ $v = .2e^{5x}$

$$= .2x^2 e^{5x} - .4\left[.2xe^{5x} - .2\int e^{5x} dx\right]$$

$$= .2x^2 e^{5x} - .08xe^{5x} + .016e^{5x} + c$$

$$= \frac{e^{5x}}{125}(25x^2 - 10x + 2) + c$$

37) $\int (\ln x)^2 dx$

$u = (\ln x)^2$ $dv = dx$
$du = 2x^{-1} \ln x dx$ $v = x$

$$= x(\ln x)^2 - 2\int \ln x dx$$

$u = \ln x$ $dv = dx$
$du = x^{-1} dx$ $v = x$

37) continued
$$= x(\ln x)^2 - 2\left[x\ln x - \int dx\right]$$
$$= x(\ln x)^2 - 2x\ln x + 2x + c$$

39) $\int x(\ln x)^3 dx$

$u = (\ln x)^3 \qquad dv = xdx$
$du = 3x^{-1}(\ln x)^2 dx \qquad v = .5x^2$
$= .5x^2(\ln x)^3 - 1.5\int x(\ln x)^2 dx$

$u = (\ln x)^2 \qquad dv = xdx$
$du = 2x^{-1}\ln x dx \qquad v = .5x^2$
$= .5x^2(\ln x)^3$
$\quad -1.5\left[.5x^2(\ln x)^2 - \int x\ln xdx\right]$

$u = \ln x \qquad dv = xdx$
$du = x^{-1}dx \qquad v = .5x^2$
$= .5x^2(\ln x)^3 - .75x^2(\ln x)^2$
$\quad + 1.5\left(.5x^2\ln x - .5\int xdx\right)$
$= .5x^2(\ln x)^3 - .75x^2(\ln x)^2$
$\quad + .75x^2\ln x - .375x^2 + c$
$= \dfrac{x^2}{8}\left(4(\ln x)^3 - 6(\ln x)^2 + 6\ln x - 3\right) + c$

41) by parts: $u = x \qquad dv = e^{-x}dx$

43) u-sub: $u = \ln x$

45) by parts: $u = x + 6 \qquad dv = e^x dx$

47) by parts: $u = 3x \qquad dv = e^{2x}dx$

49a) $\int_0^{12} 6te^{-.15t} dt$

49b) $\int_0^{12} 6te^{-.15t} dt$

$u = 6t \qquad dv = e^{-.15t}dt$
$du = 6dt \qquad v = \dfrac{-1}{.15}e^{-.15t}$

$= \left(-40te^{-.15t}\right)\Big|_0^{12} + 40\int_0^{12} e^{-.15t} dt$

$= \left(-40te^{-.15t} - \dfrac{40}{.15}e^{-.15t}\right)\Big|_0^{12}$

≈ 143.243 thousand barrels.

51) $P(t) = \int P'(t)dt = \int 2te^{.2t} dt$

$u = 2t \qquad dv = e^{.2t}dt$
$du = 2dt \qquad v = 5e^{.2t}$

$= 10te^{.2t} - 10\int e^{.2t} dt$
$= 10te^{.2t} - 50e^{.2t} + c$
$P(0) = -50 + c = 0 \Rightarrow c = 50$
$P(t) = 10e^{.2t}(t - 5) + 50$

53) $C(t) = \int C'(t)dt = \int te^{-.15t} dt$

$u = t \qquad dv = e^{-.15t}dt$
$du = dt \qquad v = \dfrac{e^{-.15t}}{-.15}$

$= \dfrac{te^{-.15t}}{-.15} + \dfrac{1}{.15}\int e^{-.15t} dt$

$= \dfrac{-20}{3}te^{-.15t} - \dfrac{400}{9}e^{-.15t} + c$

$C(0) = \dfrac{-400}{9} + c = 2 \Rightarrow c = \dfrac{418}{9}$

$C(t) = -\dfrac{20}{9}e^{-.15t}(3t + 20) + \dfrac{418}{9}$

55) $\int_0^6 4te^{-.51t}\,dt$

$u = 4t \qquad dv = e^{-.51t}\,dt$
$du = 4dt \qquad v = \dfrac{-1}{.51}e^{-.51t}$

$= \dfrac{-4}{.51}te^{-.51t} + \dfrac{4}{.51}\int e^{-.51t}\,dt$

$= \left(e^{-.51t}\left(\dfrac{-4}{.51}t - \dfrac{4}{.2601}\right)\right)\Big|_0^6$

≈ 12.451 mg

57) Given the continuous income stream function $f(t) = 2t$, we can use the formula $P = \int_0^T f(t)e^{-rt}\,dt$ to determine the present value of the continuous income stream.

$P = \int_0^5 2te^{-.06t}\,dt$

$u = 2t \qquad dv = e^{-.06t}\,dt$
$du = 2dt \qquad v = \dfrac{-1}{.06}e^{-.06t}$

$= \dfrac{-1}{.03}te^{-.06t}\Big|_0^5 + \dfrac{1}{.03}\int_0^5 e^{-.06t}\,dt$

$= e^{-.06t}\left(\dfrac{-t}{.03} - \dfrac{1}{.0018}\right)\Big|_0^5 \approx 20.52$

$\approx \$20{,}520{,}000$

59) $P = \int_0^4 (200e^{-.1t})e^{-.0575t}\,dt$

$= 200\int_0^4 e^{-.1575t}\,dt$

$= \left(\dfrac{-200}{.1575}e^{-.1575t}\right)\Big|_0^4 \approx 593.534$

Present value $\approx \$593{,}534$

61) Oil well: $\int_0^5 50te^{-.08t}\,dt$

$u = 50t \qquad dv = e^{-.08t}\,dt$
$du = 50dt \qquad v = -12.5e^{-.08t}$

$= -625te^{-.08t}\Big|_0^5 + 625\int_0^5 e^{-.08t}\,dt$

$= -625e^{-.08t}(t + 12.5)\Big|_0^5 \approx 480.874$

Gas well: $\int_0^5 (30 + 40t)e^{-.08t}\,dt$

$u = 30 + 40t \qquad dv = e^{-.08t}\,dt$
$du = 40dt \qquad v = -12.5e^{-.08t}$

$= -125(3 + 4t)e^{-.08t}\Big|_0^5 + 500\int_0^5 e^{-.08t}\,dt$

$= -125e^{-.08t}(4t + 53)\Big|_0^5 \approx 508.330$

Since the present value of the oil well is $480,874 and the present value of the gas well is $508,330, the natural gas well is the better investment.

63a) $\dfrac{1}{11-1}\int_1^{11}(2.18 + 1.1\ln x)\,dx$

$= .218x\Big|_1^{11} + .11\int_1^{11} \ln x\,dx$

$u = \ln x \qquad dv = dx$
$du = x^{-1}dx \qquad v = x$

$= 2.18 + .11\left(x\ln x - \int dx\right)\Big|_1^{11}$

$= 2.18 + .11(x\ln x - x)\Big|_1^{11} \approx 3.98$

On average, 3.98 million men enrolled in college annually in the U.S. in the 1960s.

63b) $\dfrac{1}{21-11}\int_{11}^{21}(2.18 + 1.1\ln x)\,dx$

$= .218x\Big|_{11}^{21} + .11(x\ln x - x)\Big|_{11}^{21} \approx 5.21$

On average, 5.21 million men enrolled in college annually in the U.S. in the 1970s.

63c) $\dfrac{1}{31-21}\int_{21}^{31}(2.18+1.1\ln x)dx$

$=.218x\big|_{21}^{31}+.11(x\ln x-x)\big|_{21}^{31}\approx 5.76$

On average, 5.76 million men enrolled in college annually in the U.S. in the 1980s.

65) $.2\int_{1}^{6}(35.8+1.96\ln x)dx$

$=7.16x\big|_{1}^{6}+.392\int_{1}^{6}\ln x\,dx$

$u=\ln x \qquad dv=dx$
$du=x^{-1}dx \qquad v=x$

$=35.8+.392\left(x\ln x\big|_{1}^{6}-\int_{1}^{6}dx\right)$

$=35.8+.392(x\ln x-x)\big|_{1}^{6}\approx 38.05$

Between 1991 and 1996, paper and cardboard averaged 38.05% of all solid waste in the U.S.

67) $\dfrac{1}{6}\int_{1}^{7}(2096+264\ln x)dx$

$=\dfrac{1048}{3}x\bigg|_{1}^{7}+44\int_{1}^{7}\ln x\,dx$

$u=\ln x \qquad dv=dx$
$du=x^{-1}dx \qquad v=x$

$=2096+44\left(x\ln x\big|_{1}^{7}-\int_{1}^{7}dx\right)$

$=2096+44(x\ln x-x)\big|_{1}^{7}\approx 2431.34$

United States imports of crude oil averaged about 2431 million barrels annually between January 1, 1990 and December 31, 1996.

69a) $.1\int_{1}^{11}(17.13+3.4\ln x)dx$

$=1.713x\big|_{1}^{11}+.34\int_{1}^{11}\ln x\,dx$

69a) continued

$u=\ln x \qquad dv=dx$
$du=x^{-1}dx \qquad v=x$

$=17.13+.34\left(x\ln x\big|_{1}^{11}-\int_{1}^{11}dx\right)$

$=17.13+.34(x\ln x-x)\big|_{1}^{11}\approx 22.7$

During the 1980s, annual per capita cheese consumption averaged 22.7 pounds.

69b) $.2\int_{11}^{16}(17.13+3.4\ln x)dx$

$=3.426x\big|_{11}^{16}+.68(x\ln x-x)\big|_{11}^{16}\approx 26$

During the first half of the 1990s, annual per capita cheese consumption averaged 26 pounds.

71) $.2\int_{1}^{6}(8.3+.88\ln x)dx$

$=1.66x\big|_{1}^{6}+.176\int_{1}^{6}\ln x\,dx$

$u=\ln x \qquad dv=dx$
$du=x^{-1}dx \qquad v=x$

$=8.3+.176\left(x\ln x-\int dx\right)\bigg|_{1}^{6}$

$=8.3+.176(x\ln x-x)\big|_{1}^{6}\approx 9.3$

During the first half of the 1990s, variable costs of operating an automobile averaged 9.3 cents per mile.

Chapter 7.5 Exercises

1) $\Delta x = \dfrac{3-(-1)}{4} = 1$

$\int_{-1}^{3} x^3 \, dx = \dfrac{1}{2}[-1 + 0 + 2 + 16 + 27]$

$= 22$

3) $\Delta x = \dfrac{1}{2}; \int_{1}^{3} \dfrac{5x}{5+2x^2} \, dx$

$= \dfrac{1}{4}\left[\dfrac{5}{7} + 1.579 + 1.538 + 1.429 + .652\right]$

$= 1.48$

5) $\Delta x = \dfrac{1}{4}; \int_{0}^{1} x^2 e^x \, dx$

$= \dfrac{1}{8}[0 + .161 + .824 + 2.38 + e]$

$= .76$

7) $\Delta x = \dfrac{1}{2}; \int_{3}^{5} (\ln x)^2 \, dx$

$= \dfrac{1}{4}[1.207 + 3.139 + 3.844 + 4.524$

$+ 2.590] = 3.83$

9) $a = 0, b = 2$

n trapezoids	Trap Approximation
10	6.50656
50	6.404266496
100	6.401066656

11) $a = 0, b = 2$

n trapezoids	Trap Approximation
10	.4128406644
50	.4005283873
100	.4001322241

13) $a = 1, b = 4$

n trapezoids	Trap Approximation
10	.0690011825
50	.0671679267
100	.0671104334

15) $a = 1 b = 2$

n trapezoids	Trap Approximation
10	2.453701923
50	2.451740936
100	2.451679645

17) $a = 0, b = 1$

n trapezoids	Trap Approximation
10	.4091397464
50	.396143122
100	.3957354568

19) $a = 2, b = 5$

n trapezoids	Trap Approximation
10	6.178488696
50	6.172491939
100	6.172304409

21) $a = 1, b = 3$

n trapezoids	Trap Approximation
10	4.051132552
50	4.049970734
100	4.049934358

23) Revenue for the 5 weeks after April 1st...

$\Delta x = \dfrac{5-1}{4} = 1$

$\dfrac{1}{2}[2.35 + 7.5 + 2.5 + 5 + 1.75]$

$= 9.55$ thousand dollars

25) Total cost in going from production level of 10 mugs to 50 mugs ...

288 Chapter 7 Applications of Integral Calculus

$\Delta x = \dfrac{50-10}{4} = 10$

$5[1.5 + 4 + 4.8 + 6.2 + 3.5] = \$100$

27) Total profit during the first five months of the year ... $\Delta x = \dfrac{5-1}{4} = 1$

$\dfrac{1}{2}[.62 + 1.54 + 1.56 + 1.7 + 1.03]$

$= \$3,225$

29) Revenue for the ten weeks after May 1st...

$\Delta x = \dfrac{10-1}{9} = 1$

$\dfrac{1}{2}[2.25 + 7.5 + 2.7 + 5.1 + 3.7 + 4.24$
$+ 6.22 + 4.22 + 3 + 3.12] = \$21,025$

31) Total cost of producing 20 items...

$\Delta x = \dfrac{20-0}{10} = 2$

$[7 + 22 + 26 + 32 + 36 + 42 + 48$
$+ 56 + 62 + 70 + 39] = \$44,000$

33a) Using the TI-83 program for the trapezoidal rule, enter the points for x into L_1, the points for single into L_2, start the program and enter 8 for the number of data points.

$= \dfrac{1}{36-1}[362222.5] \approx 10349.214$, so the average number of single females was about 10,349,214

33b) Using the TI-83 program for the trapezoidal rule, enter the points for x into L_1, the points for married into L_2, start the program and enter 8 for the number of data points.

33b) continued

$= \dfrac{1}{36-1}[808445] \approx 23098.429$, so the average number of married females was about 23,098,429.

35a) Using the TI-83 program for the trapezoidal rule, enter the points for x into L_1, the points for volatile organic compounds into L_2, start the program and enter 6 for the number of data points. Total amount emitted $\approx 1,223,140$ thousand tons.

35b) Average amount yearly =

$\dfrac{1}{51-1}[1,223,140] \approx 24,463$ thousand tons.

37a) Enter x into L_1, Stoppages into L_2, and start the program with 8 data points. Total work stoppages ≈ 6478.

37b) Average number of work stoppages annually $= \dfrac{1}{36-1}(6478) \approx 185$

39a) Enter x into L_1, males into L_2, and start the program with 9 data points. Total number of males who graduated $\approx 41,200,000$.

39b) Average number of males who graduated annually

$= \dfrac{1}{33-1}(41,200,000) \approx 1,287,500$.

41) Enter x into L_1, Cost into L_2, and start the program with 5 data points. Total cost in 1987 dollars spent on regulations and monitoring $\approx \$12.677$ billion.

43a) Enter x into L_1, Number into L_2, and start the program with 9 data points. Total number of pregnancies ≈ 96.5 million.

43b) Average number of pregnancies annually

$$\approx \frac{1}{17-1}[96.538] \approx 6.03 \text{ million.}$$

45) Enter x into L_1, 2-year degree into L_2, and start the program with 6 data points. The average number of first-year registering nursing students in a 2-year associate degree program

$$\approx \frac{1}{11-1}[666.9] \approx 66.7 \text{ thousand.}$$

Chapter 7.6 Exercises

1) $\lim_{b\to\infty}\int_0^b x^{-4}dx = \lim_{b\to\infty}\left[\frac{-1}{3}x^{-3}\Big|_2^b\right]$

$= \lim_{b\to\infty}\left[\frac{-1}{3b^3}+\frac{1}{24}\right] = \frac{1}{24}$

3) $\lim_{b\to\infty}\left[\int_3^b x^{-1/2}dx\right] = \lim_{b\to\infty}\left[2x^{1/2}\Big|_3^b\right]$

$= \lim_{b\to\infty}\left[2\sqrt{b}-2\sqrt{3}\right] = \infty$; diverges

5) $\lim_{b\to\infty}\left[\int_2^b x^{-1.5}dx\right] = \lim_{b\to\infty}\left[-2x^{-1/2}\Big|_2^b\right]$

$= \lim_{b\to\infty}\left[\frac{-2}{\sqrt{b}}+\frac{2}{\sqrt{2}}\right] = \frac{2}{\sqrt{2}} = \sqrt{2}$

7) $\lim_{b\to\infty}\left[\int_1^b x^{-2/3}dx\right] = \lim_{b\to\infty}\left[3x^{1/3}\Big|_1^b\right]$

$= \lim_{b\to\infty}\left[3b^{1/3}-3\right] = \infty$; diverges

9) $\lim_{b\to\infty}\left[\int_1^b e^{-x}dx\right] = \lim_{b\to\infty}\left[-e^{-x}\Big|_1^b\right]$

$= \lim_{b\to\infty}\left[-e^{-b}+e^{-1}\right] = \frac{1}{e}$

11) $\lim_{b\to\infty}\left[\int_1^b e^{-.03x}dx\right] = \lim_{b\to\infty}\left[\frac{-1}{.03}e^{-.03x}\Big|_1^b\right]$

$= \frac{-1}{.03}\lim_{b\to\infty}\left[e^{-.03b}-e^{-.03}\right] = \frac{e^{-.03}}{.03}$

≈ 32.35

13) $\lim_{b\to\infty}\left[\int_1^b \frac{\ln x}{x}dx\right]$

$u = \ln x \Rightarrow du = \frac{1}{x}dx$

13) Continued

$\int u\,du = \frac{1}{2}u^2+c$

$\lim_{b\to\infty}\left[\int_1^b \frac{\ln x}{x}dx\right] = \lim_{b\to\infty}\left[\frac{1}{2}(\ln x)^2\Big|_1^b\right]$

$= \frac{1}{2}\lim_{b\to\infty}\left[(\ln b)^2-0^2\right] = \infty$; diverges

15) $u = x^2+3 \Rightarrow du = 2x\,dx$

$\int \frac{2x}{x^2+3}dx = \int \frac{1}{u}du = \ln u + c$

$\lim_{b\to\infty}\left[\int_1^b \frac{2x}{x^2+3}dx\right] = \lim_{b\to\infty}\left[\ln(x^2+3)\Big|_1^b\right]$

$= \lim_{b\to\infty}\left[\ln(b^2+3)-\ln 4\right] = \infty$; diverges

17) $\lim_{a\to-\infty}\left[\int_a^0 e^{2x}dx\right] = \lim_{a\to-\infty}\left[\frac{1}{2}e^{2x}\Big|_a^0\right]$

$= \frac{1}{2}\lim_{a\to-\infty}\left[1-e^{2a}\right] = \frac{1}{2}$

19) $\lim_{a\to-\infty}\left[\int_a^0 e^{-x}dx\right] = \lim_{a\to-\infty}\left[-e^{-x}\Big|_a^0\right]$

$= \lim_{a\to-\infty}\left[-1+e^{-a}\right] = \infty$; diverges

21) $\lim_{a\to-\infty}\left[\int_a^0 e^{.2x}dx\right] = \lim_{a\to-\infty}\left[5e^{.2x}\Big|_a^0\right]$

$= 5\lim_{a\to-\infty}\left[1-e^{.2a}\right] = 5$

23) $\lim_{a\to-\infty}\left[\int_a^{-4} x^{-2}dx\right] = \lim_{a\to-\infty}\left[-x^{-1}\Big|_a^{-4}\right]$

$= \lim_{a\to-\infty}\left[\frac{1}{4}+\frac{1}{a}\right] = \frac{1}{4}$

25) $u = x^4 - 1 \Rightarrow \frac{1}{4} du = x^3 dx$

$\int \frac{x^3}{(x^4-1)^2} dx = \frac{1}{4} \int u^{-2} du = -\frac{1}{4} u^{-1} + c$

$\lim_{a \to -\infty} \left[\int_a^{-2} \frac{x^3}{(x^4-1)^2} dx \right]$

$= \lim_{a \to -\infty} \left[-\frac{1}{4}(x^4-1)^{-1} \Big|_a^{-2} \right]$

$= \frac{-1}{4} \lim_{a \to -\infty} \left[\frac{1}{15} - \frac{1}{(a^4-1)} \right] = \frac{-1}{60}$

27) $u = -x^2 \Rightarrow du = -2x dx$

$\int 2xe^{-x^2} dx = \int e^{-u} du = -e^{-u} + c$

$\lim_{b \to \infty} \left[\int_0^b 2xe^{-x^2} dx \right] = \lim_{b \to \infty} \left[-e^{-x^2} \Big|_0^b \right]$

$= \lim_{b \to \infty} \left[-e^{-b^2} + 1 \right] = 1$

29) $u = x^4 + 3 \Rightarrow du = 4x^3 dx$

$\int \frac{4x^3}{(x^4+3)^2} dx = \int u^{-2} du = -u^{-1} + c$

$\int_{-\infty}^{\infty} \frac{4x^3}{(x^4+3)^2} dx$

$= \int_{-\infty}^0 \frac{4x^3}{(x^4+3)^2} dx + \int_0^{\infty} \frac{4x^3}{(x^4+3)^2} dx$

$= \lim_{a \to -\infty} \left[\int_a^0 \frac{4x^3}{(x^4+3)^2} dx \right]$

$+ \lim_{b \to \infty} \left[\int_0^b \frac{4x^3}{(x^4+3)^2} dx \right]$

29) continued

$= \lim_{a \to -\infty} \left[-(x^4+3)^{-1} \Big|_a^0 \right]$

$+ \lim_{b \to \infty} \left[-(x^4+3)^{-1} \Big|_0^b \right]$

$= \lim_{a \to -\infty} \left[\frac{-1}{3} + \frac{1}{a^4+3} \right]$

$+ \lim_{b \to \infty} \left[\frac{-1}{b^4+3} + \frac{1}{3} \right] = 0$

31) $u = -x^2 \Rightarrow du = -2x dx$

$\int 2xe^{-x^2} dx = \int -e^u du = -e^{-x^2} + c$

$\int_{-\infty}^{\infty} 2xe^{-x^2} dx$

$= \int_{-\infty}^0 2xe^{-x^2} dx + \int_0^{\infty} 2xe^{-x^2} dx$

$= \lim_{a \to -\infty} \left[-e^{-x^2} \Big|_a^0 \right] + \lim_{b \to \infty} \left[-e^{-x^2} \Big|_0^b \right]$

$= \lim_{a \to -\infty} \left[-1 + e^{-a^2} \right] + \lim_{b \to \infty} \left[-e^{-b^2} + 1 \right]$

$= 1 - 1 = 0$

33) $\int_{-\infty}^{\infty} e^{-x} dx = \int_{-\infty}^0 e^{-x} dx + \int_0^{\infty} e^{-x} dx$

$= \lim_{a \to -\infty} \left[-e^{-x} \Big|_a^0 \right] + \lim_{b \to \infty} \left[-e^{-x} \Big|_0^b \right]$

$= \lim_{a \to -\infty} \left[-1 + e^{-a} \right] + \lim_{b \to \infty} \left[-e^{-b} + 1 \right]$

$= \infty$; diverges

35) $u = x^2 + 1 \Rightarrow du = 2x dx$

$\int \frac{2x}{x^2+1} dx = \int \frac{1}{u} du = \ln|u| + c$

$\int_{-\infty}^{\infty} \frac{2x}{x^2+1} dx$

$= \int_{-\infty}^0 \frac{2x}{x^2+1} dx + \int_0^{\infty} \frac{2x}{x^2+1} dx$

35) continued

$$= \lim_{a \to -\infty}\left[\ln(x^2+1)\Big|_a^0\right] + \lim_{b \to \infty}\left[\ln(x^2+1)\Big|_0^b\right]$$

$$= \lim_{a \to -\infty}\left[-\ln(a^2+1)\right] + \lim_{b \to \infty}\left[\ln(b^2+1)\right]$$

$$= -\infty + \infty \text{ ; diverges}$$

37) $\int_0^\infty \left(60e^{-.05t} - 60e^{-.1t}\right)dt$

$$= \lim_{b \to \infty}\left[\int_0^b \left(60e^{-.05t} - 60e^{-.1t}\right)dt\right]$$

$$= \lim_{b \to \infty}\left[\left(-1200e^{-.05t} + 600e^{-.1t}\right)\Big|_0^b\right]$$

$$= \lim_{b \to \infty}\left[-1200e^{-.05b} + 600e^{-.1b} + 600\right]$$

$= 600$ thousand barrels

39) $\int_0^\infty te^{-.25t}dt = \lim_{b \to \infty}\left[\int_0^b te^{-.25t}dt\right]$

$$u = t \qquad dv = e^{-.25t}dt$$
$$du = dt \qquad v = -4e^{-.25t}$$

$$\int te^{-.25t}dt = -4te^{-.25t} + 4\int e^{-.25t}dt$$

$$= -4te^{-.25t} - 16e^{-.25t} + c$$

$$\lim_{b \to \infty}\left[\int_0^b te^{-.25t}dt\right]$$

$$= \lim_{b \to \infty}\left[\left(-4te^{-.25t} - 16e^{-.25t}\right)\Big|_0^b\right]$$

$$= \lim_{b \to \infty}\left[-4be^{-.25b} - 16e^{-.25b} + 16\right]$$

$$= -4\lim_{b \to \infty}\left[be^{-.25b}\right] - 0 + 16$$

Since we don't know how to evaluate the remaining limit, we use a table to approximate it's value:

b:	1	10	100
$be^{-.25b}$	0.7788	0.8208	1.389×10^{-9}

From this it appears that the limit = 0 (this can be verified using L'Hopital's rule, but that is beyond the scope of this book), so we use this value to get an answer of
$= 16$ million cubic feet

41) $\int_0^\infty 350e^{-.3t}dt = \lim_{b \to \infty}\left[\int_0^b 350e^{-.3t}dt\right]$

$$= \lim_{b \to \infty}\left[\frac{350}{-.3}e^{-.3t}\Big|_0^b\right]$$

$$= \lim_{b \to \infty}\left[\frac{350}{-.3}e^{-.3b} + \frac{350}{.3}\right]$$

$$= \frac{3500}{3} \approx 1166.67 \text{ tons}$$

43) $\int_0^\infty \left(5e^{-.02t} - 5e^{-.035t}\right)dt$

$$= \lim_{b \to \infty}\left[\int_0^b \left(5e^{-.02t} - 5e^{-.035t}\right)dt\right]$$

$$= \lim_{b \to \infty}\left[\left(-250e^{-.02t} + \frac{5}{.035}e^{-.035t}\right)\Big|_0^b\right]$$

$$= \lim_{b \to \infty}\left[-250e^{-.02b} + \frac{1000}{7}e^{-.035b} + \frac{750}{7}\right]$$

$$= \frac{750}{7} \approx 107 \text{ ml}$$

45) $\int_0^\infty 10000e^{-.06t}dt$

$$= \lim_{b \to \infty}\left[\int_0^b 10000e^{-.06t}dt\right]$$

$$= \lim_{b \to \infty}\left[\frac{10000}{-.06}e^{-.06t}\Big|_0^b\right]$$

$$= \lim_{b \to \infty}\left[\frac{10000}{-.06}e^{-.06b} + \frac{10000}{.06}\right]$$

$$= \frac{10000}{.06} \approx \$166,667$$

47) $\int_0^\infty 30000 e^{-.0725t}\, dt$

$= \lim_{b\to\infty}\left[\int_0^b 30000 e^{-.0725t}\, dt\right]$

$= \lim_{b\to\infty}\left[\dfrac{30000}{-.0725} e^{-.0725t}\Big|_0^b\right]$

$= \lim_{b\to\infty}\left[\dfrac{30000}{-.0725} e^{-.0725b} + \dfrac{30000}{.0725}\right]$

$= \dfrac{30000}{.0725} \approx \$413{,}793$

49) $\int_0^\infty 70000 e^{-.065t}\, dt$

$= \lim_{b\to\infty}\left[\int_0^b 70000 e^{-.065t}\, dt\right]$

$= \lim_{b\to\infty}\left[\dfrac{70000}{-.065} e^{-.065t}\Big|_0^b\right]$

$= \lim_{b\to\infty}\left[\dfrac{70000}{-.065} e^{-.065b} + \dfrac{70000}{.065}\right]$

$= \dfrac{70000}{.065} \approx \$1{,}076{,}923$

51) $\int_0^\infty 12000 e^{-.055t}\, dt$

$= \lim_{b\to\infty}\left[\int_0^b 12000 e^{-.055t}\, dt\right]$

$= \lim_{b\to\infty}\left[\dfrac{12000}{-.055} e^{-.055t}\Big|_0^b\right]$

$= \lim_{b\to\infty}\left[\dfrac{12000}{-.055} e^{-.055b} + \dfrac{12000}{.055}\right]$

$= \dfrac{12000}{.055} \approx \$218{,}182$

53) $\int_0^\infty 8000 e^{-.075t}\, dt$

$= \lim_{b\to\infty}\left[\int_0^b 8000 e^{-.075t}\, dt\right]$

$= \lim_{b\to\infty}\left[\dfrac{8000}{-.075} e^{-.075t}\Big|_0^b\right]$

$= \lim_{b\to\infty}\left[\dfrac{8000}{-.075} e^{-.075b} + \dfrac{8000}{.075}\right]$

$= \dfrac{8000}{.075} \approx \$106{,}667$

55) $\int_0^\infty 3000 e^{-.066t}\, dt$

$= \lim_{b\to\infty}\left[\int_0^b 3000 e^{-.066t}\, dt\right]$

$= \lim_{b\to\infty}\left[\dfrac{3000}{-.066} e^{-.066t}\Big|_0^b\right]$

$= \lim_{b\to\infty}\left[\dfrac{3000}{-.066} e^{-.066b} + \dfrac{3000}{.066}\right]$

$= \dfrac{3000}{.066} \approx \$45{,}454$

57a) $\int_0^{75} 5000 e^{-.0625t}\, dt$

$= \left(-80000 e^{-.0625t}\right)\Big|_0^{75}$

$= -736.775 + 80000 = \$79{,}263$

57b) $\int_0^\infty 5000 e^{-.0625t}\, dt$

$= \lim_{b\to\infty}\left[\int_0^b 5000 e^{-.0625t}\, dt\right]$

$= \lim_{b\to\infty}\left[\dfrac{5000}{-.0625} e^{-.0625t}\Big|_0^b\right]$

$= \lim_{b\to\infty}\left[\dfrac{5000}{-.0625} e^{-.0625b} + \dfrac{5000}{.0625}\right]$

57b) continued

$$= \frac{5000}{.0625} = \$80{,}000$$

59a) $\int_{-\infty}^{\infty} f(x)dx = \int_{-\infty}^{0} 0 + \int_{0}^{\infty} .3e^{-.3x} dx$

$= \lim_{b\to\infty}\left[\int_{0}^{b} .3e^{-.3x} dx\right] = \lim_{b\to\infty}\left[-e^{-.3x}\Big|_{0}^{b}\right]$

$= \lim_{b\to\infty}\left[-e^{-.3b} + 1\right] = 1$

59b) $\int_{3}^{4} .3e^{-.3x} dx = -e^{-.3x}\Big|_{3}^{4} \approx 0.1054$

59c) $\lim_{b\to\infty}\left[\int_{3}^{b} .3e^{-.3x} dx\right] = \lim_{b\to\infty}\left[-e^{-.3x}\Big|_{3}^{b}\right]$

$= \lim_{b\to\infty}\left[-e^{-.3b} + e^{-.9}\right] = e^{-.9} \approx 0.4066$

61a) $\int_{-\infty}^{\infty} f(x)dx = \int_{-\infty}^{0} 0 + \int_{0}^{4} \frac{1}{8}x\,dx + \int_{4}^{\infty} 0$

$= 0 + \frac{x^2}{16}\Big|_{0}^{4} + 0 = 1$

61b) $\int_{1}^{2} \frac{1}{8}x\,dx = \frac{x^2}{16}\Big|_{1}^{2} = \frac{3}{16} = .1875$

61c) $\int_{.5}^{1} \frac{1}{8}x\,dx = \frac{x^2}{16}\Big|_{.5}^{1} = \frac{3}{64} = .046875$

63a) $\int_{-\infty}^{\infty} f(x)dx = \int_{-\infty}^{0} 0 + \int_{0}^{\infty} .1e^{-.1x} dx$

$= \lim_{b\to\infty}\left[\int_{0}^{b} .3e^{-.3x} dx\right] = \lim_{b\to\infty}\left[-e^{-.3x}\Big|_{0}^{b}\right]$

$= \lim_{b\to\infty}\left[-e^{-.3b} + 1\right] = 1$

63b) $\int_{5}^{10} .1e^{-.1x} dx = -e^{-.1x}\Big|_{5}^{10} \approx .2387$

63c) $\int_{1}^{5} .1e^{-.1x} dx = -e^{-.1x}\Big|_{1}^{5} \approx .2983$

63d) $\lim_{b\to\infty}\left[\int_{8}^{b} .1e^{-.1x} dx\right] = \lim_{b\to\infty}\left[-e^{-.1x}\Big|_{8}^{b}\right]$

$= \lim_{b\to\infty}\left[-e^{-.1b} + e^{-.8}\right] = e^{-.8} \approx .4493$

65a) $\int_{-\infty}^{\infty} f(x)dx = \int_{-\infty}^{0} 0 + \int_{0}^{\infty} .03e^{-.03x} dx$

$= \lim_{b\to\infty}\left[\int_{0}^{b} .03e^{-.03x} dx\right] = \lim_{b\to\infty}\left[-e^{-.03x}\Big|_{0}^{b}\right]$

$= \lim_{b\to\infty}\left[-e^{-.03b} + 1\right] = 1$

65b) $\int_{0}^{12} .03e^{-.03x} dx = -e^{-.03x}\Big|_{0}^{12} \approx .3023$

65c) $\int_{12}^{24} .03e^{-.03x} dx = -e^{-.03x}\Big|_{12}^{24} \approx .2109$

67a) $\int_{-\infty}^{\infty} f(x)dx = \int_{-\infty}^{0} 0 + \int_{0}^{\infty} .3e^{-.3x} dx$

$= \lim_{b\to\infty}\left[\int_{0}^{b} .025e^{-.025x} dx\right] = \lim_{b\to\infty}\left[-e^{-.025x}\Big|_{0}^{b}\right]$

$= \lim_{b\to\infty}\left[-e^{-.025b} + 1\right] = 1$

67b) $\int_{3}^{6} .025e^{-.025x} dx = -e^{-.025x}\Big|_{3}^{6} \approx .0670$

67c) $\int_{7}^{10} .025e^{-.025x} dx = -e^{-.025x}\Big|_{7}^{10} \approx .0607$

67d) $\lim_{b\to\infty}\left[\int_{10}^{b} .025e^{-.025x} dx\right] = \lim_{b\to\infty}\left[-e^{-.025x}\Big|_{10}^{b}\right]$

$= \lim_{b\to\infty}\left[-e^{-.025b} + e^{-.25}\right] = e^{-.25} \approx .7788$

Chapter 7 Review Exercises

1) $\dfrac{1}{6-0}\int_0^6 (3x+4)dx = \dfrac{1}{6}\left(\dfrac{3}{2}x^2+4x\right)\Big|_0^6$
   $= 13$

3) $\dfrac{1}{5-0}\int_0^5 (4x^3-3x^2+2)dx$
   $= \dfrac{1}{5}(x^4-x^3+2x)\Big|_0^5 = 102$

5a) RTSUM ≈ 11.8624

5b) $\dfrac{1}{8-0}[11.8624] \approx 1.4828$

7a) $\dfrac{1}{400}\int_0^{400}(1300+18x)dx$
   $= \dfrac{1}{400}(1300x+9x^2)\Big|_0^{400} = \$4900$

7b) $AC(x) = \dfrac{C(x)}{x} = \dfrac{1300}{x}+18$
   $AC(400) = \dfrac{13}{4}+18 = \$21.25$

7c) Part a is the average cost of producing a total of 0 to 400 phones, while part b is the average cost of producing each phone if 400 are produced.

9a) $\dfrac{1}{1-0}\int_0^1 2400e^{.047t}dt = \dfrac{2400}{.047}e^{.047t}\Big|_0^1$
   $= \dfrac{2400}{.047}(e^{.047}-1) \approx \$2457.29$

9b) $(.0085)(2457.29) \approx \$20.89$

11) $\int_0^5 4000e^{.09t}dt = \dfrac{4000}{.09}e^{.09t}\Big|_0^5$
   $\approx \$25,258$

13) $T=192, r=.004, P=65$
   $\int_0^{192} 65e^{.004(192-t)}dt$
   $= \int_0^{192} 65e^{.768-.004t}dt$
   $= 65e^{.768}\int_0^{192} e^{-.004t}dt$
   $= 65e^{.768}\dfrac{e^{-.004t}}{-.004}\Big|_0^{192} \approx \$18,776$

15) $\int ((2-x^2)-(x-2))dx$
   $= \int (4-x^2-x)dx$; To find limits of integration, we have to find the intersection points:
   $2-x^2 = x-2$
   $x^2+x-4 = 0$
   $x = \dfrac{-1\pm\sqrt{1^2-4(1)(-4)}}{2(1)} = \dfrac{-1\pm\sqrt{17}}{2}$
   $\int_{\frac{-1-\sqrt{17}}{2}}^{\frac{-1+\sqrt{17}}{2}} (4-x^2-x)dx$

17) $\int_{-3}^3 (9-x^2)dx = \left(9x-\dfrac{1}{3}x^3\right)\Big|_{-3}^3 = 36$

19) $\int_{-2}^3 ((x+2)-(x^2-4))dx$
   $= \int_{-2}^3 (-x^2+x+6)dx$

296 Chapter 7 Applications of Integral Calculus

19) continued

$$= \left(\frac{-x^3}{3} + \frac{x^2}{2} + 6x\right)\Big|_{-2}^{3} = \frac{125}{6}$$

21) $\int_{-2}^{6}\left((4x+6)-(x^2-6)\right)dx$

$$= \int_{-2}^{6}(12+4x-x^2)dx$$

$$= \left(12x+2x^2-\frac{x^3}{3}\right)\Big|_{-2}^{6} = \frac{256}{3}$$

23) $\int_{-2}^{5}(5e^x - 3.7^x)dx$

$$= \int_{-2}^{5}(5e^x - e^{x\ln 3.7})dx$$

$$= \left(5e^x - \frac{e^{x\ln 3.7}}{\ln 3.7}\right)\Big|_{-2}^{5} \approx 211.43$$

25) $x^2 = x+6 \Rightarrow (x+2)(x-3)=0$

$$\int_{-2}^{3}(-x^2+x+6)dx$$

$$= \left(\frac{-x^3}{3}+\frac{x^2}{2}+6x\right)\Big|_{-2}^{3} = \frac{125}{6}$$

27) $x^3 = 2x^2+8x \Rightarrow x(x+2)(x-4)=0$

$$\int_{-2}^{0}(x^3-2x^2-8x)dx$$

$$+\int_{0}^{4}(2x^2+8x-x^3)dx$$

$$= \left(\frac{x^4}{4}-\frac{2}{3}x^3-4x^2\right)\Big|_{-2}^{0}$$

$$+\left(\frac{2}{3}x^3+4x^2-\frac{x^4}{4}\right)\Big|_{0}^{4} = \frac{20}{3}+\frac{128}{3}$$

$$= \frac{148}{3}$$

29) $\int_{0}^{15}(164e^{.09t} - 1.7t^2 + 16.1t - 160)dt$

$$= \left(\frac{164}{.09}e^{.09t} - \frac{1.7}{3}t^3 + \frac{16.1}{2}t^2 - 160t\right)\Big|_{0}^{15}$$

≈ 2706; If Lucky is used for the ad campaign, total sales in the first 15 months will increase by 2706 tricycles.

31) $\int_{4}^{8}(215e^{.14x} - 80x^{3/4})dx$

$$= \left(\frac{215}{.14}e^{.14x} - \frac{320}{7}x^{7/4}\right)\Big|_{4}^{8}$$

$\approx \$765,766$

33a) Fred: $\int_{0}^{4}(-5x^2 + 23x + 31)dx$

$$= \left(\frac{-5}{3}x^3 + \frac{23}{2}x^2 + 31x\right)\Big|_{0}^{4} \approx 201$$

33b) Ginger: $\int_{0}^{4}(-4x^2 + 25x + 32)dx$

$$= \left(\frac{-4}{3}x^3 + \frac{25}{2}x^2 + 32x\right)\Big|_{0}^{4} \approx 242$$

So Ginger assembles about 41 more units than Fred during the first four hours.

35) [0,100],[0,50]

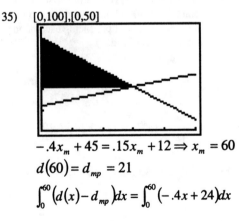

$-.4x_m + 45 = .15x_m + 12 \Rightarrow x_m = 60$

$d(60) = d_{mp} = 21$

$\int_{0}^{60}(d(x) - d_{mp})dx = \int_{0}^{60}(-.4x + 24)dx$

37) [0,100],[0,50]

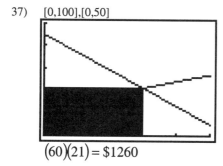

$(60)(21) = \$1260$

39) $d(100) = 370$

$\int_0^{100} (80 - .8x)dx = (80x - .4x^2)\Big|_0^{100}$
$= \$4000$

41) $d(36) = 94$

$\int_0^{36} (6 - x^{1/2})dx = \left(6x - \frac{2}{3}x^{3/2}\right)\Big|_0^{36}$
$= \$72$

43) $s(400) = 1400$

$\int_0^{400} (200 - .5x)dx = (200x - .25x^2)\Big|_0^{400}$
$= \$40,000$

45) $s(50) = 75$

$\int_0^{50} (50 - .02x^2)dx$
$= \left(50x - \frac{1}{150}x^3\right)\Big|_0^{50} = \frac{5000}{3} \approx \$1667$

47a) $250 - .3x = 60 + .2x \Rightarrow x = 380$
$d(380) = 136$; The equilibrium point is $(380, 136)$

47b) $\int_0^{380} (114 - .3x)dx$

$= \left(114x - \frac{.3}{2}x^2\right)\Big|_0^{380} = \$21,660$

47c) $\int_0^{380} (76 - .2x)dx = (76x - .1x^2)\Big|_0^{380}$
$= \$14,440$

49a) $4500 - .04x^2 = .03x^2 + 1700$
$\Rightarrow x^2 = 40000 \Rightarrow x = 200$
$d(200) = 2900$; $(200, 2900)$

49b) $\int_0^{200} (1600 - .04x^2)dx$

$= \left(1600x - \frac{.04}{3}x^3\right)\Big|_0^{200} = \frac{640000}{3}$
$\approx \$213,333$

49c) $\int_0^{200} (1200 - .03x^2)dx$

$= (1200x - .01x^3)\Big|_0^{200} = \$160,000$

51a) $14 = 22 - 2\sqrt{x} \Rightarrow \sqrt{x} = 4 \Rightarrow x = 16$

51b) $\int_0^{16} (8 - 2\sqrt{x})dx = \left(8x - \frac{4}{3}x^{3/2}\right)\Big|_0^{16}$
$= \frac{128}{3} \approx \$42.67$

53) $47 = 18e^{.007x} \Rightarrow .007x = \ln\frac{47}{18}$

$\Rightarrow x \approx 137; \int_0^{137} (47 - 18e^{.007x})dx$

$= \left(47x - \frac{18}{.007}e^{.007x}\right)\Big|_0^{137} \approx \$2301$

55a) $12.48 - .005x^2 = \sqrt{x}$

graph $0 = \sqrt{x} + .005x^2 - 12.48$ in [-30,40],[-30,30] and use the calculator's zero function to find the x-intercept of x=36. $d(36) = 6$, so the equilibrium point is (36,6)

55b) $\int_0^{36} (6.48 - .005x^2) dx$

$= \left(6.48x - \frac{.005}{3} x^3 \right) \Big|_0^{36} = \$155.52$

55c) $\int_0^{36} (6 - \sqrt{x}) dx = \left(6x - \frac{2}{3} x^{3/2} \right) \Big|_0^{36}$

$= \$72$

57) $\int 4xe^{2x} dx$

$u = 4x \qquad dv = e^{2x} dx$
$du = 4dx \qquad v = .5e^{2x}$

$= 2xe^{2x} - 2\int e^{2x} dx$

$= 2xe^{2x} - e^{2x} + c = e^{2x}(2x - 1) + c$

59) $\int x^4 \ln x\, dx$

$u = \ln x \qquad dv = x^4 dx$
$du = x^{-1} dx \qquad v = .2x^5$

$= .2x^5 \ln x - .2\int x^4 dx$

$= .2x^5 \ln x - .04x^5 + c$

$= \frac{x^5}{25} (5\ln x - 1) + c$

61) $\int 8te^{-2t} dt$

$u = 8t \qquad dv = e^{-2t} dt$
$du = 8dt \qquad v = -.5e^{-2t}$

$= -40te^{-2t} + 40\int e^{-2t} dt$

$= -40te^{-2t} - 200e^{-2t} + c$

$= -40e^{-2t}(t + 5) + c$

63) $\int x^2 e^{-x} dx$

$u = x^2 \qquad dv = e^{-x} dx$
$du = 2x\, dx \qquad v = -e^{-x}$

$= -x^2 e^{-x} + \int 2xe^{-x} dx$

$u = 2x \qquad dv = e^{-x} dx$
$du = 2dx \qquad v = -e^{-x}$

$= -x^2 e^{-x} - 2xe^{-x} + \int 2e^{-x} dx$

$= -x^2 e^{-x} - 2xe^{-x} - 2e^{-x} + c$

$= -e^{-x}(x^2 + 2x + 2) + c$

65) $\int (\ln x)^4 dx$

$u = (\ln x)^4 \qquad dv = dx$
$du = 4(\ln x)^3 x^{-1} dx \qquad v = x$

$= x(\ln x)^4 - \int 4(\ln x)^3 dx$

$u = (\ln x)^3 \qquad dv = 4dx$
$du = 3(\ln x)^2 x^{-1} dx \qquad v = 4x$

$= x(\ln x)^4 - 4x(\ln x)^3 + \int 12(\ln x)^2 dx$

$u = (\ln x)^2 \qquad dv = 12dx$
$du = 2(\ln x)x^{-1} dx \qquad v = 12x$

65) Continued
$$= x(\ln x)^4 - 4x(\ln x)^3 + 12x(\ln x)^2$$
$$- \int 24 \ln x\, dx$$

$u = \ln x \qquad dv = 24dx$
$du = x^{-1}dx \qquad v = 24x$

$$= x(\ln x)^4 - 4x(\ln x)^3 + 12x(\ln x)^2$$
$$- 24x\ln x + \int 24 dx$$

$$= x(\ln x)^4 - 4x(\ln x)^3 + 12x(\ln x)^2$$
$$- 24x\ln x + 24x + c$$

67) u-sub, $u = -x^3 \Rightarrow \dfrac{-1}{3}du = x^2 dx$

69) By parts, $u = x \qquad dv = e^{5x}dx$

71a) $\int_0^{12} 8te^{-.17t}\, dt$

71b) $\int_0^{12} 8te^{-.17t}\, dt$

$u = 8t \qquad dv = e^{-.17t}$
$du = 8dt \qquad v = \dfrac{-1}{.17}e^{-.17t}$

$$= \dfrac{-8}{.17}te^{-.17t}\Big|_0^{12} + \dfrac{8}{.17}\int_0^{12} e^{-.17t}\, dt$$

$$= \left(\dfrac{-8}{.17}te^{-.17t} - \dfrac{8}{.0289}e^{-.17t}\right)\Big|_0^{12}$$

≈ 167.395 thousand barrels

73) $\int_0^4 90te^{-.07t}\, dt$

$u = 90t \qquad dv = e^{-.07t}\, dt$
$du = 90dt \qquad v = \dfrac{-1}{.07}e^{-.07t}$

73) continued
$$= \dfrac{-90}{.07}te^{-.07t}\Big|_0^4 + \dfrac{90}{.07}\int_0^4 e^{-.07t}\, dt$$

$$= \left(\dfrac{-90}{.07}te^{-.07t} - \dfrac{90}{.0014}e^{-.07t}\right)\Big|_0^4$$

$\approx \$598.717$ thousand

75a) $\dfrac{1}{6-1}\int_1^6 (5.63x + 39.25)\, dx$

$$= \dfrac{1}{5}\left(\dfrac{5.63}{2}x^2 + 36.25x\right)\Big|_1^6$$

$= \$58{,}955$; The average sales price for a home in the U.S. between October 1, 1976 and October 1, 1981 was $\$58{,}955$.

75b) $\dfrac{1}{12-8}\left(\dfrac{5.63}{2}x^2 + 36.25x\right)\Big|_8^{12}$

$= \$95{,}550$; The average sales price for a home in the U.S. between October 1, 1983 and October 1, 1987 was $\$95{,}550$.

75c) $\dfrac{1}{21-16}\left(\dfrac{5.63}{2}x^2 + 36.25x\right)\Big|_{16}^{21}$

$= \$143{,}405$; The average sales price for a home in the U.S. between October 1, 1991 and October 1, 1996 was $\$143{,}405$.

77) $\Delta x = \dfrac{1-(-3)}{4} = 1$

$$\int_{-3}^1 x^5\, dx \approx \dfrac{1}{2}[-243 - 64 - 2 + 0 + 1]$$
$$= -154$$

300 Chapter 7 Applications of Integral Calculus

79) $\int_0^1 (x^2 - 2x)dx$

$\approx \frac{1}{8}[0 - .875 - 1.5 - 1.875 - 1]$

$\approx -.65625$

81) n = 10; ≈ 74.79658962
n = 50; ≈ 74.7981479
n = 100; ≈ 74.7981974

83) n = 10; ≈ 1.679481689
n = 50; ≈ 1.677478259
n = 100; ≈ 1.677420313

85) $\frac{1}{2}[2047 + 6248 + 4774 + 3874$
$+ 4366 + 2706] = \$12,007.50$

87) Put x in L_1 and $C'(x)$ in L_2 and start the program with n = 12. Total cost of producing 24 items = \$153.5 thousand or \$153,500

89a) Enter the data into L_1 and L_2 and run the program with n = 5. Total number of businesses that incorporated from 1985 to 1995 = 6.830 million

89b) $\frac{1}{11-1}(6,830,000) = 683,000$

91) $\lim_{b \to \infty}\left[\int_5^b x^{-2}dx\right] = \lim_{b \to \infty}\left[-x^{-1}\Big|_5^b\right]$

$= \lim_{b \to \infty}\left[\frac{-1}{b} + \frac{1}{5}\right] = \frac{1}{5}$

93) $\lim_{b \to \infty}\left[\int_{10}^b \frac{3}{x}dx\right] = \lim_{b \to \infty}\left[3\ln x\Big|_{10}^b\right]$

$= \lim_{b \to \infty}[3\ln b - 3\ln 10] = \infty$ diverges

95) $\lim_{b \to \infty}\left[\int_0^b e^{.1x}dx\right] = \lim_{b \to \infty}\left[10e^{.1x}\Big|_0^b\right]$

$= \lim_{b \to \infty}[10e^{.1b} - 10] = \infty$; diverges

97) $\lim_{a \to -\infty}\left[\int_a^0 e^{3x}dx\right] = \lim_{a \to -\infty}\left[\frac{e^{3x}}{3}\Big|_a^0\right]$

$= \lim_{a \to -\infty}\left[\frac{1}{3} - \frac{1}{3}e^{3a}\right] = \frac{1}{3}$

99) $\int_{-\infty}^\infty e^{.05x}dx = \int_{-\infty}^0 e^{.05x}dx + \int_0^\infty e^{.05x}dx$

$= \lim_{a \to -\infty}\left[20e^{.05x}\Big|_a^0\right] + \lim_{b \to \infty}\left[20e^{.05x}\Big|_0^b\right]$

$= 20\lim_{a \to -\infty}[1 - e^{.05a}] + 20\lim_{b \to \infty}[e^{.05b} - 1]$

$= 20 - 0 + \infty - 20 = \infty$ diverges

101) $\int_0^\infty (43e^{-.03t} - 43e^{-.08t})dt$

$= 43\lim_{b \to \infty}\left[\int_0^b (e^{-.03t} - e^{-.08t})dt\right]$

$= 43\lim_{b \to \infty}\left[\left(\frac{e^{-.03t}}{-.03} + \frac{e^{-.08t}}{.08}\right)\Big|_0^b\right]$

$= 43\lim_{b \to \infty}\left[\frac{e^{-.03b}}{-.03} + \frac{e^{-.08b}}{.08} + \frac{1}{.03} - \frac{1}{.08}\right]$

$= \frac{5375}{6} \approx 895.833$ thousand barrels

103) $\int_0^\infty 8000e^{-.09t}\,dt$

$= \lim_{b\to\infty}\left[\int_0^b 8000e^{-.09t}\,dt\right]$

$= \lim_{b\to\infty}\left[\dfrac{8000}{-.09}e^{-.09t}\bigg|_0^b\right]$

$= \lim_{b\to\infty}\left[\dfrac{8000}{-.09}e^{-.09b} + \dfrac{8000}{.09}\right]$

$= \dfrac{800000}{9} \approx \$88,889$

105a) $\int_{-\infty}^{\infty} f(x)\,dx$

$= \int_{-\infty}^{0} 0\,dx + \int_{0}^{21} \dfrac{2}{441}x\,dx + \int_{21}^{\infty} 0\,dx$

$= \dfrac{x^2}{441}\bigg|_0^{21} = 1$

105b) $\dfrac{x^2}{441}\bigg|_7^{14} = \dfrac{1}{3} \approx 0.333$

105c) $\dfrac{x^2}{441}\bigg|_{14}^{21} = \dfrac{5}{9} \approx 0.556$

302 Chapter 8 Calculus of Several Variables

Chapter 8.1 Exercises

1) $(2)^2 - 2(2)(3) + 3(3)^2 = 19$

3) $(-2)^2 - 2(-2)(-1) + 3(-1)^2 = 3$

5) $\dfrac{(3)^2 + (1)^2}{(3) + (1)} = \dfrac{5}{2}$

7) $\dfrac{(-2)^2 + (-3)^2}{(-2) + (-3)} = \dfrac{13}{-5}$

9) $\dfrac{3(1) - (4) + 1}{(1)^2 - (4)^2} = 0$

11) $\dfrac{3(-2) - (2) + 1}{(-2)^2 - (2)^2}$; undefined

13) $(2) + (1)\ln(1) + (1)e^{(2)} = 2 + e^2$

15) $(0) + (e)\ln(e) + (e)e^{(0)} = 2e$

17) $domf(x, y) = \{(x, y) | x, y \in R\}$

19) $domf(x, y) = \{(x, y) | x \neq -y\}$

21) $domf(x, y) = \{(x, y) | x \neq 4y\}$

23)
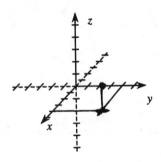

25)
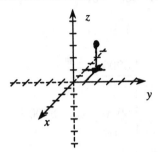

27)
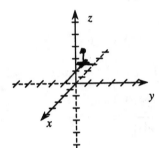

29)
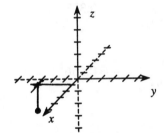

31)
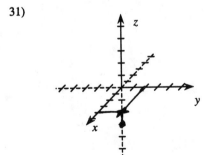

33)

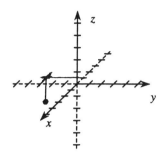

35)

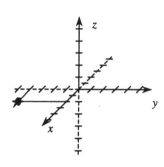

37) The point (2,2,3) is above the xy-plane, because the z value is positive.

39) The point (-3,-2,5) is above the xy-plane, because the z value is positive.

41) The point (4,-2,-1) is below the xy-plane, the z value is negative.

43) The point (0,3,-0.5) is below the xy-plane, the z value is negative.

45)

47)

49)

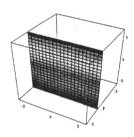

51a) $S(5,3) = 2(5)^2 + 3 = 53$

When Seamount spends $5,000 on newspaper ads and $3,000 on radio advertising each week, weekly sales are $530,000.

51b) $S(3,5) = 2(3)^2 + 5 = 23$

When Seamount spends $3,000 on newspaper ads and $5,000 on radio advertising each week, weekly sales are $230,000.

53a) $TS(1,0.5) = 1.5(1)^2 + 3.2(0.5)^3 = 1.9$

When Tube Town spends $1,000 on radio advertising and $500 on TV advertising each week, weekly sales are $19,000.

53b) $TS(0.5,1) = 1.5(0.5)^2 + 3.2(1)^3 = 3.575$

When Tube Town spends $500 on radio advertising and $1,000 on TV advertising each week, weekly sales are $35,750.

55) $R(36,1) = K \cdot \dfrac{36}{1^4} = 36K$

$R(36,2) = K \cdot \dfrac{36}{2^4} = \dfrac{9}{4}$

57) $\left(\dfrac{30-4}{4}\right)^2 \cdot 12 = 507$; A tree with a 30-inch diameter and 12 foot length yields 507 board-feet.

59) $1000 + 35(50) + 1.5(30) = 2795$; The monthly cost of producing 50 leaf blowers per month and 30 attachments per month is $2795.

61a) $P(x,y) = R(x,y) - C(x,y)$
$= 85x - 0.8x^2 - 0.25xy + 50.8y$
$+ 0.015y^2 - 1000$

61b) $85(50) - 0.8(50)^2 - 0.25(50)(30)$
$+ 50.8(30) + 0.015(30)^2 - 1000$
$= 2412.5$; The monthly profit from sales of 50 leaf blowers and 30 attachments per month is $2412.50.

63a) $C(x,y) = 5x + 4y + 50$

63b) $R(x,y) = 12x + 9.5y$

63c) $P(x,y) = R(x,y) - C(x,y)$
$= 7x + 5.5y - 50$

65) $2000\left(1 + \dfrac{0.06125}{12}\right)^{12(20)} \approx 6787.15$

Two thousand dollars invested at an annual interest rate of 6.125%, compounded monthly for 20 years, yields $6787.15.

67) $2000e^{0.0725(25)} \approx 12{,}251.49$; Two thousand dollars invested at an annual interest rate of 7.25%, compounded continuously for 25 years, yields $12,251.49.

69) $42(300)^{0.37}(100)^{0.66} \approx 7241$; When 300 units of labor and 100 units of capital are used, 7241 units of sport shoe are produced.

71a) $\dfrac{100(12)}{9} \approx 133.33$; A 9-year old person with a 12-year old mental age has an IQ of 133.33.

71b) $\dfrac{100(9)}{12} = 75$; A 12-year old person with a 9-year old mental age has an IQ of 75.

71c) $\dfrac{100(12)}{12} = 100$; A 12-year old person with a 12-year old mental age has an IQ of 100.

71d) If a=m, then $\dfrac{100m}{a} = 100$. So, an IQ of 100 represents a person with mental age equal to actual age.

73a) [0,200],[0,1000]

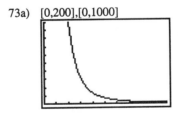

73b) [0,200],[0,1000]

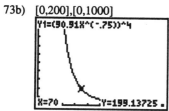

$x = 70; y = 199$. 199 units of capital are needed if 70 units of labor are used.

73c) [0,200],[0,1000]

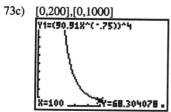

$x = 100; y = 68$. 68 units of capital are needed if 100 units of labor are used.

73d) Any point on the graph will work. Use the TRACE command to get some values. Three of the many possible answers are (119,40), (81,129), (40,1067).

Chapter 8.2 Exercises

1) $y = -x - 1 + c$
   $[-5,5], [-5,5]$

3) $y = \dfrac{-2x - 6 + c}{3}$
   $[-5,5], [-5,5]$

5) $y = \pm\sqrt{16 - x^2 - c^2}$
   $[-5,5], [-5,5]$

7) $y = c - e^x$
   $[-10,10], [-10,10]$

9) $y = \dfrac{c^2}{24^2 x}$
   $[0,5], [0,5]$

11) $y = \dfrac{c^{100/41}}{5.6^{100/41} x^{65/41}}$
    $[0,10], [0,10]$
    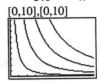

13) $y = \left(\dfrac{c}{180}\right)^{10/3} x^{-7/3}$
    $[0,5], [0,5]$
    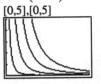

15) $y = c^2 x^{-1}$
    $[0,10], [0,10]$

17) $y = c^4 x^{-3}$
    $[0,10], [0,10]$
    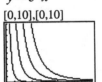

19) $y = c^{20/9} x^{-11/9}$
    $[0,10], [0,10]$
    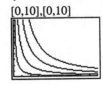

21) [0,1000],[0,1000]
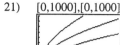

23) [0,1000],[0,1000]

25) [0,25000],[0,25000]
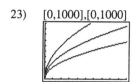

27) [0,15000],[0,350000]
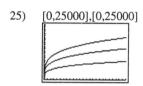

29) $y = \left(\dfrac{Q}{1.2}\right)^{10/3} x^{-7/3}$
[0,75],[0,100]
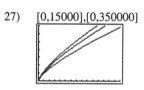

31) $y = c^3 x^{-2}$
[0,10],[0,10]
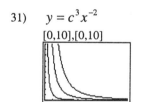

33) $y = \dfrac{100}{c} x$
[0,15],[0,15]
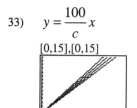

35) $y = \left(\dfrac{Q}{1.64}\right)^{5/2} x^{-3/2}$
[0,200],[0,200]
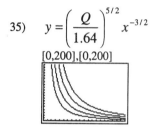

37a) vertical

37b) [0,150],[0,150]

37c) $x = 138, z = 100.203$, approximately 138 thousand labor hours would be required to produce 100.
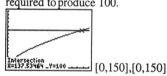
[0,150],[0,150]

37d) The graph shows the relationship between production and labor hours when capital is fixed at $18 million.

39a) vertical

39b) $x = 18 \Rightarrow z = 83.6025 y^{0.75}$

[0,50],[0,1600]

308 Chapter 8 Calculus of Several Variables

39c) 47 units of capital are required to produce 1500 units of golf clubs, with labor fixed at 100 units.
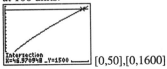 [0,50],[0,1600]

39d) The graph shows the relationship between production and capital when labor is held fixed at 100 units.

41) $y = \left(\dfrac{c}{1.7}\right)^5 x^{-4}$
[0,500],[0,500]
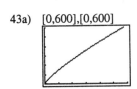

43a) [0,600],[0,600]
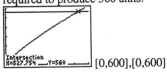

43b) About 528 thousand units of labor are required to produce 560 units.
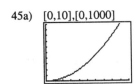 [0,600],[0,600]

45a) [0,10],[0,1000]

45b) A radius of 3" is required to get a volume of 113 in³.
[0,10],[0,1000]

45c) volume approximately 452 cubic inches
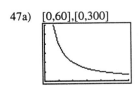 [0,10],[0,1000]

45d) The graph shows the relationship between the radius and volume when the height is held fixed at 4 inches.

47a) [0,60],[0,300]

47b) A 15-year old person, with a mental age of 20 years, has an IQ of about 133.

47c) A 16-year old person, with a mental age of 20 years, has an IQ of 125.

Chapter 8.3 Exercises

1) $f_x(x,y) = 10x$
   $f_y(x,y) = -18y^2$

3) $f_x(x,y) = 2y$
   $f_y(x,y) = 2x - 2y$

5) $f_x(x,y) = 3x^2 y^2$
   $f_y(x,y) = 2x^3 y$

7) $f_x(x,y) = 2x + 6xy^3 - y$
   $f_y(x,y) = 9x^2 y^2 - 6y^2 - x$

9) $f_x(x,y) = 6x^2 + 2$
   $f_y(x,y) = -2y$

11) $f_x(x,y) = 3(2x + 3y) \cdot 2$
    $= 6(2x + 3y)^2$
    $f_y(x,y) = 3(2x + 3y)^2 \cdot 3$
    $= 9(2x + 3y)^2$

13) $f_x(x,y) = 5.5815 x^{-0.85} y^{0.87}$
    $f_y(x,y) = 32.3727 x^{0.15} y^{-0.13}$

15) $f_x(x,y) = e^x \ln y$
    $f_y(x,y) = \dfrac{e^x}{y}$

17) $f_x(x,y) = \dfrac{-y}{x^2}$
    $f_y(x,y) = \dfrac{1}{x}$

19) $f_x(x,y) = 6x - 6x^2 y^4$
    $f_y(x,y) = -8x^3 y^3$

21) $f_x(x,y) = 2(x^2 + y^3) \cdot 2x$
    $= 4x(x^2 + y^3)$
    $f_y(x,y) = 2(x^2 + y^3) \cdot 3y^2$
    $= 6y^2(x^2 + y^3)$

23) $f_x(x,y) = \dfrac{2x}{x^2 + y^3}$
    $f_y(x,y) = \dfrac{3y^2}{x^2 + y^3}$

25) $f_x(x,y) = \dfrac{2xy(y+x) - x^2 y}{(y+x)^2}$
    $f_y(x,y) = \dfrac{x^2(y+x) - x^2 y}{(y+x)^2}$

27) $f_x(x,y) = 6xy^3 + e^{x+y}$
    $f_y(x,y) = 9x^2 y^2 + e^{x+y}$

29) $f_x(x,y) = ye^{xy}$
    $f_y(x,y) = xe^{xy} + \dfrac{1}{y}$

31) $f_x(x,y) = -10xy$
    $f_x(1,2) = -20$; If we are at the point (1,2) and move along the surface in the x-direction, parallel to the x-axis, the function is decreasing at a rate of 20 (units of f)/(units of x).

33) $f_y(x,y) = 2x^2y - 6y - 1$
$f_y(2,3) = 5$; If we are at the point (2,3) and move along the surface in the y-direction, parallel to the y-axis, the function is increasing at a rate of 5 (units of f)/(units of y).

35) $f_x(x,y) = \dfrac{-1}{y^2 - x}$
$f_x(2,5) = \dfrac{-1}{23}$; If we are at the point (2,5) and move along the surface in the x-direction, parallel to the x-axis, the function is decreasing at a rate of 1/23 (units of f)/(units of x).

37) $f_x(x,y) = 5.5815x^{-0.85}y^{0.87}$
$f_x(5,2) \approx 2.5973$; If we are at the point (5,2) and move along the surface in the x-direction, parallel to the x-axis, the function is increasing at a rate of approximately 2.6 (units of f)/(units of x).

39) $f_x(x,y) = -6x^2y - e^{x-y}$
$f_x(2,4) = -96 - e^{-2} \approx -96.136$; If we are at the point (2,4) and move along the surface in the x-direction, parallel to the x-axis, the function is decreasing at a rate of approximately 96.136 (units of f)/(units of x).

41) $f_x(x,y) = -10xy$
$f_y(x,y) = 12y^2 - 5x^2$
$f_{xx}(x,y) = -10y$
$f_{xy}(x,y) = -10x$
$f_{yy}(x,y) = 24y$

43) $f_x(x,y) = 6x - 6x^2y^2 + 6x^2$
$f_y(x,y) = -4x^3y$
$f_{xx}(x,y) = 6 - 12xy^2 + 12x$
$f_{xy}(x,y) = -12x^2y$
$f_{yy}(x,y) = -4x^3$

45) $f_x(x,y) = 5y - 18x^2$
$f_y(x,y) = 5x + 7$
$f_{xx}(x,y) = -36x$
$f_{xy}(x,y) = 5$
$f_{yy}(x,y) = 0$

47) $f_x(x,y) = -6x^2y^2 + 14x$
$f_y(x,y) = 5y^4 - 4x^3y$
$f_{xx}(x,y) = -12xy^2 + 14$
$f_{xy}(x,y) = -12x^2y$
$f_{yy}(x,y) = 20y^3 - 4x^3$

49) $f_x(x,y) = y^3e^{xy} + \dfrac{1}{x}$
$f_y(x,y) = 2y(e^{xy}) + xe^{xy}(y^2)$
$\quad = 2ye^{xy} + xy^2e^{xy}$
$f_{xx}(x,y) = y^4e^{xy} - x^{-2}$
$f_{xy}(x,y) = 3y^2e^{xy} + xy^3e^{xy}$
$f_{yy}(x,y) = 2(e^{xy}) + xe^{xy}(2y)$
$\quad + 2xye^{xy} + xe^{xy}(xy^2)$
$\quad = 2e^{xy} + 4xye^{xy} + x^2y^2e^{xy}$

51) $f_x(x,y) = 3.38x^{-0.35}y^{0.4}$
$f_y(x,y) = 2.08x^{0.65}y^{-0.6}$
$f_{xx}(x,y) = -1.183x^{-1.35}y^{0.4}$
$f_{xy}(x,y) = 1.352x^{-0.35}y^{-0.6}$
$f_{yy}(x,y) = -1.248x^{0.65}y^{-1.6}$

53) $f_x(x,y) = \dfrac{-2y}{x^2}$
$f_y(x,y) = \dfrac{2}{x}$
$f_{xx}(x,y) = \dfrac{4y}{x^3}$
$f_{xy}(x,y) = \dfrac{-2}{x^2}$
$f_{yy}(x,y) = 0$

55) $f_x(x,y) = y^2 e^{xy}$
$f_y(x,y) = e^{xy} + xye^{xy}$
$f_{xx}(x,y) = y^3 e^{xy}$
$f_{xy}(x,y) = 2ye^{xy} + xy^2 e^{xy}$
$f_{yy}(x,y) = xe^{xy} + xe^{xy} + xe^{xy} \cdot xy$
$\quad = 2xe^{xy} + x^2 ye^{xy}$

57) $f_x(x,y) = 2x - y$
$f_y(x,y) = 2y - x + 3$
$2x - y = 0 \Rightarrow y = 2x$
$2y - x + 3 = 0 \Rightarrow y = \dfrac{x-3}{2}$
$2x = \dfrac{x-3}{2} \Rightarrow x = -1 \Rightarrow y = -2$

59a) $S_x(x,y) = 4x \qquad S_y(x,y) = 1$

59b) $S_x(3,5) = 12$; If $3,000 is spent on newspaper advertising and the amount spent on radio advertising is kept fixed at $5,000 per week, then sales will be increasing at a rate of $120,000 in sales per thousand dollars spent on newspaper advertising.

$S_y(3,5) = 1$; If $5,000 is spent on radio advertising each week and the amount spent on newspaper advertising is kept fixed at $3,000 per week, then sales will be increasing at a rate of $10,000 in sales per thousand dollars spent on newspaper advertising.

61a) $TS_x(x,y) = 3x \qquad TS_y(x,y) = 6.4y$

61b) $TS_x(1,0.5) = 3$; If $1,000 is spent on radio advertising and the amount spent on television advertising is kept fixed at $500 per week, then sales will be increasing at a rate of $30,000 in sales per thousand dollars spent on radio advertising.

$TS_y(1,0.5) = 3.2$; If $500 is spent on television advertising each week and the amount spent on radio advertising is kept fixed at $1,000 per week, then sales will be increasing at a rate of $32,000 in sales per thousand dollars spent on television advertising.

63) $f_P(P,t) = e^{0.06t}$ = The rate of change of the total amount accumulated when time is held fixed and the amount of the initial investment is increased.

$f_t(P,t) = 0.06Pe^{0.06t}$ = The rate of change of the total amount accumulated when the amount of the initial investment is held fixed and time is increasing.

65a) $V_r(r,h) = \dfrac{2}{3}\pi hr \qquad V_h(r,h) = \dfrac{1}{3}\pi r^2$

65b) $V(3,5) = 15\pi$; The volume of a cone with a radius of 3 and a height of 5 is 15π.
$V_r(3,5) = 10\pi$; If a cone with a radius of 3 units and a height of 5 units has its height remain constant while the radius is increased, the volume will increase by 10π units per unit increase in radius.
$V_h(3,5) = 3\pi$; If a cone with a radius of 3 units and a height of 5 units has its radius remain constant while the height is increased, the volume will increase by 3π units per unit increase in height.

67) $C_W(W,L) = \dfrac{100}{L}$
$C_W(6,8.2) \approx 12.2$; For a head with a width of 6" and a length of 8.2", if the length of the head remains constant while the width of the head increases, the cephalic index will increase by approximately 12.2 per inch change of head width.
$C_L(W,L) = \dfrac{-100W}{L^2}$
$C_L(6,8.2) \approx -8.92$; For a head with a width of 6" and a length of 8.2", if the width stays constant while the length increases, the cephalic index will decrease by approximately 8.92 per inch change in length.

69a) $R(x,y) = x \cdot p(x,y) + y \cdot q(x,y)$
$= -4x^2 + 350x + 3xy + 450y - 3y^2$

69b) $R_x(x,y) = -8x + 350 + 3y$ = The marginal revenue from an increase in sales of racing bicycles.
$R_y(x,y) = -6y + 450 + 3x$ = The marginal revenue from an increase in sales on mountain bicycles.

69c) $R(15,20) = 13,050$; With a weekly demand of 15 racing bicycles and 20 mountain bicycles, the revenue is $13,050.
$R_x(15,20) = 290$; With a weekly demand of 15 racing bicycles and the demand for mountain bicycles fixed at 20 bicycles, the marginal revenue from an increase in sales of racing bicycles is 290 dollars per bike.

71a) $R_x(x,y) = 70 + 0.5y - 0.08x$

71b) $R_y(x,y) = 95 + 0.5x - 0.08y$

73a) $P(x,y) = R(x,y) - C(x,y)$
$= -0.5x^2 + 37x - 0.8xy + 35.5y - y^2$

73b) $P_x(x,y) = 37 - x - 0.8y$ = The marginal profit function from an increase in sales of two-stroke engines, when the sales of four-stroke engines is held fixed.
$P_y(x,y) = 35.5 - 0.8x - 2y$ = The marginal profit function from an increase in sales of four-stroke engines, when sales of two-stroke engines is held fixed.

73c) $P(18,10) = 615$; The profit from sales of 18 thousand two-stroke engines and 10 thousand four-stroke engines is $615,000
$P_x(18,10) = 11$; With sales of 18 thousand two-stroke engines and sales of four-stroke engines fixed at 10 thousand, the marginal profit from and increase in sales of two-stroke engines is 11 thousand dollars per thousand engines.

73d) $P_y(18,10) = 1.1$; With sales of 10 thousand four-stroke engines and sales of two-stroke engines fixed at 18 thousand, the marginal profit form an increase in sales of four-stroke engines is 1.1 thousand dollars per thousand engines.

75a) $f_x(x,y) = 128.8x^{-0.2}y^{0.25}$
$f_y(x,y) = 40.25x^{0.8}y^{-0.75}$

75b) Marginal productivity of labor:
$f_x(111,25) = 112.29$; If the company is now using 111 units of labor and keeps capital fixed at 25 units, production is increasing at a rate of 112.29 calculators per unit of labor.
Marginal productivity of capital:
$f_y(111,25) = 155.80$; If the company is now using 25 units of capital and keeps labor fixed at 111 units, production is increasing at a rate of 155.80 calculators per unit of capital

77a) $f_x(x,y) = 860.598x^{-0.4}y^{0.45}$
$f_y(x,y) = 645.4485x^{0.6}y^{-0.55}$

77b) Marginal productivity of labor:
$f_x(47,8) = 470.27$; If the company is now using 47 units of labor and keeps capital fixed at 8 units, production is increasing at a rate of 470.27 engines per thousand of hours of labor.
Marginal productivity of capital:
$f_y(47,8) = 2072.1$; If the company is now using 8 units of capital and keeps labor fixed at 47 units, production is increasing at a rate of 2072.1 engines per unit of capital.

79a) $f_x(x,y) = 33x^{-0.45}y^{0.5}$
$f_y(x,y) = 30x^{0.55}y^{-0.5}$

79b) Marginal productivity of labor:
$f_x(220,140) = 34.474$; If the company is now using 220 units of labor and keeps capital fixed at 140 units, production is increasing at a rate of 34.474 motorcycles per unit of labor.

79b) continued
Marginal productivity of capital:
$f_y(220,140) = 49.248$; If the company is now using 140 units of capital and keeps labor fixed at 220 units, production is increasing at a rate of 49.248 motorcycles per unit of capital.

79c) Production would increase more with increased spending on capital since the change in productivity with respect to capital is greater than the change in productivity with respect to labor (as shown in part b).

Chapter 8.4 Exercises

1) $f_x(x,y) = 2x - 4$
$f_y(x,y) = 2y + 6$
$2x - 4 = 0 \Rightarrow x = 2$
$2y + 6 = 0 \Rightarrow y = -3$
critical point: (2,-3)

3) $f_x(x,y) = 2x + 2$
$f_y(x,y) = 2y - 4$
$2x + 2 = 0 \Rightarrow x = -1$
$2y - 4 = 0 \Rightarrow y = 2$
critical point: (-1,2)

5) $f_x(x,y) = 4x - 8$
$f_y(x,y) = 6y + 6$
$4x - 8 = 0 \Rightarrow x = 2$
$6y + 6 = 0 \Rightarrow y = -1$
critical point: (2,-1)

7) $f_x(x,y) = 2x - y$
$f_y(x,y) = 2y - x - 3$
$2x - y = 0 \,\&\, 2y - x - 3 = 0$
$y = 2x \Rightarrow 2(2x) - x - 3 = 0$
$\Rightarrow x = 1 \Rightarrow y = 2$
critical point: (1,2)

9) $f_x(x,y) = 2x + y + 2$
$f_y(x,y) = -2y + x - 9$
$2x + y + 2 = 0 \,\&\, -2y + x - 9 = 0$
$\Rightarrow y = -2x - 2$
$\Rightarrow -2(-2x - 2) + x - 9 = 0$
$\Rightarrow x = 1 \Rightarrow y = -4$
critical point: (1,-4)

11) $f_x(x,y) = 2x - 6$
$f_y(x,y) = 2y - 4$
$2x - 6 = 0 \Rightarrow x = 3$
$2y - 4 = 0 \Rightarrow y = 2$
critical point: (3,2)
$f_{xx}(x,y) = 2; f_{xx}(3,2) = 2$
$f_{yy}(x,y) = 2; f_{yy}(3,2) = 2$
$f_{xy}(x,y) = 0; f_{xy}(3,2) = 0$
$D = f_{xx}(3,2) \cdot f_{yy}(3,2) - [f_{xy}(3,2)]^2$
$= 2 \cdot 2 - 0 = 4$
$D > 0; f_{xx}(3,2) > 0$
(3,2,-10) is a relative minimum.

13) $f_x(x,y) = 2x - 6$
$f_y(x,y) = 2y + 4$
$2x - 6 = 0 \Rightarrow x = 3$
$2y + 4 = 0 \Rightarrow y = -2$
critical point: (3,-2)
$f_{xx}(x,y) = 2; f_{xx}(3,-2) = 2$
$f_{yy}(x,y) = 2; f_{yy}(3,-2) = 2$
$f_{xy}(x,y) = 0; f_{xy}(3,-2) = 0$
$D = f_{xx}(3,-2) \cdot f_{yy}(3,-2) - [0]^2 = 4$
$D > 0; f_{xx}(3,-2) > 0$
(3,-2,-11) is a relative minimum.

15) $f_x(x,y) = -2x - 4$
$f_y(x,y) = -6y - 4$
$-2x - 4 = 0 \Rightarrow x = -2$
$-6y - 4 = 0 \Rightarrow y = -2/3$
critical point: (-2,-2/3)
$f_{xx}(x,y) = -2; f_{xx}(-2,-2/3) = -2$
$f_{yy}(x,y) = -6; f_{yy}(-2,-2/3) = -6$

15) continued
$f_{xy}(x,y) = 0; f_{xy}(-2, -2/3) = 0$
$D = f_{xx}\left(-2, \frac{-2}{3}\right) \cdot f_{yy}\left(-2, \frac{-2}{3}\right) - [0]^2$
$= 12 > 0, f_{xx}(-2, -2/3) < 0$
$\left(-2, \frac{-2}{3}, \frac{13}{3}\right)$ is a relative maximum.

17) $f_x(x,y) = 2x + y$
$f_y(x,y) = 2y + x - 6$
$2x + y = 0 \,\&\, 2y + x - 6 = 0$
$\Rightarrow y = -2x \Rightarrow 2(-2x) + x - 6 = 0$
$\Rightarrow x = -2 \Rightarrow y = 4$
critical point: (-2,4)
$f_{xx}(x,y) = 2; f_{xx}(-2,4) = 2$
$f_{yy}(x,y) = 2; f_{yy}(-2,4) = 2$
$f_{xy}(x,y) = 1; f_{xy}(-2,4) = 1$
$D = 2 \cdot 2 - 1^2 = 3 > 0$
$D > 0, f_{xx}(-2,4) > 0$
(-2,4,-11) is a relative minimum.

19) $f_x(x,y) = 2x - y - 3$
$f_y(x,y) = 2y - x - 3$
$2x - y - 3 = 0 \,\&\, 2y - x - 3 = 0$
$\Rightarrow y = 2x - 3$
$\Rightarrow 2(2x - 3) - x - 3 = 0 \Rightarrow x = 3$
$\Rightarrow y = 3$
critical point: (3,3)
$f_{xx}(x,y) = 2; f_{xx}(3,3) = 2$
$f_{yy}(x,y) = 2; f_{yy}(3,3) = 2$
$f_{xy}(x,y) = -1; f_{xy}(3,3) = -1$
$D = 2 \cdot 2 - [(-1)^2] = 3$
$D > 0; f_{xx}(3,3) > 0$
(3,3,-7) is a relative minimum.

21) $f_x(x,y) = 4x^2 - 8x$
$f_y(x,y) = -2y + 2$
$4x^2 - 8x = 0 \Rightarrow x = 0, x = 2$
$-2y + 2 = 0 \Rightarrow y = 1$
critical points: (0,1), (2,1)
$f_{xx}(x,y) = 8x - 8, f_{xx}(0,1) = -8$
$f_{yy}(x,y) = -2; f_{xy}(x,y) = 0$
$D_1 = f_{xx}(0,1) \cdot f_{yy}(0,1) - [0]^2 = 16$
$D_1 > 0, f_{xx}(0,1) < 0$
(0,1,0) is a relative maximum.
$f_{xx}(2,1) = 8; f_{yy}(2,1) = -2$
$f_{xy}(2,1) = 0$
$D_2 = 8 \cdot -2 - [0]^2 = -16 < 0$
$\left(2, 1, \frac{-16}{3}\right)$ is a saddle point.

23) $f_x(x,y) = 4x - 4$
$f_y(x,y) = 6y^2 - 12y - 18$
$4x - 4 = 0 \Rightarrow x = 1$
$6y^2 - 12y - 18 = 0 \Rightarrow y = -1, y = 3$
critical points: (1,-1), (1,3)
$f_{xx}(x,y) = 4; f_{xy}(x,y) = 0$
$f_{yy}(x,y) = 12y - 12, f_{yy}(1,-1) = -24$
$D_1 = f_{xx}(1,-1) \cdot f_{yy}(1,-1) - 0^2 = -96$
$D_1 < 0 \Rightarrow (1,-1,20)$ is a saddle point.
$f_{xx}(1,3) = 4; f_{yy}(1,3) = 24$
$f_{xy}(1,3) = 0; D_2 = 4 \cdot 24 - 0^2 = 96$
$D_2 > 0, f_{xx}(1,3) = 4 > 0$
(1,3,-44) is a relative minimum.

25) $f_x(x,y) = 3x^2 - 3y$
$f_y(x,y) = -3x + 3y^2$
$3x^2 - 3y = 0 \,\&\, -3x + 3y^2 = 0$
$\Rightarrow y = x^2 \Rightarrow -3x + 3(x^2)^2 = 0$
$\Rightarrow x = 0, x = 1$
$x = 1 \Rightarrow y = 1; x = 0 \Rightarrow y = 0$
critical points: (0,0),(1,1)
$f_{xx}(x,y) = 6x, f_{xx}(0,0) = 0$
$f_{yy}(x,y) = 6y, f_{yy}(0,0) = 0$
$f_{xy}(x,y) = -3$
$D_1 = 0 \cdot 0 - [-3]^2 = -9 < 0$
(0,0,0) is a saddle point.
$f_{xx}(1,1) = 6; f_{yy}(1,1) = 6$
$f_{xy}(1,1) = -3$
$D_2 = 6 \cdot 6 - [-3]^2 = 27$
$D_2 > 0, f_{xx}(1,1) = 6 > 0 \Rightarrow f(1,1)$
(1,1,-1) is a relative minimum.

27) $f_x(x,y) = -2xe^{-x^2-y^2}$
$f_y(x,y) = -2ye^{-x^2-y^2}$
$-2xe^{-x^2-y^2} = 0 \Rightarrow x = 0$
$-2ye^{-x^2-y^2} = 0 \Rightarrow y = 0$
critical point: (0,0)
$f_{xx}(x,y) = -2e^{-x^2-y^2} + 4x^2e^{-x^2-y^2}$
$f_{yy}(x,y) = -2e^{-x^2-y^2} + 4y^2e^{-x^2-y^2}$
$f_{xy}(x,y) = 4xye^{-x^2-y^2}, f_{xy}(0,0) = 0$
$f_{xx}(0,0) = -2; f_{yy}(0,0) = -2$
$D = (-2) \cdot (-2) - 0^2 = 4$
$D > 0, f_{xx}(0,0) = -2 < 0$
(0,0,1) is a relative maximum.

29) $f_x(x,y) = 6y - 12x^{-2}$
$f_y(x,y) = 6x + 3y^{-2}$
$6y - 12x^{-2} = 0 \,\&\, 6x + 3y^{-2} = 0$
$\Rightarrow y = 2x^{-2} \Rightarrow 6x + \frac{3}{4}x^4 = 0$
$= \frac{3}{4}x(8 + x^3) = 0 \Rightarrow x = 0, x = -2$
$x = 0 \Rightarrow y = undefined$
$x = -2 \Rightarrow y = \frac{1}{2}$
critical point: (-2,½)
$f_{xx}(x,y) = 24x^{-3}, f_{xx}(-2,\tfrac{1}{2}) = -3$
$f_{yy}(x,y) = -6y^{-3}, f_{yy}(-2,\tfrac{1}{2}) = -48$
$f_{xy}(x,y) = 6$
$D = (-3) \cdot (-48) - [6]^2 = 108$
$D > 0, f_{xx}(-2,\tfrac{1}{2}) < 0$
(-2,½,-18) is a relative maximum.

31a) $f_x(x,y) = 10x + 5y - 40$
$f_y(x,y) = 5x + 15y - 45$
$10x + 5y - 40 = 0$
$5x + 15y - 45 = 0$
$\Rightarrow y = 8 - 2x$
$\Rightarrow 5x + 15(8 - 2x) - 45 = 0$
$\Rightarrow x = 3 \Rightarrow y = 2$
critical point: (3,2)
$f_{xx}(x,y) = 10; f_{xx}(3,2) = 10$
$f_{yy}(x,y) = 15; f_{yy}(3,2) = 15$
$f_{xy}(x,y) = 5; f_{xy}(3,2) = 5$
$D = 10 \cdot 15 - 5^2 = 125$
$D > 0, f_{xx}(3,2) = 10 > 0$
(3,2) is a relative minimum. The company should spend $3 million on labor and $2 million on robotics equipment to minimize costs.

31b) $f(3,2) = 30$; The minimum cost is $30 million.

33a) $P_x(x, y) = -0.4x + 6y$
$P_y(x, y) = -0.9y^2 + 6x$
$-0.4x + 6y = 0$ & $-0.9y^2 + 6x = 0$
$\Rightarrow x = 15y \Rightarrow -0.9y^2 + 90y = 0$
$\Rightarrow y = 0, y = 100$
$y = 0 \Rightarrow x = 0; y = 100 \Rightarrow x = 1500$
critical point: (1500,100)
$f_{xx} = -0.4; f_{xx}(1500,100) = -0.4$
$f_{yy} = -1.8y; f_{yy}(1500,100) = -180$
$f_{xy} = 6; f_{xy}(1500,100) = 6$
$D = (-0.4) \cdot (-180) - 6^2 = 36$
$D > 0, f_{xx}(1500,100) < 0$
(1500,100) is a relative maximum. The company should sell 1500 regular sized woks and 100 jumbo woks to maximize profit.

33b) $f(1500,100) = 149,998$; maximum profit is $149,998.

35a) $P(x, y) = R(x, y) - C(x, y)$
$= -x^2 - x + xy + 11y - 2y^2 - 2$

35b) $P_x(x, y) = -2x - 1 + y$
$P_y(x, y) = x + 11 - 4y;$
$-2x - 1 + y = 0$ & $x + 11 - 4y = 0$
$\Rightarrow y = 2x + 1$
$\Rightarrow x + 11 - 4(2x + 1) = 0$
$\Rightarrow x = 1 \Rightarrow y = 3$
critical point: (1,3)
$f_{xx}(x, y) = -2; f_{xx}(1,3) = -2$
$f_{yy}(x, y) = -4; f_{yy}(1,3) = -4$

35b) continued
$f_{xy}(x, y) = 1; f_{xy}(1,3) = 1$
$D = 7 > 0, f_{xx}(1,3) = -2 < 0$
(1,3) is a relative maximum. The company should sell 1 million bags of sour cream and chives potato chips and 3 million bags of barbecue potato chips to maximize profit.

35c) $P(1,3) = 14$; Maximum profit is fourteen million dollars.

37a) $R(x, y) = xp(x, y) + yq(x, y)$
$= -4x^2 + 349x + 3xy + 466y - 3y^2$

37b) $R_x(x, y) = -8x + 349 + 3y$
$R_y(x, y) = 3x + 466 - 6y$
$-8x + 349 + 3y = 0$
$3x + 466 - 6y = 0 \Rightarrow y = \dfrac{8x - 349}{3}$
$\Rightarrow 3x + 466 - 6\left(\dfrac{8x - 349}{3}\right)$
$\Rightarrow x = \dfrac{1164}{13} \approx 89 \Rightarrow y = \dfrac{367}{3} \approx 122$
critical point: (89,122)
$f_{xx}(x, y) = -8; f_{yy}(x, y) = -6$
$f_{xy}(x, y) = 3$
$D = 39 > 0, f_{xx}(89,122) = -8 < 0$
(89,122) is a relative maximum. Sammy's Cycles should produce 89 touring and 122 mountain bicycles each week to maximize revenue.

37c) $R(89,122) = 44151$; Maximum revenue is approximately $44,151.

39a) $R(x, y) = xp(x.y) + yq(x, y)$
$= 228x - 8x^2 - 2xy + 31y - 0.15y^2$

318 Chapter 8 Calculus of Several Variables

39b) $R_x(x,y) = 228 - 16x - 2y$
$R_y(x,y) = -2x + 31 - 0.3y;$
$228 - 16x - 2y = 0$
$-2x + 31 - 0.3y = 0$
$\Rightarrow y = 114 - 8x$
$\Rightarrow -2x + 31 - 0.3(114 - 8x) = 0$
$\Rightarrow x = 8 \Rightarrow y = 50$
critical point: (8,50)
$R_{xx}(x,y) = -16$
$R_{yy}(x,y) = -0.3$
$R_{xy}(x,y) = -2$
$D = 0.8 > 0, R_{xx}(8,50) = -16 < 0$
(8,50) is a relative maximum. The company should produce and sell 8 leaf blowers and 50 attachments each month to maximize revenue.

41) $A = 4wh + 2lh + lw$; $V = lwh$
$lwh = 125 \Rightarrow h = \dfrac{125}{lw}$
$A = 4w\left(\dfrac{125}{lw}\right) + 2l\left(\dfrac{125}{lw}\right) + lw$
$A(l,w) = 500l^{-1} + 250w^{-1} + lw$
$A_l = -500l^{-2} + w = 0$
$A_w = -250w^{-2} + l = 0$
$-500l^{-2} + w = 0 \Rightarrow w = 500l^{-2}$
$\Rightarrow -250(500l^{-2})^{-2} + l = 0$
$= l(-.001l^3 + 1) \Rightarrow l = 0, l = 10$
$l = 0 \Rightarrow w = $ undefined
$l = 10 \Rightarrow w = 5$
critical point (10,5)

41) continued
$A_{ll} = 1000l^{-3}; A_{ll}(10,5) = 1$
$A_{ww} = 500w^{-3}; A_{ww}(10,5) = 4$
$A_{lw} = 1$
$D = 3 > 0; A_{ll}(10,5) = 1 > 0$
(10,5) is a relative minimum. The box should be 10"x5"x2.5" to minimize the material needed.

Chapter 8.5 Exercises

1) $F(x,y,\lambda) = 3xy + \lambda(x+y-1)$
$F_x(x,y,\lambda) = 3y + \lambda$
$F_y(x,y,\lambda) = 3x + \lambda$
$F_\lambda(x,y,\lambda) = x + y - 1$
$x + y - 1 = 0;$
$3y + \lambda = 0 \Rightarrow -\lambda = 3y$
$3x + \lambda = 0 \Rightarrow -\lambda = 3x$
$3x = 3y \Rightarrow y = x$
$x + x - 1 = 0 \Rightarrow x = \dfrac{1}{2} \Rightarrow y = \dfrac{1}{2}$
Critical point: $\left(\tfrac{1}{2}, \tfrac{1}{2}\right)$; The maximum value of $f(x,y) = 3xy$, subject to $x + y = 1$, is $\tfrac{3}{4}$.

3) $F(x,y,\lambda) = xy + \lambda(2x+y-12)$
$F_x(x,y,\lambda) = y + 2\lambda$
$F_y(x,y,\lambda) = x + \lambda$
$F_\lambda(x,y,\lambda) = 2x + y - 12$
$2x + y - 12 = 0;$
$y + 2\lambda = 0 \Rightarrow \lambda = \dfrac{-y}{2}$
$x + \lambda = 0 \Rightarrow \lambda = -x$
$-x = \dfrac{-y}{2} \Rightarrow y = 2x$
$2x + (2x) - 12 = 0 \Rightarrow x = 3$
$y = 6$
Critical point: (3,6); The maximum value of $f(x,y) = xy$, subject to $2x + y = 12$, is 18.

5) $F(x,y,\lambda) = xy + \lambda(x-y+8)$
$F_x(x,y,\lambda) = y + \lambda$
$F_y(x,y,\lambda) = x - \lambda$
$F_\lambda(x,y,\lambda) = x - y + 8$
$x - y + 8 = 0;$
$y + \lambda = 0 \Rightarrow \lambda = -y$
$x - \lambda = 0 \Rightarrow \lambda = x$
$-y = x \Rightarrow y = -x$
$x - (-x) + 8 = 0 \Rightarrow x = -4 \Rightarrow y = 4$
Critical point: (-4,4); The minimum value of $f(x,y) = xy$, subject to $x - y = -8$, is -16.

7) $F(x,y,\lambda) = x^2 + 2y^2 + \lambda(2x+y-8)$
$F_x(x,y,\lambda) = 2x + 2\lambda$
$F_y(x,y,\lambda) = 4y + \lambda$
$F_\lambda(x,y,\lambda) = 2x + y - 8$
$2x + y - 8 = 0;\ \ 4y + \lambda = 0$
$2x + 2\lambda = 0 \Rightarrow \lambda = -x$
$4y + (-x) = 0 \Rightarrow y = \dfrac{1}{4}x$
$2x + \left(\dfrac{1}{4}x\right) - 8 = 0 \Rightarrow x = \dfrac{32}{9}$
$y = \dfrac{8}{9}$
Critical point: $\left(\tfrac{32}{9}, \tfrac{8}{9}\right)$; The minimum value of $f(x,y) = x^2 + 2y^2$, subject to $2x + y = 8$, is $\dfrac{128}{9}$.

320 Chapter 8 Calculus of Several Variables

9) $F(x, y, \lambda) = 2xy - 4x + \lambda(x + y - 12)$
$F_x(x, y, \lambda) = 2y - 4 + \lambda$
$F_y(x, y, \lambda) = 2x + \lambda$
$F_\lambda(x, y, \lambda) = x + y - 12$
$x + y - 12 = 0$
$2y - 4 + \lambda = 0 \Rightarrow \lambda = -2y + 4$
$2x + \lambda = 0 \Rightarrow \lambda = -2x$
$-2y + 4 = -2x \Rightarrow y = x + 2;$
$x + (x + 2) - 12 = 0 \Rightarrow x = 5$
$y = 7$
Critical point: (5,7); The maximum value of $f(x, y) = 2xy - 4x$, subject to $x + y = 12$, is 50.

11) $F(x, y, \lambda) = 1.64 x^{0.6} y^{0.4}$
$+ \lambda(12,000x + 5,000y - 1,100,000)$
$F_x(x, y, \lambda) = 0.984 x^{-0.4} y^{0.4} + 12,000\lambda = 0$
$F_y(x, y, \lambda) = 0.656 x^{0.6} y^{-0.6} + 5,000\lambda = 0$
$F_\lambda(x, y, \lambda) = 12,000x + 5,000y - 1,100,000$
$12,000x + 5,000y - 1,100,000 = 0$
$\dfrac{0.984 x^{-0.4} y^{0.4}}{12,000} = -\lambda = \dfrac{0.656 x^{0.6} y^{-0.6}}{5,000}$
$\left(x^{0.4} y^{0.6}\right) \dfrac{0.984 x^{-0.4} y^{0.4}}{12,000} = \dfrac{0.984 y}{12,000}$
$= \dfrac{0.656 x^{0.6} y^{-0.6}}{5,000} \left(x^{0.4} y^{0.6}\right) = \dfrac{0.656 x}{5,000}$
$\Rightarrow y = 1.6x$
$12,000x + 5,000(1.6x) - 1,100,000 = 0$
$\Rightarrow x = 55 \Rightarrow y = 88$
Critical point: (55,88); The maximum value of $f(x, y) = 1.64 x^{0.6} y^{0.4}$, subject to $12,000 + 5,000y = 1,100,000$, is approximately 108.86.

13) $F(x, y, \lambda) = x^2 + y^2 - xy - 4$
$+ \lambda(x + y - 6)$

13) continued
$F_x(x, y, \lambda) = 2x - y + \lambda = 0$
$F_y(x, y, \lambda) = 2y - x + \lambda = 0$
$F_\lambda(x, y, \lambda) = x + y - 6 = 0$
$2x - y + \lambda = 0 \Rightarrow \lambda = y - 2x$
$2y - x + \lambda = 0 \Rightarrow \lambda = x - 2y$
$y - 2x = x - 2y \Rightarrow y = x$
$x + (x) - 6 = 0 \Rightarrow x = 3 \Rightarrow y = 3$
Critical point: (3,3); The minimum value of $f(x, y) = x^2 + y^2 - xy - 4$, subject to $x + y - 6 = 0$, is 5.

15) $F(x, y, \lambda) = xy + \lambda(x^2 + y^2 - 8)$
$F_x(x, y, \lambda) = y + 2x\lambda = 0$
$F_y(x, y, \lambda) = x + 2y\lambda = 0$
$F_\lambda(x, y, \lambda) = x^2 + y^2 - 8 = 0$
$y + 2x\lambda = x + 2y\lambda \Rightarrow y = x$
$x^2 + (x)^2 - 8 = 0 \Rightarrow x = \pm 2$
$x = -2 \Rightarrow y = -2, \ x = 2 \Rightarrow y = 2$
Critical points: (-2, -2), (2, 2); Each point will produce a maximum value of 4 for $f(x, y) = xy$, subject to $x^2 + y^2 = 8$.

17) $F(x, y, \lambda) = e^{xy} + \lambda(x^2 + y^2 - 8)$
$F_x(x, y, \lambda) = ye^{xy} + 2\lambda x = 0$
$F_y(x, y, \lambda) = xe^{xy} + 2\lambda y = 0$
$F_\lambda(x, y, \lambda) = x^2 + y^2 - 8 = 0$
$ye^{xy} + 2\lambda x = xe^{xy} + 2\lambda y \Rightarrow y = x$
$x^2 + (x)^2 - 8 = 0 \Rightarrow x = \pm 2$
$x = -2 \Rightarrow y = -2, \ x = 2 \Rightarrow y = 2$
Critical points: (-2,-2), (2,2); Each point will produce a maximum value of

17) continued
$e^4 \approx 54.6$ for $f(x,y) = e^{xy}$, subject to $x^2 + y^2 = 8$.

19a) $F(x,y,\lambda) = 0.1x^2 + 0.2y^2$
$\quad + \lambda(x+y-180)$
$F_x(x,y,\lambda) = 0.2x + \lambda = 0$
$F_y(x,y,\lambda) = 0.4y + \lambda = 0$
$F_\lambda(x,y,\lambda) = x + y - 180 = 0$
$0.2x + \lambda = 0 \Rightarrow \lambda = -0.2x$
$0.4y + \lambda = 0 \Rightarrow \lambda = -0.4y$
$-0.2x = -0.4y \Rightarrow y = 0.5x$
$x + (0.5x) - 180 = 0 \Rightarrow x = 120$
$\Rightarrow y = 60$

Critical point: (120, 60); To minimize cost, the company should produce 120 regular models and 60 joggers' model each day.

19b) $C(120,60) = \$2160.00$.

21a) $F(x,y,\lambda) = 0.1x^2 + 0.2y^2$
$\quad + \lambda(x+y-120)$
$F_x(x,y,\lambda) = 0.2x + \lambda = 0$
$F_y(x,y,\lambda) = 0.4y + \lambda = 0$
$F_\lambda(x,y,\lambda) = x + y - 120 = 0$
$0.2x + \lambda = 0 \Rightarrow \lambda = -0.2x$
$0.4y + \lambda = 0 \Rightarrow \lambda = -0.4y$
$-0.2x = -0.4y \Rightarrow y = 0.5x$
$x + (0.5x) - 120 = 0 \Rightarrow x = 80$
$y = 40$

Critical point: (80, 40); To minimize cost, the company should produce 80 pepper spray devices and 40 siren devices each day.

21b) $C(80,40) = \$960.00$.

23a) $F(x,y,\lambda) = 2x^2 - xy + y^2 + 250$
$\quad + \lambda(x+y-104)$
$F_x(x,y,\lambda) = 4x - y + \lambda = 0$
$F_y(x,y,\lambda) = -x + 2y + \lambda = 0$
$F_\lambda(x,y,\lambda) = x + y - 104 = 0$
$4x - y + \lambda = 0 \Rightarrow \lambda = y - 4x$
$-x + 2y + \lambda = 0 \Rightarrow \lambda = -2y + x$
$y - 4x = -2y + x \Rightarrow y = \dfrac{5x}{3}$
$x + \left(\dfrac{5x}{3}\right) - 104 = 0 \Rightarrow x = 39$
$\Rightarrow y = 65$

Critical point: (39, 65); To minimize cost 39 units should be produced by the old production line and 65 units should be produced by the new production line.

23b) $C(39,65) = \$4{,}982.00$.

25a) $F(x,y,\lambda) = 13x^{0.75} y^{0.3}$
$\quad + \lambda(15x + 50y - 18{,}000)$
$F_x(x,y,\lambda) = 9.75x^{-0.25} y^{0.3} + 15\lambda = 0$
$F_y(x,y,\lambda) = 3.9x^{0.75} y^{-0.7} + 50\lambda = 0$
$F_\lambda(x,y,\lambda) = 15x + 50y - 18{,}000 = 0$
$\dfrac{9.75x^{-0.25} y^{0.3}}{15} = -\lambda = \dfrac{3.9x^{0.75} y^{-0.7}}{50}$
$(x^{0.25} y^{0.7}) \dfrac{9.75x^{-0.25} y^{0.3}}{15} = \dfrac{9.75y}{15}$
$= \dfrac{3.9x^{0.75} y^{-0.7}}{50} (x^{0.25} y^{0.7}) = \dfrac{3.9x}{50}$

322 Chapter 8 Calculus of Several Variables

25a) continued
$\Rightarrow y = 0.12x$
$15x + 50(0.12x) - 18,000 = 0$ Critical
$\Rightarrow x \approx 857.14 \Rightarrow y \approx 102.86$
point: (857.14, 102.86); Approximately 857.14 units of labor and 102.86 units of capital are required to maximize production.

25b) $f(857.14, 102.86) \approx 8268$ golf carts per week.

27a) $F(x, y, \lambda) = x^{0.6} y^{0.45}$
$+ \lambda(16,000x + 4,000y - 1,700,000)$
$F_x(x, y, \lambda) = 0.6x^{-0.4} y^{0.45} + 16,000\lambda = 0$
$F_y(x, y, \lambda) = 0.45x^{0.6} y^{-0.55} + 4,000\lambda = 0$
$F_\lambda(x, y, \lambda) = 16,000x + 4,000y$
$- 1,700,000 = 0$

$\dfrac{0.6x^{-0.4} y^{0.45}}{16,000} = -\lambda = \dfrac{0.45x^{0.6} y^{-0.55}}{4,000}$

$(x^{0.4} y^{0.55}) \dfrac{0.6x^{-0.4} y^{0.45}}{16,000} = \dfrac{0.6y}{16,000}$

$= \dfrac{0.45x^{0.6} y^{-0.55}}{4,000} (x^{0.4} y^{0.55}) = \dfrac{0.45x}{4,000}$

$\Rightarrow y = 3x$
$16,000x + 4,000(3x) - 1,700,000 = 0$
$\Rightarrow x \approx 60.7 \Rightarrow y \approx 182.1$
Critical point: (60.7, 182.1);
Approximately 60.7 units of labor and 182.1 units of capital are required to maximize production.

27b) $-\lambda = \dfrac{0.6 \cdot x^{-0.4} \cdot y^{0.45}}{16,000} \approx 0.000075$

For each additional dollar value available to the budget, production increases by about 0.000075 units.

27c) $-\lambda(200,000) = 0.000075(200,000)$
≈ 15.1; Optimal increase in production with an additional $200,000 budgeted is about 15.1 units.

29a) $F(x, y, \lambda) = 5.6x^{0.65} y^{0.41}$
$+ \lambda(2,000x + 1,500y - 800,000)$
$F_x(x, y, \lambda) = 3.64x^{-0.35} y^{0.41} + 2,000\lambda$
$F_y(x, y, \lambda) = 2.296x^{0.65} y^{-0.59} + 1,500\lambda$
$F_\lambda(x, y, \lambda) = 2,000x + 1,500y - 800,000$
$F_x = F_y = F_\lambda = 0$

$\dfrac{3.64x^{-0.35} y^{0.41}}{2,000} = -\lambda = \dfrac{2.296x^{0.65} y^{-0.59}}{1,500}$

$(x^{.35} y^{.59}) \dfrac{3.64x^{-0.35} y^{0.41}}{2,000} = \dfrac{3.64y}{2,000}$

$= \dfrac{2.296x^{0.65} y^{-0.59}}{1,500} (x^{0.35} y^{0.59}) = \dfrac{2.296x}{1,500}$

$\Rightarrow y = 0.841x$
$2,000x + 1,500(0.841x) - 800,000 = 0$
$\Rightarrow x \approx 245.3 \Rightarrow y \approx 206.3$
Critical point: (245.3, 206.3)
Approximately 245.3 units of labor and 206.3 units of capital should be used to maximize production.

29b) $-\lambda = \dfrac{3.64(245.3)^{-0.35} (206.3)^{0.41}}{2000}$

≈ 0.002358; For each additional dollar value available to the budget, production increases by about 0.002358 units.

29c) $0.002358(50,000) \approx 117.9$; Optimal increase in production with an additional $50,000 budgeted is about 118 units.

31a) $F(x, y, \lambda) = 100x^{0.6} y^{0.4}$
$+ \lambda(100x + 150y - 60,000)$

31a) continued
$$F_x(x,y,\lambda) = 60x^{-0.4}y^{0.4} + 100\lambda$$
$$F_y(x,y,\lambda) = 40x^{0.6}y^{-0.6} + 150\lambda$$
$$F_\lambda(x,y,\lambda) = 100x + 150y - 60{,}000$$
$$100x + 150y - 60{,}000 = 0$$
$$60x^{-0.4}y^{0.4} + 100\lambda = 0$$
$$40x^{0.6}y^{-0.6} + 150\lambda = 0$$
$$\frac{60x^{-0.4}y^{0.4}}{100} = -\lambda = \frac{40x^{0.6}y^{-0.6}}{150}$$
$$\left(x^{0.4}y^{0.6}\right)\frac{60x^{-0.4}y^{0.4}}{100} = \frac{60y}{100}$$
$$= \frac{40x^{0.6}y^{-0.6}}{150}\left(x^{0.4}y^{0.6}\right) = \frac{40x}{150}$$
$$\Rightarrow y = \frac{4}{9}x$$
$$100x + 150\left(\frac{4}{9}x\right) - 60{,}000 = 0$$
$$\Rightarrow x = 360 \Rightarrow y = 160$$
360 units of labor and 160 units of capital are required to maximize production.

31b) $-\lambda \approx 0.43379$; For each additional dollar value available to the budget, production increases by about 0.43379 units

31c) $-\lambda(10{,}000) \approx 4338$; Optimal increase in production with an additional \$10,000 budgeted is about 4338 units.

33) $F(x,y,\lambda) = xy + \lambda(2x + y - 625)$
$$F_x(x,y,\lambda) = y + 2\lambda$$
$$F_y(x,y,\lambda) = x + \lambda$$
$$F_\lambda(x,y,\lambda) = 2x + y - 625$$

33) continued
$$2x + y - 625 = 0$$
$$y + 2\lambda = 0 \Rightarrow y = -2\lambda$$
$$x + \lambda = 0 \Rightarrow x = -\lambda$$
$$2(-\lambda) + (-2\lambda) - 625 = 0$$
$$\Rightarrow \lambda = \frac{-625}{4}$$
$$x + \left(\frac{-625}{4}\right) = 0 \Rightarrow x = \frac{625}{4}$$
$$y + 2\left(\frac{-625}{4}\right) = 0 \Rightarrow y = \frac{625}{2}$$
The dimensions of the parking lot should be $\frac{625}{2}$ feet by $\frac{625}{4}$ feet.

35) Maximize $f(x) = x \cdot y \cdot z$, subject to
$$2x + 2z + y = 300$$
$$F(x,y,z,\lambda) = xyz + \lambda(2x + 2z + y - 300)$$
$$F_x(x,y,z,\lambda) = yz + 2\lambda = 0$$
$$F_y(x,y,z,\lambda) = xz + \lambda = 0$$
$$F_z(x,y,z,\lambda) = xy + 2\lambda = 0$$
$$F_\lambda(x,y,z,\lambda) = 2x + 2z + y - 300 = 0$$
$$\Rightarrow -\lambda = \frac{yz}{2} = xz = \frac{xy}{2}$$
$$\frac{yz}{2} = xz \Rightarrow x = \frac{y}{2};\ \frac{xy}{2} = xz \Rightarrow z = \frac{y}{2}$$
$$2\left(\frac{y}{2}\right) + 2\left(\frac{y}{2}\right) + y - 300 = 0 \Rightarrow y = 100$$
$$x = \frac{y}{2} = 50,\ z = \frac{y}{2} = 50$$
The dimensions of the box should be 100 feet by 50 feet by 50 feet.

Chapter 8.6 Exercises

1) $\int_1^3 3x^2 y^2 dy = x^2 y^3 \Big|_1^3$
$= x^2(3)^3 - x^2(1)^2 = 26x^2$

3) $\int_1^3 3x^2 y^2 dx = x^3 y^2 \Big|_1^3$
$= (3)^3 y^2 - (1)^3 y^2 = 26y^2$

5) $\int_0^2 (2x+3y)dy = 2xy + \dfrac{3y^2}{2}\Big|_0^2$
$= \left[2x(2) + \dfrac{3(2)^2}{2}\right] - [0] = 4x+6$

7) $\int_0^2 (2x+3y)dx = x^2 + 3yx \Big|_0^2$
$= \left[(2)^2 + 3(2)y\right] - [0] = 4+6y$

9) $\int_1^2 (x^3 y^2 - 2xy)dy = \dfrac{x^3 y^3}{3} - xy^2 \Big|_1^2$
$= \left[\dfrac{2^3 x^3}{3} - 2^2 x\right] - \left[\dfrac{1^3 x^3}{3} - 1^2 x\right]$
$= \dfrac{7x^3}{3} - 3x$

11) $\int_1^2 (x^3 y^2 - 2xy)dx = \dfrac{x^4 y^2}{4} - x^2 y \Big|_1^2$
$= \left[\dfrac{2^4 y^2}{4} - 2^2 y\right] - \left[\dfrac{1^4 y^2}{4} - 1^2 y\right]$
$= \dfrac{15y^2}{4} - 3y$

13a) [-1,3],[-1,4]

13b) $\int_0^2 \int_0^3 2xy\, dy\, dx$
$= \int_0^2 xy^2 \Big|_0^3 dx = \int_0^2 9x\, dx$
$= \dfrac{9x^2}{2}\Big|_0^2 = 18$

13c) $\int_0^3 \int_0^2 2xy\, dx\, dy$
$= \int_0^3 x^2 y \Big|_0^2 dy = \int_0^3 4y\, dy$
$= 2y^2 \Big|_0^3 = 18$

15a) [-1,2],[-1,3]

15b) $\int_0^1 \int_0^2 (3x+y)dy\, dx$
$= \int_0^1 3xy + \dfrac{y^2}{2}\Big|_0^2 dx = \int_0^1 (6x+2)dx$
$= 3x^2 + 2x \Big|_0^1 = 5$

15c) $\int_0^2 \int_0^1 (3x+y)dx\, dy$
$= \int_0^2 \dfrac{3x^2}{2} + xy \Big|_0^1 dy = \int_0^2 \left(\dfrac{3}{2}+y\right)dy$
$= \dfrac{3y}{2} + \dfrac{y^2}{2}\Big|_0^2 = 5$

17a) [0,20],[-2,7]

17b) $\int_1^{16}\int_1^4 \sqrt{xy}\,dy\,dx$

$= \int_1^{16} \frac{2x^{1/2}y^{3/2}}{3}\bigg|_1^4 dy = \int_1^{16} \frac{14x^{1/2}}{3}dy$

$= \frac{28x^{3/2}}{9}\bigg|_1^{16} = 196$

17c) $\int_1^4 \int_1^{16} \sqrt{xy}\,dx\,dy$

$= \int_1^4 \frac{2x^{3/2}y^{1/2}}{3}\bigg|_1^{16} dy = \int_1^4 42y^{1/2}dy$

$= 28y^{3/2}\bigg|_1^4 = 196$

19a) [0,3],[-1,2]

19b) $\int_1^2 \int_0^1 x^2 y^2\,dy\,dx$

$= \int_1^2 \frac{x^2 y^3}{3}\bigg|_0^1 dx = \int_1^2 \frac{x^2}{3}dx = \frac{x^3}{9}\bigg|_1^2$

$= \frac{7}{9}$

19c) $\int_0^1 \int_1^2 x^2 y^2\,dx\,dy$

$= \int_0^1 \frac{x^3 y^2}{3}\bigg|_1^2 dy = \int_0^1 \frac{7y^2}{3}dy$

$= \frac{7y^3}{9}\bigg|_0^1 = \frac{7}{9}$

21a) [-1,2],[-1,2]

21b) $\int_0^1 \int_0^1 \left(2 - \frac{1}{2}x^2 + y^2\right)dy\,dx$

$= \int_0^1 2y - \frac{1}{2}x^2 y + \frac{1}{3}y^3\bigg|_0^1 dx$

$= \int_0^1 \left(\frac{7}{3} - \frac{1}{2}x^2\right)dx$

$= \frac{7}{3}x - \frac{1}{6}x^3\bigg|_0^1 = \frac{13}{6}$

21c) $\int_0^1 \int_0^1 \left(2 - \frac{1}{2}x^2 + y^2\right)dx\,dy$

$= \int_0^1 2x - \frac{1}{6}x^3 + xy^2 \bigg|_0^1 dy$

$= \int_0^1 \left(\frac{11}{6} + y^2\right)dy$

$= \frac{11}{6}y + \frac{1}{3}y^3\bigg|_0^1 = \frac{13}{6}$

23) $\int_0^1 \int_0^2 xy\,dy\,dx = \int_0^1 \frac{xy^2}{2}\bigg|_0^2 dx = \int_0^1 2x\,dx$

$= x^2\bigg|_0^1 = 1$

[-5,5],[-5,5]

326 Chapter 8 Calculus of Several Variables

25) $\int_1^2 \int_1^3 (x^2 y + y)\,dx\,dy$

$= \int_1^2 \left.\dfrac{x^3 y}{3} + xy \right|_1^3 dy = \int_1^2 \dfrac{32y}{3}\,dy$

$= \left.\dfrac{16y^2}{3}\right|_1^2 = 16$

[-5,5],[-5,5]

27) $\int_{-1}^1 \int_0^2 x^2 e^y\,dx\,dy$

$= \int_{-1}^1 \left.\dfrac{x^3 e^y}{3}\right|_0^2 dy = \int_{-1}^1 \dfrac{8e^y}{3}\,dy$

$= \left.\dfrac{8e^y}{3}\right|_{-1}^1 = \dfrac{8e}{3} - \dfrac{8}{3e} \approx 6.268$

[-5,5],[-5,5]

29) $\dfrac{1}{(1)(1)} \int_0^1 \int_0^1 (x^2 + y^2)\,dx\,dy$

$= \int_0^1 \left.\dfrac{x^3}{3} + xy^2\right|_0^1 dy = \int_0^1 \dfrac{1}{3} + y^2\,dy$

$= \left.\dfrac{y}{3} + \dfrac{y^3}{3}\right|_0^1 = \dfrac{2}{3}$

31) $\dfrac{1}{(1)(1)} \int_0^1 \int_0^1 (9 - x^2 - y^2)\,dy\,dx$

$= \int_0^1 \left.9y - x^2 y - \dfrac{1}{3}y^3\right|_0^1 dx$

$= \int_0^1 \dfrac{26}{3} - x^2\,dx = \left.\dfrac{26x}{3} - \dfrac{x^3}{3}\right|_0^1 = \dfrac{25}{3}$

33) $\dfrac{1}{(2)(2)} \int_0^2 \int_0^2 2xy\,dy\,dx$

$= \dfrac{1}{4}\int_0^2 \left.xy^2\right|_0^2 dx = \dfrac{1}{4}\int_0^2 4x\,dx$

$= \left(\dfrac{1}{4}\right)\left.2x^2\right|_0^2 = 2$

35) $\dfrac{1}{1 \cdot 2} \int_0^1 \int_0^2 x^2 e^y\,dx\,dy$

$= \dfrac{1}{2}\int_0^1 \left.\dfrac{x^3 e^y}{3}\right|_0^2 dy = \dfrac{1}{2}\int_0^1 \dfrac{8e^y}{3}\,dy$

$= \left(\dfrac{1}{2}\right)\left.\dfrac{8e^y}{3}\right|_0^1 = \left(\dfrac{4e}{3} - \dfrac{4}{3}\right) \approx 2.291$

37) $V = \int_0^1 \int_0^2 (9 - x^2 - y^2)\,dy\,dx$

$= \int_0^1 \left.9y - x^2 y - \dfrac{1}{3}y^3\right|_0^2 dx$

$= \int_0^1 \dfrac{46}{3} - 2x^2\,dx = \left.\dfrac{46x}{3} - \dfrac{2x^3}{3}\right|_0^1$

$= \dfrac{44}{3}$ cubic units

39) $V = \int_0^2 \int_0^1 (x^2 + 2y^2) dx dy$

$= \int_0^2 \frac{x^3}{3} + 2xy^2 \Big|_0^1 dy = \int_0^2 \frac{1}{3} + 2y^2 dy$

$= \frac{y}{3} + \frac{2y^3}{3} \Big|_0^2 = 6$ cubic units

41) $V = \int_0^1 \int_0^2 3xy \, dx \, dy = \int_0^1 \frac{3x^2 y}{2} \Big|_0^2 dy$

$= \int_0^2 6y \, dy = 3y^2 \Big|_0^2 = 3$ cubic units

43) $V = \int_0^2 \int_0^1 x^2 e^y \, dy \, dx$

$= \int_0^2 x^2 e^y \Big|_0^1 dx = \int_0^2 (x^2 e - x^2) dx$

$= \frac{x^3 e - x^3}{3} \Big|_0^2 = \frac{8e - 8}{3} \approx 4.582$

4.582 cubic units

45) $\frac{1}{30 \cdot 10} \int_{30}^{40} \int_{150}^{180} 250 x^{0.7} y^{0.3} dy \, dx$

$= \frac{1}{300} \int_{30}^{40} 192.308 x^{0.7} y^{1.3} \Big|_{150}^{180} $

$= \frac{1}{300} \int_{30}^{40} 34688.2 x^{0.7} dx$

$= \frac{1}{300} \left[20404.8 x^{1.7} \Big|_{30}^{40} \right] \approx 13918.6$

47) $\frac{1}{10 \cdot 10} \int_{70}^{80} \int_{20}^{30} 161 x^{0.8} y^{0.25} dy \, dx$

$= \frac{1}{100} \int_{70}^{80} 128.8 x^{0.8} y^{1.25} \Big|_{20}^{30}$

$= \frac{1}{100} \int_{70}^{80} 3595.53 x^{0.8} dx$

$= \frac{1}{100} \left[1997.52 x^{1.8} \Big|_{70}^{80} \right] \approx 11,370$

49) $\frac{1}{6} \int_0^2 \int_0^3 (36 + 2x - 3y) dx \, dy$

$= \frac{1}{6} \int_0^2 36x + x^2 - 3xy \Big|_0^3 dy$

$= \frac{1}{6} \int_0^2 (117 - 9y) dy$

$= \frac{1}{6} \left[117y - \frac{9y^2}{2} \Big|_0^2 \right] = 36$

51) $\frac{1}{25} \int_0^5 \int_0^5 (30,000 e^{-y}) dx \, dy$

$= \frac{1}{25} \int_0^5 (30,000 e^{-y} x) \Big|_0^5 dy$

$= \frac{1}{25} \int_0^5 (150,000 e^{-y}) dy$

$= \frac{1}{25} \left[-150,000 e^{-y} \Big|_0^5 \right] \approx 5959.57$

The average population density for Cedar Falls is approximately 5956.6 people per square mile.

53) $\int_0^5 \int_0^5 (30,000 e^{-y}) dx \, dy$

$= \int_0^5 (30,000 e^{-y} x) \Big|_0^5 dy$

$= \int_0^5 (150,000 e^{-y}) dy$

$= \left[-150,000 e^{-y} \Big|_0^5 \right] \approx 148,989$

328   Chapter 8 Calculus of Several Variables

53) continued
The total population of Cedar Falls is approximately 148,989.

55) $\dfrac{1}{2}\displaystyle\int_0^2\int_0^1 (125-10x^2-10y^2)\,dydx$

$= \dfrac{1}{2}\displaystyle\int_0^2 \left(125y-10x^2y-\dfrac{10y^3}{3}\right)\Bigg|_0^1 dx$

$= \dfrac{1}{2}\displaystyle\int_0^2 \left(\dfrac{365}{3}-10x^2\right)dx$

$= \dfrac{1}{2}\left[\dfrac{365x}{3}-\dfrac{10x^3}{3}\right]\Bigg|_0^2 = \dfrac{325}{3}$

57) $\dfrac{1}{9\cdot 6}\displaystyle\int_{19}^{25}\int_{23}^{32}\dfrac{100m}{a}\,dmda$

$= \dfrac{1}{54}\displaystyle\int_{19}^{25}\dfrac{50m^2}{a}\Bigg|_{23}^{32} da$

$= \dfrac{1}{54}\displaystyle\int_{19}^{25}\dfrac{24750}{a}\,da$

$= \dfrac{1}{54}\left[24750\ln|a|\right]_{19}^{25} \approx 125.784$

Chapter 8 Review Exercises

1) $f(3,5) = \dfrac{-6}{11}$

3) $f(10,3) = 103$

5) $domf = \{(x,y) | x + y \geq 0\}$

7)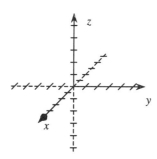

9)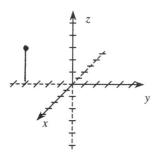

11) (-3, -4, 2) is above the xy-plane (the z value is positive.)

13) (6, -2, -1) is below the xy-plane (the z value is negative.)

15)

17) $P(4000, 3000) = 800$, With sales of 4000 cups of coffee and 3000 cups of cappuccino per month, the monthly profits are $800.

19a) $C(x,y) = 7x + 13y + 5000$

19b) $R(x,y) = 12x + 21y$

19c) $P(x,y) = R(x,y) - C(x,y)$
$= 5x + 8y - 5000$

21) $f(4500, 0.045) \approx 7057.40$; If $4,500 is invested at an annual rate of 4.5%, compounded continuously for 10 years, the total amount accumulated is approximately $7057.40.

23) $y = 2x + 3 - c$
[-5,5],[-5,5]

25) $y = \pm\sqrt{16 - x^2 - c}$
[-5,5],[-5,5]

330 Chapter 8 Calculus of Several Variables

27) $y = \left(\dfrac{c}{10}\right)^2 x^{-1}$
[0,10],[0,10]
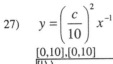

29) $y = \left(\dfrac{c}{8.9}\right)^{50/21} x^{-31/21}$
[0,50],[0,50]

31) $y = c^{3/2} x^{-1/2}$
[0,10],[0,10]

33) $y = c^{20/17} x^{3/17}$
[0,10],[0,10]

35) [0,10],[0,200]

37) [-5,5],[-10,10]

39) $y = \left(\dfrac{Q}{1.87}\right)^{10/3} x^{-7/3}$
[0,100],[0,100]

41a) Vertical

41b) [0,15],[0,200]
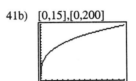

41c) [0,15],[0,200]

The point (6.2, 150) shows that $6.2 million in capital is required to produce 150,000 units.

43) $f_x(x,y) = 5 \qquad f_y(x,y) = -6y$

45) $f_x(x,y) = 5(8y^2 - 3x)^4(-3)$
$= -15(8y^2 - 3x)^4$
$f_y(x,y) = 5(8y^2 - 3x)^4(16y)$
$= 80y(8y^2 - 3x)^4$

47) $f_x(x,y) = \dfrac{3x - 3(x+y)}{(3x)^2} = \dfrac{-y}{3x^2}$
$f_y(x,y) = \dfrac{3x(1) - (x+y)\cdot 0}{(3x)^2} = \dfrac{1}{3x}$

49) $f_x(x,y) = e^{x-y} + 15x^2 y$
$f_y(x,y) = -e^{x-y} + 5x^3$

51) $f_y(x,y) = 12xy^3 - 10y^4$

$f_y(3,5) = -1750$; If we are at the point (3,5) and move along the surface in the y-direction, parallel to the y-axis, the function is decreasing by 1750 units of f per unit of y.

53) $f_y(x,y) = 7.967x^{0.38}y^{-0.38}$

$f_y(3,2) \approx 9.29$; If we are at the point (3,2) and move along the surface in the y-direction, parallel to the y-axis, the function is increasing at approximately 9.29 units of f per unit of y.

55) $f_x(x,y) = -12x^2 + 10y + y^5$
$f_y(x,y) = 10x + 5xy^4$
$f_{xx}(x,y) = -24x$
$f_{xy}(x,y) = 10 + 5y^4$
$f_{yy}(x,y) = 20xy^3$

57) $f_x(x,y) = 2(x+2y)\cdot 1 = 2x + 4y$
$f_y(x,y) = 2(x+2y)\cdot 2 = 4x + 8y$
$f_{xx}(x,y) = 2; f_{xy}(x,y) = 4;$
$f_{yy}(x,y) = 8$

59) $f_x(x,y) = 12x - 4y + 1 = 0$
$f_y(x,y) = -4x + 10y - 3 = 0$
$12x - 4y + 1 = 0 \Rightarrow y = 3x + \dfrac{1}{4}$
$-4x + 10\left(3x + \dfrac{1}{4}\right) - 3 = 0$
$\Rightarrow x = \dfrac{1}{52} \Rightarrow y = \dfrac{4}{13}$

61a) $R(x,y) = xp(x,y) + yq(x,y)$
$= -0.6x^2 + 480x + 0.5xy + 720y$
$-0.7y^2$

61b) $R_x(x,y) = -1.2x + 480 + 0.5y$; This gives us the amount that each additional 15-inch monitor sold will add to the total revenue.

$R_y(x,y) = 0.5x + 720 - 1.4y$; This gives us the amount that each additional 17-inch monitor sold will add to the total revenue.

61c) $R_x(640,850) = 137$; If 640 15-inch monitors and 850 17-inch monitors are sold weekly, the company receives an additional \$137 for the 641st 15-inch monitor sold.

$R_y(640,850) = -150$; If 640 15-inch monitors and 850 17-inch monitors are sold weekly, the company loses \$150 profit for the 851st 17-inch monitor sold.

63a) $P(x,y) = R(x,y) - C(x,y)$
$= -3.5x^2 + 110x - 18.5xy + 130y$
$+ 1.2y^2 - 600$

63b) $P_x(x,y) = -7x + 110 - 18.5y$. This function gives us the marginal profit from an increase in sales of regular VCR's in hundreds of dollars per hundred VCR's.

$P_y(x,y) = -18.5x + 2.4y + 130$.

This function gives us the marginal profit from an increase in sales of stereo VCR's in hundreds of dollars per hundred VCR's.

63c) $P_x(8,11) = -149.5$; With sales of 800 regular VCR's and 1100 stereo VCR's, the marginal profit from an increase in sales of regular VCR's is –14,950 dollars per hundred VCR's.

63c) continued
$P_y(8,11) = 8.4$; With sales of 800 regular VCR's and 1100 stereo VCR's, the marginal profit from an increase in sales of stereo VCR's is 840 dollars per hundred VCR's.

63d) $P(8,11) = 3.2$; The profit from sales of 800 regular VCR's and 1100 stereo VCR's is $320.

65) $f_x(x,y) = 2x + 8 = 0 \Rightarrow x = -4$
$f_y(x,y) = 2y - 6 = 0 \Rightarrow y = 3$
Critical point: (-4,3)

67) $f_x(x,y) = 2x - 5y + 6 = 0$
$f_y(x,y) = 2y - 5x - 8 = 0$
$2x - 5y + 6 = 2y - 5x - 8$
$y = x + 2 \Rightarrow 2x - 5(x+2) + 6 = 0$
$\Rightarrow x = \dfrac{-4}{3} \Rightarrow y = \dfrac{2}{3}$
Critical point: $\left(\dfrac{-4}{3}, \dfrac{2}{3}\right)$

69) $f_x(x,y) = 2x - 2; f_y(x,y) = 2y + 8$
$2x - 2 = 0 \Rightarrow x = 1$
$2y + 8 = 0 \Rightarrow y = -4$
Critical point: (1,-4,-24)
$f_{xx}(x,y) = 2; f_{yy}(x,y) = 2$
$f_{xy}(x,y) = 0; D = 2 \cdot 2 - 0^2 = 4 > 0$
$D > 0, f_{xx}(1,-4) > 0$
(1,-4,-24) is a relative minimum.

71) $f_x(x,y) = 2x + y + 3 = 0$
$f_y(x,y) = -4y + x - 3 = 0$
$2x + y + 3 = 0 \Rightarrow y = -2x - 3$
$-4(-2x-3) + x - 3 = 0$
$\Rightarrow x = -1 \Rightarrow y = -1$
Critical point: (-1,-1,0)
$f_{xx}(x,y) = 2; f_{yy}(x,y) = -4$
$f_{xy}(x,y) = 1;$
$D = 2(-4) - 1^2 = -9 < 0$
(-1,-1,0) is a saddle point.

73) $f_x(x,y) = 6x^2 - 4x + 4y = 0$
$f_y(x,y) = 4x + 2y = 0$
$4x + 2y = 0 \Rightarrow y = -2x$
$6x^2 - 4x + 4(-2x) = 0 = 6x(x-2)$
$x = 0, x = 2$
$x = 0 \Rightarrow y = 0, x = 2 \Rightarrow y = -4$
Critical points: (0,0,0), (2,-4,-8)
$f_{xx}(x,y) = 12x - 4; f_{xx}(0,0) = -4$
$f_{yy}(x,y) = 2; f_{xy}(x,y) = 4$
$D_1 = (-4)(2) - 4^2 = -24 < 0$
(0,0,0) is a saddle point.
$f_{xx}(2,-4) = 20; f_{yy}(2,-4) = 2$
$f_{xy}(2,-4) = 4$
$D_2 = (20)(2) - 4^2 = 24 > 0$
$D_2 > 0, f_{xx}(2,-4) > 0$
(2,-4,-8) is a relative minimum.

75) $f_x(x,y) = 2xe^{x^2-y^2+4y}$
$f_y(x,y) = (-2y+4)e^{x^2-y^2+4y}$
$2xe^{x^2-y^2+4y} = 0 \Rightarrow x = 0$
$(-2y+4)e^{x^2-y^2+4y} = 0 \Rightarrow y = 2$
Critical point: $(0,2,e^4)$

75) continued
$f_{xx}(x,y) = 2e^{x^2-y^2+4y} + 4x^2 e^{x^2-y^2+4y}$
$f_{yy}(x,y) = -2e^{x^2-y^2+4y}$
$\quad + (-2y+4)(-2y+4)e^{x^2-y^2+4y}$
$f_{xy}(x,y) = 2x(-2y+4)e^{x^2-y^2+4y}$
$f_{xx}(0,2) \approx 109; f_{yy}(0,2) \approx -109$
$f_{xy}(0,2) = 0$
$D = 109(-109) - 0^2 \approx -11881 < 0$
$(0,2,e^4)$ is a saddle point.

77a) $R(x,y) = xp(x,y) + yq(x,y)$
$= -0.3x^2 + 12x + 0.2xy + 20y$
$\quad - 0.2y^2$

77b) $R_x(x,y) = -0.6x + 12 + 0.2y = 0$
$R_y(x,y) = 0.2x + 20 - 0.4y = 0$
$\Rightarrow y = 3x - 60$
$0.2x + 20 - 0.4(3x - 60) = 0$
$\Rightarrow x = 44 \Rightarrow y = 72$
Critical point: (44, 72)
$f_{xx}(x,y) = -0.6; f_{yy}(x,y) = -0.4$
$f_{xy}(x,y) = 0.2$
$D = (-0.6)(-0.4) - 0.2^2 = 0.2$
$D > 0, f_{xx}(44,72) < 0$
(44,72) is a relative maximum. To maximize revenue, the salon should give 44 basic haircuts and 72 deluxe haircuts daily.

77c) $R(44,72) = 984$; Maximum daily revenue is $984.

79a) $P_x(x,y) = -6x + 2y + 800 = 0$
$P_y(x,y) = -4y + 2x + 1200 = 0$
$-6x + 2y + 800 = 0 \Rightarrow y = 3x - 400$
$-4(3x - 400) + 2x + 1200 = 0$
$\Rightarrow x = 280 \Rightarrow y = 440$
Critical point: (280,440)
$P_{xx}(x,y) = -6; P_{yy}(x,y) = -4$
$P_{xy}(x,y) = 2$
$D = (-6)(-4) - 2^2 = 20$
$D > 0, f_{xx}(280,440) < 0$
(280,440) is a relative maximum. To maximize profits, 280,000 fine-tip pens and 440,000 large-tip pens should be sold each day.

79b) $P(280,440) = 375,600$; Maximum daily profit is $375,600.

81) $F(x,y,\lambda) = 5xy + \lambda(x + y - 10)$
$F_x(x,y,\lambda) = 5y + \lambda = 0$
$F_y(x,y,\lambda) = 5x + \lambda = 0$
$F_\lambda(x,y,\lambda) = x + y - 10 = 0$
$5y + \lambda = 0 \Rightarrow \lambda = -5y$
$5x + (-5y) = 0 \Rightarrow y = x$
$x + x - 10 = 0 \Rightarrow x = 5 \Rightarrow y = 5$
$f(5,5) = 125$

83) $F(x,y,\lambda) = 16 - 2x^2 + 5xy - 18y^2$
$\quad + \lambda(x + 3y - 6)$
$F_x(x,y,\lambda) = -4x + 5y + \lambda = 0$
$F_y(x,y,\lambda) = 5x - 36y + 3\lambda = 0$
$F_\lambda(x,y,\lambda) = x + 3y - 6 = 0$
$-4x + 5y + \lambda = 0 \Rightarrow \lambda = -5y + 4x$
$5x - 36y + 3(-5y + 4x) = 0$

83) continued

$$\Rightarrow y = \frac{1}{3}x$$

$$x + 3\left(\frac{1}{3}x\right) - 6 = 0 \Rightarrow x = 3 \Rightarrow y = 1$$

$$f(3,1) = -5$$

85) $F(x,y,\lambda) = 3.8x^{0.35}y^{0.6}$
$\quad + \lambda(120x + 260y - 2500)$
$F_x(x,y,\lambda) = 1.33x^{-0.65}y^{0.6} + 120\lambda$
$F_y(x,y,\lambda) = 2.28x^{0.35}y^{-0.4} + 260\lambda$
$F_\lambda(x,y,\lambda) = 120x + 260y - 2500$
$F_x = F_y = F_\lambda = 0$

$$\frac{1.33x^{-0.65}y^{0.6}}{120} = -\lambda = \frac{2.28x^{0.35}y^{-0.4}}{260}$$

$$\left(x^{0.65}y^{0.4}\right)\frac{1.33x^{-0.65}y^{0.6}}{120} = \frac{1.33y}{120}$$

$$= \frac{2.28x^{0.35}y^{-0.4}}{260}\left(x^{0.65}y^{0.4}\right) = \frac{2.28x}{260}$$

$$\Rightarrow y = \frac{72x}{91}$$

$$120x + 260\left(\frac{72x}{91}\right) - 2500 = 0$$

$$\Rightarrow x \approx 7.675 \Rightarrow y \approx 6.073$$

$$f(7.675, 6.073) \approx 22.888$$

87a) $F(x,y,\lambda) = 28x^{0.6}y^{0.4}$
$\quad + \lambda(500x + 800y - 4,000)$
$F_x(x,y,\lambda) = 16.8x^{-0.4}y^{0.4} + 500\lambda = 0$
$F_y(x,y,\lambda) = 11.2x^{0.6}y^{-0.6} + 800\lambda = 0$
$F_\lambda(x,y,\lambda) = 500x + 800y - 4,000 = 0$

$$\frac{16.8x^{-0.4}y^{0.4}}{500} = -\lambda = \frac{11.2x^{0.6}y^{-0.6}}{800}$$

87a) continued

$$\left(x^{0.4}y^{0.6}\right)\frac{16.8x^{-0.4}y^{0.4}}{500} = \frac{16.8y}{500}$$

$$= \frac{11.2x^{0.6}y^{-0.6}}{800}\left(x^{0.4}y^{0.6}\right) = \frac{11.2x}{800}$$

$$\Rightarrow y \approx 0.417x$$

$$500x + 800(0.417x) - 4000 = 0$$

$$\Rightarrow x \approx 4.8 \Rightarrow y \approx 2$$

Approximately 4.8 units of labor and 2 units of capital are needed to maximize production.

87b) $f(4.8, 2) = 94.692$; Maximum production is approximately 94,692 telephones.

89) Let c = the cost of the fence side, $5c$ = the cost of the wall side. Total cost:
$5c \cdot y + c(2x + y) = 6cy + 2xc$
Minimize $f(x,y) = 6cy + 2xc$, subject to $xy = 300$.
$F(x,y,\lambda) = 6cy + 2xc + \lambda(xy - 300)$
$F_x(x,y,\lambda) = 2c + y\lambda = 0$
$F_y(x,y,\lambda) = 6c + x\lambda = 0$
$F_\lambda(x,y,\lambda) = xy - 300 = 0$

$$2c + y\lambda = 0 \Rightarrow \lambda = \frac{-2c}{y}$$

$$6c + x\left(\frac{-2c}{y}\right) = 0 \Rightarrow x = 3y$$

$$(3y)y - 300 = 0 \Rightarrow y = 10 \Rightarrow x = 30$$

The dimensions of the garden should be 10 feet by 30 feet.

91) $$\int_5^8 (2x^2 y^3)\,dx = \left.\frac{2x^3 y^3}{3}\right|_5^8 = 258y^3$$

93) $\int_2^6 (x^2y - 4y^3)dy$

$= \frac{1}{2}x^2y^2 - y^4 \Big|_2^6 = 16x^2 - 1280$

95a) [-2,7],[-2,6]

95b) $\int_0^5 \int_0^2 (3x - 4xy)dydx$

$= \int_0^5 3xy - 2xy^2 \Big|_0^2 dx$

$= \int_0^5 -2xdx = -x^2 \Big|_0^5 = -25$

95c) $\int_0^2 \int_0^5 (3x - 4xy)dxdy$

$= \int_0^2 \frac{3}{2}x^2 - 2x^2y \Big|_0^5 dy$

$= \int_0^2 \left(\frac{75}{2} - 50y\right)dy$

$= \frac{75y}{2} - 25y^2 \Big|_0^2 = -25$

97a) [-2,6],[-1,5]

97b) $\int_0^4 \int_1^3 (3xy^2)dydx = \int_0^4 xy^3 \Big|_1^3 dx$

$= \int_0^4 (27x - x)dx = 13x^2 \Big|_0^4 = 208$

99) $V = \int_0^3 \int_0^4 (25 - x^2 - y^2)dydx$

$= \int_0^3 25y - yx^2 - \frac{1}{3}y^3 \Big|_0^4$

$= \int_0^3 \left(100 - 4x^2 - \frac{64}{3}\right)dx$

$= \frac{236}{3}x - \frac{4}{3}x^3 \Big|_0^3 = 200$

101) $\frac{1}{12}\int_0^3 \int_0^4 (x^2 + 3xy - y^2)dydx$

$= \frac{1}{12}\int_0^3 x^2y + \frac{3}{2}xy^2 - \frac{1}{3}y^3 \Big|_0^4 dx$

$= \frac{1}{12}\int_0^3 \left(4x^2 + 24x - \frac{64}{3}\right)dx$

$= \frac{1}{12}\left[\frac{4}{3}x^3 + 12x^2 - \frac{64}{3}x\right]_0^3 = \frac{20}{3}$

103) $\frac{1}{16}\int_1^5 \int_3^7 x^3 y \, dydx$

$= \frac{1}{16}\int_1^5 \frac{x^3 y^2}{2} \Big|_3^7 dx = \frac{1}{16}\int_1^5 20x^3 dx$

$= \frac{1}{16}\left[5x^4 \Big|_1^5\right] = 195$

105) $\frac{1}{96}\int_{18}^{30} \int_{16}^{24} (11.8x^{0.68} y^{0.45})dydx$

$= \frac{1}{96}\int_{18}^{30} \frac{11.8}{1.45}x^{0.68} y^{1.45} \Big|_{16}^{24} dx$

$\approx \frac{1}{96}\int_{18}^{30} (362.838 x^{0.68})dx$

$\approx \frac{1}{96}\left[215.975 x^{1.68} \Big|_{18}^{30}\right] \approx 392.8$

107) $\dfrac{1}{48}\displaystyle\int_0^8\int_0^6 (150 - x^2 - 2y^2)\,dy\,dx$

$= \dfrac{1}{48}\displaystyle\int_0^8 \left[150y - x^2 y - \dfrac{2y^3}{3}\right]_0^6 dx$

$= \dfrac{1}{48}\displaystyle\int_0^8 (756 - 6x^2)\,dx$

$= \dfrac{1}{48}\left[756x - 2x^3 \Big|_0^8\right] \approx 104.667$

The average concentration of air pollution is approximately 104.667 parts per million.

Appendix A Exercises

1) $\dfrac{-4}{x^6} = -4x^{-6}$

3) $1.6\sqrt{x} = 1.6x^{1/2}$

5) $\dfrac{8}{\sqrt[3]{x^2}} = 8x^{-2/3}$

7) $x^{3/7} = \sqrt[7]{x^3}$

9) $6.3x^{4/5} = 6.3\sqrt[5]{x^4}$

11) $2x^{-2/3} = \dfrac{2}{\sqrt[3]{x^2}}$

13) $y = x + 2$
$x + 2(x+2) = 16$
$x = 4$
$y = 6$

15) $x + 2y = 8$
$x = 8 - 2y$
$3(8-2y) - 4y = 9$
$y = \dfrac{3}{2}$
$x = 5$

17) $4x + 3z = 4$
$x = 1 - \dfrac{3}{4}z$
$2y - 6z = -1$
$y = 3z - \dfrac{1}{2}$
$8\left(1 - \dfrac{3}{4}z\right) + 4\left(3z - \dfrac{1}{2}\right) + 3z = 9$
$z = \dfrac{1}{3}$
$x = 1 - \dfrac{3}{4}z = 1 - \dfrac{3}{4}\left(\dfrac{1}{3}\right) = \dfrac{3}{4}$
$y = 3z - \dfrac{1}{2} = 3\left(\dfrac{1}{3}\right) - \dfrac{1}{2} = \dfrac{1}{2}$

19) $3x - 2y = -16$
$y = \dfrac{3}{2}x + 8$
$2x + 5y = 21$
$y = \dfrac{-2x}{5} + \dfrac{21}{5}$

Graph each equation solved for y and find the intersection point to get the solution to the system of equations. The solution is the point (-2, 5).

21) $\log_b\left(\dfrac{2x}{5y}\right)^3 = 3\log_b 2x - 3\log_b 5y$
$= 3\log_b 2 + 3\log_b x - 3\log_b 5 - 3\log_b y$
$= \log_b 8 + 3\log_b x - \log_b 125 - 3\log_b y$

23) $\log\left(\dfrac{xy^2}{y+4}\right) = \log xy^2 - \log(y+4)$
$= \log x + 2\log y - \log(y+4)$

25) $\ln\sqrt{\dfrac{5x^3}{y^9}} = \ln\left(\dfrac{5x^3}{y^9}\right)^{1/2}$

$\quad = \dfrac{1}{2}\ln 5 + \dfrac{3}{2}\ln x - \dfrac{9}{2}\ln y$

27) $\log_b(x-2) = \log_b(x-7) + \log_b 4$

$\log_b(x-2) = \log_b[(x-7)(4)]$

$x - 2 = 4(x-7)$

$3x = 26$

$x = \dfrac{26}{3}$

29) $\ln(4x+5) - \ln(x+3) = \ln 3$

$\ln\left(\dfrac{4x+5}{x+3}\right) = \ln 3$

$\dfrac{4x+5}{x+3} = 3$

$4x + 5 = 3(x+3)$

$x = 4$

31) $3^x = 6$

$\ln(3^x) = \ln 6$

$x \ln 3 = \ln 6$

$x = \dfrac{\ln 6}{\ln 3} \approx 1.631$

33) $2^{5x+2} = 8$

$\ln(2^{5x+2}) = \ln 8$

$(5x+2)\ln 2 = \ln 8$

$5x + 2 = \dfrac{\ln 8}{\ln 2}$

$5x = \dfrac{\ln 8}{\ln 2} - 2$

$x = \left(\dfrac{\ln 8}{\ln 2} - 2\right)\cdot\dfrac{1}{5} = \dfrac{1}{5}\left(\dfrac{\ln 2^3}{\ln 2} - 2\right)$

$= \dfrac{1}{5}\left(\dfrac{3\ln 2}{\ln 2} - 2\right) = \dfrac{1}{5}(3-2) = \dfrac{1}{5}$